Natural Capital

Natural Capital

Theory & Practice of Mapping Ecosystem Services

EDITED BY

Peter Kareiva
The Nature Conservancy and Santa Clara University, USA

Heather Tallis
Natural Capital Project, Stanford University, USA

Taylor H. Ricketts
World Wildlife Fund, USA

Gretchen C. Daily
Stanford University, USA

Stephen Polasky
University of Minnesota, USA

OXFORD
UNIVERSITY PRESS

OXFORD

UNIVERSITY PRESS

Great Clarendon Street, Oxford OX2 6DP

Oxford University Press is a department of the University of Oxford.
It furthers the University's objective of excellence in research, scholarship,
and education by publishing worldwide in

Oxford New York

Auckland Cape Town Dar es Salaam Hong Kong Karachi
Kuala Lumpur Madrid Melbourne Mexico City Nairobi
New Delhi Shanghai Taipei Toronto

With offices in

Argentina Austria Brazil Chile Czech Republic France Greece
Guatemala Hungary Italy Japan Poland Portugal Singapore
South Korea Switzerland Thailand Turkey Ukraine Vietnam

Oxford is a registered trade mark of Oxford University Press
in the UK and in certain other countries

Published in the United States
by Oxford University Press Inc., New York

British Library Cataloguing in Publication Data

Data available

Library of Congress Cataloging in Publication Data

Library of Congress Control Number: 2010942945

Typeset by SPI Publisher Services, Pondicherry, India
Printed in Great Britain
on acid-free paper by
CPI Antony Rowe, Chippenham, Wiltshire

ISBN 978-0-19-958899-2 (Hbk.)
       978-0-19-958900-5 (Pbk.)

3 5 7 9 10 8 6 4 2

Contents

Section II: Multi-tiered models for ecosystem services

Contributors

W. L. (Vic) Adamowicz—Department of Rural Economy, University of Alberta, Edmonton, Alberta, Canada T6G 2H1

Edward B. Barbier—Department of Economics and Finance, University of Wyoming, Laramie, WY 82071-3985, USA

Karen Bennett—World Resources Institute, 10 G Street NE, Suite 800, Washington, DC 20002, USA

Kenneth Brooks—Department of Forest Resources, University of Minnesota, 1530 Cleveland Avenue North, Saint Paul, MN 55108-6112, USA

Berry Brosi—Department of Environmental Studies, Emory University, 400 Dowman Dr., Ste. E510, Atlanta, GA 30322, USA

Lauretta Burke—World Resources Institute, 10 G Street NE, Suite 800, Washington, DC 20002, USA

Giorgio Caldarone—Land Assets Division, Kamehameha Schools, 567 South King Street, Suite 200, Honolulu, HI 96813, USA

D. Richard Cameron—The Nature Conservancy—California, 201 Mission Street, 4th Floor, San Francisco, CA 94105, USA

Karen Carney—Stratus Consulting, Inc., 1920 L Street NW, Suite 420, Washington, DC, 20036, USA

Stephen R. Carpenter—Center for Limnology, University of Wisconsin, 680 North Park Street, Madison WI 53706, USA

Kai M. A. Chan—Institute for Resources, Environment, and Sustainability, University of British Columbia, Vancouver, BC, Canada

Marc Conte—Department of Biology and Natural Capital Project, Woods Institute for the Environment, 371 Serra Mall, Stanford University, Stanford, CA 94305-5020, USA

Emily Cooper—World Resources Institute, 10G Street NE, Suite 800, Washington, DC 20002, USA

Gretchen C. Daily—Department of Biology and Natural Capital Project, Woods Institute for the Environment, Stanford University, 371 Serra Mall, Stanford, CA 94305-5020, USA

Brett Day—Centre for Social and Economic Research on the Global Environment (CSERGE), School of Environmental Sciences, University of East Anglia, Norwich, NR4 7TJ, UK

Eric Dinerstein—Conservation Science Program, World Wildlife Fund-US, 1240, 24th Street NW, Washington, DC 20037, USA

T. Ka'eo Duarte—Land Assets Division, Kamehameha Schools, 567 South King Street, Suite 200, Honolulu, HI 96813, USA

Paul R. Ehrlich—Department of Biology, 371 Serra Mall, Stanford University, Stanford, CA 94305-5020, USA

Driss Ennaanay—Department of Biology and Natural Capital Project, Woods Institute for the Environment, 371 Serra Mall, Stanford University, Stanford, CA 94305-5020, USA

Cinzia Fissore—Department of Soil, Water, and Climate, University of Minnesota, 439 Borlaug Hall, Saint Paul, MN 55108, USA

David Freyberg—Department of Civil and Environmental Engineering, 473 Via Ortega, Stanford University, Stanford, CA 94305, USA

Nicola Gallai—INRA, UMR406 Abeilles & Environnement, 84914 Avignon Cedex 9, France and INRA, UMR LAMETA, 2 place Viala, 34060 Montpellier Cedex 1, France

Edward Game—The Nature Conservancy, PO Box 5681, West End, QLD 4101, Australia

Renzo Giudice—Centre for Ecology, Evolution and Conservation (CEEC), School of Biological Sciences, University of East Anglia, Norwich, NR4 7TJ, UK

Joshua Goldstein—Department of Human Dimensions of Natural Resources, Colorado State University, Fort Collins, CO 80523, USA

Lawrence H. Goulder—Environmental and Resource Economics, Stanford University, Stanford, CA 94305, USA

Sarah Greenleaf—Department of Biological Sciences, California State University, 6000 J Street, Sacramento, CA 95819, USA

Craig Groves—The Nature Conservancy, 520 E. Babcock Street, Bozeman, MT 59715, USA

Anne D. Guerry—Department of Biology and Natural Capital Project, Woods Institute for the Environment, 371 Serra Mall, Stanford University, Stanford, CA 94305-5020, USA AND Conservation Biology Division, NOAA Northwest Fisheries Science Center, 2725 Montlake Boulevard E, Seattle, WA 98112, USA

Rajendra Gurung—World Wildlife Fund Nepal Program, Baluwatar, Nepal

Neil Hannahs—Kamehameha Schools, 567 South King Street, Suite 200, Honolulu, HI 96813, USA

Craig Hanson—World Resources Institute, 10 G Street NE, Suite 800, Washington, DC 20002, USA

David Harrison—The Nature Conservancy, USA

Chris J. Harvey—Conservation Biology Division, NOAA Northwest Fisheries Science Center, 2725 Montlake Blvd E, Seattle, WA 98112, USA

Lauren Hay—USGS Denver Federal Center, Lakewood, CO, USA

Norbert Henninger—World Resources Institute, 10 G Street NE Suite 800, Washington DC 20002, USA

Frances Irwin—World Resources Institute, 10 G Street NE, Suite 800, Washington, DC 20002, USA

Peter Kareiva—The Nature Conservancy, USA

Donald Kennedy—Environmental Sciences, Stanford University, Stanford, CA 94305, USA

Kekuewa Kikiloi—Land Assets Division, Kamehameha Schools, Honolulu, HI, USA

Christopher Kirkby—State Key Laboratory of Genetic Resources and Evolution; Ecology, Conservation and Environment Center (ECEC), Kunming Institute of Zoology, Chinese Academy of Science, Kunming, Yunnan, 650223, China, AND Centre for Ecology, Evolution and Conservation (CEEC), School of Biological Sciences, University of East Anglia, Norwich, NR4 7TJ, UK, AND Centre for Social and Economic Research on the Global Environment (CSERGE), School of Environmental Sciences, University of East Anglia, Norwich, NR4 7TJ, UK

Jawoo Koo—International Food Policy Research Institute, 2033 K Street NW, Washington, DC 20006, USA

Carolyn Kousky —Resources for the Future, 1616 P Street NW, Washington, DC 20036, USA

Claire Kremen—Department of Environmental Science, Policy and Management, University of California, 217 Wellman Hall, Berkeley, CA 94720-3114, USA

Florence Landsberg—World Resources Institute, 10 G Street NE Suite 800, Washington, DC 20002, USA

Joshua J. Lawler—School of Forest Resources, University of Washington, Box 352100, Seattle, WA 98195, USA

David Lobell—Environmental Earth System Science, Stanford University, 473 Via Ortega, Stanford CA 94305, USA

Eric Lonsdorf—Urban Wildlife Institute, Lincoln Park Zoo, 2001 North Clark Street, Chicago, IL 60614, USA

Jianzhong Ma—The Nature Conservancy, China Program, Yunnan Provincial Meteorological Building, 8th Floor, 77 Xi Chang Road, Kunming, Yunnan Province, People's Republic of China 650034

Andrew R. Marshall—Research Fellow, Environment Department, University of York, Heslington, York YO10 5DD, UK

Emily McKenzie—World Wildlife Fund and The Natural Capital Project, 1250 24th Street NW, Washington, DC 20009, USA

Guillermo Mendoza—National Research Council—Research Associateships Program Fellow, 500 5th Street NW, Washington, DC 20001 USA

Harold Mooney—Department of Biology, Stanford University, 371 Serra Mall, Stanford, CA 94305-5020, USA

Claire Montgomery—Department of Forest Resources, Oregon State University, 205 Peavy Hall, Corvallis, OR 97331, USA

P.K.T. Munishi—Department of Forest Biology, Faculty of Forestry and Nature Conservation, Sokoine University of Agriculture (SUA), PO Box 3010, Morogoro, Tanzania

Robin Naidoo—Conservation Science Program, World Wildlife Fund-US, 1250 24th Street NW, Washington, DC 20009, USA

Erik Nelson—Department of Economics, Bowdoin College, 9700 College Station, Brunswick, Maine 04011–8497, USA

John Nieber—Department of Bioproducts and Biosystems Engineering, University of Minnesota, 1390 Eckles Avenue, St Paul, MN 55108-3005, USA

Hermann Oliveira-Rodrigues—CSR, Universidade Federal de Minas Gerais, Belo Horizonte, 31270-901, MG, Brazil

Nasser Olwero—Conservation Science Program, World Wildlife Fund-US, 1250 24th Street NW, Washington, DC 20037-1193, USA

Jouni Paavola—School of Earth and Environment, University of Leeds, Leeds, LS2 9JT, UK

Stefano Pagiola—Economics Unit, Socially Sustainable Development Department, Latin America and Caribbean, World Bank

Liba Pejchar—Warner College of Natural Resources, Colorado State University, Fort Collins, CO 80534-1401, USA

Andrew J. Plantinga—Department of Agricultural and Resource Economics, Oregon State University, 213 Ballard Extension Hall, Corvallis, OR 97331-4501, USA

Mark L. Plummer—Conservation Biology Division, NOAA Northwest Fisheries Science Center, 2725 Montlake Blvd E, Seattle, WA 98112, USA

Stephen Polasky—Department of Applied Economics, Department of Ecology, Evolution and Behavior, University of Minnesota, 1994 Buford Avenue, St. Paul, MN 55108, USA

Simon G. Potts—Centre for Agri-Environmental Research, University of Reading, RG6 6AR, UK

Jai Ranganathan—National Center for Ecological Analysis and Synthesis, 735 State Street, Suite 300, Santa Barbara, CA 93101, USA

Janet Ranganathan—World Resources Institute, 10 G Street NE, Suite 800, Washington, DC 20002, USA

James Regetz—National Center for Ecological Analysis and Synthesis (NCEAS), University of California, 735 State Street, Suite 300, Santa Barbara, CA 93101, USA

Taylor H. Ricketts—World Wildlife Fund-US, 1250 24th Street NW, Washington, DC 20090, USA

Lee D. Ross—Department of Psychology, Bldg 420-Rm 380, Stanford University, Stanford, CA 94305-2130, USA

Mary H. Ruckelshaus—Natural Capital Project, Woods Institute for the Environment, Stanford University, 371 Serra Mall, Stanford, CA 94305-5020, USA

Susan Ruffo—The Nature Conservancy, 4245 North Fairfax Drive, Suite 100, Arlington, VA 22203, USA

Jean-Michel Salles—CNRS, UMR LAMETA, 2 place Viala, 34060 Montpellier Cedex 1, France

James Salzman—Law School and Nicholas School of the Environment and Earth Sciences, Duke University, P.O. Box 90360, Durham, NC 27708, USA

M. Sanjayan—The Nature Conservancy, 1011 Poplar Street, Missoula, MT 59802, USA

Terre Satterfield—Institute for Resources, Environment, and Sustainability, University of British Columbia, Vancouver, BC, Canada

Sarah L. Shafer—U.S. Geological Survey, 3200 SW Jefferson Way, Corvallis, OR 97331, USA

Sabina Shaikh—Public Policy Studies and Program on Global Environment, University of Chicago and RCF Economic Consulting, USA

Manu Sharma—Natural Capital Project, Woods Institute for the Environment, 371 Serra Mall, Stanford University, Stanford, CA 94305-5020, USA

Priya Shyamsundar—South Asian Network for Development and Environmental Economics

Britaldo Silveira Soares-Filho—State Key Laboratory of Genetic Resources and Evolution; Ecology, Conservation and Environment Center (ECEC), Kunming Institute of Zoology, Chinese Academy of Science, Kunming, Yunnan, 650223, China

Luis Solorzano—Gordon and Betty Moore Foundation, 1661 Page Mill Road, Palo Alto, CA 94304, USA

Bill Stanley—Climate Change Team, The Nature Conservancy, 4245 North Fairfax Drive, Arlington, VA 22203, USA

Charlotte Stanton—Emmett Interdisciplinary Program in Environment and Resources, 397 Panama Mall, Stanford University, Stanford, CA 94305, USA

Heather Tallis—Natural Capital Project, Woods Institute for the Environment, 371 Serra Mall, Stanford University, Stanford, CA 94305, USA

Christine Tam—Natural Capital Project, Woods Institute for the Environment, 371 Serra Mall, Stanford University, Stanford, CA 94305, USA

C. Peter Timmer—Harvard Professor of Development Studies, emeritus, P.O. Box 1402, Kenwood, CA 95452, USA

R. Kerry Turner—Centre for Social and Economic Research on the Global Environment (CSERGE), School of Environmental Sciences, University of East Anglia, Norwich, NR4 7TJ, UK

Bernard E. Vaissière—INRA, UMR406 Abeilles & Environnement, 84914 Avignon Cedex 9, France

Nathan Vadeboncoeur—Institute for Resources, Environment, and Sustainability, University of British Columbia, Vancouver, BC, Canada

Nicole Virgilio—Climate Change Team, The Nature Conservancy, 4245 North Fairfax Drive, Arlington, VA 22203, USA

Michael Todd Walter—Department of Biological and Environmental Engineering, Riley Robb Hall, Cornell University, Ithaca, NY 14853-5701, USA

Yukuan Wang—Institute of Mountain Hazards, Chinese Academy of Sciences, No.9, Block 4, Renminnan Road, Chengdu, China

Sue White—School of Applied Sciences, Building 53, Cranfield University, Cranfield, Bedfordshire, UK

Eric Wikramanayake—Conservation Science Program, World Wildlife Fund-US, 1240 24th Street NW, Washington, DC 20037, USA

Neal Williams—Department of Biology, Bryn Mawr University, Park Science Building, 101 N. Merion Avenue, Bryn Mawr, PA 19010, USA

Rachel Winfree—Department of Entomology, State University New Brunswick, 119 Blake Hall 93 Lipman Drive, Rutgers, NJ 08901, USA

Stacie Wolny—Department of Biology and Natural Capital Project, Woods Institute for the Environment, 371 Serra Mall, Stanford University, Stanford, CA 94305, USA

Stanley Wood—International Food Policy Research Institute, 2033 K Street NW, Washington, DC 20006, USA

Ulalia Woodside—Land Assets Division, Kamehameha Schools, Honolulu, HI, USA

Hazel Wong—The Nature Conservancy, USA

Douglas W. Yu—State Key Laboratory of Genetic Resources and Evolution; Ecology, Conservation and Environment Center (ECEC), Kunming Institute of Zoology, Chinese Academy of Science, Kunming, Yunnan, 650223, China, AND Centre for Ecology, Evolution and Conservation (CEEC), School of Biological Sciences, University of East Anglia, Norwich, NR4 7TJ, UK

Jing Zhang—Department of Bioresource Policy, Business and Economics, University of Saskatchewan, Saskatoon, Saskatchewan, Canada, S7N 5A8

Wei Zhang—International Food Policy Research Institute

Foreword

Getting there

In 1997 two books were published that focused on the significance of biological diversity for the welfare of humankind. One was Yvonne Baskin's delightful book on *The Work of Nature: How the Diversity of Life Sustains Us* (Baskin 1997) and the second, the highly influential volume edited by Gretchen Daily on *Nature's Services: Societal Dependence on Natural Ecosystems* (Daily 1997). These volumes provided a compelling rationale for conserving biological diversity as not only a social responsibility of society but also as a necessity for human prosperity and survival. The crucial interface between biological diversity and ecosystem services to human well-being had finally been made explicit in these seminal publications. These books marked a turning point in ecological science and conservation.

The idea that ecosystem services provide an imperative for conservation became the launching pad for the monumental Millennium Ecosystem Assessment (MA). Using the ecosystem service paradigm, the MA took a global view of the status and trends of ecosystems and the services they provide, plausible scenarios of the capacity of ecosystems to deliver services in the future, and the response options available to society that would lead to the continuance of the delivery of the vital services that underpin human endeavors.

The MA was developed under the auspices of the UN, and guided by wide representation including those from intergovernmental conventions, NGO's, and industry. Over 1300 scientists were involved in the production of a number of products that received wide distribution. The social process that led to its development, as well as the resultant publications, sparked the interest of a wide audience including the policy-making community, one of the targeted audiences. Throughout the world ecosystem service concepts are now being incorporated into development and strategic planning. The concept that ecosystem services benefit society has resonated with an extraordinary breadth of constituencies, including the development community that has traditionally viewed environmental priorities as an impediment to development.

So, the case has been made, but the means to practical utilization of the concepts and findings of the MA are not sufficiently developed for easy implementation. It was noted in the MA summary that, "the scientific and assessment tools and models available to undertake a cross-scale integrated assessments and to project future changes in ecosystem changes in ecosystem services are only now being developed" (MA 2005). The tools needed to carry out assessments at local levels, in the framework of the MA, have been aided greatly by the recent publication of *A Manual for Assessment Practitioners* (Ash *et al.* 2010).

Now, this volume represents a major leap forward in providing tools to utilize ecosystem service concepts in decision-making. It has been produced by a team with an unusual history. In 2006, three veterans of the MA formed a unique partnership to take the next step toward bringing ecosystem services science into practice. Gretchen Daily (Stanford University), Peter Kareiva (The Nature Conservancy), and Taylor Ricketts (World Wildlife Fund) founded the Natural Capital Project, dedicated to producing quantitative tools for spatially explicit valuation of ecosystem services, and applying them in major resource decisions worldwide. The project has blended the muscle of a research university with the practical perspectives and global networks of the two largest conservation organizations. From the outset, Stephen Polasky (University of Minnesota) and Heather Tallis (Stanford University) have co-led the modeling efforts, and Minnesota has recently joined as a fourth formal partner.

This team has produced a series of models for an array of key ecosystem services that can be used in concert to provide scenarios of the land-use decisions on the subsequent delivery of a bundle of services. These models were designed for use by a wide range of practitioners and have the capacity to utilize input data of differing levels of resolution. The Natural Capital Project has not only developed these models, but has been applying them in many regions in the world.

There is no doubt that the application of the material in this volume will provide a major step forward in the search for practical approaches that will become mainstream and will further the goal of society of conserving the biological diversity that produces the ecosystem services vital for our future well-being.

Hal Mooney

Ash, N., Blanco, H. *et al.*, Eds. (2010). Ecosystems and Human Well-Being. A Manual for Assessment Practioners. Island Press, Washington, DC.

Baskin, Y. (1997). The Work of Nature. How the Diversity of Life Sustains Us. Island Press, Washington, DC.

Daily, G.D., Ed. (1997). Nature's Services. Societal Dependence on Natural Ecosystems. Island Press, Washington, DC.

MA (2005). Millennium Ecosystem Assessment. Ecosystems and Human Well-being. Synthesis. Island Press, Washington, DC.

How to read this book

This book is an outgrowth of the Natural Capital Project, which seeks to ensure that nature's value is accounted for in all of our business, policy, and development decisions. In this sense, the book depicts the "state of science" relevant to the Natural Capital Project. Together, the chapters lay the scientific foundation for our project, and are linked by that common goal. More importantly, however, each chapter represents a general contribution to modeling ecosystem services and connecting them to resource management. Each is written to "stand alone" as much as possible. The book therefore need not be read front to back—read the chapters of interest to you.

In order to cover a wide array of topics concerning nature's benefits and how those benefits might shape public and private decisions, we have had to ask authors to leave out a lot of important detail. To find more details on the models one can go to http://naturalcapitalproject.org/InVEST.html, and to find publications in technical journals about various applications of these models go to the Natural Capital website: http://naturalcapitalproject.org/publications.html.

We recognize that thousands of scientists and policy makers around the world are striving to incorporate nature's value into their work. There are hundreds of stories that reveal personal and institutional discoveries surrounding nature's benefits. In order to capture some of this exciting diversity of experience regarding the value of nature, we include essays throughout the book—set off in boxes—that tell some story apart from the technical details of the main text. These boxes can be read on their own. They are not intended to be part of the main course—they are appetizers.

Finally, we encourage all readers to join the community of scientists and policy makers working to create a world molded by better-informed decisions that factor in the services nature provides for human well-being. We invite you all to go to the Natural Capital Project website and join a community of users for the models described in this book.

The Editors

Acknowledgments

This book was possible only because of the generous support of several foundations and donors. We are grateful for the support of the National Science Foundation, the John D. and Catherine T. MacArthur Foundation, the David and Lucille Packard Foundation, the Gordon and Betty Moore Foundation, the Google Inc. Charitable Giving Fund of Tides Foundation, the Leverhulme Trust, the Resources Legacy Fund Foundation, the Winslow Foundation, and the National Center for Ecological Analysis and Synthesis. Several individuals have also donated their time, advice, and financial support. In particular, we could not even have begun this project without start-up funding from Vicki and Roger Sant, and Peter and Helen Bing.

Advice is cheap, but good advice is hard to come by. We are lucky to have received advice from some of the leading thinkers in conservation and science, including Paul Ehrlich, Steve McCormick, Walt Reid, Hal Mooney, Jane Lubchenco, Mary Ruckelshaus, Buzz Thompson, Jeff Koseff, Jim Salzman, Chuck Katz, Kerry Turner, Joshua Galdstein, Neil Hannahs, Dennis White, Joshua Lawler, Jimmy Kagan, Stacie Wolny, Sue White, Andrew Balmford, Neil Burgess, Mattieu Rouget, Kai Chan, Rebecca Shaw, Nasser Olwero, Jim Regetz, Yukuan Wang, Hua Zheng, Zhiyun Ouyang, Li Shuzhuo, Mark Plummer, Robin Naidoo, George Jambiya, Silvia Benitez, Eric Lonsdorf, Dick Cameron, and many others. We are also deeply grateful to the practitioners at WWF and TNC who continue to field test our ideas and give critical, real-world feedback to our efforts. Finally, our employers have given us the freedom to occasionally ignore our day jobs and get this book done and for their tolerance and patience we are grateful to Stanford University, WWF, The Nature Conservancy, and the University of Minnesota.

The Editors

SECTION I

A vision for ecosystem services in decisions

CHAPTER 1

Mainstreaming natural capital into decisions

Gretchen C. Daily, Peter M. Kareiva, Stephen Polasky, Taylor H. Ricketts, and Heather Tallis

1.1 Mainstreaming ecosystem services into decisions

The past several decades have produced tremendous change in how people think about the environment and human development. The focus of environmental issues in the 1960s and 1970s was on air and water pollution with an immediate impact on the local surroundings. Actions to reduce pollution occurred primarily in relatively wealthy countries able to afford it.

Now the focus has expanded to encompass the benefits from (and losses to) living natural capital: Earth's lands and waters and their biodiversity. Food and fiber production, provision of clean water, maintenance of a livable climate, security from floods, the basis for many pharmaceuticals, and appreciation of the wonders and beauty of the natural world are a few of the many dimensions of human well-being that hinge on living natural capital (Daily 1997).

The importance of maintaining natural capital for the ecosystem service benefits that flow from it is increasingly seen as vital in both poor and rich countries alike. Indeed, declining natural capital poses a direct threat to rural poor since they depend closely on the environment for their livelihood (Dasgupta 2010). After spending decades struggling to fence off nature from people, conservation is emerging on the global stage with a new vision that emphasizes the importance of connecting nature and people (Kareiva and Marvier 2007).

One of the largest efforts to date, the Millennium Ecosystem Assessment, illustrated the many ways in which natural systems are vital assets critical for human well-being (MA 2005). The Millennium Ecosystem Assessment took a giant step forward in developing a widely shared vision, a conceptual framework, and a synthesis of existing knowledge. It spawned a suite of further efforts, including an Intergovernmental Science-Policy Platform for Biodiversity and Ecosystem Services (Mooney and Mace 2009; Larigauderie and Mooney 2010). By almost any measure—scientific papers published, media mentions, Google search trends—awareness of natural capital and efforts to sustain it have skyrocketed since the Millennium Assessment.

The Millennium Ecosystem Assessment's vision is starting to take hold. China, for instance, has invested over 700 billion yuan (approximately $100 billion) in ecosystem service payments over 1998–2010 (Zhang *et al.* 2000; Liu *et al.* 2008). In addition, China has established a new system of "ecosystem function conservation areas," spanning 25% of the nation's land area where the most vital elements of living natural capital will be protected for securing and harmonizing human and natural well-being.

It is not just giant modernizing nations that are bringing a new view to nature. The value of natural capital is being included in decisions taken by community leaders, traditional cultures, and global corporations. For example, payments for watershed service projects make up a significant portion of existing ecosystem services schemes (many others relate to carbon) (Goldman *et al.* 2010). These schemes typically involve water users paying upstream land managers for the delivery of clean, consistent water supplies (Brauman *et al.* 2007; Porras *et al.* 2007; Wunder *et al.* 2008), and have in some places become more extensive and sophisticated in design (e.g., Nel *et al.* 2009). In Hawai`i, policies and payments for a wide array of services is being promoted through local watershed agreements, the state's House Concurrent Resolution on Ecosystem Services (passed in 2006), the state's Climate Bill (passed in 2007), and leadership of the state's largest private landowner, Kamehameha Schools (Chapter 14). Companies including Coca-Cola, LaFarge, and Mondi are evaluating the role of ecosystem services within their supply chains and working to invest in them (Varga 2009, WRI 2010, McKenzie *et al.*, Chapter 19 this volume).

Including the value of ecosystem services in the decisions of governments, corporations, traditional cultures, and individuals does not replace or undermine the intrinsic value of nature, nor the moral imperative to conserve it (e.g., Leopold 1949; Norton 1987; Ehrenfeld 1988; Rolston 2000). Instead, valuing ecosystem services and natural capital complements these moral concerns, broadening our understanding of the roles nature plays in our lives and the reasons for conserving it. If we can add how nature contributes to human well-being to the arguments for conservation, why wouldn't we?

While the recent transformation in the way people think about nature and human development has been productive, the urgent challenge now is in moving from ideas to action on a broad scale (Carpenter *et al.* 2006, 2009). Mainstreaming ecosystem services into everyday decisions requires a systematic method for characterizing their value—and the change in value resulting from alternative polices or human activities. Unlike the well-established accounting tools we apply to measure the value of traditional economic goods and services, we have no ready set of accounting tools to measure the value of ecosystem services (MA 2005; NRC 2005; Mäler *et al.* 2008). Absent these, ecosystem services are invariably undervalued or not valued at all—by governments, businesses, and the public (Daily *et al.* 2000; Balmford *et al.* 2002; NRC 2005; Dasgupta 2010). The result is continued losses in natural capital and biodiversity. Often, it is only after their loss that we recognize the importance of ecosystem services, such as in the wake of Hurricane Katrina or cyclones in India (Stokstad 2005; Das and Vincent 2009).

But who and what will catalyze the next giant step forward? Part of the answer lies with improving science. The natural and social science communities need to attack a set of difficult and compelling issues: How can such complex processes as the role of forests in flood control or crop pollination be quantified accurately? How can such diverse values as are embodied in cultural services be characterized meaningfully? How can we make credible projections of natural capital under scenarios of change, such as in population, climate, or resource management? And how can we build the capacity in civil society and institutions—and in deep aspects of human beliefs and behavior—to take account of ecosystem services and natural capital?

This book tackles these science issues, while acknowledging the many other social and political elements to the problem. It is intended to supply one of the catalysts required for a new approach that harmonizes conservation and development.

1.2 What is new today that makes us think we can succeed?

An appreciation of ecosystems as valuable capital assets traces back to Plato and doubtless much earlier (Mooney and Ehrlich 1997), and the current research agenda on ecosystem services continues long-standing lines of work. For example, renewable resources have been an active area of study in economics since at least the 1950s, when Gordon (1954), Scott (1955), and Schaefer (1957) characterized harvesting a biological stock and the problems of open-access fisheries. In the 1960s and 1970s, economists set out to measure "the value of services that natural areas provide" (Krutilla and Fisher 1975, p. 12) that included the value of renewable resources (Krutilla 1967; Clark 1990), non-renewable resources (Dasgupta and Heal 1979), and environmental amenities (Freeman 1993). More recent advances in a broad range of areas, such as in ecology and global change, economics, policy and institutions, and especially their integration, have broadened this work to include a wider set of ecosystem services and an examination of the set of human actions needed to maintain the flow the services (e.g., Dasgupta 2001; MA 2005; NRC 2005; Ruhl *et al.* 2007).

Four big advances of the past decade promise to make an old good idea a new beacon for real change. First, the Millennium Ecosystem Assessment represented a visionary and seminal step in global science—it was the first comprehensive global assessment of the status and trends of all of the world's major ecosystem services. The key finding of this assessment was that two-thirds of the world's ecosystem services were declining, a finding that captured the attention of world leaders (MA 2005).

Second, the science of ecosystem functions and processes has made huge advances so that we can now model (albeit with uncertainty) the impacts of land use and resource management decisions on a wide variety of ecosystem processes. Ecological science has also become adept at spatially explicit modeling, which is essential for mapping ecosystem services and their flows to people (e.g., Chan *et al.* 2006; Rokityanskiy *et al.* 2007; Bennett *et al.* 2009; Nelson *et al.* 2009; Harrison *et al.* 2010).

Third, economic valuation methods have been applied to the spatial provision of ecosystem services to estimate the monetary value of benefits and the distribution of those benefits to various segments of society (NRC 2005; Naidoo and Ricketts 2006). In addition, qualitative and quantitative methods from other social sciences have been applied to gain better understanding of the social and cultural importance of ecosystem services (e.g., MA 2005; US EPA 2009).

Lastly, experiments in payments for ecosystem services (Pagiola *et al.* 2002; Pagiola and Platais 2007; Wunder *et al.* 2008), in ecosystem-based management (Barbier *et al.* 2008), and in regional planning give us the empirical data for evaluating approaches to valuing ecosystem services and incorporating values into decision-making. There is a growing recognition that bundling together of ecosystem services and explicit attention to trade-offs will both better inform decisions, and help diverse stakeholders to appreciate the perspectives of others (e.g., Boody *et al.* 2005; Naidoo and Ricketts 2006; Egoh *et al.* 2008; Bennett *et al.* 2009; Nelson *et al.* 2009).

Our challenge today is to build on this foundation and integrate ecosystem services into real decisions. Doing so requires understanding the interlinked; joint production of services; quantify-

ing the multitude of benefits derived from services to various segments of society; understanding the decision-making process of individuals, corporations, and governments; integrating research with institutional design and policy implementation; and crafting policy interventions that are designed for learning and improvement through time. Each of these alone is a complex task; together they form a daunting but critically important agenda requiring a global collaboration.

1.3 Moving from theory to implementation

In moving from theory to practical implementation, Figure 1.1 presents a framework of the role that ecosystem services can play in decision-making (Daily *et al.* 2009). This framework connects the science of quantifying services with valuation and policy to devise payment schemes and management actions that take account of ecosystem services.

Though the framework is a continuous loop, we start with the **decisions** oval to emphasize our focus. After all, the main point of understanding and valuing natural capital and ecosystem services is improving natural resource decisions. So we start—and end—there. These **decisions** encourage and constrain **actions** relating to the use of land, water, and other elements of natural capital.

Continuing clockwise around Figure 1.1, "biophysical sciences" are central to understanding the

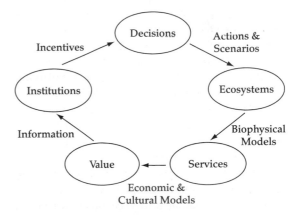

Figure 1.1 A framework showing how ecosystem services can be integrated into decision-making. One could link any two ovals, in any direction; we present the simplest version here.

link between **decisions** and **ecosystems**, and along with economics and social science, the links between **ecosystems** and **services**. We study the former link with classic ecology and conservation biology to, for example, estimate impacts of land-use change on biodiversity (e.g., Daily *et al.* 2001; Steffan-Dewenter *et al.* 2007). And we pursue the latter link with "ecological production functions" that relate, say, forest condition and management practices to the supply of carbon storage, pollination, and other ecosystem services (e.g., Ricketts *et al.* 2004).

Social sciences are also central to understanding the **value** of services to people ("economic and cultural models"). Economic valuation techniques are commonly used for this link, to place monetary value on natural capital. Value is often not fully captured in monetary terms, though, so it is important to characterize value in multiple dimensions, including, for example, health, livelihood support, cultural significance (e.g., Dasgupta 2001). This will help ensure that valuation and broader decision-making approaches are inclusive of the range of benefits and people concerned (Heal 2000a, 2000b).

Finally, valuing ecosystem services provides useful information that can help shape **institutions** (e.g., agricultural markets, subsidies, land-use policies, conservation NGOs) to guide resource management and policy. Having the right **institutions** can create incentives so that the **decisions** of individuals, communities, corporations, and governments promote widely shared values. The links between the **value**, **institutions**, and **decisions** ovals are much more the art and politics of social change than science, though scientists can inform these debates if they target specific decisions and are attuned to the social and political contexts.

This idealized framework is helpful in clarifying the many frontiers of research and implementation en route to operationalizing ecosystem services into decisions (see also Carpenter *et al.* 2009). This includes continued biophysical research on the impacts of human actions on ecosystems, all the way to studies on the way landowners respond to conservation incentives. Chapters in this book touch on all ovals and all arrows within the framework, but the core chapters focus on moving from ecosystems to services, and from services to value, using production functions and valuation techniques.

1.4 Using ecosystem production functions to map and assess natural capital

There are several methods for mapping and assigning value to ecosystem services, each with its own advantages and limitations. The initial valuation work in the field of ecosystem services primarily used what is called the benefit transfer approach (e.g., Costanza *et al.* 1997). This approach typically uses empirical estimates of the value of goods produced from some habitat type and transfers those benefits to similar habitats elsewhere, including anywhere in the world (Costanza *et al.* 1997). Local knowledge can be used to adjust the benefits because one knows, for instance, that the west coast marshes of North America are less productive than the east coast marshes and so forth. The general idea, however, is to use lookup tables of benefits per unit area of habitat type, and thereby quantify overall natural capital.

An alternative method favored in this book is called a "production function approach." Instead of relying on lookup tables, we build models that predict local ecosystem service supply based on land cover, land use, ecosystem attributes, human demand, etc. These functions are analogous to those long used in agriculture, which relate amounts of water, fertilizer, and labor to resulting crop yield. In our view, production functions have key advantages over benefits transfer, and we delve into these further in Chapter 3.

1.5 Roadmap to the book

Our book begins with three chapters that introduce the core approach and hypotheses of our work on natural capital. Chapter 2 examines the philosophical bases for ecosystem service value and explores ways of measuring such value, distinguishing alternative approaches and highlighting some ethical issues underlying the choices among them. It also explores the strengths and weaknesses of these measurement approaches, and indicates which approaches are best suited to the different types of value conferred by ecosystem services. Chapter 3 then introduces the modeling approach we have developed, which strives to integrate many different ecosystem services, to do so over scales appropriate to important resource decisions, and to assess trade-offs among services on real landscapes. All resource decisions involve these trade-offs (e.g., between biological carbon sequestration and stream flow; Jackson *et al.* 2005). Yet, all too often, the importance of trade-offs among services is lost in decision-making, with the result that unintended consequences arise while pursuing what at first seems like a good idea.

The middle section of the book delves into details for each of the core models of ecosystem services. The specific services we model are water supply for hydropower and irrigation (Chapter 4), flood damage avoidance (Chapter 5), water pollution regulation (Chapter 6), carbon storage and sequestration (Chapter 7), production of timber and non-timber forest products (Chapter 8), agricultural production (Chapter 9), crop pollination (Chapter 10), enhancement of recreation and tourism (Chapter 11), and provision of cultural services (Chapter 12). We also model biodiversity, as an ecosystem attribute (Chapter 13).

Like all early efforts in modeling, we try to strike a balance between scientific rigor, data availability, and practical usability. Some will object that the models are overly simplistic; others will find them hopelessly complicated (indeed, reviewers have made both arguments for almost every chapter). We offer two tiers of models for each ecosystem service. Tier 1 is the simplest credible model we could devise, with data needs that can be met even in data-poor regions that are often so fundamental to both conservation and human livelihoods. Tier 2 models offer more complexity, specificity, and realism for users and places with the data to support them. We include enough math in each chapter to make the modeling approach clear. And we have implemented tier 1 equations into a modeling tool available for free download at http://invest.ecoinformatics.org.

The final section of the book is based on potential applications of our approach to modeling and mapping natural capital. Applications are messy and demanding, and require links to other fields of science as well as policy. In this "getting real" section of the book, we discuss trade-offs (Chapter 14), difficult choices about how complicated or detailed models need to be (Chapter 15), the implicit but rarely quantified link between ecosystem services and poverty (Chapter 16), the challenge of extending our approach to marine ecosystems (Chapter 17), assessing the impacts of climate change on ecosystem services (Chapter 18), and ideas for how all of this science might actually enter into decision-making and policy (Chapter 19).

In all chapters we include short essays by contributors who are using concepts of natural capital in their conservation and policy work. We include these essays to emphasize that our models play only one small part in a world of innovation surrounding natural capital. There has never been a more exciting time for conservation and ecosystem science than now—but some of that excitement is shrouded in equations and modeling details. It is in our essays that one can find evidence of the tipping point that is before us, in contemplating the African boy cooking a monkey (Box 1.1, this chapter); the first national exchange for carbon storage credits (Chapter 7); the hopes we pin on agroecosystems as highways out of poverty (Chapter 16); vastly different options for a sustainable future (Box 1.2, this chapter); and many others.

Box 1.1 The everyday meaning of natural capital to the world's rural poor

M. Sanjayan

The boy is no more than 10 years old, bare legs scarred by tropical parasites, clad only in dingy shorts despite the threatening rain (Figure 1.A.1). He is engrossed in his task of carefully burning the fur off a dead monkey. I have stumbled onto this unfolding scene in a village in Sierra Leone, West Africa, whose inhabitants are amongst the poorest in the world despite being surrounded by a wealth of biodiversity. The monkey is a *Cercopithecus* of some sort, perhaps a white-nosed Guenon, a relatively common crop-raiding monkey in these parts. Holding it carefully in

continues

Box 1.1 *continued*

Figure 1.A.1   Young boy preparing to eat a monkey, rural Sierra Leone. Photo by M. Sanjayan.

both hands, the boy slowly turns it as the pelt singes and curls into soft gray ash. It is clearly a delicate task, with the flames struggling to catch the rain-soaked pieces of stick fed into the weak fire. Occasionally the boy pauses and, with a piece of tin sheet metal, furiously fans the smoky mess. An acrid odor hangs in the heavy air.

The monkey will soon be food. It will be dismembered, every bit from nose to tail, thrown into a pot with some okra, peppers, or other meager vegetables, and a few drops of palm oil—a stew ultimately yielding, based on the small crush of spectators, what I estimate to be about two tablespoons of meat protein for each person. A small monkey in a big pot. Fascinated by the boy's handiwork, I pull out my camera and snap a photo—and immediately feel a little shabby about it. The boy just giggles.

Spend any amount of time quietly observing the daily rituals of rural village life in any tropical country and you cannot but be impressed by the magnitude and diversity of services people derive from their immediate surroundings; nature if you will. However, these services provided by nature are nearly always taken for granted. Local communities are usually myopic in their understanding of similar communities, encircling what are to them virtually endless resources, and governments who can see the big picture are reluctant to acknowledge it lest they expose their shortcomings. Plus, changes are usually slow and accumulate over a long period of time, thus hiding cumulative impacts in the imperfect memories of the elders.

As I see it, six basic services provide most of the daily needs of extremely poor rural people. Fresh water is the most obvious and its procurement is taxing, particularly to women, who bear most of the load. While there are taps sprouting in many rural villages in Africa, few connect to sustainable water sources. Fuel wood, collected from forests, plantations, or local groves, is indispensible for the heating and cooking needs of 40% of rural homes. Gathering it is the second most taxing chore (after water) that impacts daily life. Fisheries provide protein to 20% of the world's population. On the Ganges River in India, for example, ten million people in 2000 villages depend on fishing to both meet their daily needs and provide jobs. Fertility of soils, and its natural renewal through processes of nutrient cycling, is essential to places untouched by the Green Revolution and does not involve the consequences of industrial fertilization in terms of energy use and nutrient overload. Forest products, like meat for protein from forest animals, fruit, honey, medicinal plants, and fiber, have a myriad of vital uses. Fodder, in terms of grass and browse for livestock, is important in rural communities because it is one of the few ways through which the poor can access the global economy. Livestock is the common bank for rural populations. These "6 Fs" of nature (six free services) are part of the staple packet of goods and services that virtually every rural community depends upon and that governments conveniently ignore, and non-profits underestimate the importance of. Lose them and people will suffer.

Later, when I look through my day's pictures on my camera, it is not the dead monkey that draws my attention. It's the sheet of tin metal, the one the boy used to fan the flames that I focus on. It's a piece of a signboard—one of millions that line roads all over Africa, from charities, governments, and religious organizations. On this one, framed against a white background, are the blue letters—WFP—suspended over the image of a bundle of grain. The World Food Programme. Here is this kid, fanning the flames with a sign board for a humanitarian organization, to burn the fur off a dead monkey from the forest, which is what he is actually going to eat. For the poor, nature often provides when governments and institutions don't suffice, and that is a powerful lesson.

As we begin to better map, quantify, and assign monetary value on nature's services, from carbon to water, to more difficult-to-assess services like fodder, we need to ensure that we are adequately capturing the roughly one-sixth of the world's population who live on the very margins of national or global economies but whose needs for such services are not just dire, but virtually irreplaceable. The value of water to a woman in West Africa trying to prepare a meal is far higher (though not in strictly monetary terms) than that to a woman filling up her swimming pool in California. Valuation of ecosystem services must properly recognize, and incorporate, this vast social net that nature provides to the poor.

I think about returning to the village and bartering for the sign; hanging in my office, it would be a powerful reminder of this insight. But I have nothing useful to trade. Nothing sustainable. For now, the boy needs it more than I do.

1.6 Open questions and future directions

This book and the modeling approaches we introduce are only a beginning. We anticipate the research community adding other ecosystem services over time, as well as continually improving the models and data for those presented in this book. These additions and improvements will come from confronting these and other models with a variety of real-world data and challenges. Here we mention two of the many key arenas in which further understanding is crucial.

First, major advances in methods and tools are needed to incorporate dynamic effects, as well as shocks and uncertainty. Dynamic changes (e.g., in climate and in the nitrogen cycle) and changes arising through economic development and evolving human preferences over time are important to include. The possibility of feedbacks within ecosystems, and between ecosystem services and human behavior, are important areas for further development. Feedback effects can give rise to thresholds and rapid changes in systems that can fundamentally alter system states (Scheffer *et al.* 2001). The ability to incorporate shocks and the possibility of surprises is another area where further development is needed. Fires, droughts, and disease all can have major influences on ecosystems and affect the services produced. Changes in economic conditions or fads in human behavior can similarly cause major changes in systems (e.g., financial crises). The occurrence of each of these and other potential disturbances is difficult to predict but virtually certain to come about. Understanding their likely impacts on ecological and social systems will help us prepare for them.

A second major area for further development is in relating ecosystem condition to human health. The relationships between biophysical attributes of ecosystems and human communities are complex (Myers and Patz 2009). Destruction of ecosystems can at times improve aspects of community health. For example, draining swamps can reduce habitat for the mosquito vector that transmits the parasite that causes malaria. On the other hand, ecosystems provide many services that sustain human health, for which substitutes are not available at the required scale, such as purification and regulation of drinking water flow; regulation of air quality; nutrition (especially of protein and micronutrients); psychological benefits; and, in complex ways, regulation of vector-borne disease (Levy *et al.* in press). To date, there is little rigorous research establishing the links between ecosystem conditions and human health.

Box 1.2 Sorting among options for a more sustainable world

Stephen R. Carpenter

"What gives you the most hope for the environment over the next 50 years?" When we asked that question of 59 global leaders in 2003, we expected great variability in the answers. To our surprise, the answers fell rather cleanly into three clusters. Some respondents were not worried and simply had faith in economic growth, being convinced that a sustainable environment would follow automatically from economic development. Two clusters stood out, however, because they envisioned futures that were less automatic, and that would need some guidance if we were to achieve a hopeful outcome. The first of these "we need change" clusters thought that innovation and investments in environmentally friendly technology was the key. The second "we need change" group felt that governance should be restructured to motivate local innovation and learning, and thereby create sustainable landscapes from the bottom up.

These two clusters of ideas became the *Technogarden* and *Adapting Mosaic* scenarios of the Millennium Ecosystem Assessment (MA 2005). We evaluated the condition of 24 global ecosystem services from 2000 to 2050 under these and other scenarios. *Technogarden* and *Adapting Mosaic* were the most successful scenarios for maintaining ecosystem services. However, these two scenarios represent very different policies that lead to different bundles of global ecosystem services by 2050.

In *Technogarden*, society addresses global environmental problems such as climate heating, materials cycles, and nutrient mobilization through innovations in energy production, buildings, transportation, and agriculture (Figure 1.B.1). Improvements in agriculture and urban design make it possible to feed and house humanity without extensive new conversion of wild lands. Market mechanisms and sophisticated economic instruments are deployed to manage ecosystem services. International cooperation on incentives for better technology lays the foundation for improved cooperation on other problems of the global commons, such as pelagic marine fisheries, disease containment, and conservation of antibiotics. Expanded access to education leads to smaller family sizes and thereby slows the pace of population growth. Even though many ecosystem services are in sustainable condition by 2050, there are some downsides. Some aesthetic, cultural, and spiritual aspects of ecosystems are

Figure 1.B.1   Depiction of the *Technogarden* scenario of the Millennium Ecosystem Assessment. From *Ecosystems and Human Well-Being: Scenarios*, by the Millennium Ecosystem Assessment. Copyright © 2005 Millennium Ecosystem Assessment. Reproduced by permission of Island Press, Washington, D.C.

lost or irreversibly changed. Local ecological knowledge is sometimes lost as management becomes more centralized. Unexpected consequences of technology lead to some big accidents. Nonetheless, *Technogarden* offers many successes in management of ecosystem services.

Adapting Mosaic begins with reorganization of governance around institutions tailored to naturally occurring clusters of ecosystem services (Figure 1.B.2). For example, the Headwaters of the Missisippi River in North America (Minnesota and Wisconsin, plus parts of eastern Iowa and northern Illinois) organizes around sustainable agriculture mostly for local consumption, ecosystem management for abundant clean freshwater, and urban areas known for environmentally friendly high-technology and biotechnology industries. Within the overarching Headwaters region, responsibility for ecosystem management organizes around sub-water-sheds at the smallest spatial extent, and major ecore-gions at an intermediate spatial extent. Governance of most other regions of the world undergoes similar adjustments to accommodate natural patterns of ecosystem services.

Adapting Mosaic stresses local innovation and learning by doing to improve ecosystem services. Even though global economic linkages are sparse, information linkages are strong. Innovations and news of failed experiments can spread rapidly. The global network makes rather fast progress on improving practices for ecosystem services. Ironically, however, the withering of global institutions hinders progress on problems of the global commons such as climate heating and pelagic marine fisheries.

The scenarios of the Millennium Ecosystem Assessment are not a prescription for solving the world's problems. They are more like hypotheses to be tested. So why not combine the best of *Technogarden* and *Adapting Mosaic*? The global commons problems are critical, and improved technology will be needed to create better energy sources, agriculture, transportation, and infrastructure. Half of humanity lives in cities. The USA alone will replace most of its infrastructure by 2030, and in the 21st century the world will erect more buildings than in the entire history of our species before 2000. This reconstruction is an opportunity for lowering the impact of cities on the global commons and on the rural regions that feed the cities and absorb their waste. At the same time we are addressing the global commons, there are many benefits available from multiscale adaptive management of landscapes. These benefits can be financed through appropriate pricing of the ecosystem services that rural regions and wild regions provide.

Figure 1.B.2 Depiction of the *Adapting Mosaic* scenario of the Millennium Ecosystem Assessment. From *Ecosystems and Human Well-Being: Scenarios*, by the Millennium Ecosystem Assessment. Copyright © 2005 Millennium Ecosystem Assessment. Reproduced by permission of Island Press, Washington, D.C.

continues

Box 1.2 *continued*

Even though the Millennium Ecosystem Assessment did not compute an optimal path to 2050, it is likely that the best mix of options is some combination of *Technogarden* and *Adapting Mosaic*. The challenge is to understand which combinations of ecosystem services are biophysically possible, the trade-offs among different bundles of ecosystem services possible from a given region, and the institutional frameworks that enable ongoing flows of ecosystem services.

1.7 A general theory of change

Mainstreaming natural capital into decisions is a long-term proposition, requiring co-evolving advances in knowledge, social institutions, and culture. Certainly a single book is not sufficient for achieving this. We propose instead that our book contributes to an overall theory of change (Bradach *et al.* 2008) involving three key, broad elements.

First, businesses, governments, and individuals must find it easy to inculcate ecosystem services and natural capital into their decisions, and the methods for doing so must be transparent, credible, and predictable. In many cases, sectors of society are open to the idea of ecosystem services and natural capital, but simply do not know how to take the idea and use it in a concrete way.

Second, there need to be examples of projects or enterprises that—as a result of properly valuing ecosystem services and natural capital—end up with improved decisions, institutions, and human well-being. These examples both test our science against real-world problems and produce compelling stories of how an ecosystem services approach made a difference.

Lastly, political and thought leaders must appreciate these examples of success and spread the word. This is where the lessons of a set of examples can be mainstreamed into the myriad decisions—by businesses, governments, farmers, and banks—that are made every year and that impact our natural world. This is where the impact of scattered projects can be magnified into worldwide change.

None of these steps are complicated, and our theory of change does not require a brilliant and novel strategy. In fact, we are convinced that all three ingredients are within striking distance. The environmental movement has a much bigger and more diverse and powerful community behind it now than ever before (Daily and Matson 2008). Science is beginning to provide tools and methods that will reduce the transaction costs. And there are enough policy experiments underway that compelling examples of natural capital stewardship enhancing human well-being should be forthcoming.

Our book targets the first element of our theory: to make quantifying and valuing natural capital straightforward and routine. Science is not everything, but both modeling and empirical science provide the foundation for action. The models we rely on are not a *fait accompli*—they are the first step in an iterative process between basic science and application to real-world problems. That is why we highlight case studies in which valuation of natural capital is being used to influence land and water management. Science by itself cannot change the world, but science plus the vision and action of leaders can—and that is what we seek.

References

Balmford, A., Bruner, A., Cooper, P., *et al.* (2002). Economic reasons for conserving wild nature. *Science*, **297**, 950–3.

Barbier, E., Koch, E., Silliman, B., *et al.* (2008). Coastal ecosystem-based management with non-linear ecological functions and values. *Science*, **319**, 321–3.

Bennett, E., Peterson, G., and Gordon, L. (2009). Understanding relationships among multiple ecosystem services. *Ecology Letters*, **12**, 1394–1404.

Boody, G., Vondracek, B., Andow, D., *et al.* (2005). Multifunctional agriculture in the United States. *BioScience*, **55**, 27–38.

Bradach, J., Tierney, T., and Stone, N. (2008). Delivering on the promise of nonprofits. *Harvard Business Review*, **Dec.**, 88–97.

Brauman, K., Daily, G. C., Duarte, T. K., and Mooney, H. A. (2007). The nature and value of ecosystem services: An overview highlighting hydrologic services. *Annual Review in Environment and Resources*, **32**, 67–98.

Carpenter, S., DeFries, R., Dietz T., *et al.* (2006). Millennium Ecosystem Assessment: research needs. *Science*, **314**, 257–8.

Carpenter, S., Mooney, H., Agard, J., *et al.* (2009). Science for managing ecosystem services: Beyond the Millennium Ecosystem Assessment. *Proceedings of the National Academy of Sciences, USA*, **106**, 1305–12.

Chan, K. M. A., Shaw, M. R., Cameron, D. R., *et al.* (2006). Conservation planning for ecosystem services. *PLoS Biol*, **4**, e379.

Clark, C. (1990). *Mathematical Bioeconomics: The Optimal Management of Renewable Resources*, 2nd edn. Wiley, New York.

Costanza, R., d'Arge, R., de Groot, R., *et al.* (1997). The value of the world's ecosystem services and natural capital. *Nature*, **387**, 253–60.

Daily, G. C., Ed. (1997). *Nature's Services: Societal Dependence on Natural Ecosystems*, Island Press, Washington, DC.

Daily, G. C., Söderqvist, T., Aniyar, S., *et al.* (2000). The value of nature and the nature of value. *Science*, **289**, 395–6.

Daily, G. C., Ehrlich, P. R., and Sánchez-Azofeifa, G. (2001). Countryside biogeography: Use of human-dominated habitats by the avifauna of southern Costa Rica. *Ecological Applications*, **11**, 1–13.

Daily, G. C., and Matson, P. (2008). Ecosystem services: From theory to implementation. Special Feature. *Proceedings of the National Academy of Sciences, USA*, **105**, 9455–6.

Daily, G. C., Polasky, S., Goldstein, J., *et al.* (2009). Ecosystem services in decision-making: time to deliver. *Frontiers in Ecology and the Environment*, **7**, 21–8.

Das, S., and Vincent, J. (2009). Mangroves protected villages and reduced death toll during Indian super cyclone. *Proceedings of the National Academy of Sciences, USA*, **106**, 7357–60.

Dasgupta, P., and Heal, G. (1979). *Economic Theory and Exhaustible Resources*. Cambridge University Press, Cambridge.

Dasgupta, P. (2001). *Human Well-Being and the Natural Environment*, Oxford University Press, Oxford.

Dasgupta, P. (2010). Nature's role in sustaining economic development. *Philosophical Transactions of the Royal Society B*, **365**, 5–11.

Egoh, B., Reyers, B., Rouget, M., *et al.* (2008). Mapping ecosystem services for planning and management. *Agriculture Ecosystems and Environment*, **127**, 135–40.

Ehrenfeld, D. (1988). Why put a value on biodiversity? In *Biodiversity*, E. O. Wilson, Ed., National Academy Press, Washington, DC.

Freeman, A. M., III (1993). *The Measurement of Environmental and Resource Values: Theory and Methods*. Resources for the Future, Washington, DC.

Goldman, R., Benítez, S., Calvache, A., and Rámos, A. (2010). *Water funds: protecting watersheds for nature and people*. The Nature Conservancy, Arlington.

Gordon, H. (1954). The economic theory of a common-property resource: the fishery. *Journal of Political Economy*, **62**, 124–42.

Harrison, J., Bouwman, A., Mayorga, E., and Seitzinger, S. (2010). Magnitudes and sources of dissolved inorganic phosphorus inputs to surface fresh waters and the coastal zone: A new global model. *Global Biogeochemical Cycles*, **24**, 271–9.

Heal, G. (2000a). *Nature and the Marketplace: Capturing the Value of Ecosystem Services*. Island Press, Washington, DC.

Heal, G. (2000b). Valuing ecosystem services. *Ecosystems*, **3**, 24–30.

Jackson, R. B., Jobággy, E., Avissar, R., *et al.* (2005) Trading water for carbon with biological carbon sequestration. *Science*, **310**, 1944–7.

Kareiva, P., and Marvier, M. (2007). Conservation for the people. *Scientific American*, **297**, 50–7.

Krutilla, J. (1967). Conservation reconsidered. *American Economic Review*, **47**, 777–86.

Krutilla, J., and Fisher, A. (1975). *The Economics of Natural Environments: Studies in the Valuation of Commodity and Amenity Resources*. Johns Hopkins University Press, Baltimore.

Larigauderie, A., and Mooney, H. (2010). The Intergovernmental Science-Policy Platform on Biodiversity and Ecosystem Services: moving a step closer to an IPCC-like mechanism for biodiversity. *Current Opinion in Environmental Sustainability*, **2**, 9–14.

Leopold, A. (1949). *A Sand County Almanac*. Oxford University Press, Oxford.

Levy, K., Daily, G. C., and Myers, S. (in press). Ecosystem services and human health: A conceptual framework.

Liu, J., Li, S., Ouyang, Z., Tam, C. and Chen, X. (2008). Ecological and socioeconomic effects of China's policies for ecosystem services. *Proceedings of the National Academy of Sciences, USA*, **105**, 9489–94.

MA (Millennium Ecosystem Assessment). (2005). *Ecosystems and Human Well-being: The Assessment Series (Four Volumes and Summary)*. Island Press, Washington, DC.

Mäler K.-G., Aniyar, S., and Jansson, A. (2008). Accounting for ecosystem services as a way to understand the requirements for sustainable development. *Proceedings of the National Academy of Sciences USA*, **105**, 9501–06.

Mooney, H., and Ehrlich, P. (1997). Ecosystem services: A fragmentary history. In G. Daily, Ed., *Nature's Services*. Island Press, Washington, DC.

Mooney, H. and Mace, G. (2009). Biodiversity policy challenges. *Science*, **325**, 1474.

Myers, S. and Patz, J. (2009). Emerging threats to human health from global environmental change. *Annual Review of Environment and Resources*, **34**, 223–52.

Naidoo, R. and Ricketts, T. H. (2006). Mapping the economic costs and benefits of conservation. *PLoS Biology*, **4**, e360.

Nel, D., Marais, C., and Blignaut, J. (2009). Water neutrality: A first quantitative framework for investing in water in South Africa. *Conservation Letters*, **2**, 11–18.

Nelson, E., Mendoza, G., Regetz, J., *et al.* (2009). Modeling multiple ecosystem services, biodiversity conservation, commodity production, and tradeoffs at landscape scales. *Frontiers in Ecology and the Environment*, **7**, 4–11.

Norton, B. (1987). *Why Preserve Natural Variety?* Princeton University Press, Princeton.

NRC (National Research Council) (2005). *Valuing Ecosystem Services: Toward Better Environmental Decision-Making*. National Academies Press, Washington, DC.

Pagiola, S., Bishop, J., and Landell-Mills, N. (2002). *Selling Forest Environmental Services*. Earthscan, London.

Pagiola, S. and Platais, G. (2007). *Payments for Environmental Services: From Theory to Practice*. World Bank, Washington, DC.

Porras, I., Grieg-Gran, M., and Neves, N. (2007). *All That Glitters: A Review of Payments for Watershed Services in Developing Countries*. International Institute of Environment and Development, London.

Ricketts, T. H., Daily, G. C., Ehrlich, P. R., and Michener, C. (2004). Economic value of tropical forest to coffee production. *Proceedings of the National Academy of Sciences, USA*, **34**, 12579–82.

Rokityanskiy, D., Benítez, P., Kraxner, F., *et al.* (2007). Geographically explicit global modeling of land-use change, carbon sequestration, and biomass supply. *Technological Forecasting and Social Change*, **74**, 1057–82.

Rolston, H., III (2000). The land ethic at the turn of the millennium. *Biodiversity and Conservation*, **9**, 1045–58.

Ruhl, J., Kraft, S., and Lant, C. (2007). *The Law and Policy of Ecosystem Services*. Island Press, Washington, DC.

Schaefer, M. B. (1957). Some considerations of population dynamics and economics in relation to the management of marine fishes. *Journal of the Fisheries Research Board of Canada*, **14**, 669–81.

Scheffer, M., Carpenter, S. R., Foley, J. *et al.* (2001). Catastrophic shifts in ecosystems. *Nature*, **413**, 591–6.

Scott, A. (1955). The fishery: the objectives of sole ownership. *Journal of Political Economy*, **63**, 116–24.

Steffan-Dewenter, I., Kessler, M., Barkmann, J., *et al.* (2007). Tradeoffs between income, biodiversity, and ecosystem functioning during tropical rainforest conversion and agroforestry intensification. *Proceedings of the National Academy of Sciences USA*, **104**, 4973–8.

Stokstad, E. (2005). Louisiana's wetlands still struggle for survival. *Science*, **310**, 1264–6.

US EPA (United States Environmental Protection Agency) Science Advisory Board. (2009). *Valuing the Protection of Ecological Systems and Services*. EPA-SAB-09-012. US EPA, Washington, DC.

Varga, A. (2009). *Payments for Ecosystem Service: An Analysis of Cross Cutting Issues in Ten Case Studies*. Columbia University, New York.

WRI (World Resources Institute). (2010). *Lafarge Presque Island Quarry: Corporate Ecosystem Services Review*. World Resources Institute, Washington, DC.

Wunder, S., Engel, S., and Pagiola, S. (2008). Taking stock: a comparative analysis of payments for environmental service programs in developed and developing countries. *Ecological Economics*, **65**, 834–52.

Zhang, P., Shao, G., Zhao, G., *et al.* (2000). China's forest policy for the 21st century. *Science*, **288**, 2135–6.

CHAPTER 2

Interpreting and estimating the value of ecosystem services

Lawrence H. Goulder and Donald Kennedy

2.1 Introduction: why is valuing nature important?

Many of the critical ecosystem services generated by natural capital (such as pollination services, flood control, water filtration, and provision of habitat for biodiversity) are externalities—they are not given a price in markets. As a result, unfettered markets often lead to the compromising or collapse of ecosystems, much to the detriment of human welfare. Oftentimes society would benefit from greater protection of ecosystems and their services than results from unregulated markets.

Public policy has a crucial role to play in regulating or influencing markets so as to prevent them from producing unfortunate societal outcomes. Yet decisions about such public policies are often contentious. Agricultural interests will vie for greater ability to purchase wetlands and convert them through drainage to agricultural land. Urban developers will push to convert open space to new housing tracts.

Perhaps the most important basis for supporting a policy that would protect otherwise threatened ecosystem services is evidence that society gains more value from such protections than it gives up. Providing such evidence requires an understanding of the biophysical processes involved, that is, the various services offered by the ecosystem in question. It also requires an assessment of the benefits to well-being—or values to society—of these ecosystem services.

This chapter clarifies how such an assessment of ecosystem services can be made. It has two main components. One is to examine the philosophical basis of ecosystem service value. In considering this basis, we contrast competing approaches to value and bring out some ethical issues underlying the choice among different approaches.

The other component is to lay out various methods for measuring the values of ecosystem services, and to consider the strengths and limitations of these approaches. Quantitative assessments of ecosystem service value have become much more widespread in recent years. The expanding literature now includes estimates of the value of such ecosystem services as pollination, pest control, and water purification. These assessments are beginning to play a significant role in the formulation of land-use policies.

Setting out the values of ecosystem services to society provides a basis for making public policy decisions. However, it is not the only basis. As we discuss briefly below, it may make sense to consider as well whether a policy decision is consistent with preserving the intrinsic rights of the various organisms or ecosystems that might be affected by the decision. If intrinsic rights are involved, it is reasonable to restrict the set of serious alternatives to those that are consistent with these rights.

The chapter is organized as follows. Section 2.2 examines alternative philosophical foundations for valuing living things and ecosystems. It also considers how attention to intrinsic rights might supplement or even offer an alternative to a consideration of values. This philosophical discussion lays the groundwork for Section 2.3's examination of empirical valuation methods. In Section 2.4 we indicate some valuation problems that arise in a few specific real-world cases. Our final section draws conclusions.

2.2 Philosophical issues: values, rights, and decision-making

2.2.1 Competing philosophical approaches to value

2.2.1.1 *The anthropocentric approach*

From what do nature's values derive? When we claim that a given living thing or species or habitat is worth such-and-such, what is the basis of that claim?

Among US policy analysts, the prevailing approach to value is anthropocentric. This approach claims that natural things (indeed, all things) have value to the extent that they confer satisfactions to humans. It stipulates that value is based on the ability to give *utility* (or well-being) to humans. Economists tend to support this viewpoint which, as we discuss below, is inherent in benefit–cost analysis.

At first blush, this anthropocentric approach might seem inconsistent with safeguarding the planet or protecting non-human forms of life. But the approach does not necessarily imply a ruthless exploitation of nature. On the contrary, it can be consistent with the fervent protection of non-human things, both individually and as collectivities. After all, we may feel that the protection of nature or particular non-human forms of life is important to our satisfaction or well-being, and thus we may place a high value on these forms. The anthropocentric approach does not rule out our making substantial sacrifices to protect and maintain other living things. However, it asserts that we should assign value (and therefore help other forms of life) *only insofar as we humans gain satisfaction or well-being from doing so*. The notion of satisfaction here should be interpreted broadly, to encompass not only mundane enjoyments (as with consuming plants or animals for food) but also more lofty pursuits (such as marveling at the beauty of an eagle).

Anthropocentric value can be categorized according to the way the satisfaction is generated. *Use value* refers to satisfaction that involves (directly or indirectly) a physical encountering with the object in question. There are *direct* use values (for example, the satisfaction from catching or eating trout) as well as *indirect* use values (for example, the value that can be attached to plankton because it provides

nutrients for other living things that in turn feed humans). The anthropocentric approach does not restrict value to forms of nature that are consumed: there are both *consumptive* and *non-consumptive* use values. Examples of the former are the values that might be attached to ducks insofar as they provide food. Examples of the latter are the values ducks provide in the form of pleasure to bird-watchers.

Satisfactions also include non-use values: values that involve no actual direct or indirect physical involvement with the natural thing in question. The most important value of this type may be *existence* value (or passive use value)—the satisfaction one enjoys from the pure contemplation of the existence of some entity. For example, a New Jersey resident who has never seen the Grand Canyon and who never intends to visit it can derive satisfaction simply from knowing it exists. As another example, many people experienced a loss of satisfaction or well-being simply from learning of the ecological damage resulting from the Bluewater Horizon oil spill in the Gulf of Mexico in 2010. This was a loss of existence value.

The array of services provided by ecosystems spans all of these categories of values. The pest control and flood control services they offer have direct use value to nearby agricultural producers. Their provision of habitat for migratory birds confers an indirect use value for people who enjoy watching them (non-consumptive) or hunting them (consumptive). Ecosystems also yield an existence value: wetlands, for example, provide such value to people who simply appreciate the fact that wetlands or their services exist.

The fundamental assertion of the anthropocentric approach is that the value of a given species or form of nature to an *individual* is entirely based on its ability to yield satisfaction to that person (directly or indirectly). Benefit–cost analysis invokes the anthropocentric approach, while introducing a further assumption—that the value to *society* of the natural thing is the sum of the values it confers to persons.

Benefit–cost analysis offers a rather convenient way of measuring the overall social values of alternative policies. Thus it provides a basis for making difficult policy decisions. It seeks to ascertain in monetary terms the gain or loss of satisfaction to different groups of humans under each of various policy alternatives. Under each alternative, it adds up the gains and subtracts the losses, and then compares the net gains across policy options. Importantly, benefit–cost analysis often does not differentiate between one person's valuation of a given species and another's—that is, each person's valuation receives the same weight as another's. Many times, no attempt is made to correct for differences in awareness, education, or "enlightenment" among individuals. The preferences of people who have no concern for future generations, or who have no sense of the ecological implications of their actions, are often counted equally with those of people who are more altruistic or who recognize more fully the fragility of ecosystems.

Such benefit–cost analyses are non-discriminating, perhaps to a fault. Consider the fact that preferences change. They may change for a given person over his or her lifetime, or from generation to generation. To impute values for future generations (such as the value that future generations might place on certain ecosystem functions), benefit–cost analysis must impute preferences to these generations. Clearly, this can only involve guesswork. Usually benefit–cost analyses assume that future generations' preferences are similar to those of the current generation. Costanza *et al.* (1995) indicate that preferences seem to evolve toward an increasing concern for sustainability. They consider the notion that this natural evolution of preferences ought to be accounted for in social decisions—that more evolved, developed preferences deserve greater weight in analyses of policy options. However, some benefit–cost analyses do in fact give special attention to the assessments offered by experts.

Many ecologists are uneasy with the tendency of benefit–cost assessments to give considerable weight to valuations made by relatively uninformed individuals. There is a basic appeal to the idea that the preferences of some individuals—particularly those who are better informed or have more relevant expertise—should count more. But it is very difficult to arrive at an objective standard for "relevant expertise." Philosophers offer varying viewpoints as to what's appropriate here (NRC 2004).

2.2.1.2   *A biocentric approach*

The biocentric approach offers another basis for value. It asserts that value consists in the ability to provide well-being or utility to humans *and* to other species. Under the anthropocentric approach, the well-being of other species counts only indirectly: such well-being is important only to the extent that it contributes to human well-being. In contrast, the biocentric approach gives weight directly to the well-being of other species. Thus, it allows for the possibility that another species will have value even if it does not confer satisfaction directly or indirectly to humans. This independent value is sometimes referred to as intrinsic value.

Defenders of the anthropocentric approach point out that since human beings are the dominant species on the planet; they are obliged to define ethical principles in terms of human wants and needs. However, biocentrists can counter by pointing out the following implication of anthropocentric logic. Suppose that representatives of another species should arrive from outer space, a species clearly superior to human beings in intelligence, perceptiveness, and technological know-how. To the extent that defenders of anthropocentrism have invoked the "dominant species" argument, consistency would require humans to allocate some decision-making authority to this other species, no matter whether humans like their decisions or not. Human well-being would count only insofar as it served the well-being of the superior species. This may seem troubling to many of us! What if the dominant species felt it was best to exterminate humans? This *reductio ad absurdum* argument has been invoked to support a biocentric approach that gives weight directly to a range of species.

While the biocentric approach may have some appeal, it is difficult to implement. As discussed below, "willingness to pay" offers a measure of the change in well-being to humans generated by a given policy change to protect nature or environmental quality. No comparable measure is currently available for assessing changes in satisfaction to other species or communities of them. Also, it is difficult to draw a clear line between biocentric value and certain anthropocentric values. When individuals call for a biocentric approach, they may actually be expressing the anthropocentric satisfaction they would gain if that approach were followed. For example, when someone calls for the preservation of a given habitat on the grounds that the species residing there has intrinsic value, that individual may really be revealing the (anthropocentric) existence value that the species provides. It thus becomes difficult to distinguish biocentric intrinsic value from existence value, which suggests that the biocentric approach may be superfluous.

2.2.2   Intrinsic rights: a further consideration

Under the value-based approaches just discussed, social decisions are to be made based on a comparison of values. If Policy A generates greater value than Policy B, then Policy A should be given preference over Policy B. Consider in particular the anthropocentric approach to value. If a given a species or other element of nature does not convey satisfaction to human beings directly or indirectly, then according to this approach it should be given no value. It must produce no use value, either directly or indirectly. Thus, it must be something we do not enjoy eating (there is no consumptive use value) and something we do not enjoy observing (there is no non-consumptive use value). In addition, the organism must not serve any positive ecosystem function (there must be no indirect use value). Also it must be the case that we are certain that humans' tastes and ecosystem function will not change to give rise to a future use value. To complete the picture, the organism must also have zero existence value—humans must not enjoy contemplating this thing. Is there any real-world organism that fits this picture? Perhaps some lowly species of cockroach comes close. Whether it exactly fits the picture is not important. The key point is that such a creature would be given virtually no value in a benefit–cost analysis. This means that if we are considering a development project that threatens its existence, this threat does not cause us to refrain from undertaking the project. As long as there are some benefits from the project and no other, "significant" form of life is put at risk, we would not prevent the loss of this particular species. If destroying the habitat and putting the area involved to an alternative use (e.g., resi-

dential housing) had any value at all, then according to the anthropocentric approach this is the best option for society.

Based on examples of this sort, some philosophers argue that the fate of other species becomes too precarious when it must depend on a link to human values or satisfactions. (See, for example, Skidmore 2001.) An *intrinsic rights* approach provides an entirely different basis for decision-making. When intrinsic rights are involved, then the appropriate social decision must respect those rights. Attention to intrinsic rights can in some cases complement the weighing of values. In such cases, policy makers would first restrict the set of options to be considered to those that respect intrinsic rights. Within this restricted set, policy makers would then choose the option that yielded the highest value.

In other cases, an attention to intrinsic rights is fully dispositive. This applies, for example, when any change to a given habitat would violate the claimed intrinsic rights of the species that currently reside there. In such circumstances, a defender of intrinsic rights could argue that the value-based approach is inappropriate: any comparison of benefits (values gained) and costs (values sacrificed) is not justified. Many analysts argue that species and natural communities have intrinsic rights to exist and prosper. They claim that, consequently, society should uphold these rights irrespective of the values gained (benefits) or sacrificed (costs) in the process.

Arguments for intrinsic rights are not entirely independent of references to well-being or satisfactions of other species. For example, in *Animal Liberation*, ethicist Peter Singer argues that nonhuman animals have the basic right to be spared of suffering deliberately caused by humans (Singer 1975). This argument is grounded in the notion that, like humans, other animals are sentient creatures, capable of experiencing pleasure and pain, and that there is something fundamentally wrong about causing pain to any creature. However, even though the call for intrinsic rights may be based on a concern for well-being, it proposes a very different basis for decision-making: the appropriate social decision must respect intrinsic rights. A policy that satisfies a benefit–cost test should be rejected if it violates intrinsic rights.

2.2.3 Public policy's inconsistent approach to decision-making

When should a value-based approach be employed, and when should attention to intrinsic rights supply the primary basis for decision-making? And which of the two approaches do societies in fact adopt? United States environmental policy adopts both the anthropocentric value approach (via benefit–cost analysis) and an intrinsic rights approach, and often acts inconsistently. Oftentimes the mandate for a particular environmental law will embrace the intrinsic rights approach, but actual implementation yields to a value-based approach.

A key example is the US Endangered Species Act, passed in 1973 after a previous Act was brought up to date and linked to the Convention on International Trade in Endangered Species. In addition to defining the status of "endangered" and "threatened," it made eligible for protection all plants and invertebrates, and prohibited the "taking" of all endangered animals. "Taking" included destruction of essential habitat. Federal agencies were required to use their authority to conserve listed species and were prohibited from undertaking actions that would jeopardize listed species or modify their critical habitats.

There is an assumption here that certain species under threat have an intrinsic right to exist. However, when it comes to actual implementation of the Endangered Species Act, the Congress only allocated funds sufficient for protecting a small fraction of species that may qualify for the designation of threatened or endangered. Based on threat criteria and the availability of appropriated funds, the Interior Department decided that some species are more worthy of protection than others. In effect, it adopted an anthropogenic, value-based approach, with the priorities reflecting the range of values that people place on different species. Charismatic megafauna like the wolf or the peregrine falcon get more protection than the white-footed mouse. The anthropocentric, value-based approach involved in implementation contradicts the intrinsic rights basis of the mandate declared by Congress.

One might be tempted to fault Congress for failing to allocate enough funds to protect all species. However, the allocation reflects the preferences of the broader public: to protect every species would require far more funds than the public generally wishes to devote to this purpose. People want species protection, but they also want funds for other things (e.g., education, defense, and their own consumption).

The case of the Endangered Species Act is not unusual. The US Clean Air offers another example. The mandate is broad: setting air quality standards tight enough to assure an "adequate margin of safety" to all individuals. There is no reference to a comparison of benefits and costs. Yet in the actual establishment of the standards, the EPA paid close attention to costs, and the ultimate standards imposed were not tight enough to prevent serious health problems or premature mortality to the most pollution-sensitive individuals.

In many instances there is a fundamental inconsistency between the stated objectives of environmental laws and the way the laws are implemented. Lawmakers and the public may experience rewards from establishing broad mandates that declare intrinsic rights. At the same time, they are free to implement the laws much more restrictively; so that people need not sacrifice as many other things as they would had they enforced intrinsic rights fully.

We do not suggest that society must choose between the universal application of an intrinsic rights approach and the across-the-board adoption of a value-based approach. In certain circumstances, it may be best to invoke intrinsic rights. In the Clean Water Act, Congress essentially found that the population of the United States had an intrinsic right to clean water. In other circumstances the appeals to intrinsic rates are fairly rare; for example, there are few claims that farmers enjoying the pollination services provided to agriculture by bees inhabiting nearby natural habitats have an intrinsic right to such services. Are these pollination services somewhat less "fundamental" than the various services offered by clean water? Does this explain the relative infrequency of claims that woodland-based pollination services are an intrinsic right?

Even if one adopts a value-based approach rather than an intrinsic rights approach in making policy decisions, this does not necessarily preclude invoking additional evaluation criteria in the decision process. Benefit–cost analysis considers the aggregate values gained (benefits) and values sacrificed (costs) of a given policy option. It usually does not focus on how the benefits and costs are distributed across members of society (rich vs. poor, current generations vs. future generations, etc.; see Chapter 16 for further discussion). The distribution of policy impacts is important and deserves attention as well. Other evaluative criteria (minimization of risk and political feasibility) can also be important. Thus, the results of a benefit–cost study may not be sufficient to settle the question of which policy is best.

2.3   Measuring ecosystem values

Although attention to values need not be the sole criterion for decision-making, we believe it is sufficiently important to justify a focus on how to measure various values. Here we describe central methods for measuring anthropocentric value. Considerable progress has been made over the years in developing such methods. However, the science is far from perfect. Controversies persist.

Ecosystem services are especially difficult to measure for the same reason that ecosystems are threatened. Many of the services provided by ecosystems are positive externalities. The flood control benefits, water filtration services, and species-sustaining services offered by ecosystems are usually external to the parties involved in the market decision as to whether and at what price a given habitat will be sold. As a result, the habitats that support complex ecosystems tend to be sold too cheaply in the absence of public intervention, since important social benefits are not captured in the price. Public attention to the values of these (largely external) benefits is important for providing support for reasonable public policies to protect important habitats. This makes it all the more important to determine the values of these services.

2.3.1   Willingness to pay

As indicated, under the anthropocentric approach the value of a given living thing is the amount of human satisfaction that thing provides. How could such satisfaction be measured? Nearly every empirical approach assumes that the value of a given

natural amenity is revealed by the amount that people would be willing to pay or sacrifice in order to enjoy it. Willingness to pay is thus the measure of satisfaction.

It is important to be clear as to what is meant by "willingness to pay." It is not always an actual, expressed willingness; it is not restricted to what we observe from people's actual payments in market transactions. Rather, it represents a kind of psychological equivalence. Suppose a project would improve water quality an individual enjoys. That individual's willingness to pay is the income sacrifice that just brings the individual back to his or her original utility level after the improvement in water quality. More formally, the willingness to pay W for a given improvement to environmental quality Q is the value W that leads to the following equality:

$$U(Q + \Delta Q, Y - W) = U(Q, Y), \qquad (2.1)$$

where U stands for the individual's utility and Y is the individual's income. Willingness to pay expresses the maximum payment an individual would make that just compensates for (or undoes) the utility gain from the environmental improvement. It is the size of the payment that, if made, would keep the person's utility from changing. It is not necessarily what people say they are willing to pay. In some cases, markets indicate individuals' true willingness to pay as defined above: for example, the market price of a tomato might indicate what consumers are willing to pay at the margin for this product. But in other circumstances researchers need to rely on other, more indirect methods to fathom it.

2.3.2   Methods for measuring the values of ecosystem services

Above, we have distinguished various types of ecosystem service values. The ecosystem services themselves can be placed in various categories as well. As in the Millennium Ecosystem Assessment (2003), we will distinguish provisioning, regulating, and other services offered by ecosystems. Below, we consider the types of values associated with each of these major categories of service. Different types of

Table 2.1 Ecosystem services and valuation methods

Services	Types of values offered	Valuation method
Provisioning services		
Sustenance of plants and animals	Direct use values	
	—Consumptive	Direct valuations based on market prices
	—Non-consumptive	Indirect valuations (revealed expenditure methods, contingent valuation method)
	Indirect use values	(No valuations necessary if plants/animals with direct values are counted)
Regulating services		
Water filtration, flood control, pest control, pollination, climate stabilization	Direct and indirect use values	Estimation of service's contribution to profit (holding all else constant)
Other services		
Generation of spiritual, esthetic, and cultural satisfaction	Existence value	Indirect valuations (contingent valuation method)
	Direct, non-consumptive use value	Indirect valuations (revealed expenditure methods, contingent valuation method)
Recreational services	Non-consumptive direct use value (e.g., from bird-watching)	Indirect valuations (revealed expenditure methods, contingent valuation method)
Generation of option value*	Option value	Empirical assessments of individual risk-aversion

* Option value represents a component of the overall value offered by a potential future ecosystem service, supplementing other values attributed to this potential service. See discussion text.

valuation techniques are called for, depending on the category of service involved. Table 2.1 shows the relationships between service types and valuation methods.

2.3.2.1 Valuing the provisioning services of ecosystems

As suggested by Table 2.1, a general type of service provided by ecosystems is the sustenance of plants and animals. In choosing a method for valuing this type of service, it helps to distinguish plants and animals with *direct use values* from those with *indirect use values*. Examples of the former are plants or animals that are consumed as food or that directly offer recreational values (sightseeing, nature-watching, etc.). Examples of the latter are plants and animals (such as organisms that are lower on the food chain) that help sustain other plants and animals that we enjoy directly. To give specific examples: ecosystems generate direct use values by supporting the various types of birds that we either enjoy non-consumptively as bird-watchers or consumptively as bird-hunters. They generate indirect use values by supporting the life of various plants or insects that in turn enable birds to thrive.

Direct, consumptive use values. When direct use values are involved, two main valuation methods may apply. In the case of direct *consumptive* use values, one can employ direct valuation methods based on market prices. When natural ecosystems provide a habitat for animals that are harvested and sold commercially, the commercial market value provides a gauge of the value of the habitat services. For example, part of the value of marine ecosystems is conveyed by the value of the commercial fish that they help sustain. Of course, this only represents a portion of the value of the ecosystem—namely, the value of the ecosystem's potential to sustain those fish that have a market value.

There is an important difference between the *marginal* and *total* value associated with market prices or the willingness to pay of consumers in markets. Economists regard the prices that people are willing to pay as indicators of the marginal value—the value they place on the *last unit* purchased. Consider what a homeowner would be willing to pay for residential water in a given month. He might be willing to pay a huge sum for the privilege of consuming the first ten cubic feet, because doing without them would deprive him of even the most fundamental (and valuable) uses of water for that month: drinking water, the occasional shower, etc. The *next* ten cubic feet would probably not be worth quite as much. They would allow him additional opportunities to fill a glass from the faucet, and an extra shower or two, but these would not be as critical to him (or to the people with whom he associates!) as the first ten cubic feet. Thus the marginal value of water—the amount one is willing to pay for each successive increment—falls steadily.

Figure 2.1 displays a typical willingness-to-pay schedule. The first cubic foot is shown to be worth a great deal more than the fiftieth, which in turn is worth much more than the hundredth. In reality, of course, households do not have to purchase each unit of water at its marginal value. If they did, they would be charged larger amounts for the first increments than for later ones. Instead, utilities charge households a given price per unit of water, regardless of how much they consume.

In Figure 2.1, the horizontal line at $0.02 represents the price charged for the water. (We use this number arbitrarily.) The standard economic assumption is that users will continue to purchase water until the marginal value of the water (or marginal willingness to pay) is equal to the marginal sacrifice (or price). In these circumstances, the price

is an expression of the marginal willingness to pay, or marginal value. (In the example of Figure 2.1, the user would demand 400 cubic feet of water per month at this price.)

The *total* value of the water consumed is much more than the price, however. The total value is the area under the marginal willingness-to-pay schedule (the sum of areas I and II in the diagram). Note that to ascertain total value (as opposed to marginal value), researchers need to have information on the entire marginal willingness-to-pay schedule (or demand curve), not just the price paid. A main challenge of empirical valuation techniques is to trace out marginal willingness-to-pay schedules.

In the context of commercial products of ecosystems, this means that market prices represent only the *marginal* value of these products. The value of the total sales of these products corresponds to area II in Figure 2.1. Note that this is less than the total value to consumers, which is the sum of areas I and II. Thus market sales understate the overall value of the commercially viable forms of life supported by ecosystems.

Direct, non-consumptive use values. Within the category of direct use values from living things maintained by ecosystems, we have another case to consider: the case where the life forms are used *non-consumptively*. For such uses, the relevant markets do not usually arise, and thus it is not possible to gauge values directly by observing market prices. Markets tend to arise for goods or services that are excludable: the failure to pay for the good or service implies an inability to enjoy or consume the good. For non-consumptive use values (like bird-watching), it is difficult to establish a market because people cannot easily be excluded from enjoying the good or service. In these cases, it is necessary to apply more inferential methods to ascertain the relevant values.

Revealed expenditure methods represent a broad category of inferential approaches (NRC 2004). Revealed expenditure methods have been applied to ascertain some of the values provided by parks, lakes, and rivers—or, equivalently, the costs that results from the loss of these elements of nature. Here we describe one of the first and simplest revealed expenditure methods: the travel cost

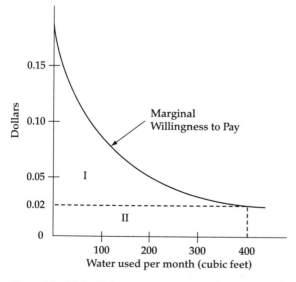

Figure 2.1 Relationship between water use and marginal willingness to pay. From Daily (1997) reproduced by permission of Island Press, Washington, DC.

method (Freeman 1993). In recent years several more general and sophisticated approaches have tended to supplant the travel cost method, but the basic logic of the newer approaches is the same as that of the travel cost method.

Non-consumptive uses are not directly bought or sold in markets; prices are not usually charged for their use. In those instances when use prices *are* charged (through entry fees, etc.), the prices are unlikely to be good indicators of (marginal) value. That is because the users of these resources actually "pay" more than the entry fees to use them. For example, the cost of a family visit to Yosemite National Park is much greater than the daily use fee. The travel cost method recognizes that by adding to the entry fee (if any) the transportation cost and time cost expended to visit a particular site, one can ascertain the overall travel cost. This method regards the overall travel cost as a measure of the marginal willingness to pay by a visitor to the park; this is considered to be the same as the marginal value of the park to the visitor. The underlying assumption is that people will continue to visit the park until the value of the last unit (that is, the marginal value) is just equal to the travel cost.

It is also possible to employ survey methods, such as the contingent valuation method, to determine how much value people place on the non-consumptive uses. In contingent valuation assessments of value, interviewees are asked what they would be willing to pay in order to prevent some real or hypothetical amenity. For an exposition of this approach, see Mitchell and Carson (1989). Many economists distrust results from survey approaches, claiming that individuals' asserted preferences in the hypothetical circumstances posed by surveys bear no systematic relationship to their true preferences. Defenders of survey methods counter that, in many cases, surveys are the only method available. This "only game in town" argument may have force when existence values are involved, as discussed below.

Indirect use values. Ecosystems contain many living organisms that support other, often "higher" forms of life that provide direct or indirect value to humans. It could be assumed that the value of ecosystem services should include the values of the services provided by these life forms. In fact, there is no need to include the values of these services in an accounting of the overall value of an ecosystem! These values are already captured in the values attached to the life forms that humans enjoy. Consider the value of certain plants whose fruits are eaten by birds and other "higher" life forms; assume humans obtain no direct use value from these plants. If we abide by the anthropocentric approach to value, then there is no value to these plants over and above the value that we attach to the higher life forms to which they contribute. To add their indirect use values to the direct use values would be double counting. The accounting here is perfectly analogous to the economic valuation of net economic output, which disregards the value of intermediate inputs, that is, inputs used up in the process of producing final goods such as consumer goods and capital goods.

2.3.2.2 *Valuing the regulating services of ecosystems*

Table 2.1 lists four examples of production inputs from ecosystems: water filtration, flood control, pollination, and climate stabilization. These services are inputs to the sustained production of agricultural products in the sense that it would be difficult to maintain agricultural production without relatively pure water, flood control, pest control, or a stable climate.

An appropriate measure of the value of productive inputs is the additional economic income or profit that they provide, holding everything else constant. Thus, for example, the value of pest control services provided by ecosystems is their contribution to profits. To assess these values, agronomists and other researchers develop models in which the profitability of various agricultural products is assessed in the presence and absence of ecosystem-provided pest control, a key production input. The difference is the value of the pest control services. Similarly, one can gauge the value of flood control services by comparing profitability in the presence and absence of such services. A favorable climate can be considered a productive input. Numerous studies have aimed to assess the damage from climate change to agriculture by comparing yields and agricultural profits under current climate with those that would apply after predicted future climate change (e.g. Mendelsohn *et al.* 1994; Schlenker *et al.* 2005). The damage from a changed climate is equivalent to the monetized benefit or value from avoiding this change.

Pollination services are another example of an important production input. In the Central Valley of northern California, various specialty crops, including melons, nuts, and tree fruits, depend upon the pollination services supplied by wild bees whose population is maintained by breeding sites in nearby "natural" areas such as undeveloped forestland. The value of these pollination services is the additional profit generated by the populations of wild bees.

These pollination services have declined over time as larger and larger proportions of the region have been developed for agricultural purposes. One can only assume that at some point in the developmental history of this unusually productive agriculture, wild insect populations alone were sufficient to guarantee some base level of pollination services and thus guarantee yields adequate to keep the farmers in business. Since that is clearly no longer the case, farmers now have to substitute a costly alternative—pollination services from the bees supplied by commercial beekeepers. This is now an economically significant activity in these regions. In this example, the avoided cost is the difference in cost between the case where farmers enjoyed free pollination services from wild bees and the case where they must pay for the services of bees husbanded by commercial beekeepers.

It is sometimes suggested that one can place a value on production inputs by examining what costs or expenditures agricultural producers manage to avoid by having these inputs and thus not having to substitute other inputs for them. For example, where ecosystems provide effective pest control, farmers can avoid having to pay for alternative pest control methods such as the use of synthetic pesticides. In fact, avoided cost is not a theoretically valid indicator of value. To see this, consider the following (extreme) situation. Suppose it were infinitely costly for farmers to find an alternative to wetlands in providing flood control. If avoided cost indicated value, then the value of the wetlands' flood control services would be considered infinite. In fact, although the loss of flood control services would cause a significant loss of profit to farmers, the loss would not be infinite.

Although avoided cost is not a measure of value, it is still important information. It indicates the *net* advantage of having access to the productive input provided by ecosystems, as opposed to having to achieve the same input through an alternative. It provides a rationale for preserving the ecosystem service. For example, when the New York City Water District struggled with how to preserve water quality in the Catskills, it determined that it was far less costly to do so by restoring the ecosystems surrounding the city's upstate reservoirs rather than by constructing a new water treatment plant. The very high avoided cost motivated the decision to pursue ecosystem-generated water quality control (Daily and Ellison 2002).

2.3.2.3 *Valuing ecosystem services offering non-use values*

Other important services include the generation of spiritual, esthetic, and cultural satisfaction, the provision of recreational services, and the generation of option value. Recreational services provide a non-consumptive direct use value. For example, a National Park offers opportunities for hiking, swimming, and bird-watching. Park visitors engaging in these activities physically encounter the ecosystems involved (implying a use value), but (hopefully) do not use up the hiking trails, lakes, or birds in the process of enjoying them. The non-consumptive use values from these activities can be estimated using the methods described for such values under Section 2.1 above.

Ecosystems also provide services with values other than use values—values that do not derive from a physical encounter with the item of nature in question. The values provided here are *non-use* values. There are two main types of non-use value.

Existence value. This is the value that derives from the sheer contemplation of the existence of ecosystems. While much of our enjoyment of biodiversity involves use value—that is, it derives from a physical encountering with various plants and animals—we also derive satisfaction from simply recognizing that these forms of nature exist. Thus existence value is an important element of the value people attach to nature or the functioning of ecosystems. It may reflect the spiritual, esthetic, or cultural satisfaction we obtain when we contemplate the diversity, beauty, complexity, or power of nature.

Survey approaches such as contingent valuation assessments may be the only way of ascertaining existence value, since actual market and non-market behavior gives little hint of its magnitude. As mentioned, survey approaches are controversial. Yet, when it comes to existence values, surveys may be the only way of ascertaining values because people's actions do not leave a "behavioral trail" from which their valuations can be inferred. In this limited space we cannot offer an appraisal of survey approaches. However, we can point out what seems to be the key underlying question: Is the information obtained from surveys, however imperfect, better than no information at all?

Option value. As developed in the economics literature (e.g., Bishop 1982), the term "option value" refers to a premium that people are willing to pay to preserve an environmental amenity, over and above the mean value (or expected value) of the use values anticipated from the amenity. Suppose, for example, a habitat is threatened with destruction. And suppose that, if the habitat is preserved, there is a 50% chance you would visit it, and a 50% chance you would not. If you were to visit it, you would derive a use value of 10; if you didn't you would enjoy no use value. In this case the expected value of the use value is 5. But you might be willing to pay, say, 7 to ensure the preservation of the habitat. If so, your option value is 2 (7–5). This premium reflects individual risk-aversion: in the absence of risk-aversion, people's willingness to pay would equal the mean use value (its expected value), and option value would be 0.

We follow general practice in subsuming option value under the general category of non-use values. However, the case can be made that option value is so closely connected with (potential) use that it should be placed in the use-value category.

It is much easier to define option value than to measure it. Its measurement requires a gauging of individuals' risk-aversion, and this may depend on the specific context: persons are not equally averse to different types of risk. For an empirical assessment of option value, see Cameron (1992).

2.3.3 Marginal vs. total value

In discussions of ecosystems, one often might have in mind their *total* value. However, in many real-world circumstances, the policy debate concerns the *change in value* or *marginal loss of value* that results from alteration or conversion of a *part* of the region that occupies an ecosystem. In benefit–cost analyses, when a portion of the ecosystem is threatened with conversion, it may be more important to know the change or loss of ecosystem value associated with such conversion than to know the total value of the entire original ecosystem. Does a "minor" encroachment on the land area of an ecosystem generate small losses in ecosystem value, or do small encroachments precipitate large damages?

To examine this issue, we can begin with a very large area of a (relatively) undisturbed ecosystem. The value we place on a given amount of area lost to other uses depends on the area of this system. Let A represent the land area of our ecosystem, and suppose that the initial area is A_0. This ecosystem, valued for its natural beauty and its biological diversity, is being decreased marginally in area through conversion to farmland. Suppose first (counter to fact) that this decrease takes place *without changing the ecosystem's character through species loss.* Since a larger area is worth more than a small one, the marginal value of each withdrawn unit rises gradually as the area (A) decreases. But in the limit, an area of size 0 is worthless, and tiny areas are less attractive because they have a rather zoo-like character. Thus at small values of A, the marginal value begins to fall again. This relationship is shown in the path labeled "1" in Figure 2.2. The relationship between area and value expresses the pure *ecosystem-scale* effect.

In fact we know that the biological diversity of the ecosystem—one of the features contributing to

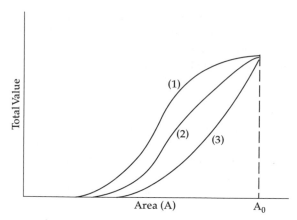

Figure 2.2   Relationship between loss of habitat area and loss of ecosystem value viewed through three different dimensions: ecosystem scale effect (1), diversity effect (2), and species effects (3). A_0 denotes the initial area of the ecosystem and the curves show how value declines as that initial area is reduced. In all cases there is some threshold area that is too small to support any value.

its value to nature lovers—is *not* area-independent. The relationship, established mainly in studies on islands and (to a more limited extent) on tropical forests, is a non-linear one. The precise form varies, but in a variety of studies the number of species lost is slight until a quarter to a half of the area is lost, and rises precipitously after about three-quarters of the area is lost. The effect on marginal value is to exaggerate the loss of ecosystem value as A is reduced. The impact of the loss in numbers of species as A is reduced may be termed the *diversity* effect. This effect is taken into account in the path labeled "2" in Figure 2.2. As indicated by the differences between paths 1 and 2, this intensifies the marginal loss of value from a given reduction in A.

A third effect needs to be considered. The species in ecosystem A are not considered to be of equal value to humans. People seem to care more about eagles and panthers than about mosses and bacteria. We also know that species are related to one another in a complex, co-evolved web of dependencies: prey and predator, plant and pollinator. Trophic relationships are also vitally important. Often, higher order species on the food chain have the most exacting environmental requirements, and are thus valuable indicators of the health of the entire ecosystem; they or others may also be critical "keystone" species because they are located at the center of a network of interdependencies. Thus, as a

practical matter, species values become proxies for ecosystem values: the Endangered Species Act in the United States is an embodiment of this principle in policy. Of course we regularly justify large expenditures to save some species (e.g., the Black Rhinoceros) but not others (there is no Save the Furbish Lousewort Society).

On what basis do we assign value to species? The following are some axes along which different people make selections.

Taxonomic proximity. We like animals that are like us. Primates attract human attention not only because there may be utility in the relationship ("animal models" for human disease) but because we respond to their quasi-human qualities.

Rarity. All other things being equal, we have more interest in rare things than in common ones. This is not simply a matter of vulnerability, although it is true that rare organisms are more vulnerable to extinction than abundant ones. Rarity itself can be the attraction; in some sense animals and plants in nature are "collectibles," if only in the sense of finding and listing them, and collections of the rare are more desired than collections of the commonplace. Indeed, "collection" in the form of listing is a motive with powerful economic consequences. Many birdwatchers will undertake extreme expenditures to visit ecosystems harboring rare species for the purpose of expanding their "life-lists."

Genetic uniqueness. If a species represents a unique evolutionary line—is, for example, the only extant member of its genus or family—then it may be entitled to higher value than it would otherwise. Scientists especially would favor the use of this criterion.

Importance to ecosystem function. Certain species (often called "keystone" species) create conditions that permit the maintenance of the entire ecosystem. The dominant trees in a forest, or birds that dig nest-holes in trees that are used by other species, or insects vital to the pollination of a dominant plant, would be examples.

How can these preferences be related to the marginal value calculation? Biological diversity is reduced as A shrinks, but species do not fall out randomly; certain kinds tend to drop out relatively early, others only when A becomes quite small. For conservation biologists and others, this means that wise policies cannot be made unless some value is attached to the different kinds. If, for example, the ones we view as most valuable did well in relatively small areas, we might argue for a patchwork of little parks, whereas if the opposite were true we would insist on large refuges.

Obviously the number of possible criteria is large enough to prohibit development of a precise relationship among area, species loss, and value. However, larger organisms with broad ranges that are especially area-sensitive would be likely to be rarer, on average closer taxonomically to humans, favored for "charm," and important to ecosystem function. Thus it is reasonable to assume a *species-composition effect*: that as A is reduced, the species lost early in the reduction are more valuable than those lost later. When this effect is taken into account, the marginal loss from a reduction in species area is even greater than indicated by path 2. Path 3 incorporates this effect (and the others). Indeed, our analysis applies specifically to the simple case in which A is reduced by shrinkage from the outside edges. In many situations, the reduction occurs by fragmentation—a patch here, a patch there, leading to a checkerboard of "natural" and "modified" areas. The new habitats provided by "edge effects" can raise local biodiversity (at least transiently). In the longer run the area/diversity rule will apply over the entire region, but the value of species lost may differ. In recent studies of plant diversity in grassland patches, the first species lost are the most effective, narrow-niche competitors: fragmentation gives an advantage to those species adept at dispersal and at rapid colonization (Tilman 1997).

Clearly some of these relationships are uncertain, and the exercise could be applied to real natural areas only after substantial research. However, it points up the importance of thinking about value in marginal rather than aggregate terms, and suggests a discipline that could be applied in the framing of general conservation policy.

2.4 Some case studies

2.4.1 Wetlands in the United States

Wetlands provide important ecosystem services, including flood control, water purification, and pro-

vision of habitat for numerous species. The values of these and many other services are very difficult to quantify. Perhaps even more important to the measurement challenge is the complexity of the network that links wetlands to groundwater and thence to streams and lakes and other "navigable waters."

In theory, society could decide on the desired level of protection of wetlands and their ecosystem services by calculating the values of the numerous ecosystem services, determining the extent of wetlands that maximizes the net benefits from these services minus the opportunity cost to society, and then implementing laws that protect just this amount of wetland. In fact, wetland protection has largely ignored valuation. In the United States, the law does not invoke an explicit comparison of benefits or costs as a basis for the protection offered; this may partly reflect the measurement difficulties just mentioned. Indeed, US law does not even acknowledge cost as a consideration in determining the extent to which wetlands or their services are to be protected. Rather, these broad ecosystem services are treated like a public "right," something to be safeguarded irrespective of the cost of protection.

That right is protected by two public agencies under several laws. The Clean Water Act, administered through the Environmental Protection Agency, has several sections devoted to wetlands protection, and these refer to the general set of the "navigable waters" referred to above. The Food Quality Protection Act has a provision colloquially known as the "swampbuster" clause that prohibits the drainage or alteration of wetlands for farming purposes. That provision is administered by the US Army Corps of Engineers. These two agencies are required to issue permits when a landowner undertakes measures that would contribute fill or pollutants to wetlands that lay within the drainage system of the owner's property. In an important wetlands case called Borden Ranch, the Supreme Court ruled that a California farmer could not be issued a permit for a technique of plowing called "deep ripping," on the grounds that it violated provisions of the Clean Water Act. In short, the Court found that the connection of the groundwater under the farmer's plow to the navigable waters could not be disturbed or interrupted.

In a much later case, the Supreme Court again split about a proposal to fill some wetlands near Lake St. Claire in Michigan. This particular wetland was some distance from the lake, which clearly fit any ordinary definition of a "navigable water." Once again, the issue rested on the question on whether a wetland that is distant from clearly navigable surface waters is nevertheless entitled to protection. The responsibility, again shared by the Environmental Protection Agency and the Army Corps of Engineers, was complicated once again by the ambiguity of the Clean Water Act's language. A four-four split on the Court was eventually decided by Justice Kennedy, who wrote in his opinion that the Clean Water Act intended to apply its provisions to the nation's waters generally, not restricted to surface waters. However, he also argued that the case should be ultimately decided scientifically by the federal agencies responsible for applying the protection.

This case exemplifies the difficulty of interpreting Congressional intention. It also highlights the difficulties of measuring ecosystem values in a network of rivers, lakes, wetlands, and groundwater that is diffuse, interconnected, and complex. In the face of such difficulties, policy-makers might prefer simply to establish a broad right to protection, rather than aim to compare values gained and values sacrificed under alternative levels of protection.

2.4.2 Vegetation and coastal protection

About one-third of the world's population lives in coastal areas or small islands, and they are at risk from buffeting and damage from storms and extreme weather events, like the hurricanes that regularly visit Caribbean and Gulf of Mexico coastlines and the recent tsunamis that swept across Indonesia and coastlines in the south Pacific and Indian Oceans. In coastal areas, societies often must make difficult choices between economic development activities and risk-reducing conservation measures. Decision-makers practicing ecosystem-based management are required to address both of these competing needs in a way that balances the welfare of residents. In order to do this, they must be able to measure the values associated with the ecosystem services provided by conservation measures (see Box 2.1).

This problem has been analyzed by Barbier and a team of fifteen authors (Barbier *et al.* 2008; Box 17.1 in this book) from various international institutions. Their analysis examined the trade-offs between conservation of coastal mangroves in Thailand and the conversion of mangrove lands to shrimp aquaculture. The authors of the study started with the reasonable assumption that the buffering capacity of the mangroves would depend on their area. On measuring the ocean wave attenuation at various distances inland from the shore, they showed that the storm buffering service provided by the mangroves was in fact nonlinear owing to the shape of the declining wave attenuation.

Using the estimated nonlinear relationship, Barbier *et al.* (2008) found that the policy that provides the greatest overall benefit to the local population is one that prohibits shrimp aquaculture in much, but not all, of the area in question, and thus reserves some area for shrimp farming ponds. Some economic benefits to aquaculture investors were retained. (In contrast, the assumption of a linear relationship between area conserved and buffering capacity would have suggested that prohibition of all shrimp farming was optimal.) This significant effort at ecosystem service valuation provided the basis for a solution favoring both stakeholders—both conservation groups and investors in aquaculture.

2.4.3 The Galapagos Islands

A third example, international in character, is provided by the Galapagos Islands. This archipelago, located 600 miles west of the Ecuador coast, consists of thirteen large islands and a number of smaller ones. All are of recent volcanic origin (100 000 to a 1 000 000 years old), and they contain a unique assemblage of plants and animals. They were visited by Charles Darwin during the voyage of the *Beagle*, and now are an important site for contemporary studies of evolutionary biology.

Managed as a National Park by Ecuador since the 1950s, the islands have also become a favorite destination for tourists, who explore the islands from boats and debark on the islands to follow carefully marked trails in the company of trained naturalist-guides. With the growth in popularity of

Box 2.1 Designing coastal protection based on the valuation of natural coastal ecosystems

R. K. Turner

Depending on the precise definition used, coastal zones occupy around 20% of the earth's surface but host more than 45% of the global population and 75% of the world's megacities (> 10 million inhabitants). The zone's underpinning coastal ecosystems—coral reefs, mangroves, salt marshes and other wetlands, sea grasses and seaweed beds, beaches and sand dunes, estuaries and lagoons, forests and grasslands—necessary to sustain human occupation, are highly diverse, productive, and biocomplex. These ecosystems provide a range of services, such as, nutrient and sediment storage, water flow regulation and quality control, and storm and erosion buffering, summarized in Table 2.A.1 (Crossland et al. 2005). Coastal zones are impacted by dynamic environmental changes that occur both ways across the land–ocean boundary. The natural and anthropogenic drivers of change (including climate change) cause impacts ranging from erosion, siltation, eutrophication, and over-fishing to expansion of the built environment, and inundation due to sea level rise. All coastal zone natural capital assets have suffered significant losses over the past three decades (e.g., 50% of marshes lost or degraded, 35% of mangroves, and 30% of reefs) (MA 2005). The consequences for services and economic value of this loss at the margin are considerable but have yet to be properly recognized and more precisely quantified and evaluated (Daily 1997; Turner et al. 2003; Barbier et al. 2008).

The ecosystem services (storm buffering) valuation example illustrated below is drawn from the European coastal zone context and in particular the east coast of England, which is one of UK's most "at-risk" areas from climate change and other impacts. The European coastal zone is around 600 000 km² (within 10 km of shorelines) and home to 80 million people and 280 major cities. The annual value of coastal tourism alone is 75 billion EUR (*The Changing Faces of Europe's Coastal Areas* (2006): http://reports.eea.europa.eu/eea_report_2006_6/en). In response to the multitude of pressures bearing down on coastal areas, coastal protection and sea defense policy in the UK and Europe is being reappraised and reoriented toward a more flexible and adaptive strategy anchored to an ecosystem services approach and decision support system (Turner and Daily 2008; Turner et al. 2008).

The Millennium Ecosystem Assessment (MA 2005) states that "ecosystem services are the benefits people obtain from ecosystems" and subdivides them into supporting, regulating, provisioning, and cultural services. Building on this platform, it can be argued that when the focus is on national accounting (Boyd and Banzhaf 2007), or landscape management (Wallace 2007), or in our example valuation of service benefits (Fisher et al. 2008), further elaboration is required for actual choice-making involving human welfare. A key step is the separation of ecosystem processes and functions into intermediate and final services, with the latter yielding welfare benefits (see Table 2.A.1).

The generation of services and the enjoyment of benefits is spatially conditioned and therefore a key step in any evaluation process must be the setting of the ecosystems in their appropriate contexts. The valuation process must also be restricted to marginal gains/losses and should avoid double counting and should note possible threshold effects and any nonlinearity between change in ecosystem services and habitat variables such as size of area. Failure to account for these limitations will lead to under-/overestimated economic values and unnecessarily polarized cost–benefit decision choices (Turner 2007; Barbier et al. 2008).

Traditional sea defense and coastal protection strategies in Europe have sought to provide rigid engineered "hold-the-line" protection for people, property, and other assets against the vagaries of dynamic coastal environments. Given the growing awareness of the consequences of climate change a policy switch is taking place toward a "coping strategy" based on a mixed approach with engineered protection focused on high commercial value areas, and the rest of the coastline left to adapt to change more naturally. Measures such as "managed realignment," which involves the deliberate breaching of engineered defenses to allow coastal migration and the creation of extended intertidal marshes and mudbanks, at the expense of agricultural land, are now being tested. Testing sites have been carefully chosen to minimize impacts on existing people, property, and environmental assets that enjoy engineered protection from the sea. The question is do they represent cost-effective strategies for society?

Managed realignment schemes yield benefits in terms of ecosystem services. They generate carbon storage benefits (via saltmarsh creation) that can be valued in terms of the damage costs avoided per tonne of CO_2. The sites also serve to improve fisheries' productivity via nursery areas and this gain can be valued via market prices for commercial species. There are also general recreation and amenity benefits related to walking, bird-watching, and other recreational activities, as well as biodiversity

continues

Box 2.1 *continued*

Table 2.A.1 Coastal zone services* and benefits

Intermediate services	Final services	Benefits	Econ valuation methods
e.g:	e.g:	e.g:	e.g:
• Geodynamics: sediment and nutrient cycling and transport	• Creation of beaches, dunes, estuaries	• Flood/storm buffering	• Market prices/damage cost avoided
• Primary production	• Sediment, nutrients, contaminants retention/storage	• Shoreline stabilization/erosion control	• Production function market prices
• Water cycling		• Fish production	• Survey-based contingent valuation/ choice experimentation
• Climate mitigation	• Biomass export	• Biodiversity maintenance	• Damage cost avoided
	• Maintenance of fish nurseries and refuges	• Carbon storage	• Travel cost, hedonic pricing, survey based CV or CE
	• Regulation of water flow and quality	• Amenity and recreation activities	• Survey based CV or CE
	• Carbon sequestration	• Cultural/heritage conservation	
	• Recreation and amenity		
	• Cultural heritage		

*. European coastal areas including estuaries and saltmarshes.

maintenance and existence value benefits. An indication of the composite value of some of these amenity and related benefits can be estimated by transferring benefits data from the published literature if the spatial and other contextual variables are similar, or, more properly, by conducting site-specific contingent valuation/choice experiment studies to estimate willingness-to-pay values. Finally, the maintenance costs of the existing engineered defenses will be saved as realignment schemes are implemented. On the costs side, secondary defenses may be required further inland and there are opportunity costs associated with any agricultural land that is sacrificed as the old defenses are breached.

Cost–benefit analysis of managed realignment schemes took the following approach.

The "status quo" existing protection system is appraised as follows:

$$PV_t^{sq} = -\sum_{t=0}^{T} \frac{1}{(1+r)^t}[(I^{sq}C_{m,t}^{sq})], \qquad (2.A.1)$$

where PV_t^{sq} is the present value of total costs of current defenses (£million), r is the discount rate, I^{sq} is the length of defenses and $C_{m,t}^{sq}$ is the maintenance costs (£km^{-1} yr^{-1}).

The costs and benefits of managed realignment is given as follows:

$$PV_t^{mr} = \sum_{t=0}^{T} \frac{1}{(1+r)^t} \begin{bmatrix} I^{mr}(C_{k,t}^{mr} + C_{m,t}^{mr}) \\ -(a_t^{mr}L_{agr,t}^{agr}) - (a_h^{mr}B_{e,t}) \end{bmatrix} \qquad (2.A.2)$$

where PV_t^{mr} is the present value of managed realignment schemes (£million), I^{mr} is the length of managed realignment (km), $C_{k,t}^{mr}$ is the capital cost of realignment, $C_{m,t}^{mr}$ is the maintenance costs, a_t^{mr} is the agricultural land lost, $L_{agr,t}^{agr}$ is the forgone agricultural land value, a_h^{mr} is the area of intertidal habitat created, and $B_{e,t}$ is the ecosystem value benefits (£ha^{-1}).

Finally, the overall CBA result is found as:

$$NPV_t^{mr} = (PV_t^{sq} - PV_t^{mr}) \qquad (2.A.3)$$

where NPV_t^{mr} is the net present value of managed realignment compared to hold-the-line for a given stretch of coastline at time t (£million).

The analysis was carried out with data from a number of different estuaries, and indicates that appropriately sited schemes do represent gains in economic efficiency if declining discount rates are applied over a 100-year time horizon (see Turner *et al.* 2007 and Luisetti *et al.* 2008 for further details).

"ecotourism," the Galapagos now attract over 150 000 visitors each year.

There is a resident population on several of the larger islands, with a few service industries and a subsistence economy that depends on agriculture and fishing. These have been augmented by other direct uses that compete with the "natural" state of the larger islands, whereas recent reports suggest that the less occupied islands are doing better than they did 25 years ago. A significant fishery for sea cucumbers, a delicacy prized in Asian and French cuisine, developed in the 1990s and still exists, although the catch is declining. Illegal long-line shark fishing continues to create a problem. Not only do these activities threaten the intertidal fauna, they pose significant risks to the terrestrial ecosystem through the introduction of "exotic" species and destructive camping on some islands. Fortunately, the Park's protection system is much more effective now than in the past, and efforts to eradicate goats, cats, and other invasive species are continuing, most effectively on the four smaller, less-inhabited islands.

Arrayed against these direct, consumptive use values are two other values. The first is the direct, non-consumptive use value from ecotourism, which brings significant revenue. A sample calculation of this value would be that the average visitor (a week on a boat is a typical excursion) spends well more than $10 000. If the visitor is from the USA, additional revenue will accrue to the Ecuadorian economy through accommodations on the mainland, the flight to the islands, and (if a national carrier is used) the flight to Quito or Guayaquil. A total per-visit value of $15 000 would be a reasonable figure for the "overseas" visitor: if half were Ecuadorian nationals and half from elsewhere, the value of the industry would approach one billion annually.

Local residents, however, would make quite a different calculation. The shops and restaurants at Puerto Ayora collect some money, and the support of the Darwin Station by tourists flows into the local economy. Some boat operators are islanders, and some services for all vessels are locally provided. However, the vast majority of the revenue flows to tour operators, many of them non-Ecuadorian, and to other off-island entities.

Thus it is not surprising that a sometimes violent controversy has arisen over the protection of the islands. When the government closed the sea cucumber fishery in 1994 because the catch limit was being vastly exceeded, fishermen and some other local residents seized the Darwin Station and took scientists hostage. In a political controversy over a bill that would have given the islanders more local autonomy (and relaxed many of the ecological protections) there was another takeover. The tense historical contest between extraction and conservation in the Galapagos is, at least with respect to this particular indirect use value, the result of distributional effects. The economic potential of ecotourism is almost certainly greater than that of the resource-extraction uses. Yet the residents retain most of the rents from the second, and little from the first.

A second use value stems from the (uncertain) future benefits that would emanate from the scientific research underway on the Galapagos. The large number of endemic species found there, and the recentness of their evolutionary divergence from mainland relatives, make the islands a living laboratory for studies of species formation. Important recent work (see Grant 1986) depends on the integrity of the ecosystems of certain islands. Calculating its value, of course, would be extremely difficult.

Finally, there are two important non-use values. First, as in the case of the wetland example, people who have never been to the Galapagos and never expect to may experience a loss of existence value that they would willingly pay to avoid. The unique quality of the islands and the considerable publicity they have received as a mecca for naturalists gives this consideration a weight it might lack in less special areas. In addition, in the presence of uncertainty, people might be willing to pay a premium (over and above the expected future use value) to ensure the preservation of the unique flora and fauna of the islands. This is the option value.

2.5 Conclusions

Society must often make difficult choices about how and how much to protect natural capital and the ecosystem services such capital generates. Perhaps the most important basis for supporting a policy that would protect otherwise threatened ecosystem services is evidence that society gains more value from such protections than it gives up. This requires an assessment of the values that human beings place on such services—values that often are not expressed in markets.

In this chapter we have aimed to clarify the philosophical underpinnings of these values. The most prevalent and perhaps most workable philosophical basis is anthropocentric—value consists in the ability to provide satisfaction or well-being to humans. Although anthropocentric, this approach is consistent with society's making great sacrifices in order to protect valued species and ecosystem services.

We have also indicated the various types of value generated by ecosystem services, and laid out principal empirical methods for measuring these values. None of the empirical approaches is perfect; the uncertainties in measurement can be vast. However, even with the imperfections the methods generally are good enough to provide a basis for public policies to protect ecosystem services. The InVEST models described throughout this book represent the frontier in valuing ecosystem services. These models exemplify the substantial progress of the past decade in researchers' abilities to depict the gains and losses associated with the protection of a wide range of ecosystems and their services.

Many of the benefits from ecosystem services are not captured by unregulated markets. An individual's private gain from protecting ecosystem services falls short of the value of such protection to society. Hence private markets tend to fail to provide sufficient protection, and there is an important role for public policy to protect these services.

Two types of public policy could stem from the information offered by the InVEST models and other empirical studies. One is the introduction of quantitative restrictions that restrict the way natural capital gets used or converted and thereby protect the services such capital generates. Limits on wetland conversion, for example, help protect the various ecosystem services (flood control, water purification, and habitat provision) that wetlands offer. Another approach is the introduction of prices for ecosystem services—prices that the market would not generate on its own. For example, a tax on wetland conversion would serve to reduce the

rate of such conversion. According to economic theory, the tax rate should be set equal to the marginal value of the ecosystem services provided by the natural capital or ecosystem in question. This tax rate would lead to a lowered frequency of conversion that maximizes the net gain to society from intact wetlands—the benefit to agriculture minus the lost ecosystem service value. In many cases, the tax would prevent any conversion from taking place. These are instances in which the marginal value of the existing ecosystem services exceeds the marginal value generated by any conversion or alternative use of the natural capital.

Although our discussion acknowledges a key role for benefit–cost analysis in the valuation of ecosystem services, we would emphasize that such analysis is usually not sufficient for deciding policy. Benefit–cost analyses yield useful information on aggregate net benefits under alternative policy scenarios, but usually ignore issues of fairness or distribution. These analyses need to be accompanied by an assessment of the distribution of the gains and losses, both across the current generation and between current and future generations. If a proposed policy clearly would lead to serious inequities, it is reasonable to reject the policy, even if it passes a benefit–cost test.

The topics of valuation and policy choice raise a number of imponderables. In arriving at the social value of a given option, should every person's willingness to pay count equally, or should some members be given more weight than others? Are the preferences of sophisticated ecologists worth more than those of city-dwellers who evidence neither knowledge of nor interest in "nature"? How much weight should we give to the preferences or well-being of future generations, as compared to that of current inhabitants of the planet? How can we gauge the preferences of future generations in attempts to ascertain the gains or losses they might experience under different policies?

The fact that these questions have no easy answers need not make us pessimistic about the prospects for sensible public policy. We can go a long way toward improving policy-making by calling attention to the underlying philosophical questions, by developing empirical methods that generate better information about the gains and losses at stake under alternative public policies, and by developing channels for communicating this information to the general public.

References

Barbier, E., Koch, E., Silliman, B., *et al.* (2008). Coastal ecosystem-based management with nonlinear ecological functions and values. *Science*, **319**, 321–3.

Bishop, R. C. (1982). Option value: an exposition and extension. *Land Economics*, **58**, 1–15.

Boyd, J., and Banzhaf, S. (2007). What are ecosystem services? *Ecological Economics*, **63**, 616–26.

Cameron, T. (1992). Nonuser resource values. *American Journal of Agricultural Economics*, **74**, 1133–7.

Costanza, R., Norton, B., and Bishop. R. C. (1995) *The evolution of preferences: Why sovereign preferences may not lead to sustainable policies and what to do about it.* SCASSS workshop on Economics. Ethics and the Environment, Sweden.

Crossland, C. J., Kremer, H. H., Lindeboom, H. J., *et al.*, Eds. (2005). *Coastal fluxes in the anthropocene*, IGBP Series. Springer, Berlin.

Daily, G. C., Ed. (1997). *Nature's services: Societal dependence on natural ecosystems.* Island Press, Washington, DC.

Daily, G. C., and Ellison, K. (2002). *The New Economy of Nature.* Island Press, Washington, DC.

Fisher, B., Turner, K., Zylstra, M., *et al.* (2008). Ecosystem services and economic theory: integration for policy-relevant research, *Ecological Applications*, **18**, 2050–67.

Freeman, A. M. (1993). *The Measurement of Environmental and Resources Values: Theory and Methods.* Resources for the Future, Washington, DC.

Grant, P. (1986). *Ecology and Evolution of Darwin's Finches.* Princeton University Press, Princeton.

Luisetti, T., Turner, R. K., and Bateman, I. J. (2008) *An ecosystems' services approach to assess managed realignment coastal policy in england*, CSERGE Working Paper, ECM 2008–04,CSERGE, University of East Anglia, Norwich.

Mendelsohn, R., Nordhaus, W., and Shaw, D. (1994) The impact of global warming on agriculture: a Ricardian analysis. *American Economic Review*, **84**, 753–71.

Millennium Ecosystem Assessment (MA) (2003). Concepts of ecosystem value and valuation approaches. In *Ecosystems and Human Well-being, A Framework for Assessment; A Report of the Conceptual Framework Working Group of the Millennium Ecosystem Assessment.* Island Press, Washington, DC.

Mitchell, R. C., and Carson, R. T. (1989). *Using Surveys to Value Public Goods: The Contingent Valuation Method.* Resources for the Future, Washington, DC.

National Research Council (NRC) (2004). *Valuing Ecosystem Services: Toward Better Environmental Decision-Making.* National Academies Press, Washington, DC.

Schlenker, W., Hanemann, W. M., and Fisher, A. (2005). Will US agriculture really benefit from global warming? *American Economic Review*, **95**, 395–406.

Singer, P. (1975). *Animal Liberation.* Random House, New York.

Skidmore, J. (2001). Duties to animals: The failure of Kant's Moral Theory. *Journal of Value Inquiry*, **35**, 541–59.

Tilman, D. (1997). Biodiversity and ecosystem functioning. In G. C. Daily, Ed., *Nature's Services: Societal Dependence on Natural Ecosystems.* Island Press, Washington, DC.

Turner, R. K. (2007). Limits to CBA in UK and European environmental policy: Retrospects and future prospects. *Environmental and Resource Economics*, **37**, 253–69.

Turner, R. K., Burgess, D., Hadley, D., *et al.* (2007). A cost-benefit appraisal of coastal managed realignment policy. *Global Environmental Change*, **17**, 397–407.

Turner, R. K., and Daily, G. C. (2008). The ecosystem services framework and natural capital conservation. *Environmental and Resource Economics*, **39**, 25–35.

Turner, R. K., Georgiou, S., and Fisher B. (2008). *Valuing ecosystem services: The case of multifunctional wetlands.* Earthscan, London.

Wallace, K. J. (2007). Classification of ecosystem services: Problems and solutions. *Biological Conservation*, **139**, 235–46.

Watson, R. A. (1983) A critique of anti-anthropocentric biocentrism. *Environmental Ethics*, **5**, 245–56.

CHAPTER 3

Assessing multiple ecosystem services: an integrated tool for the real world

Heather Tallis and Stephen Polasky

3.1 Today's decision-making: the problem with incomplete balance sheets

Conservation and natural resource management have been dominated by approaches that focus on a single sector and a single objective. These approaches often fail to include a wider set of consequences of decision-making. For example, maximizing profit from industrial production often leads to negative impacts on air quality and human health. Maximizing agricultural production often leads to poor water quality and in some cases losses of productivity in downstream fisheries. Maximizing biodiversity conservation can come at the cost of local jobs, food production, and other important benefits.

Many of these consequences are the result of management decisions that overlook the broad suite of ecosystem conditions and processes that sustain and enrich human life. Ecosystem services, defined as the contribution of ecosystem conditions and processes to human well-being, include the production of goods (such as agricultural crops, seafood, timber, and natural pharmaceuticals), processes that control variability and support life (climate regulation, flood mitigation, pollination, and the provision of soil fertility and clean water), enrich cultural life (recreational opportunities, and satisfaction of aesthetic and spiritual needs), and preserve options (Ehrlich and Ehrlich 1981; Daily 1997; MA 2005).

In most cases and for most services, there is little incentive for business managers and local landowners to account for the provision of ecosystem services in their decision-making. Landowners receive financial rewards for producing crops or for developing their land as real estate. They typically do not receive financial rewards for providing public goods from ecosystems, such as pollution filtration or flood mitigation. For example, in tropical coastal ecosystems, mangroves are routinely cleared for shrimp aquaculture. People clearing the mangroves receive high market prices for the shrimp they produce but they do not bear the full costs associated with the loss of habitat for coastal fisheries, storm surge protection (Das and Vincent 2009), pollution filtration, or the loss of other ecosystem services provided by mangrove ecosystems. A more complete accounting shows that maintaining mangroves generally provides greater benefits for society than shrimp aquaculture provides (Sathirathai and Barbier 2001; Barbier 2007). A single-sector approach, which ignores the multitude of connections among components of natural and social systems, generally fails to provide as high a value to society as would management that accounted for the full range of social benefits.

3.2 The decision-making revolution

In this chapter, we outline the major challenges in taking an integrated approach to decision-making, and present a new spatially explicit modeling tool that takes critical steps toward addressing these challenges. The inclusion of ecosystem services in decision-making provides a framework that enables managers to broaden their perspectives by considering the multiple, interlinked consequences of their decisions. Our approach builds from the Millennium Ecosystem Assessment (MA), which contributed substantially to our understanding of how to take such an approach at a global scale (MA 2005). Within a year of its completion, findings from the MA were incorporated into the Convention on Biological Diversity, the RAMSAR Convention on Wetlands, and the Convention to Combat Desertification (Boerner 2007). Despite the overall success of the MA at the global scale, we are still left with the grand challenge of bringing useful models and information to bear at local, regional, and national scales where most decisions are made. Although ecosystem service assessment has been attempted at sub-global scales (Imperial 1999) there are no systematic tools that can be applied in a general, consistent way across sites at the spatial scales and time frames relevant to major decisions affecting ecosystems.

One of the most challenging aspects of creating such tools is integrating robust ecological models and understanding in "ecological production functions" that define how the spatial extent, structure, and functioning of ecosystems determine the production of ecosystem services. This challenge is particularly acute with ecosystem functions and services that operate across ecosystem boundaries (such as nutrient transport from land to sea) and across spatial scales (Engel et al. 2008).

A second major challenge of including ecosystem services in specific decisions is generating estimates of the value of ecosystem services in economic and other terms. This task requires linking ecological models and understanding with social and economic methods to reveal the values people hold for different ecosystem services (NRC 2005). This task is especially hard for the many ecosystem services that generate global public benefits, such as climate regulation or existence value of species, for which there are no market prices or other readily available signals of value. Over the past 40 years or so, economists have developed a number of methods and tools for "non-market valuation" that can be applied to estimate the value of ecosystem services (Freeman 2003, NRC 2005). Whether for market or non-market values, appropriately linking social and economic valuation with ecological production functions is necessary to ensure that values reflect underlying ecological conditions. We describe these challenges further below and offer plausible solutions.

3.3   The ecological production function approach

An ecological production function specifies the output of ecosystem services provided ("produced") by an ecosystem given its condition and processes. These functions vary spatially due to site- and ecosystem-specific relationships. Once an ecological production function is specified, researchers can quantify the impact of landscape change on the level of ecosystem service outputs. In the twentieth century, human alteration of ecosystems on a large scale, such as the conversion of native ecosystems to monoculture agriculture, led to an increase in some provisioning services (e.g., food production) at the expense of many regulating, supporting, and cultural services (Vitousek et al. 1997; MA 2005).

Most applications of an ecological production function modeling approach have been done at small scales or for a small set of services. There are a growing number of such studies (e.g., Ellis and Fisher 1987; Barbier and Strand 1998; Wilson and Carpenter 1999; Barbier 2000; Kaiser and Roumasset 2002; Ricketts et al. 2004) and useful overviews and summaries have been compiled (Pagiola et al. 2004; NRC 2005; Barbier 2007). One of the most challenging tenets of production functions is to integrate modeling across multiple services. The essential next step toward informing decision-making is a systematic approach that combines the rigor of the small-scale studies with the breadth of broad-scale assessments. Recent work has taken strides in this vein (e.g., Boody et al. 2005; Jackson et al. 2005; Antle and Stoorvogel 2006; Naidoo and Ricketts 2006; Nelson et al. 2008, 2009).

There are some cases where an understanding of ecological production functions alone is sufficient. Many government agencies make decisions about what activities will be allowed based on whether or not they meet an environmental standard. For example, an agency may assess how activities will likely affect the ability of an ecosystem to meet water quality standards. Their decision is not based on how much it would cost to treat that water for consumption or the value of access to clean drinking water, but rather on the expected change in contaminant levels or the likelihood of crossing a contamination threshold. In these cases, simply knowing how ecosystem services will change in biophysical terms is informative and useful.

Other decisions are tied to financial costs and benefits, and many decision-makers are conditioned to analyzing policy alternatives in terms of the net benefits measured in monetary terms. One concern about keeping measurement of ecosystem services in biophysical units is that services not measured in monetary terms may not be given full weight in decision-making or may be ignored altogether (Daily 1997). In these cases it can be very useful to combine ecological production functions with economic valuation methods to estimate and report the monetary value of ecosystem services.

Some ecological production function approaches have been combined with appropriate market prices and non-market valuation methods to estimate economic value, and illustrate the change in the monetary value of services with changing environmental conditions (e.g., Swallow 1994; Naidoo and Ricketts 2006; Barbier 2007). Ecological production functions can be used to determine how the provision of various market goods and services change as ecosystem conditions or processes change. The value of the changes in output of marketed goods and services can be evaluated using market prices for marginal changes or by using changes in consumer and producer surplus for non-marginal changes (Just et al. 2004; Barbier 2007). Non-market valuation methods, including revealed preference and stated preference methods, can be used for ecosystem services that are not traded in markets (Freeman 2003; NRC 2005). At present, however, we lack comprehensive studies that tie together economic valuation methods with ecological production functions to estimate the monetary value of ecosystem services both for a broad range of ecosystem services and at a broad geographic scale (NRC 2005, but see Naidoo and Ricketts 2006).

Detailed non-market valuation studies appropriate to a particular service in a particular place tend to be time- and-resource intensive, limiting the applicability to broad-scale assessments. Benefit transfer provides a less time-intensive approach than production functions for generating broad-scale monetary estimates of ecosystem services. Benefit transfer studies of the value of ecosystem services have garnered significant inter-

national attention especially following Costanza et al. (1997), which estimated the monetary value of ecosystem services for the entire planet. The most common application of the benefit transfer approach uses estimates of the value of services per unit area from a single or small number of locations and applies this value to other locations with the same ecosystem type (e.g., Costanza et al. 1997; Konarska et al. 2002; Troy and Wilson 2006; Turner et al. 2007). This approach relies on existing estimates and does not require any additional analysis, which is a distinct advantage if decisions are imminent and primary data collection is not feasible (Wilson and Hoehn 2006).

The assumption of constant per hectare values that makes the benefit transfer approach so tractable, however, has significant disadvantages that limit its social, economic, and ecological realism. In some cases, the benefits transfer approach provides an estimate of total economic value rather than estimates of value for individual services (e.g., Konarska et al. 2002). When limited to estimates of total economic value, we cannot analyze how the provision and value of each individual service will change under new conditions. If a wetland is converted to agricultural land, how does this subsequently affect the provision of clean drinking water, floods downstream, climate regulation, or soil fertility? Without service-specific information, it is impossible to design effective policies or payment programs that ensure the continued provision of ecosystem services.

Further, assuming that every hectare of a given habitat type is of equal value ignores well-demonstrated differences between sites in terms of scarcity, spatial configuration, size, quality of habitat, number of nearby people, or their social practices and preferences, all of which may be crucial in determining the value of ecosystem services. These simplifications mean that area-based benefit estimates that rely on transferring values from a study site (where original valuation took place) to a policy site cannot consider important spatial aspects of land use, habitat distribution and geometry, or economic insights on the importance of proximity and value. For these reasons, we do not believe that application of benefits transfer based on value per hectare by habitat type is a good direction to pursue.

A more promising approach to benefit transfer is to use a value function approach, which estimates value as a function of ecological and socio-economic conditions, rather than a unit value approach that depends only on habitat type. Value functions can be estimated from data at a single site or from a meta-analysis of a number of sites (Rosenberger and Phipps 2007). Although the field of benefit transfer is moving toward value function approaches that include socio-demographic characteristics, environmental attitudes, and biophysical contexts at study and policy sites (Wilson and Hoehn 2006; Rosenberger and Stanley 2006; Eshet *et al.* 2007), examples of sound applications are still hard to find (Plummer 2009).

The creation of general tools capable of integrating ecological production functions with valuation for multiple services would allow a major advance toward integrated decision-making in diverse contexts across scales. At present, the lack of such tools makes ecosystem approaches to most resource decisions expensive, time-consuming, unwieldy, and difficult to replicate. In the next section, we describe the development of a new general tool aimed at filling this gap.

3.4   InVEST: mapping and valuing ecosystem services with ecological production functions and economic valuation

The Natural Capital Project (http://www.naturalcapitalproject.org) has developed a new tool designed to facilitate integrated decision-making, bringing together credible, useful models based on ecological production functions and economic valuation methods. The intention is to incorporate biophysical and economic information about ecosystem services into conservation and natural resource decisions at an appropriate scale. The tool is called InVEST, for Integrated Valuation of Ecosystem Services and Tradeoffs. We have built in several key features that make this a flexible yet scientifically grounded tool. InVEST is a set of computer-based models that:

• Focuses on ecosystem services themselves, rather than on the underlying biophysical processes alone;

• Is spatially explicit;
• Provides output in both biophysical and monetary terms;
• Is scenario driven;
• Clearly reveals relationships among multiple services; and
• Has a modular, tiered approach to deal flexibly with data availability and the state of system knowledge.

Several of these features are discussed in greater detail below.

3.4.1   A multiple ecosystem service approach

Managers are often forced to make trade-offs among sectors and goals. A fundamental socio-economic truth is that a manager cannot simultaneously maximize returns for all sectors of society at once. As the old saying goes, we cannot have our cake and eat it too. Despite the ubiquity of trade-offs, managers frequently lack a set of tools to inform them about the trade-offs they face. Often, mental assessments of the existence and magnitude of trade-offs are wrong and lead to decisions that result in poor outcomes. Although management actions that strike a balance among goals may be plausible, these actions are often hard to identify in a highly charged political environment. Arguments commonly based on qualitative assumptions are difficult to balance in a systematic and clearly understood way. A formal modeling framework that can reveal the likely relationship among services can help dispel incorrect assumptions and identify management options that provide a high level of a range of ecosystem services. There is growing evidence that decision-makers are ready for this kind of approach, if only they had tools to help them move forward (see examples in Section 3.4.4).

Trade-offs among sectors, or among ecosystem services exist because services are not perfectly correlated. Using data from the Willamette Basin in Oregon, Nelson *et al.* (2008) found that targeting policies to provide carbon sequestration, by limiting enrollment to landowners who would grow forests on their land, was effective at increasing carbon storage but not effective for species conservation. Alternatively, targeting policies to meet species conservation objectives, by limiting enrollment to landowners who would restore rare habitat types (e.g., oak savannah and prairie), was effective at increasing species conservation but not effective for carbon sequestration. More generally, the MA (2005) found pervasive trade-offs between provisioning services (e.g., food and timber production) and other types of services (regulating, supporting, and cultural services). However, some trade-offs are more a consequence of past land-use decisions than of the underlying potential of the socio-ecological system. Polasky *et al.* (2008) demonstrated that higher levels of both biodiversity conservation and value of marketed commodities could be achieved by rearranging the spatial pattern of activities on a landscape. A modeling framework that allows assessments of biodiversity and multiple ecosystem services can identify policies or geographies that can lead to win–win outcomes, where all objectives can be increased relative to the status quo, and to those situations where outcomes necessarily lead to trade-offs.

We address the need to reveal and quantify synergies and trade-offs by providing models for a suite of ecosystem services in InVEST. The services we currently model are provision of hydropower and irrigation water, mitigation of storm peak flows, avoidance of reservoir sedimentation, regulation of water quality, climate regulation through carbon storage and sequestration, timber production, production of non-timber forest products, agriculture

production, provision of pollination for agricultural crops, recreation and tourism, and provision of cultural values and non-use values (Table 3.1). We also provide models for terrestrial biodiversity, as an attribute of natural systems that underpins the delivery of ecosystem services (Chapter 13).

Additional ecosystem services besides those listed above should be considered in many natural resource decision-making processes (see Box 3.1 for example). We have focused initially on the above-mentioned subset because of their global importance, relevance to major decisions being made currently, and proximity of many of these services to markets. As the modeling effort progresses our aim is to include other services that likely provide value to society.

3.4.2   Ecological processes vs. ecosystem services: a critical distinction

Ecological processes are essential for the provision of ecosystem services but processes are not synonymous with services. Until there is some person somewhere benefiting from an ecological process, it is only a process and not an ecosystem service (Luck *et al.* 2009). This distinction is critical yet often overlooked. Extensive research has been applied to the modeling and measurement of ecological processes, and it is tempting to simply apply those models to ecosystem service-related decisions. However, ecological processes tell us only about the ecological

Table 3.1   Classification of ecosystem services modeled in InVEST

Service	MA classification	Valuation technique	Chapter
Provision of water for hydropower	Supporting	Market valuation	4
Provision of water for irrigation	Supporting	Market valuation	4
Storm peak mitigation	Regulating	Avoided damages	5
Water purification: nutrient retention	Regulating	Avoided damages	6
Avoided reservoir sedimentation	Regulating	Avoided damages	6
Carbon sequestration	Regulating	Social value, Market valuation	7
Timber production	Provisioning	Market valuation	8
Non-timber forest product production (NTFP)	Provisioning	Market valuation	8
Agricultural production	Provisioning	Market valuation	9
Pollination of agricultural crops	Supporting	Market valuation	10
Recreation/tourism	Cultural/aesthetic	Market and non-market valuation	11
Cultural/aesthetic	Cultural/aesthetic	None	12

Valuation technique identifies the general approach used to derive value once InVEST has generated the amount of ecosystem service.

Box 3.1 Unsung ecosystem service heroes: seed dispersal and pest control

Liba Pejchar

At dusk in Mexico, a bat flashes low over a coffee plantation, swiping a moth off a leaf and disappearing into the night. In a large city park halfway around the world, a jay screeches shrilly as it snaps up an acorn and swoops away, burrowing it neatly for the winter. In Texas, 100 million bats pour out from caves and from under bridges, feeding in a frenzy over 10 000 acres of cotton plantations. And in a Hawaiian forest, a thrush flutters down, nabs a red berry, and swoops to a perch, dropping the seed in alarm when the shadow of a hawk passes overhead. What do all of these actions have in common? They all involve the feeding habits of birds and bats, our winged cousins. But they also illustrate two frequently unrecognized ecosystem services: seed dispersal and pest control, services that are provided for free every day, all over the world.

Understanding how bird and bat species provide key functions in ecosystems opens up new opportunities to value the provision of little-known ecosystem services. The acorn-toting Eurasian Jay (*Garrulus glandarius*) is a neat example of such an opportunity (Figure 3.A.1). A city park near Stockholm supports one of the largest populations of giant oaks (*Quercus* spp.) in Europe—a keystone species that harbors high biodiversity and plays an important role in the cultural landscape of Sweden. Natural regeneration and long-distance dispersal is largely dependent on the

Eurasian Jay, which bury 4500–11 000 acorns per year at depths ideal for germination and in areas where light conditions are perfect for growth. The fraction of acorns that are buried and forgotten have a far greater chance of germinating than passively dispersed acorns that typically suffer 100% predation by mammals. In the absence of the jays, the replacement cost of this service (using human labor to plant oak seedlings) for the park alone would be 1.5–6.7 million SEK per pair of jays (approximately US$200 000–950 000 per pair (Hougner *et al.* 2006)).

In this and other rare cases, seed dispersal and pest control services have been quantified in monetary terms by comparing costs and benefits to the human equivalent (i.e., hand-planting seeds or applying pesticides). For example, breeding colonies of Brazilian free-tailed bats (*Tadarida brasiliensis*) feed on extraordinary numbers of insects in south-central Texas (Figure 3.A.1). Lactating females consume the insect equivalent of two-thirds of their body mass every hot summer night. Taking into account the private and social costs of pesticides that would be required in the absence of these insectivores, cotton farmers benefit from bats to the tune of approximately US$750 000 per year on a harvest worth US$4–6 million (Cleveland *et al.* 2006). Building on ecological understanding to demonstrate the economic value of wild nature is crucial for incorporating conservation of these organisms and processes into decision-making.

Figure 3.A.1 (a) Eurasian Jays (*Garrulus glandarius*) and (b) Brazilian free-tailed bats (*Tadarida brasiliensis*) provide valuable seed dispersal and pest control services.

---

production function, which is only the **supply** of ecosystem services. It is critical to include the **demand** for services as well. Where are the people who enjoy services, and how much do they use? The combination of supply and demand generates "use"

of ecosystem services. We define the term "use" quite broadly. "Use" includes not only the consumption of physical goods (e.g., agricultural crops, fish), but also the recreational, cultural, spiritual, and aesthetic appreciation of nature (non-consumptive use value), as well as option, existence, and bequest values that do not generate a benefit from any type of consumptive or non-consumptive use ("non-use" values). To model use of ecosystem services, we need to integrate analysis of the supply of ecosystem services (ecological processes) with analysis of the location, type, and intensity of demand for services. This is true for all categories of ecosystem service (Table 3.1).

Water-related services provide good examples for how to think about the distinction between ecological processes (supply) and ecosystem services (supply and demand). Consider water purification for drinking, a regulating service. Many useful ecological production function models exist that can help us predict the concentration of contaminants in waterways (e.g., Soil and Water Assessment Tool (SWAT), Gassman *et al.* 2007; Annualized Agricultural Non-Point Source (AnnAGNPS), Yuan *et al.* 2006). However, the provision of clean drinking water is not a service unless there is someone there who wishes to drink it. This does not mean that a natural system providing clean water in a remote area with no people does not provide *any* services. Clean water in remote areas can maintain biodiversity or provide ecological functions that underpin other ecosystem services, and may also be a value as clean water *for drinking in the future*. However, if no one currently makes use of the water for drinking, then there is no clean drinking water ecosystem service in that particular place at the current time. Ecological processes (the production function) must be connected to beneficiaries to gain an accurate picture of the level of use of the service provided.

Consider another example involving the pollination of agricultural crops, a supporting service. Many agricultural crops require insect pollination (e.g. almonds, strawberries), but many other crops do not (e.g., rice, corn). A patch of native habitat in an agricultural landscape may house bee populations, but if there are no agricultural fields within foraging distance with a crop in need of pollination, then that native habitat patch does not provide pollination benefits for crops at that time. So, in this case, a model of native pollinator meta-population dynamics could give us a very clear sense of how much pollination service *could be supplied* by patches of native habitat in an agricultural landscape. Until that information is paired with information on the identity of crops grown, their distribution in the area, and the crop-specific yield benefits of pollination, we cannot estimate the amount of pollination service actually being provided at a given time.

InVEST deals with this challenge by using a three-step modeling process. First, the ecological production function, or the **supply** side of ecosystem services, is modeled (Table 3.2). These models require biological, physical, geological, and other kinds of inputs, and draw heavily from existing knowledge. For example, our water-related service models start out with similar fundamental hydrologic processes as those included in models such as SWAT (Gassman *et al.* 2007). Our model for avoided reservoir sedimentation draws heavily from the Universal Soil Loss Equation (USLE) (Brooks *et al.* 1982). Our biodiversity model is based on species-area relationships in extensive native habitats and in countrysides (Connor and McCoy 1979; Pereira and Daily 2006). The outputs from this step of modeling are in biophysical units and represent the level of each ecological process supported by each part of the landscape (Table 3.2). For example, our model of irrigation first predicts the total amount of surface runoff from each parcel on the landscape. This represents the supply of water for all potential consumptive uses (Table 3.3). Outputs like this can be useful for model calibration or for understanding the maximum potential level of service.

Table 3.2 Three-step structure of InVEST ecosystem service models

Modeling step	Model inputs	Model outputs	Units
Ecological process	Geological, morphological, biological, etc.	Supply	Biophysical
Use	Socio-economic, management characteristics, etc.	Level of use—intermediate service	Biophysical
	Socio-economic, management characteristics, etc.	Level of use—final service	Final product
Valuation	Financial	Value	Monetary

Table 3.3 Examples of ecosystem service outputs and units for one final service (timber production) and two intermediate services (crop pollination and provision of irrigation water)

Output	Service		
	Timber production	Crop pollination	Provision of irrigation water
Supply	Standing stock of wood (ft^3 ha^{-1})	Insect abundance (# insects ha^{-1})	Surface runoff (vol ha^{-1})
Use—intermediate service	None	Insect abundance contributing to crop pollination (# of insects ha^{-1})	Runoff available and used for irrigation (vol ha^{-1})
Use—final service	Harvested wood (ft^3 ha^{-1})	Crop yield due to insects (kg crop ha^{-1})	Crop yield due to runoff used for irrigation (kg ha^{-1})
Value	NPV of harvested timber ($ha^{-1})	NPV of additional crop yield ($ha^{-1})	NPV of additional crop yield ($ha^{-1})

NPV = net present value.

The second step of modeling determines the use of ecosystem services. This step incorporates socio-economic, management, and other kinds of data on demand for ecosystem services with information on supply. **Use** of an ecosystem service is the level of supply in an area actually demanded by people for the service of interest (Table 3.2). For example, in our irrigation model, we first consider how much water supply is available, which is determined by surface runoff generated upstream of agricultural fields with subtractions for withdrawals upstream for other consumptive uses such as for drinking or industry. We then consider how much water demanded by crops in the region is not met by rainfall. There may be a large amount of water available for irrigation (after other consumptive uses), but if there are no crops left with a water deficit after factoring in rainfall, there is no demand for irrigation. In such a case, although there is a supply of irrigation water, because there is no demand there is no resulting use and therefore no ecosystem service provided. This example illustrates why it is so important to combine supply and demand in the use step in modeling. It is only by combining supply and demand to determine use that we quantify the level of outputs of ecosystem services.

If the service of interest is an intermediate service, there are two possible model outputs: one for the use of the intermediate service and one for the use of the final service (Table 3.2). In our irrigation example, we can report the amount of water used for crop irrigation, which would have the units of volume per hectare (vol ha^{-1}). This is the level of use of the intermediate service. Irrigation water is an input to agricultural production (final service), so we can also use our agricultural production model to estimate how much additional crop yield can be expected given that additional amount of water available from irrigation. This output (additional yield from irrigation, in kg ha^{-1}) is the level of use of the final service (the final product) (Table 3.3).

In addition to mapping and quantifying the supply and use of ecosystem services, InVEST also has the capacity to estimate their value. As discussed in Section 3.3, we use market and non-market valuation approaches to arrive at monetary values for ecosystem service provision (Table 3.2). Our focus in estimating value is on the social value of each service, which captures the total value of the service to society as opposed to the value it offers to the owner of the service-providing land. While the social value of services for which markets exist (e.g., provisioning services) can be estimated using market prices, estimating the social value of services for which there are not markets requires alternative methods. Numerous techniques have been developed and refined in recent decades to estimate the value of non-market goods (e.g., hedonic price models, travel cost models (random utility models), choice experiments including both contingent valuation and conjoint analysis). To reduce the data collection and analytical burden on InVEST users, the default valuation methodologies we use in InVEST are typically associated with market prices for commodities traded in markets, or the damages avoided by the maintenance of service provision. In each case, the valuation methodologies described in the following chapters for services represent only one of the many viable valuation options. Users with greater sophistication and access to valuation studies may wish to utilize other valuation approaches.

Our irrigation example demonstrates the InVEST valuation approach for many provisioning services. The monetary value of increased water for irrigation requires identifying the increase in the revenue from agricultural production that would result from an increase in available water input (Table 3.3). This additional value can arise because more land can be irrigated or because existing cropland can receive more water, which could increase yields of crops or allow more water-intensive but higher value crops to be grown.

Consider another example of avoided reservoir sedimentation that demonstrates the methodology of damages avoided for valuing non-market services. The role that vegetation and management practices play in keeping sediment out of waterways can provide services to society including avoided infrastructure maintenance costs (as reservoirs silt in and require dredging) and avoided flood risk (as rivers or reservoirs silt in and lose their capacity to control or buffer floods). To estimate the value of avoided siltation in reservoirs, we calculate supply by modeling how much erosion control is provided by a landscape based on enhanced USLE equations. We then calculate use by adding demand, represented by the location of reservoirs and their characteristics (dead volume, remaining lifetime of the dam, etc.). Finally, we derive the value of this ecosystem service through avoided dredging cost calculations.

Strictly speaking, damages avoided (or replacement costs) are estimates of costs not estimates of benefits and so need to be used with caution. Avoided damages (replacement cost) can be used as a measure of value of ecosystem services only when there are at least two ways of providing an equivalent quantity and quality of an ecosystem service (one supplied by ecosystems and an alternative supply via a human-engineering approach), and where the benefits of the service exceed the costs providing the service via the human-engineering approach (Shabman and Batie 1978; NRC 2005). If these conditions are violated then damages avoided

is not an appropriate method for valuing an ecosystem service.

The details of how supply, demand, and economic valuation are combined in models for these and other services can be found in later chapters of this volume (see also Table 3.3 for other examples).

3.4.3 Spatially explicit ecosystem service modeling

Because the value of ecosystem services is determined by both the location of ecological processes that provide services (supply) and the location of people who demand and use the services, any ecosystem service modeling effort should be spatially explicit. We consider two key elements of space in the application of InVEST: the role of spatial pattern and heterogeneity in the landscape in determining the provision of services, and the scale across which different services act. Often, decision-makers want to know where to invest or how to target programs to get the greatest return from their investment. For instance, where should protected areas be located to gain the largest biodiversity and climate regulation co-benefits? Should a new agricultural subsidy program to control water quality be targeted at riparian areas in headwaters or further downstream? Will a tree planting program in a poor district help with flood control? All of these questions have a spatial element, but many existing biophysical process models are non-spatial and do not allow analysis of the landscape locations best for investment. All of the models in InVEST focus on identifying how much each parcel on the landscape contributes to each service.

Secondly, we must consider the scale across which services are provided. Some services, such as pollination and some water-related services, are provided at a very local scale, while other services, such as climate regulation, are provided at a global scale. Trees fixing carbon in the Amazon forest are providing a global benefit. Each model in InVEST looks across the appropriate scale for the service of interest. For example, the pollination model uses the foraging distance of pollinators to delineate the landscape for assessment, while the carbon sequestration model assumes that tree growth on any parcel provides a benefit no matter where it is located since the global atmosphere is well-mixed.

Considerations of scale raise two important issues, one related to modeling and one related to policy. It may be more difficult to apply models for local services since input data on land use and cover patterns need to be at a high enough resolution to capture important features of the service. One would not learn much from the crop pollination model if native pollinators at the site of interest foraged 1.5 km, but land use and cover data were only available at 10 km² resolution. When modeling multiple services, the scale of data resolution should correspond to the finest scale of resolution for the ecosystem services of interest.

In terms of policy, the scale and location of the provision of ecosystem services and the scale and location of beneficiaries of the services are often disconnected. This is the case for trees fixing carbon in the Amazon rainforest that are enjoyed (and "used") by people all around the globe. For other services, such as pollination or provision of clean water, benefits are fairly local. However, even where provision and benefits of services are local, supply and demand may be disconnected in space. For example, upstream landowners may divert water or increase nutrient loading that harms downstream water users. Such spatial disconnects between provision and benefits have important implications for policy. Explicit policies may be needed to give incentives to people who control the provision of services so that they can recognize the benefits that their actions provide to others. Such policies can be explored through the development of scenarios.

3.4.4 Scenario-driven modeling: making a decision-relevant tool

To be effective in a decision-making or policy arena, analyses should be relevant to the needs and questions of managers and decision-makers. To apply InVEST in such situations, we envision embedding the modeling within a stakeholder engagement process that allows managers to identify the choices of interest to them (Figure 3.1). InVEST is designed to work with many different kinds of scenarios derived through many types of stakeholder engagement processes.

InVEST can take input from stakeholders to consider a wide range of land use and resource management alternatives. Each ecosystem service model uses land use and land cover (LULC) patterns as inputs to predict biodiversity and the production of ecosystem services across a landscape. This means that we can consider choices that affect the type of land cover (urban, wetland, closed-canopy deciduous forest, etc.) and choices that keep land cover the same but alter management practices on any

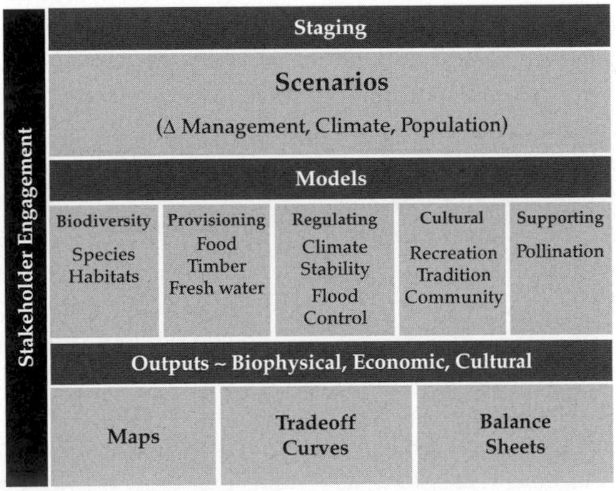

Figure 3.1 Conceptual model for applying InVEST as part of a stakeholder process. Stakeholders give input to the process at every step. They produce future options that are turned into scenarios, they identify the ecosystem services of interest and help determine the level of model complexity needed for the questions of interest, they provide input data for the models, and they request particular types of outputs and then assess those outputs. If results spur further questions or ideas for alternative scenarios, the entire process can be repeated.

particular part of the landscape (change in release pattern from an existing dam, change in crop type planted in existing agricultural areas, change in fertilizer type or amount used, change in rotation time in existing plantation forests, etc.). Most natural resource management decisions will have effects on land use and cover patterns, either directly or indirectly, so sensitivity of InVEST models to LULC patterns translates into broad sensitivity to management choices.

For InVEST to assess the choices that managers or other stakeholders want to consider, we need to translate those choices into likely future land use and cover patterns. There are many different methods for turning choices into LULC patterns. When decision-makers hold full control over the area of interest, their own planning processes usually include the development of scenarios that can be assessed by InVEST for likely ecosystem service provision. When the area of interest is a more complex landscape with multiple ownership and multiple drivers of change, more complex scenario generation options are available. One can develop predictions of how landowners will react given the various market forces, institutions, and incentives that they face at regional (e.g., Nelson *et al.* 2008; Sohl and Sayler 2008), national (e.g., Veldkamp and Fresco 1996), or global scales (e.g., Alcamo *et al.* 1994). For example, we can use models of landowner decision-making to predict how landowners would react to changes in crop prices or to government policies to generate scenarios of land use and land cover that can be input into InVEST models. In more demonstrative applications, users may want to explore what is possible on a particular landscape given the fundamental ecosystem service relationships in place. In these cases, landscape optimization modeling can be used to find solutions that maximize the provision of a combination of services. We have used this approach to maximize a measure of species conservation for a given value of commodity production (and vice versa) using early versions of InVEST models (Polasky *et al.* 2008).

We have emphasized management decisions as drivers of landscape change, but there are obviously other factors at work. Climate change and human population growth will strongly influence LULC patterns and climate conditions in the future. Scenarios can be built to include these drivers of change in addition to management practices. Many efforts have now down-scaled global climate models and used regional predictions to drive vegetation patterns, giving us maps of likely future land cover and climate. Similarly, a variety of research groups are in the process of turning numeric projections of population change into spatially explicit human population density estimates or urban/rural area extents (e.g., Salvatore *et al.* 2005). When models or maps of these drivers are available, they can be combined with any approaches that project management impacts, giving scenarios that represent all three major drivers of future change. We demonstrate one such application of InVEST with climate scenarios in Chapter 18. In the next section we explore the specifics of several real-world decisions being addressed through scenario analysis with InVEST.

3.4.5 Examples of scenario development and InVEST application

In many cases, managers want to know the likely outcomes of a proposed program, policy, or management action before they decide how to proceed (Ghazoul 2007). Maps of ecosystem service provision can be used for planning, priority setting, determining compensation or offset levels, designing policies or monitoring programs, or identifying which members of society are controlling the provision of ecosystem services and which members are receiving benefits. Here, we consider several management questions and explore how InVEST can help answer them. These are all real cases where InVEST and its precursors are being applied.

> *The Chinese national government has a program to set up conservation areas to protect the natural capital that supports human well-being. Where should these areas be located?*

This is an optimization question, like those commonly asked in the conservation arena. To answer this question, we need to know which areas of the landscape provide the highest level of services at the least cost. We can first use production function and economic valuation models to estimate ecosystem service levels and values. We can then feed these maps into optimization algorithms that determine which parts of the landscape, if protected or managed in a certain way, meet goals the policy maker sets for ecosystem service provision at the lowest cost (e.g., number of hectares, purchase value of the land, opportunity cost of foregone activities). For those familiar with conservation planning, this could be done by running MARXAN, or a similar program, with ecosystem service targets. While InVEST does not include an optimization routine, outputs of InVEST easily can be used in standard optimization packages to answer questions of this sort.

The Nature Conservancy has completed an analysis of priority sites for conservation action in the Willamette Basin, Oregon (USA). They want to work first in the places that also give the highest possible benefit for human well-being. Where should they work?

This is also a priority setting question, but one that places biodiversity conservation above ecosystem service provision. In this case, we do not have to apply production function models to the whole landscape but can instead focus analyses on the priority sites for biodiversity conservation. The options for configurations of priority areas become scenarios and InVEST can be used to estimate the levels and values of services that society will receive from conservation of each site. Planners can then choose where to act.

Colombia's Ministry of the Environment is responsible for permitting and licensing all major production sectors in the country. They are considering requiring offsets and mitigation for biodiversity and ecosystem service damages caused by development in sectors such as mining, agriculture, oil and gas, and infrastructure. If the expanded approach becomes law, how much mitigation should be required for a given development plan or set of permits, and where should the mitigation be done to adequately offset the damages?

InVEST has great potential for application in the mitigation and offsetting arena because it explicitly considers the connection between ecosystem processes and people. In many mitigation cases, offsets may be designed with the intention of providing services, but the actual connection to people is not considered explicitly. In Florida for example, a wetland mitigation bank was created on a nearshore island to compensate for development in a coastal watershed. People in this area get drinking water from groundwater wells. The wetland bank was meant to replace biodiversity and water filtration services for clean drinking water lost through development, but the island housing the bank was located over a different aquifer with no inhabitants. There was technically no net loss of wetland area or function, but the drinking water-related services provided to people were lost entirely.

InVEST models can be used to identify where on a landscape services are produced and used. Colombia's Ministry has generated scenarios that represent proposed mines and other development projects and InVEST in being used to identify how much service will be gained or lost, and where. InVEST is also being used to identify other areas in the same landscape that provide similar levels of service to the same people, thus allowing targeted offsetting that will come closer to ensuring no net loss of ecosystem services.

A state agency in Oregon (USA) could design a subsidy that would motivate conservation of natural habitat on private lands. The objective would be to achieve biodiversity protection and climate regulation (through carbon sequestration) with this single program. How should they target payments to get the best returns for both goals on a fixed budget? How much more return could they get if they increased the budget?

In data-rich environments like Oregon, scenarios can be developed by computer-based land transition models. This was the approach taken by Nelson *et al.* (2008). Scenarios were created to represent five possible subsidy programs, each designed to target a different set of landowners: (i) all landowners, (ii) owners of land in riparian zones, (iii) owners who could restore certain rare types of habitat, (iv) owners who could restore forests with large potential to sequester carbon, and (v) owners with parcels with high species conservation value. Researchers projected land use change trajectories based on historic land change data in the region and altered the land change patterns based on an econometric model that determined which landowners would likely participate in each subsidy program. InVEST and its precursor models were used to project changes in biodiversity and service provision under each policy scenario, clearly revealing returns associated with each policy design and showed the trade-off in biodiversity conservation and ecosystem service provision under alternative policies.

Kamehameha Schools, the largest private landowner in Hawai'i, seeks to manage their lands to balance economic, environmental, educational, cultural, and community values. They are considering several possible management plans for providing these values on an iconic 26 000-ha parcel of land, the North Shore of O`ahu. Which plan will yield the best results?

In some cases, like this one, the key stakeholder has control over the whole area of analysis. In this case, hand-drawing scenarios is an alternative to computer simulation of scenarios since there is little in the land use and cover patterns that is left to outside social drivers or chance. We worked with Kamehameha Schools to mark up a map of the parcel, identifying which parts of the landscape would be under different uses in the alternate plans. We then rendered these ideas with GIS software, creating landscape scenarios that were assessed with InVEST. Managers will use the multiple service outputs of InVEST to identify how well each possible scenario measures up against their multiple objectives. These results will help inform the type of management plan implemented.

The Pacific Northwest Ecosystem Research Consortium, a group of representatives from government agencies, non-government organizations, and production sectors, is currently trying to align the policy trajectory of the Willamette Basin (Oregon, USA) with the desires of its inhabitants. Which future will provide the greatest benefits to society, both in terms of current market commodities and in terms of ecosystem services not presently valued? Do some alternatives offer more equitable distribution of benefits to residents?

Most natural resource management decisions have major impacts on the balance of society, influencing who receives higher income streams, who has access to markets, who must follow restrictive regulations, or even who has the right of access or ownership. In this case, the policies implemented in the Basin will determine the balance of revenues among agriculture, timber, and real estate sectors, and which areas will be developed or conserved. To better understand the kinds of policies people living in the Basin would like to see implemented, the Consortium led a multi-stakeholder process to create three plausible scenarios for the future that fell within the bounds of society's comfort level. The three scenarios were assessed with InVEST to reveal the trade-offs and synergies among commodities and ecosystem services that can be expected under each possible future (Nelson *et al.* 2009).

Going one step further, we can combine production function models such as InVEST with socio-economic data to understand how each possible future will affect social equity. Such analyses can help identify options that may lead to inequities before social groups or sectors are marginalized. Preliminary analyses of this type are discussed in Chapter 16.

3.4.6 Tiered modeling: flexibility for a data-limited world

There is always a trade-off in modeling between making a model more complex and detailed and keeping it simple. Simple models require fewer data, are often less prone to parameter estimation errors and subsequent error propagation, and can be easier to explain and understand. Complex models require more information, but they often inspire greater confidence because they more faithfully depict the details and underlying intricacies of processes. Different applications and different users will have specific needs for either complicated or simple models of ecosystem services and valuation. For this reason, we have developed a tiered system of models in InVEST (Figure 3.2).

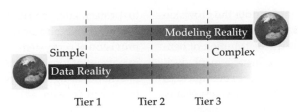

Figure 3.2 A tiered approach to modeling ecosystem services. Given the difficulty of matching models with the desired level of complexity with often sparse data, we created InVEST with different types of models. Tier 1 models are simple and require few data. Tier 2 models are more complex and require more data. Tier 3 models add even greater confidence and will often be site-specific models created by other research groups.

Tier 1 models are the simplest models. We developed these models to require few data and be easy to understand and explain to others, yet to retain sufficient credibility to guide management decisions. Their distinguishing feature is a reliance on readily available data that are generally accessible everywhere in the world. Tier 1 models can draw information from the published literature, global data sets, site-specific data sources, local traditional knowledge, or expert opinion. Because of their simplicity, these models will be most appropriately applied in scoping and planning activities where the purpose is to understand the general lay of the land. They may also be used in planning stages of payment for ecosystem service programs or policy design to estimate likely returns from alternative management options or to place focus on different parts of a landscape.

In many applications the predictions of tier 1 models may be too crude, or too prone to errors of averaging or aggregating, to meet the needs of decision-makers. For these cases, we provide more detailed tier 2 models. These models require more data, have more parameters, and are more time-consuming and difficult to apply. However, tier 2 models are likely to be seen as "better" in the sense of addressing more ecological complexity, allowing for greater spatial and temporal heterogeneity and generating more refined estimates. Instead of representing a world of "average trees" and the "average pollinator" or habitat types sans species lists, tier 2 models disaggregate groups or time steps to include age structure of trees, a variety of pollinator guilds or species, monthly precipitation patterns, and so on. We discuss when tier 1 or tier 2 models are most appropriate in Chapter 15. Further work along these lines will help clarify when greater model complexity is essential in decision-making and when it is unnecessary.

InVEST provides a general framework within which one can mix and match tier 1 and tier 2 models depending on differing data availability or need for precision among services. By mixing tiers, InVEST allows users to customize its application to specific problems. While we have developed only tier 1 and tier 2 models, it is possible to also use InVEST with what we call tier 3 models, research level, state-of-the-art models (e.g., the CENTURY model (Parton *et al.* 1994); SWAT model (Gassman *et al.* 2007)).

3.5 Future directions and open questions

InVEST provides a means for addressing many of the current challenges of mapping and valuing multiple ecosystem services to create change in conservation and natural resource management, but it is not a panacea. Many of the components of InVEST are relatively new and untested. Methods for assessing the validity and reliability of model predictions at landscape levels are needed. The models will continue to undergo modification and updating as experience and applications increase.

A large unmet challenge in ecosystem service assessment and integrated decision-making is to understand the distribution of ecosystem services among different groups in society and how this distribution will likely change as a consequence of management decisions. While it is important to know the total amount of ecosystem services provided and their overall value to society, it is also important to know who benefits from the provision of services and their social and economic status. Without information about the distribution of benefits from ecosystem services, management decisions can lead to serious unintended consequences for equity and well-being. This concern is especially strong for management actions that negatively affect underprivileged segments of society (Pagiola *et al.* 2005). In developing countries, for example, establishing a new national park or conservation area has in some cases resulted in removing and separating people and their work (Kareiva and Marvier 2007), leading to a decline in their livelihood or well-being.

Whether a particular social group wins, loses, or remains unaffected by a decision is determined by several factors. Of utmost importance is access. Underprivileged members of society will not benefit or lose from changes in ecosystem services if they cannot access those services. Access has two critical components in this context: physical overlap in space and legal rights (and their enforcement). InVEST can show clearly where services will be provided on a landscape and how their provision is likely to change in space. This ability can give insights into the spatial overlap part of access. Rapid advances are being made in the mapping of social indicators of poverty (CIESIN 2006; World Resources Institute 2007), and we can draw from these approaches to ask where ecosystem services and the poor overlap on the landscape. We provide some examples of this kind of exercise in Chapter 16. However, we currently do not have a standardized way for bringing information about institutions and their level of enforcement into a mapping and valuation context. Developing ways to represent and predict the interactions among ecosystem services, people, and institutions will be critical to the assessment of the distributional effects of management decisions.

A related challenge lies in altering financial and institutional infrastructure to give incentives that maintain and enhance the provision of ecosystem services. Often there is a spatial or temporal mismatch between those who control the provision of ecosystem services and those who benefit from the services. Without the ability to connect the demand for services with the supply from those who control it, there will be insufficient incentive for suppliers to protect and maintain ecosystems. Addressing this mismatch and providing proper incentives for provision of ecosystem services will require changes in policies of local and national governments and in international agreements. Some progress on these fronts can be seen, as with the recent emergence of carbon markets, international policy discussions on Reduced Emissions from Deforestation in Developing Countries (REDD), and expansion of programs of Payments for Ecosystem Services (PES). Successfully linking the science of mapping and valuing ecosystem services with proper institutions and policies will likely remain a major challenge for decades to come.

Finally, there is the need to expand InVEST to include models for other key terrestrial (see Box 3.1), freshwater, and marine ecosystem services. Useful groundwork is being laid for comprehensive modeling of marine ecosystem services (Chapter 17) and the Natural Capital Project is now developing InVEST models for fisheries production, shoreline protection, marine tourism and recreation, and marine biodiversity.

A user's guide for InVEST can be found at http://invest.ecoinformatics.org and there is a growing community that seeks to develop and apply multiobjective ecosystem service models to decision-making. Whether one uses InVEST or any of several other ecosystem service modeling approaches (e.g., Artificial Intelligence for Ecosystem Services (ARIES; ARIES 2008), EcoMetrix (Primozich 2008), Multiscale Integrated Models of Ecosystem Services (MIMES; unpublished)), the greatest challenge is not in developing the models, but in linking the models pragmatically to everyday land- and resource-use decisions.

References

ARIES. (2008). *ARIES white paper*. University of Vermont, Burlington.

Alcamo, J., Kreileman, G. J. J., Krol, M., *et al.* (1994). Modeling the global society-biosphere-climate system, 1. Model description and testing. *Water Air and Soil Pollution*, **76**, 1–35.

Antle, J. M., and Stoorvogel, J. J. (2006). Incorporating systems dynamics and spatial heterogeneity in integrated assessment of agricultural production systems. *Environment and Development Economics*, **11**, 39–58.

Barbier, E. B., and Strand, I. (1998). Valuing mangrove-fishery linkages: A case study of Campeche, Mexico. *Environmental and Resource Economics*, **12**, 151–66.

Barbier, E. B. (2000). Valuing the environment as input: Applications to mangrove-fishery linkages. *Ecological Economics*, **35**, 47–61.

Barbier, E. B. (2007). Valuing ecosystem services as productive inputs. *Economic Policy*, **22**, 177–229.

Boerner, B. (2007). *Ecosystem services, agriculture, and rural poverty in the Eastern Brazilian Amazon: Interrelationships and policy prescriptions.* Amsterdam, The Netherlands.

Boody, G., Vondracek, B., Andow, D. A., *et al.* (2005). Multifunctional agriculture in the United States. *Bioscience*, **55**, 27–38.

Brooks, K. N., Gregersen, H. M., Berglund, E. R., *et al.* (1982). Economic evaluation of watershed projects—an overview methodology and application. *Water Resources Bulletin*, **18**, 245.

Center for International Earth Science Information Network (CIESIN). (2006). *Where the poor are: An atlas of poverty.* The Earth Institute at Columbia University, New York.

Cleveland, C. J., Betke, M., Federico, P. *et al.* (2006). Economic value of the pest control service provided by the Brazilian free-tailed bats in south-central Texas. *Frontiers in Ecology and Evolution*, **4**, 238–43.

Connor, E. F., and McCoy, E. D. (1979). The statistics and biology of the species–area relationship. *American Naturalist*, **113**, 791–833.

Costanza, R., d'Arge, R., de Groot, R., *et al.* (1997). The value of the world's ecosystem services and natural capital. *Nature*, **387**, 253–60.

Daily, G. C. (1997). *Nature's Services.* Island Press, Washington, DC.

Das, S., and Vincent, J. R. (2009). Mangroves protected villages and reduced death toll during Indian super cyclone. *Proceedings of the National Academy of Sciences of the United States of America.* **Early Edition**, 1–4.

Ehrlich, P. R., and Ehrlich, A. H. (1981). *Extinction: The Causes and Consequences of the Disappearance of Species.* Random House, New York.

Ellis, G. M., and Fisher, A. C. (1987). Valuing the einvronment as an input. *Journal of Environmental Management*, **25**, 149–56.

Engel, S., Pagiola, S., and Wunder, S. (2008). Designing payments for environmental services in theory and practice: An overview of the issues. *Ecological Economics*, **65**, 663–74.

Eshet, T., Baron, M. G., and Shechter, M. (2007). Exploring benefit transfer: disamenities of waste transfer stations. *Environmental and Resource Economics*, **37**, 521–47.

Freeman, A. M. I. (2003). *The Measurement of Environmental and Resource Values: Theory and Methods.* Resources for the Future, Washington, DC.

Gassman, P. W., Reyes, M. R., Green, C. H., *et al.* (2007). The soil and water assessment tool: historical development, applications and future research directions. *Transactions of the American Society of Agricultural and Biological Engineers*, **50**, 1211–50.

Ghazoul, J. (2007). Recognising the complexities of ecosystem management and the ecosystem service concept. *Gaia*, **16**, 215–21.

Hougner, C., Colding, J., and Soderqvist, T. (2006). Economic valuation of a seed dispersal service in the Stockholm National Urban Park, Sweden. *Ecological Economics*, **59**, 364–74.

Imperial, M. T. (1999). Institutional analysis and ecosystem-based management: The institutional analysis and development framework. *Environmental Management*, **24**, 449–65.

Jackson, R. B., Jobbagy, E. G., Avissar, R., *et al.* (2005). Trading water for carbon with biological sequestration. *Science*, **310**, 1944–7.

Just, R. E., Hueth, D. L., and Schmitz, A. (2004). *The Welfare Economics of Public Policy: A Practical Approach to Project and Policy Evaluation.* Edward Elgar, Cheltenham, UK and Northampton, MA.

Kaiser, B., and Roumasset, J. (2002). Valuing indirect ecosystem services: the case of tropical watersheds. *Environment and Development Economics*, **7**, 701–14.

Kareiva, P., and Marvier, M. (2007). Conversation for the people—Pitting nature and biodiversity against people makes little sense. Many conservationists now argue that human health and well-being should be central to conservation efforts. Scientific American, 297, 50–7.

Konarska, K. M., Sutton, P. C., and Castellon, M. (2002). Evaluationg scale dependence of ecosystem service valuation: a comparison of NOAA-AVHRR and Landsat TM datasets. Ecological Economics, 41, 491–507.

Luck, G. W., Harrington, R., Harrison, P. A. *et al.* (2009). Quantifying the contribution of organisms to the provision of ecosystem services. *BioScience*, **59**, 223–35.

Millennium Ecosystem Assessment (MA). (2005). *Ecosystems and Human Well-being: Synthesis.* Island Press, Washington, DC.

Naidoo, R., and Ricketts, T. H. (2006). Mapping the economic costs and benefits of conservation. *PLoS Biology*, **4**, 2153–64.

National Research Council (NRC). (2005). *Valuing ecosystem services: Toward better environmental decision-making.* National Academies Press, Washington, DC.

Nelson, E., Polasky, S., Lewis, D. J., *et al.* (2008). Efficiency of incentives to jointly increase carbon sequestration and species conservation on a landscape. *Proceedings of the National Academy of Sciences of the United States of America*, **105**, 9471–6.

Nelson, E. N., Mendoza, G. M., Regetz, J., *et al.* (2009). Modeling multiple ecosystem services, biodiversity conservation, commodity production and tradeoffs at landscape scales. *Frontiers in Ecology and the Environment*, **7**, 4–11.

Pagiola, S., von Ritter, K., and Bishop, J. (2004). *How Much is an Ecosystem Worth? Assessing the Economic Value of Conservation*. The World Bank, Washington, DC.

Pagiola, S., Arcenas, A., and Platais, G. (2005). Can payments for environmental services help reduce poverty? An exploration of the issues and the evidence to date from Latin America. *World Development*, **33**, 237–53.

Parton, W. J., Schimel, D. S., Ojima, D. S., *et al.* (1994). Quantitative modeling of soil forming processes. In R. B. Bryant and R. W. Arnold *et al.*, Eds. *Special Publication*, pp. 147–67. Soil Science Society of America, Madison, WI.

Pereira, H. M., and Daily, G. C. (2006). Modeling biodiversity dynamics in countryside landscapes. *Ecology*, **87**, 1877–85.

Plummer, M. (2009). Assessing benefit transfer for the valuation of ecosystem services. *Frontiers in Ecology and the Environment*, **7**, 38–45.

Polasky, S., Nelson, E., Camm, J., *et al.* (2008). Where to put things? Spatial land management to sustain biodiversity and economic returns. *Biological Conservation*, **141**, 1505–24.

Primozich, D. (2008). *Developing the Willamette ecosystem marketplace*. Willamette Partnerhsip, Salem.

Ricketts, H. T., Daily, G. C., Ehrlich, P. R., *et al.* (2004). Economic value of tropical forest to coffee production. *Proceedings of the National Academy of Sciences of the United States of America*, **101**, 12579–82.

Rosenberger, R. S., and Stanley, T. D. (2006). Measurement, generalization, and publication: Sources of error in benefit transfers and their management. *Ecological Economics*, **60**, 372–8.

Rosenberger, R. S. and Phipps, T. T. (2007). Correspondence and convergence in benefit transfer accuracy: A meta-analytic review of the literature. In: S. Navrud, and R. Ready, Eds., *Environmental Values Transfer: Issues and Methods*. Springer, Dordrecht.

Salvatore, M., Pozzi, F., Ataman, E., *et al.* (2005). *Mapping global urban and rural population distributions*. Food and Agriculture Organization, Rome.

Sathirathai, S. and Barbier, E. B. (2001). Valuing mangrove conservation in southern Thailand. *Contemporary Economic Policy*, **19**, 109–22.

Shabman, L. A., and Batie, S. S. (1978). Economic value of natural coastal wetlands: a critique. *Coastal Zone Management Journal*, **4**, 231–47.

Sohl, T. and Sayler, K. (2008). Using the FORE-SCE model to project land-cover change in the southeastern United States. *Ecological Modelling*, **219**, 49–65.

Swallow, S. K. (1994). Renewable and nonrenewable resource theory applied to coastal agriculture, forest, wetland and fishery linkages. *Marine Resource Economics*, **9**, 291–310.

Troy, A. and Wilson, M. A. (2006). Mapping ecosystem services: Practical challenges and opportunities in linking GIS and value transfer. *Ecological Economics*, **60**, 435–49.

Turner, W. R., Brandon, K., Brooks, T. M., *et al.* (2007). Global conservation of biodiveristy and ecosystem services. *Bioscience*, **57**, 868–73.

Veldkamp, A., and Fresco, L. O. (1996). CLUE-CR: An integrated multi-scale model to simulate land use change scenarios in Costa Rica. *Ecological Modeling*, **91**, 231–48.

Vitousek, P. M., Aber, J. D., Howarth, R. W., *et al.* (1997). Human alteration of the global nitrogen cycle: Sources and consequences. *Ecological Applications*, **7**, 737–50.

Wilson, M. A. and Carpenter, S. R. (1999). Economic valuation of freshwater ecosystem services in the United States: 1971–1997. *Ecological Applications*, **9**, 772–83.

Wilson, M. A. and Hoehn, J. P. (2006). Valuing environmental goods and services using benefit transfer: The state-of-the-art and science. *Ecological Economics*, **60**, 335–42.

World Resources Institute. (2007). *Nature's Benefits in Kenya: An atlas of ecosystem services and human well-being*. World Resources Institute, Washington, DC.

Yuan, Y., Bingner, R. L., and Boydstun, J. (2006). Development of TMDL watershed implementation plan using Annualized AGNPS. *Land Use and Water Resources Research*, **6**, 2.1–2.8.

SECTION II

Multi-tiered models for ecosystem services

CHAPTER 4

Water supply as an ecosystem service for hydropower and irrigation

Guillermo Mendoza, Driss Ennaanay, Marc Conte, Michael Todd Walter, David Freyberg, Stacie Wolny, Lauren Hay, Sue White, Erik Nelson, and Luis Solorzano

4.1 Introduction

Water is necessary for all life. Precipitation through stream runoff and groundwater that is tapped via wells provide the world's supply of water for drinking, irrigation, and hydropower generation. The spatial and temporal availability of this water is strongly influenced by watershed geomorphology, vegetation, and land and water management practices. As such, natural capital (in this case, vegetation and soil) can support the provision of water services by regulating the amount and timing of water availability. In this chapter, we presents models that link land use and land cover (LULC) and several other key attributes to the quantity of surface water available for irrigation and for hydropower production.

We focus on irrigation and hydropower because of their global economic significance. The vast majority of water used worldwide is used for irrigation, accounting for up to 85% of fresh water use in developing nations (IWMI 2001). The benefit of access to irrigation is the associated increase in crop yield: irrigated agriculture provides 40% of the global food production (FAO, 2003). Water supply for hydropower is also an economically significant service, and one likely to become even more important because of its presumed low carbon emissions. For example, hydropower supplies 9% of the US electricity and 49% of all renewable energy used in the USA (Edwards 2003). Some countries depend almost exclusively on hydropower, such as Tanzania where hydropower accounts for more than 60% of total generated capacity (Lyimo 2005).

Although there is a lot still to be learned about the connections between land management, vegetation cover, and water yield (e.g., Chomitz and Kumari 1998; Bruijnzeel 2001; Bosch and Hewlett 1982; Oyebande 1998), there is a demand for science-based decision-making regarding policies, payments, or activities that can alter water use. Decision-makers need a credible and convenient methodology that explicitly links land use to water delivery (Tallis *et al.* 2008). Several complex models are available for simulating water yield—most notably Hydrological Simulation Program—Fortran (HSPF; Donigian *et al.* 1984) and Soil and Water Assessment Tool (SWAT; Arnold *et al.* 1998). SWAT and HSPF are quasi-process-based hydrology models that are data- and time-intensive; they are difficult to apply in data-poor regions of the world, or in situations lacking technically sophisticated support staff to calibrate the models. For this reason there is a need for simpler models that can be more easily applied, especially in a context of examining many ecosystem services at once, where trade-offs and relative comparisons may be sufficient.

We present two models that explicitly connect land use and land cover to the regulation of surface water flows. We describe methods for converting modeled flows into the level and value of two ecosystem services: the regulation of water flow for hydropower production and the regulation of water flow for agricultural crop irrigation. We provide two sets of models called tier 1 and tier 2 models. The tier 1 models are the simplest and require the fewest data. They provide annual average outputs and do not require daily precipitation data, yet produce predictions that closely match more complex models. The tier 2 models are more complex and provide predictions on a daily time step.

4.2 Tier 1 water supply model

4.2.1 Modeling water yield

The tier 1 water yield model is designed to evaluate how land use and land cover affect annual surface water yield across a landscape. We define the water yield on a landscape as all precipitation that does not evaporate or transpire. In developing a model designed to accommodate areas with minimal access to data, we utilize a water balance model that is drawn from globally available data on annual precipitation and dryness indices that partition the water balance for any place in the world (Budyko and Zubenok 1961; Milly 1994; Zhang *et al.* 2001). The model we describe is for surface water and does not separate groundwater, which will require another approach (see Box 4.1).

Our water balance model is based on the hypothesis that water yield can be approximated solely by

---

Box 4.1 Can we apply our simple model where groundwater really matters?

Heather Tallis, Yukuan Wang, and Driss Ennaanay

Groundwater makes up 100 times more of the world's freshwater than surface water does (30 and 0.3%, respectively) (Gleick 1996). Groundwater also constitutes 30% of streamflow, on average, around the globe (Zektser and Loaiciga 1993), although this percent varies dramatically by region. In regions of the USA where groundwater–surface water interactions are significant, groundwater contribution to streamflow can reach up to 90% (Winter *et al.* 1998). The tier 1 water yield model does not account for such interactions, instead predicting the total water depth generated from a parcel (combined surface water, shallow groundwater, and deep groundwater) based on characteristics of the watershed of interest. As part of the model development process, we wanted to test whether our simplified approach could be useful in regions where groundwater–surface water interactions are significant. We selected a watershed in Boaxing County, China (part of the Upper Yangtze River Basin) where an 11-year time series of river discharge showed a high contribution of baseflow and groundwater discharge to total streamflow. In other words, we picked a watershed where we expected our model to perform poorly.

The basin is midsized (3240 km^2), and is located in the subtropical monsoon eco-region with an annual rainfall average of 1172 mm. Rainfall is distributed over the year, but there is an intense rainy season from May to September (with flooding from July to August). The basin is rather steep (avg. slope 32°), with elevation ranging from 750 to 5328 m. Land cover and land use is relatively diverse, with more than 47% of the area in forest, 21% in natural

grasslands, and 15% in shrubs. Woodlands, paddy fields, and residential and agricultural lands account for the remaining area. Soils are also diverse, including yellow, yellow brown, mountain brown, dark brown, sub-alpine meadow, alpine meadow, and limestone soils.

We compared tier 1 annual average yield estimates to observed streamflow data (summarized to annual average runoff) for the years 1995 to 2005. Model inputs included average annual precipitation, potential evapotranspiration (calculated using Hamon method, grids generated from five weather stations within and surrounding the watershed), soil data, and a land use land cover map from 2005. Un-calibrated model runs showed poor agreement with observed streamflow levels, with our tier 1 estimates falling 32% below observed levels. This was expected, given the high contribution of baseflow apparent from monthly hydrograph and precipitation analyses. It is likely that the groundwater aquifer contributing to streamflow in this basin extends well beyond the borders of the modeled watershed, so our estimates of yield based solely on precipitation and watershed characteristics are missing a key water source. To account for this additional source, we conservatively took the lowest runoff depth in the driest period of the observed time series as the baseflow and groundwater discharge depth generated outside the modeled watershed. Adding this value to our model estimate of annual average yield gave us 92% of the observed annual average streamflow.

These results suggest that our simple, tier 1 model can be useful in regions of high groundwater–surface water interaction, but only if time series data are available for calibration. In the USA, the US Geological Survey has classified all watersheds into 24 regions that can help identify when the use of our tier 1 model would require calibration. Each region represents an area of similar physiography, climate, and ground water–surface water interactions (Winter *et al.* 1998). In six of ten regions analyzed for groundwater contribution to streamflow, more than 50% of streamflow comes from ground water, and we would strongly recommend using our tier 1 model only under careful scrutiny and calibration. The following rivers are representative of conditions in those six regions:

Dismal River, NE; Duckabush River, WA; Dry Frio River, TX; Brushy Creek, CA; Sturgeon River, MI; and Ammonoosuc River, NH. In the other four regions, less than 50% of streamflow comes from ground water and tier 1 applications could proceed more readily. Rivers characteristic of those regions are Homochitto River, MS; Santa Cruz River, AZ; Orestimba Creek, CA; and Forest River, ND. Similar kinds of classifications, or a close look at time series data, will help identify where the tier 1 water yield model can be applied most readily around the globe.

the local interaction of fluctuating precipitation and potential evapotranspiration given the water storage properties of the soil (Milly 1994). The relationship between potential and actual evapotranspiration is described by the Budyko curve, which is based on over 2000 water balance observations representing catchments of different climates and eco-regions worldwide (Budyko and Zubenok 1961; Zhang *et al.* 2001). In particular, we determine the annual amount of precipitation that does not evapotranspire, more simply called water yield, for each parcel on the landscape (indexed by $x = 1, 2, \ldots, X$):

$$Y_{jx} = \sum_j \left(1 - \frac{AET_{xj}}{P_{xj}}\right) \cdot P_{xj} \cdot A_x, \qquad (4.1)$$

where AET_{xj} is the annual actual evapotraspiration on parcel x with LULC category j, P_{xj} is the annual precipitation on parcel x with LULC j, and A_{xj} is the area of x in LULC j. Annual precipitation can be modified upward for j that have significant fog drip contributions (Bruijnzeel 2000).

The evapotranspiration portion of the water balance, AET_{xj}/P_{xj}, is an approximation of the Budyko curve developed by Zhang *et al.* (2001):

$$\frac{AET_{xj}}{P_{xj}} = \frac{1 + \omega_{xj}R_{xj}}{1 + \omega_{xj}R_{xj} + \frac{1}{R_{xj}}}, \qquad (4.2)$$

where R_{xj} is the dimensionless ratio of potential evapotranspiration to precipitation, known as the Budyko dryness index (Budyko 1974), on parcel x with LULC j, and ω_{jx} is a dimensionless ratio of plant accessible water storage to expected precipitation during the year. ω_{jx} characterizes water balance in distinctive plant communities, given prevailing

climatic and soil conditions (Milly 1994; Potter *et al.* 2005; Donohue *et al.* 2007), and is given by

$$\omega_{xj} = Z\left(\frac{AWC_x}{P_{xj}}\right), \qquad (4.3)$$

where AWC_x is a volumetric (mm) measure of the water content (mm) in the soil available to plants, and Z is a parameter applied to each homogeneous basin in the landscape and is found with calibration. Generally, ω_x varies between 0.5 and 2 (Zhang *et al.* 2001), with the 0.5 typical of pasture biomes and the 2 typical of forest biomes.

The above equations depict how the Budyko dryness index (R_{xj}) and the ratio of water available to plants relative to annual precipitation (AWC_x/P_{xj}) affect the annual water balance. However, the model does not explicitly incorporate the impact of the frequency of annual events, the sub-parcel spatial variability of soil water storage capacity, and synchronicity of the energy-precipitation cycles on the water balance, which all influence the water balance (see Milly 1994). In order to adjust for these neglected effects, Z is used as a calibration constant. To determine Z, the user must have information on the annual water balance partition, which is generally obtained as the difference of observed annual precipitation and observed annual streamflow, corrected for groundwater recharge and important consumptive losses. The parameter Z is adjusted until results from Eq. (4.2) reasonably correspond to observed water balance partitions.

Finally, we define the Budyko dryness index as

$$R_{xj} = \frac{k_j \cdot ETo_x}{P_{xj}} = \frac{PET_{xj}}{P_{xj}}, \qquad (4.4)$$

where $R_{xj} > 1$ denote parcels that are potentially arid (Budyko 1974), ETo_x is the reference evapotranspiration on parcel x, and k_j is the plant evapotranspiration coefficient associated with the LULC j on parcel x. ETo_x is an index of climatic demand with k_j largely determined by j's vegetative characteristics (Allen *et al.* 1998).

4.2.2 Water retention index

When the water retention properties of the landscape are reduced, such as due to deforestation, water yield is increased but without a means for storage; larger portions can flow to streams too quickly for beneficial use. Our annual average yield estimation does not consider the exchange between surface water and groundwater via infiltration. However, infiltration is an important process linked to the timing of water flows and the availability of groundwater. We do not attempt to provide a simple model of quantitative groundwater recharge rates. Instead, we provide a way to rank the landscape to identify areas where water can infiltrate or leak into groundwater and thus be drawn on for extended periods of time from baseflow or wells. Our water retention index does not adjust the availability of surface water based on groundwater recharge rates, nor does it estimate the amount of groundwater recharge provided by a landscape. It does provide a simple approach for identifying high and low areas of groundwater recharge, allowing managers to see how management options will affect the location and relative magnitude of recharge.

Although many factors contribute to retention, we consider three primary elements. First, because of geomorphology or artificial drainage systems, some parcels will be more hydrologically connected to streams. Secondly, vegetation and other surface features can facilitate the infiltration of surface water into the soil. Thirdly, if soil infiltration capacity is high relative to rainfall intensity or snowmelt, water will be more likely to infiltrate the soil than to runoff to streams. Soil moisture, effective soil depth, and mean rainfall depth and frequency interact to determine runoff and leakage (Porporato *et al.* 2002). We use the formulation of the topographic index as a proxy for soil moisture and unsaturated soil depth,

and use the ratio of saturated hydrologic conductivity to precipitation as a proxy for the relative effects of rainfall intensity and frequency.

We include these variables in the water retention index (v_{jx}) as:

$$v_{jx} = \begin{cases} (1 - HSS_x) \cdot \min\left\{1, \dfrac{r_j}{r_{\text{for}}}\right\} \cdot \min\left\{1, \dfrac{ksat_{jx}}{p_{\text{d}}}\right\} \\[2em] 1 \quad \text{when the water yield regulation properties of the landscape are not applicable} \end{cases}, \quad (4.5)$$

where $HSS_x \, \varepsilon \, [0,1]$ is a normalized topographic index from Lyon *et al.* 2004 of x that provides an index of hydraulic connectivity of a parcel to a stream, r_j denotes j's roughness coefficient (Kent 1972), r_{for} is a normalizing roughness coefficient that represents the roughness of a natural forest coverage (Kent 1972), $ksat_{jx}$ is the saturated hydraulic conductivity, a standard soil property, of x with LULC j scaled to a daily time step, and p_d is the mean daily rainfall depth at the area of interest. For most applications p_d is likely to be constant across parcels. HSS_{ix} is based on topographic wetness index (Beven and Kirkby 1979) that predicts areas in the landscape prone to saturation and associated runoff due to drainage area, slope, depth of soil, and permeability (Steenhuis *et al.* 1995; Lyon *et al.* 2004). We argue that areas prone to saturation will tend to have higher soil moisture and the least favorable characteristics for water retention.

The water retention index varies between 0 and 1. If v_{jx} tends to 1, parcel x with LULC j has high water retention characteristics, whereas if v_{jx} tends to 0 it implies that parcel x has poor water retention properties and any precipitation will likely runoff immediately. If data are lacking to define any of the product terms in Eq. (4.5) that define v_{jx}, one can simply use the terms for which data are available, realizing that the characterization will be less accurate. Obviously, water retention properties of the landscape are largely irrelevant if annual water yield is regulated by large reservoir storage. Using the index of water retention, we identify parcels x that contribute to the landscape regulating function as those that exceed a threshold of water retention index score as $\tilde{Y}_{xj}$, where

$$\tilde{Y}_{xj} = I\left(v_{xj} > \alpha\right)Y_{xj} \tag{4.6}$$

where I is an indicator function equal to 1 when $v_{jx} > \alpha$ and equal to 0 otherwise. The constant α is the threshold value of v_{jx} that establishes whether parcel x is a water-regulating parcel.

Our predictions of annual water yield using Eqs. (4.1)–(4.4), and of identifying high infiltration areas using Eqs. (4.5) and (4.6) are admittedly highly simplified approximations. The basic idea is to use spatial hydrology and land-use or land cover information to reveal impacts of changes in land use and land cover. Our simple models predict higher annual water yields but lower retention properties for urban areas than for forests, especially if the forests lie on permeable soils and have a low hydraulic connectivity to a water body. This agrees with field and more complicated modeling studies. Later in the chapter we compare this tier 1 approach to more traditional detailed models such as SWAT.

4.2.3   Water allocation in tier 1

Land use and land cover on a given parcel are not the only determinants of how much water arrives downstream for use. Consumptive uses also play a role in determining downstream water supply. To quantify the water available for use in irrigation or the production of electricity, the model tracks water consumption and water use along flow paths. Let v be the set of parcels that drain to parcel x. The water available for use during period p at parcel x is defined as S_{xp},

$$S_{xp} = \beta_p \cdot \sum_{v \in x} \tilde{Y}_{jv} - \sum_{v \in x} C_{jvp}, \tag{4.7}$$

where $S_{xp} \geq 0$, C_{jvp} is consumption at x with LULC j during demand period p, $\beta_p \cdot \sum \tilde{Y}_{jv}$ is the yielded water available for consumption from v given LULC j during p, and β_p is a constant, less than 1, to account for the fraction of annual water yield available during p. For example, an irrigation district without a reservoir may require water during the dry season

months that account for about 5% of annual water yield. β_p can be determined by using studies of regional streamflow distribution or by accounting for water (Eq. (4.7)) at a representative monitored catchment. When there are significant interbasin groundwater transfers, the value of β_p can be greater than 1, and the user will have to make corrections. In the following sections, except the section on valuation for irrigation, we will assume an annual period p.

By contributing surface water to parcels downstream for consumptive use, "source" parcels (the v parcels in Eq. (4.7)) provide an ecosystem service. In this section we calculate how much a source parcel contributes to the benefits derived from surface water use downstream. First, we define B_{Dp} as the benefit obtained from productive use of surface water at demand point D in period p. Second, we define B_{MDp} as the portion of benefit B_{Dp} provided by parcel x's surface water yield or the yield of a collection of parcels defined by $x \subset M$ (x or M's contributive value will be 0 by definition if x or M does not drain into demand point D).

$$B_{MDp} = \sum_{x \subset M} \frac{\beta_p \tilde{Y}_{xjp}}{S_{Dp}} \left(\prod_{y \subset D} \frac{S_{yp} - C_{yjp}}{S_{yp}} \right) B_{Dp}, \tag{4.8}$$

where $y \subset D$ is the set of parcels along the flow path between x or M and D, S_{Dp} and S_{yp} is the supply of surface water to D and parcel y in period p as defined by Eq. (4.7), and C_{yjp} is the volume of water consumed at y in year p. If there are no consumption parcels along x and D's flow path then

$$B_{MDp} = \sum_{x \subset M} \frac{\beta \tilde{Y}_{xjp}}{S_{Dp}} B_{Dp}. \tag{4.9}$$

The management unit M will likely have a wide range of sizes, ranging across the size of sub-watersheds, protected areas, riparian buffers, and the smallest size of landholdings. At the smaller size of management parcel, M, our approach likely generates greater errors because non-hydraulic boundaries neglect the effects of hydrologic interdependence within the landscape and small scales remove the benefits of averaging over heterogeneous landscape properties. Our experience indicates that tier 1 annual yield models are most useful when applied to evaluate trade-offs between management units of similar size and located within similar watershed zones.

For the sake of simplicity in defining the equations in the subsequent valuation sections, we define the contribution of water from x to support an ecosystem service at D as F_{jxD}:

$$F_{jxD} = \beta \cdot \tilde{Y}_{jx} \cdot \left(\prod_{y \in D} \frac{S_y - C_{jy}}{S_y} \right). \qquad (4.10)$$

The user can define benefits B_D as the power generated by a hydropower station, as crop yields by an irrigated plot, or simply as the water contribution from x potentially supplied for productive use at D. In each case a biophysical map of ecosystem services can be derived using Eq. (4.8) whereby the contribution of parcel x in achieving the power generated, crop yield, or water supply at D can be mapped.

In many cases, conservation planners or managers for payments of ecosystem services schemes simply want rules to effectively allocate a fixed source of revenue from royalties or user fees, such as in hydropower royalties for watershed protection in Nepal (Winrock 2004), watershed protection fees from utility bills in Quito, or sugar cane growers in the Valle del Cauca of Colombia (Pagiola *et al.* 2002). Equation (4.8), which can be modified for water quality protection or soil retention practices, provides a tool for helping spatially allocate revenue in proportion to the services provided by different sections of a watershed. In this case of proportional allocation, Eq. (4.8) becomes

$$B_M = \sum_{x \subset M} \frac{F_{jxD}}{S_D} \cdot B_B, \qquad (4.11)$$

where the benefits from x due to LULC j are not dependent on the production benefits at a water use point D, but rather a fixed value or budget, B_B, that in this case corresponds to willingness to pay (contingent valuation) a proportion of royalties or utility fees for conservation, or a total "score" value to help prioritize the landscape. B_M might correspond to an allocation of the fixed budget for protection, conservation, or restoration activities.

In the next sections we propose formulations of B_D for non contingent valuation.

4.2.4 Linking water supply to hydropower production

We modify Eq. (4.7) to define the average rate of water flow available to generate hydropower at point H in year p as

$$S_{Hp} = \frac{\beta \sum_{v \subset H} \tilde{Y}_{vjp} - \sum_{v \subset H} C_{vjp} \tilde{Y}_{vjp}}{n}, \qquad (4.12)$$

where H indexes the hydropower generating station at demand point H, $v \subset H$ is the set of parcels that drains to demand point H, and n represents the time steps to define flow rate.

Dams are not only used for electricity production. In the USA and Central America dam use is almost equally divided between irrigation, hydropower, water supply, flood control, and recreation (see Table 4.1). To value water used for electricity generation, the user must know the average releases of water that go through the generating turbines, which can depend on seasonal multi-use demands and the capacity of diversion infrastructure and of turbines. This model assumes that a constant fraction, γ, of $\bar{S}_H$ is released through the turbines to generate energy.

At hydropower station H, practically available power generated is calculated using the following equation, slightly modified from standard form (e.g., see Edwards 2003 for typical formulation):

$$\varepsilon_H = \mu \cdot \kappa \cdot (\gamma \cdot \bar{S}_H) \cdot h_H, \qquad (4.13)$$

where ε_H is power generated in kilowatt-hours, h_H is the effective average head, μ is the turbine efficiency (generally varying between 0.75 and 0.95), and κ is a constant determined by the product of water density, gravity, and a conversion factor. The product, $\gamma \cdot \bar{S}_H$, represents the average water released from turbines, which must be established from hydropower operators. ε_H represents the benefits from the ecosystem functions for the generation of electricity due to water provision and regulation. The contribution of each pixel upstream of H to ε_H is given by substituting ε_H for B_D in Eq. (4.8).

Table 4.1. Global breakdown (%) of large dams by purpose (McMahon and Mein 1986)

	Europe	Asia	North and Central America	South America	Africa	Australia
Flood control	3	2	13	18	1	2
Hydropower	33	7	11	24	6	20
Irrigation	19	63	11	15	50	13
Multipurpose	25	26	40	26	21	14
Recreation	0	0	9	0	0	0
Water supply	17	2	10	13	20	49
Other	3	0	6	4	2	2

4.2.5 Linking water supply to irrigation

Predicting the provision of water supply for irrigation poses several modeling challenges. Spatially explicit modeling of irrigation supply and use requires information on watershed characteristics as well as data on diversion systems and wells that might be used to obtain water from neighboring drainage catchments or aquifers. Unlike the demand for hydropower, irrigation water is usually required at specific periods within a year—when water is scarce. Therefore, the landscape water yield regulation properties discussed above are most critical when reservoir storage options on the landscape are limited and the water needs of crops are not being met fully by precipitation (high water yields have no value for irrigation if this water is delivered when crop growth is not limited by water). Therefore, detailed crop and water scheduling information is required for an accurate depiction of irrigation demand and use.

Equation (4.7) provides the framework to determine surface water used for irrigation. Unlike hydropower, the user must correctly account for water diversions that may not be apparent in a land-use map. The surface water available for irrigation at use point D during irrigation period p (as opposed to year p) is S_{Dp} given by a modified version of Eq. (4.7),

$$S_{Dp} = \max\left\{\beta_p \sum_{v \subset D} \tilde{Y}_{vjp} - \sum_{v \subset D} \sum_k \sigma_{kvp} A_{kvp} W_{kvp} - \sum_{v \subset D} G_{vjp},\ 0\right\}$$

(4.14)

where $\beta_p \sum_{v \subset D} \tilde{Y}_{vjp}$ is the sum of surface water yield during p from the set of parcels v that drain to D, $\sum_{v \subset D} \sum_k \sigma_{kvp} A_{kvp} W_{kvp}$ is the surface irrigation water (in volumetric units) consumed in the set of parcels v, and $\sum_{v \subset D} G_{vjp}$ is the sum of other water consumption in the set of parcels v that drain into D. In other words,

$$\sum_{v \subset D} C_{vjp} = \sum_{v \subset D} \sum_k \sigma_{kvp} A_{kvp} W_{kvp} + \sum_{v \subset D} G_{vjp}\ .$$

(4.15)

The irrigation water consumed in any parcel v is a function of the water needs per hectare of crop k in v in period p not met by precipitation in period p (W_{kvp} in depth units), the area devoted to crop k in v during period p (A_{kvp}), and a management decision of how much water needed by k in v during p is actually delivered ($\sigma_{kvp} \in [0,1]$).

4.3 Tier 1 valuation

Each parcel on the landscape is assigned a value due to its contribution to each of the services profiled in this chapter. In the case of hydropower, a parcel's value is related to the value of the energy produced. In the case of irrigation, value is determined from additional crop productivity due to the irrigation at point D. A parcel that yields water used downstream for hydropower or irrigation is assigned a share of the downstream production values according to its relative contribution to the utilized water flow. By taking this approach we can understand both the total value of each service delivered to users on a landscape and the location of high-value service supply regions.

4.3.1 The value of water flow for hydropower

Albery (1968) estimates the maximum willingness to pay for water by comparing the cost of electricity production from hydropower with that of the cheapest alternative source of electricity. The difference between these two costs can be interpreted as the economic rent to the water resource. In order to estimate this value, the user must have information about the cost of hydropower production as well as the next cheapest alternative cost of power generation. While this approach is theoretically accurate, it is also data intensive, requiring that users have information about the costs of hydropower and other sources of energy production. For this reason, we choose to value water in hydropower generation based on the price of hydropower alone. While this approach will not fully capture the social value of hydropower production (such as reduced greenhouse gas emissions compared to coal-fired power plants), it will provide a lower bound estimate of this value.

Under this framework, the net present value of hydropower production is given by

$$NPVH_H = \sum_{t=0}^{T-1} \frac{(p_e \varepsilon_H - c_H)}{(1+r)^t} \qquad (4.16)$$

where p_e is the market price of electricity (per unit of energy) provided by hydropower plant at dam d, ε_H represents the annual energy generated by hydropower station H, c_H represents the average annual cost of operating hydropower station H and should include the external environmental damages of dam construction, T indicates the number of years we expect present landscape conditions to persist or the expected remaining lifetime of the station at H (set T to the smallest value if the two time values differ), and r is the market discount rate.

We rely on a parcel's relative contribution of water used for hydropower to distribute the above value to the parcels upon which the electricity-generating water is generated. Let F_{xHj} represent the amount of water that originates on parcel x in LULC j that is available for use for hydropower production at demand point H (see Eqs. (4.8) and (4.10)). Then, the contribution of water values for hydropower, $NPVHC_{xH}$ derived from management unit, M, is obtained as

$$NPVHC_{HM} = \sum_{x \subset M} \frac{F_{xHj}}{S_H} NPVH_{H'} \qquad (4.17)$$

where S_H represents the total water used in electricity generation at point H.

Application example. Let us consider the value of water available for electricity generation in five hydropower plants of the Willamette river basin (Figure 4.1; see Plate 1). Given specific power generation ratings, and a LULC pattern from 1990, Eq. (4.1) simulates annual water yield at a 30 × 30 m parcel resolution. In this example, the estimate of the net present value of water for hydropower is based on a price of electricity of $0.01 per kilowatt, with a discount rate of 5% for a 100-year productive lifespan. Our model assigns the highest landscape values to the Detroit and Green Peter hydropower plant catchments because they have high power ratings with respect to their drainage areas (Figure 4.1a). The landscape draining into Lookout Point has lower per-parcel values despite its high energy rating because the contributing area is large. To examine how LULC change might impact the quantity and value of water available for hydropower production, we developed a scenario in which all the forested area below 1000 m in elevation was converted into pasture. Using our tier 1 surface water yield model, the increase in water yield of our deforested landscape enhanced annual runoff by up to 105 mm per parcel (Figure 4.1b). Since we were only evaluating total annual yield in this example, our tier 1 approach assigned greater landscape gains for hydropower provision to the deforested parcels (Figure 4.1c). Notice, that while deforestation is likely to negatively impact biodiversity and carbon emissions, in this situation it enhances potential hydropower production thru increased water yield. If deforestation also enhances sediment discharge into the reservoir, the net impact on hydropower generation might be reduced (see Chapter 6). This example is a good illustration of the trade-offs inherent in most land-use decisions, but glossed over when considering one service at a time.

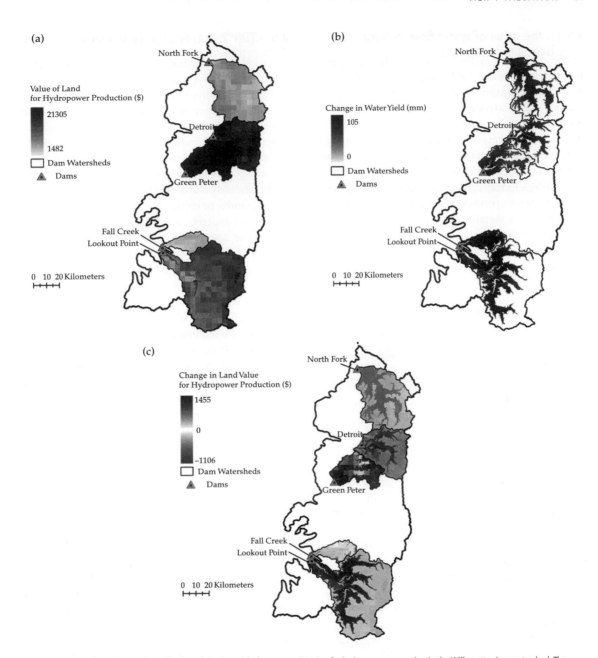

Figure 4.1 Hypothetical example application of tier 1 model of water provisioning for hydropower generation in the Willamette river watershed. The example evaluates five sub-catchments of hydropower stations at North Fork (41MW), Detroit (115MW), Green Peter (92MW), Fall Creek (6.4MW), and Lookout Point (138MW). (a) The net present value of landscape water provision services for hydropower; (b) changes in water yield as a result of hypothetical deforestation of all land below 1000 m above sea level; and (c) the changes in landscape value for hydropower under the deforestation scenario. (See Plate 1.)

4.3.2 The value of water flow used for irrigation at the crop field

The monetary value of the amount of water used at D for irrigation purposes is given by B_D. The water available for irrigation, S_{Dp}, at D is given by Eq. (4.14). Let Irr_{Dp} indicate the annual amount of surface water required for consumption by annual crop irrigation demand at D that is required during crop cycle with period, p, in an average year ($Irr_{Dp} \leq S_{Dp}$). In this case B_D is equal to the net value of agriculture production in D with Irr_{Dp}. See Chapter 9 for details on determining B_D at irrigated points D. The contribution of each management parcel in the landscape to B_D is approximated using Eq. (4.8).

4.4 Limitations of the tier 1 water yield models

As a general rule, the tier 1 water yield models are most physically realistic at watershed scales, specifically, for areas that are hydrologically coherent. A hydrologically coherent area maintains the integrity of surface and subsurface water flow paths or drainage to a point of discharge, such as up to rivers, streams, creeks, or springs. When a watershed is subdivided without accounting for the potential flows between them, our tier 1 modeling approach presented can estimate relative contributions of different parts of the landscape to water flow, retention, and consumption but will likely misrepresent the integrated watershed response. Second, we do not explicitly model groundwater, especially in terms of pumping from shallow aquifers for irrigation. A third major shortcoming of tier 1 water yield models is its annual timeframe. Many critical hydrological events or demands for water use occur on the timescale of days, and representing hydrology as an annual output can lead to large errors of interpretation. One way around this is to build seasonal models, so that as a compromise between annual and daily, one breaks the year up into different seasons, within which seasonal averages provide approximations that can get much closer to the degree of realism one seeks. Tier 2 addresses this shortcoming in that it is a daily time step model.

4.5 Tier 2 water supply model

Our tier 2 water models simulate hydrologic variability, incorporate the temporal nature of management, and model additional hydrologic processes that influence the ecosystem services discussed in this chapter. In addition, tier 2 provides the tools for characterizing water resources infrastructure. Irrigation water is diverted and often stored to support or permit agricultural production when water is lacking at crop fields. Similarly, infrastructure for hydropower plants diverts, stores, or enhances energy production.

The soil matrix, vegetation roots, and shallow aquifers of a watershed provide a regulatory function by enhancing opportunities for baseflow or groundwater recharge. In effect, this provides a storage value to the watershed landscape. Tier 2 provides the means to simulate daily hydrology and climate variability, water resources infrastructure, and all the rules of operation, management, and rights. This additional functionality provides greater accuracy in absolute and temporal terms. However, the tier 2 approach requires far more effort in compiling necessary data, and calibrating the models, and may not be feasible unless one has formal hydrological and modeling training and relatively long time windows for analysis.

The hydrology in the tier 2 models is driven by a modular system that allows users to replace or add different hydrology modules based on site-specific conditions, existing data, and hydrologic understanding. In addition, the modular approach allows one to adjust model complexity as needed to explain site-specific phenomena (Farmer *et al.* 2003). Our tier 2 approach uses the Precipitation Runoff Modeling System (PRMS) developed by the US Geological Survey (Leavesley *et al.* 1983) to estimate water supply. Supply then becomes an input to a water resources systems model that incorporates demand, or use. Outputs of this model are then linked back to the watershed, such that the service provisioning is a function of hydrology, infrastructure, and management.

4.5.1 Tier 2 modeling of physical hydrology

Our tier 2 hydrology model evaluates the impacts of various combinations of precipitation, climate, and LULC on streamflow, sediment yields, and general basin hydrology. In PRMS the spatial modeling units are defined as Hydrologic Response Units (HRUs), which represent watershed parcels of homogenous geomorphology, vegetation, management, and hydraulic connectivity. We simulate the hydrologic regime on a daily time step and per storm event using historical records or probabilistic hydrology based on stochastic methods. By incorporating inter- and intra-annual variability, we can effectively examine the nonlinear response of hydrology to climate or management practices (i.e., dry years often lead to disproportionate scarcity). Within each HRU modeling parcel, the model keeps track of critical hydrological variables such as infiltration, soil moisture, and canopy interception of water—all of which will impact water supply. In the tier 2 equations that follow, parcels or landscape units, x, are HRUs.

4.5.2 Modeling of water resource systems: water accounting

Once water supply is estimated with the PRMS model, the tier 2 approach accounts for water resource systems and their impacts on flows. In particular, we recommend using the Water Evaluation and Planning System (WEAP), which is a readily available tool that links hydrology to water resources systems (Siebert and Purkey 2007). In tier 2, the water from parcel x available for use, S_d, at demand point d is described in its most generic form for any time, t, as

$$S_d(t) = \sum_{x \in d} Q(t) - \sum_{u \in d} D_u(t, \sigma_u) + \sum_{u \in d} R_{u,STO}(t, \sigma_u)$$
$$- \sum_{sto \in d} V_{sto}(t, \sigma_u) - \sum_{u \in d} \frac{Irr_u(t, \sigma_u)}{WRR_u} \quad (4.18)$$

where $\sum_{x \in d} Q(t)$ represents streamflow (including both storm runoff and shallow groundwater discharge components) from parcels x in the watershed available for use at d during time step t. $\sum_{x \in d} Q(t)$ results from interaction of each hydrology module, such as canopy interception, infiltration and percolation, transpiration and evaporation, groundwater processes, and vegetation growth. Each of the biophysical processes represented by the modules can be represented at a level of complexity commensurate

with the data available and the user's understanding and needs. $D_u(t, \sigma_u)$ represents water withdrawals during t to satisfy demand u, which, like x, are part of the watershed landscape benefitting d. Different demands for water u will have different priorities, or management policies, σ_u, that can be function of t and $Q(t)$. $R_{u,STO}(t, \sigma_u)$ are flow releases from reservoirs STO, or return flows from demands u during t. Return flows vary depending on the type and efficiency of water use and transport. They usually represent the difference between water diverted to satisfy a demand and the amount of water actually consumed, which is often 80 to 90% of demand for municipal and industrial uses (Loucks and Van Beek 2005). $V_{sto}(t)$ is the water stored during t in a lake or reservoir, or unavailable water being routed through the stream channel network. $Irr_u(t, \sigma_u)$ is that irrigation water required at u, and WRR_u is the water requirement ratio, a fraction to account for transmission losses or other irrigation requirements for excess demand of irrigation water. See Chapter 9 for information on $Irr_{ud}(t, \sigma_u)$.

The temporal dimension of Eq. (4.18) denoted by t introduces seasonality and interannual variability, such as dry and wet seasons or wet and dry years, which are often key to meeting demand for ecosystem services. In addition, in tier 2, the last four terms of Eq. (4.18) incorporate human effects on water availability, such as storage infrastructure, diversions, management strategies, and water rights policies. As might be evident, the greater is the alteration of flow regimes by human influence during t; the lower is the relative importance of landscape hydrology. The temporal and engineering system management dimensions of tier 2 provide the tools for broader examination of ecosystem services and trade-off options. The following sections describe how we link the ecosystem provision of water, $S_d(t)$, to hydropower production and how we perform the valuation of the landscape. See Chapter 9 for the approach used in valuing the landscape for irrigation.

4.5.3 Linking water supply to hydropower production

Our tier 2 biophysical model allows us to calculate the daily water yield and use on a parcel. This fine-scale temporal resolution lets the user model the effects of time-sensitive management decisions on energy production. This ability should add realism to the model predictions associated with temporal variability in electricity production and price throughout the year. Societal demand for electricity can be decomposed into two segments: peak and off-peak. In temperate zones, peak demand for electricity occurs during the summer when electricity use increases to cope with higher temperatures. Because the price of electricity is higher during periods of peak demand, we expect managers of impoundment systems (reservoirs and dams) to maximize production during peak periods. To do this, system managers store water in the reservoir during off-peak periods.

The impoundment system manager must choose releases through generating turbines $r_{dp}(t)$ and $r_{do}(t)$ for each time step t, where there are T time steps in a year. $r_{dp}(t)$ and $r_{do}(t)$ represent peak and off-peak release volumes, respectively. The net revenue-maximizing manager will choose $r_{dp}(t)$ and $r_{do}(t)$ for each time t according to an objective function based on Eq. (4.12) similar to

$$\max_{\substack{r_{dp1},\dots,r_{dpT} \\ r_{do1},\dots,r_{doT}}} \sum_{t=1}^{T} \kappa \cdot \mu \cdot \left(\begin{matrix} p_{pe} \cdot r_{dp}(t) \cdot h_d(V_d(t)) + \\ p_{oe} \cdot r_{do}(t) \cdot h_d(V_d(t)) \end{matrix} \right) - c_{dt} \quad (4.19)$$

Subject to

$$\begin{aligned} r_{dp}(t) + r_{do}(t) &\leq V_d(t) - \overline{V}_d & t = 1,\dots,T \\ r_{dp}(t) \leq \overline{r}_{dp}(t) \text{ and } r_{do}(t) &\leq \overline{r}_{do}(t) & t = 1,\dots,T, \\ V_d(t+1) &= V_d - r_{dp}(t) - r_{do}(t) + S_d(t), \end{aligned}$$

where the objective function indicates the annual net revenue generated at d, p_{pe} and p_{oe} are the prices of electricity during peak and off-peak periods, respectively, $p_{pe} > p_{oe}$, and c_{dt} indicates the marginal cost of maintaining impoundment system d (we assume c_{dt} is the same across each t). $h_d(V_d(t))$ is the generating hydraulic head that is a function of volume of water, $V_d(t)$, in reservoir d during each time step t determined by accounting for all inflows from the watershed, $S_d(t)$, and direct precipitation, and subtracting all releases and other losses.

The last four equations place constraints on the choices of $\{r_{dp}(1),\dots,r_{dp}(T)\}$ and $\{r_{do}(1),\dots,r_{do}(T)\}$. The

first constraint states that the release of water through dam d in t (given by $r_{dp}(t) + r_{do}(t)$) cannot be greater than the amount of water in the reservoir at the beginning of t (given by $V_d(t)$) less some minimum amount of water that the manager wants to maintain in the reservoir at all times (given by $\overline{V}_d$). $\overline{V}_d$ is unique to each impoundment system and is assumed to be a function of the manager's risk preferences, d's other uses (e.g., recreation, source of drinking water, source of irrigation water), and other reservoir management policies. The second and third constraints limit the rate of release during the peak and off-peak periods to what is physically possible, given by $\overline{r}_{dp}(t)$ and $\overline{r}_{do}(t)$, which are functions of the dam's capacity and the duration of peak and off-peak times in period t. The final equation is used to track time-specific reservoir water levels where $S_d(t)$ gives the recharge rate of the impoundment d. See Eq. (4.18) for the calculation of $S_d(t)$.

4.5.4 The challenge of calibrating tier 2 water models

Calibration of physical hydrology parameters is usually part of a three-step process of calibration, verification, and validation of a model. Calibration establishes model parameters using optimizing objective functions to simulate specific hydrologic processes. Verification is generally associated with evaluating performance of a calibrated model using a different data set or time series at the same calibrated site. Validation of the model occurs when performance of the model continues to be adequate using optimized parameters and setup during calibration and verification at different yet representative sites. Having access to representative data (Yew et al. 1997) correctly configuring the processes (Farmer et al. 2003) and optimizing parameters, such that they can be considered "validated", are the biggest limitations to using tier 2 hydrology models in ecosystem service prediction.

The purpose of model calibration is to better identify model parameter values that cannot be accurately determined based on physical data. The traditional approach to calibration of a distributed-hydrologic model has been comparison of measured and simulated runoff at the outlet of the basin. Farmer et al. (2003) recommend to begin a calibration process by first adequately simulating the general annual water balance, that is, calibrate for interannual yields of a basin to adequately account for the impact of dry and wet years on ecosystem services. Then, the modeler must adequately simulate the seasonal variability of the water balance (intra-annual performance). Once the inter- and intra-annual components of the simulated water balance correspond within acceptable ranges of observed values, the calibration process should continue to match flow duration curves. Finally, the modeler can make adjustments to generate a suitable time series of predicted hydrology. Clearly, how one assesses model performance should be dictated by the ecosystem services of interest. For example, if ecosystem services for irrigation depend on drought flows during the months before a rainy season, then calibration efforts should focus on adequately modeling the relevant baseflow processes and reporting performance statistics during those dry months.

4.6 Tier 2 valuation model

The tier 2 valuation models are implemented at applicable management time steps instead of aggregating on an annual basis. Water yield in the tier 2 supply and use models is measured at a time scale that matches the temporal scale of water use decisions.

4.6.1 The value of water flow for hydropower

In order to assess the value of hydropower generation, we assume the net revenue has been maximized over some time period. In other words, given operation rules defined in Eq. (4.19) that maximize net revenue subject to policy constraints, we estimate expected annual energy production, $\bar{\varepsilon}_{hyd}$, which is used to determine the net present value of energy. Let T indicate the number of years we expect present landscape conditions to persist or the expected lifetime of d (set T to the smallest value if the two time values differ).

Let $\{r^*_{dp}(1),\ldots,r^*_{dp}(T)\}$ and $\{r^*_{do}(1),\ldots,r^*_{do}(T)\}$ indicate the peak and off-peak releases that maximize the net present value of water used in the production of hydropower, $NPVH_d$. Then $NPVH_d$ is given by

$$NPVH_d = \sum_{t=1}^{T} \Psi^* \cdot \left(\sum_{t=1}^{T} \frac{1}{(1+\gamma)^t} \right)$$

$$\Psi^* = \kappa \cdot \mu \cdot \begin{pmatrix} p_{pe} \cdot r^*_{dp}(t) \cdot h_d(V_d(t)) + \\ p_{oe} \cdot r^*_{do}(t) \cdot h_d(V_d(t)) \end{pmatrix} - c_{dt} \tag{4.20}$$

We also provide a model for predicting the contribution of water yield to run of the river hydropower production. Managers of diversion systems (run-of-river dams) do not have limited capability to store water. The total annual value of hydroelectric production at diversion system dam z with negligible storage is determined by its unmanaged flow's temporal relationship with peak and off-peak periods. A diversion system powered by flows that tend to be higher during peak hours than off-peak will generate higher revenues than other diversion systems with the opposite flow–peak relationship, all else equal. The net present value of energy produced by dam d (point of use value) until time T is given by $NPVH_z$,

$$NPVH_z = \sum_{t=1}^{T} \Psi \cdot \left(\sum_{t=1}^{T} \frac{1}{(1+\gamma)^t} \right)$$

$$\Psi = \kappa \cdot \mu \cdot \begin{pmatrix} p_{pe} \cdot r_{zp}(t) \cdot h_z(V_d(t)) + \\ p_{oe} \cdot r_{zo}(t) \cdot h_z(V_d(t)) \end{pmatrix} - c_{zt} \tag{4.21}$$

where $r_{zp}(t)$ indicates the amount of water that flows through z during the peak period of t, $r_{zo}(t)$ indicates the amount of water that flows through z during the off-peak period of t, and c_{zt} indicates the average cost of maintaining dam z (assume c_{zt} is the same across each t).

4.7 Sensitivity analyses and testing of tier 1 water supply models

4.7.1 Sensitivity analysis of tier 1 models

Because of the many simplifying assumptions of the tier 1 water models, it is essential to gain an understanding of their sensitivity to different parameters. To accomplish this, we conducted a formal multivariate sensitivity analysis (Nearing *et al.* 1989) that calculates global sensitivity using the maximum, minimum, and average values for each parameter, i, in the model as

$$SI_i = \frac{(OUT_{i,\max} - OUT_{i,\min})}{(IN_{i,\max} - IN_{i,\min})} \cdot \frac{IN_{i,\text{aver}}}{OUT_{i,\text{aver}}}, \quad (4.22)$$

where $IN_{i,\max}$, $IN_{i,\min}$, and $IN_{i,aver}$ are respectively the maximum, minimum, and average values for model input parameter i, $OUT_{i,\min}$ and $OUT_{i,\max}$ are the model outputs corresponding to input parameter $IN_{i,\min}$ and $IN_{i,\max}$ respectively while holding all other parameters at their average value, and $OUT_{i,\text{aver}}$ is the average of $OUT_{i,\max}$ and $OUT_{i,\min}$.

A sensitivity index of 1 indicates that the model average output varies with the same magnitude and direction as the model average input. A negative value means that the input and output change inversely. Sensitivity analysis for the tier 1 water yield model is presented by Table 4.2, which summarizes the sensitivity index values for different parameters of the water yield model. The parameters in Table 4.2 are listed in descending order of sensitivity. These values are all negative, meaning that any increase in parameter value corresponds to a decrease in model output. The most sensitive parameters of our tier 1 model were associated with vegetation properties: the evapotranspiration coefficient and root depth. This has two important implications. First, it may indicate that vegetation is extremely important in regulating ecosystem services. Secondly, it means that users should pay special attention to uncertainty associated with vegetation properties.

Solving Eq. (4.2) for all dryness index ratios provides insight on how sensitive the model is to climate, i.e., the energy–rainfall relationship. In general, Eq (4.2) has a greater sensitivity to land-use parameters the closer the PET–rainfall ratio is equal to 1. In particular, very dry regions are least sensitive to climate factors in the tier 1 model.

We do not present sensitivity analysis for the water retention index because it contains a hydraulic connectivity parameter that is dependent on specific watershed geomorphology, shape, and land-use configurations. Thus, these factors are somewhat invariant for each watershed, although there may be issues associated with the scale of available data or how "parcels" are defined. A valuable tool for managing uncertainty is to explore several simulations using the reasonable ranges of possible parameter values and examining how rankings of ecosystem services provision or regulation values differ.

4.7.2 Testing of tier 1 models

Given the major simplifications we have made to provide tier 1 models that can be run with minimal data inputs, it is important to know how well our simple models agree with more complex, widely accepted models. However, there are no (or very few) models that predict a landscape's contribution to particular uses such as hydropower and irrigation. Much more common are models for the first step of our modeling process: the biophysical supply step. As such we focused on verifying this biophysical modeling approach by testing it against the popular hydrology model, SWAT, in climatically diverse eco-regions of the USA: California, Texas Gulf, Tennessee, the Willamette river basin, and the lower Colorado basin (Figure 4.2). We aggregate our model results into catchment areas within each eco-region, and compare spatial variation in water yields produced by our tier 1 model to spatial variation in water yield produced by the much more detailed and data-intensive SWAT model (Arnold *et al.* 1998).

Table 4.2 Sensitivity index for the tier 1 yield model and the principal drivers that describes parameters

Parameter, i	Parameter description	Driver	SI_i
k_{xj}	Evapotranspiration coefficient	Land use	-97184
RD	Root depth	Land use	-7324
Z	Water balance calibration constant	Climate	-45
AWC_x	Available water capacity	Soil property	-0.161

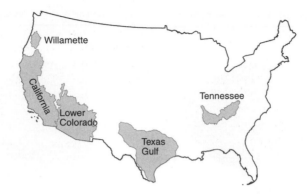

Figure 4.2 Basins across the continental USA representing different eco-regions used in testing tier 1 water yield model rankings with respect to SWAT.

Our measure of model concordance is the non-parametric Kendall tau test (e.g., Helsel and Hirsch 2002), which quantifies the degree of agreement or disagreement between two non-normal ranked sets of data; a tau coefficient value (τ) of 1 means perfect agreement, while a τ of 0 means the sets are completely independent. The Kendall tau correlation statistics for all tests are shown in Table 4.3. We evaluate performance for annual water yield, and compare our water retention index with SWAT annual groundwater percolation outputs. Figure 4.3 provides a basin scale inspection of comparative model outputs for water yield and water recharge index in the Texas Gulf and Willamette river basins, respectively. Figure 4.4 provides scatter plot comparison between outputs from our non-calibrated tier 1 water yield model and that from SWAT model applied to average annual water yield in the Texas Gulf as presented in Table 4.3.

Results show that our limited data models can predict trends and rankings fairly well for annual water yield. With respect to absolute values we overestimate higher and underestimate lower results when compared with SWAT (Figure 4.4); in other words, our tier 1 model somewhat exaggerates water yield at both the high end (overestimates) and low end (underestimates). In simulating water yield, we found the best agreement in model results in Tennessee ($\tau = 0.89$, $p < 0.01$). The lower agreement between models in the Willamette river basin is likely due to the smaller scale.

Moreover, when our tier 1 average annual water yield model was calibrated and then compared with observed data in five watersheds in the Hainan Island of China, performance is greatly improved in absolute terms (Figure 4.5). However, it is important to note that research is needed to evaluate performance within management units that do not

Table 4.3 Kendall tau ranking statistics for non-calibrated tier 1 water yield and water retention index outputs with respect to annual yield and groundwater percolation simulation by SWAT at sub-catchments in five eco-regions/river basins in the USA

Eco-region	N	Water yield			Water retention index		
		R^2	Kendall tau (τ)	p value	R^2	Kendall tau (τ)	p value
Willamette	111	nl	0.38	< 0.01	0.59	0.50	< 0.01
Texas	122	0.89	0.59	< 0.01	nl	0.25	< 0.01
Tennessee	32	0.96	0.89	< 0.01	nl	0.15	0.25
Lower Colorado	85	nl	0.59	< 0.01	nl	0.23	< 0.01
California	135	0.95	0.42	< 0.01	nl	0.18	0.003

n = Number of sub-catchments per eco-region.
nl = nonlinear

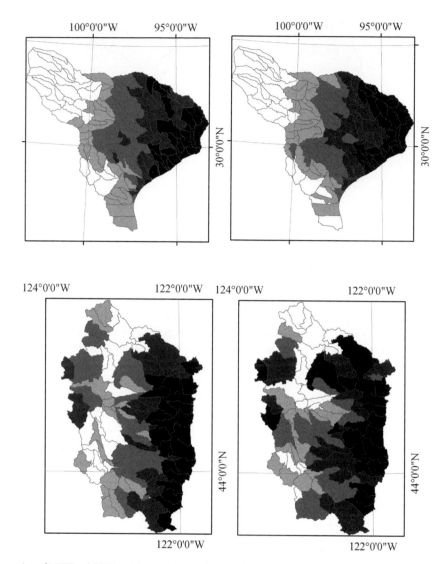

Figure 4.3   Comparison of InVEST and SWAT spatial patterns in annual water yield. Five quantile rankings of mean annual water yields for the Texas Gulf basin between our tier 1 water yield approach (top left) and the SWAT model (top right), and our tier 1 water retention index for the Willamette River basin (bottom left) and the SWAT model's mean annual groundwater percolation rate (bottom right).

necessarily have distinct hydrologic boundaries, since trade-off analysis often occurs within the boundaries of a hydrologic region.

The performance statistics of our tier 1 water retention index model were not as good. In this case, we compared our water retention index, which is based on surface roughness, hydraulic connectivity, and surface permeability, with the groundwater

percolation rate estimated by SWAT (Eq. (4.5)). Comparisons between our water retention index and groundwater fluxes in SWAT yielded significantly positive, but often relatively low measures of concordance (Table 4.3).

To preliminarily test our water retention index against observed data, rather than another model, we compared streamflow recession coefficients

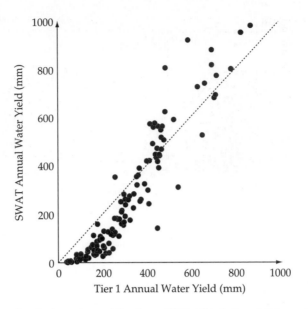

Figure 4.4   Scatter plot comparison of our tier 1 annual water yield and annual SWAT yield in the Texas Gulf. A non-calibrated average water yield tier 1 model is compared to a calibrated SWAT study to illustrate model application with limited data can provide useful guidance for landscape ranking. The dashed line corresponds to the 1:1 line.

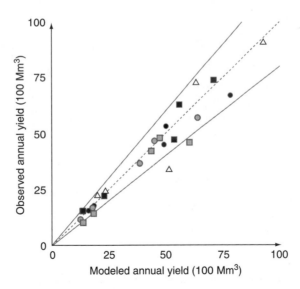

Figure 4.5   Scatter plot comparison of a calibrated tier 1 annual water yield model to observed annual water yield for 1980, 1985, 1990, 1995, and 2000, in five watersheds of Hainan Island, China. Solid circles correspond to 1980, empty circles to 1985, solid squares to 1990, empty squares to 1995, and triangles to 2000. The solid bounding lines correspond to a 40% uncertainty range and the dashed line corresponds to the 1:1 line. Units are in 100 000 m³.

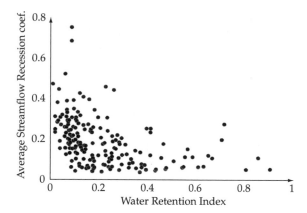

Figure 4.6   Scatter plot of water retention index and average streamflow recession coefficient for 180 unregulated stream gauges across the continental USA. Average streamflow recession coefficients are derived from unregulated streamflow (selected for minimal anthropogenic effects in the HCDN data set by Slack *et al.* 1993). Receding streamflow is sampled from data time series with at least 4 consecutive days of decrease in flow, from sub-catchments between 200 and 600 square miles, occurring between June and September, after 1970, and with aridity index greater than 1 (see Eq. (4.4)). For each of the 180 stream gauge stations, on average, 122 time series were used to approximate a recession flow coefficient (*RFC*) by a simplifying assumption of an exponential decay in flow, i.e., $RFC = \dfrac{\ln\left(\dfrac{Q_{t=0}}{Q_t}\right)}{t}$). An average water retention index was estimated by applying Eq. (4.5) for the sub-catchments corresponding to the stream gauges. The streamflow recession coefficient is usually inversely proportional to the water regulation function of the landscape. The higher the RFC, the lower the water regulation properties of the landscape, and vice versa.

from streamflow data from across the continental USA to respective values of the water retention index. The recession coefficient index (similar to an index of the rate of decay) is an indicator of the rate that baseflow decreases after a rainfall period. A high value for the recession coefficient signifies that the rate of decrease of stream baseflow is high, which might be expected to result from a landscape with poor water retention properties, and vice versa. Figure 4.6 compares streamflow recession coefficient data with water retention index. We found that, in general, watersheds with higher streamflow recession coefficient values were more likely to have lower water retention index values. That is, stream baseflow that decrease the fastest can be associated with landscape properties that score lowest on our water retention index. This finding suggests that it is reasonable to define a threshold value of water retention, such as in Eq. (4.6), to evaluate whether the landscape effectively regulates streamflow.

4.8   Next steps

Watersheds can support the provision of water-related ecosystem services by regulating the amount and timing of water availability. However, more scientific guidance is needed on how to support conservation practitioners in developing and managing watersheds to preserve or enhance these services, and how to tailor the approaches to different biomes (Rodriguez-Iturbe 2000). We have presented simple (tier 1) and more complex (tier 2) approaches for valuing the general landscape properties for water quantity and timing as applied to hydropower generation and irrigation. These tools may not necessarily provide definitive answers but they do provide a structured framework for linking hydrologic science with decision-makers and planners. These tools need to be continuously improved, used cautiously and transparently, while presenting uncertainty ranges as a starting point for informed decision-making in the spirit of adaptive management.

Improvements to our models will include the incorporation of uncertainty using two possible approaches. The first involves drawing parameter values from a distribution for a given LULC across a landscape for parameters whose values might vary with LULC type. This approach is insightful because it can provide an uncertainty range of average service provision by LULC characteristics. The second approach would assign confidence intervals directly to non-landscape specific parameters whose values might vary between regions and affect the impact of LULC characteristics.

Hydrologic improvements to tier 1 will include improvements to our water regulation function; addition of soil moisture dynamics for improving estimates of likely infiltration and leakage, and reducing errors due to modeling scales used; a proper definition of water partition parameter Z in Eq. (4.3) that is most likely a nonlinear function; improvements to hydropower generation by defining an average hydraulic head that is dependent on inflows, such as by adapting Gould-Dincer solutions (see McMahon et al. 2007); and a stochastic function to relate the timing of irrigation requirement with the timing of water availability. In addition, our water regulation and provision functions can be modified and coupled differently to support other services, such as fresh water fisheries or drinking water. Although tier 2 can better account for the complexity of natural systems at any time step, it needs to be better integrated with the spatial sophistication of tier 1.

Finally, we need to learn from field application of our models in collaboration with planners and service providers. Adaptive management, which recognizes that humans do not know enough to efficiently manage ecosystems and that decision-makers cannot afford to consequently postpone action (Lee 1999), is an important next step for integrated model improvement. Adaptive management, often described as a process of learning by doing (Lee 1999), provides a framework where our tier 1 models become incredibly valuable for helping early decision-making and dialogue, but can also be iteratively improved. At the onset of a project, tier 1 can be used to rank and prioritize the landscape to efficiently allocate scarce management resources. As a knowledge base increases through hydrologic monitoring and analysis, tier 1 market-based valuation can be more reliably applied; tier 1 hydrologic algorithms can be improved and catered to specific needs, and ultimately lead to improved valuation with the use of tier 2 models if desired.

References

Albery, A. C. (1968). Forecasting demand for instream uses. In W. R. D. Sewell and B. T. Bower, Eds. *Forecasting the Demands for Water*. Department of Engineering, Mines, and Resources, Ottawa, Canada.

Allen, R. G., Pereira, L. S., Raes, D., *et al.* (1998). *Crop evapotranspiration guidelines for computing crop water requirements*. FAO Irrigation and drainage Paper 56. Food and Agriculture Organization of the United Nations. Rome, Italy.

Arnold, J. G., Srinivasan, R., Muttiah, R. S., *et al.* (1998). Large area hydrologic modeling and assessment, Part I: Model development. *Journal of the American Water Resources Association*, **34**(1), 73–89.

Beven, K. J., and Kirkby, M J. (1979). A physically based variable contributing area model of basin hydrology. *Hydrologic Science Bulletin*, **24**(1), 43–69.

Bosch, J. M., and Hewlett, J. D. 1982, A review of catchment experiments to determine the effect of vegetation changes on water yield and evapotranspiration. *Journal of Hydrology*, **55**, 3–23.

Bruijnzeel, L.A. 2000, *Hydrology of tropical montane cloud forests: A reassessment*. Tropical Environmental Hydrology Programme, Amsterdam.

Bruijnzeel, L. A. (2001). Hydrological functions of tropical forests: not seeing the soil for the trees? *Agriculture, Ecosystems and Environment*, **104**(1), 185–228.

Budyko, M. I. (1974). *Climate and Life*, Academic Press, San Diego, CA.

Budyko, M. I., and Zubenok, L. I. (1961). The determination of evaporation from the land surface. *Izvestiya Akademii Nauk. SSR Seriya Geografiya*, **6**, 6–17.

Chomitz, K. M., and Kumari, K. (1998). The domestic benefits of tropical forest preservation: A critical review emphasizing hydrological functions. *World Bank Research Observer*, **13**(1), 13–35.

Donigian, A. S., Jr., Imhoff, J. C., Bicknell, B. R., *et al.* (1984). *Application guide for Hydrological Simulation Program—Fortran (HSPF)*. U.S. Environmental Protection Agency, Environmental Research Laboratory, Athens, GA, EPA-600/3-84-065.

Donohue, R. J., Roderick, M. L., and McVicar, T. R. (2007). On the importance of including vegetation dynamics in Budyko's hydrological model. *Hydrology and Earth System Sciences*, **11**, 983–95.

Edwards, B. K. (2003). *The economics of hydroelectric power.* Edward Elgar, Northampton, MA.

Farmer, D., Sivalapan, M., and Jothityangkoon, C. (2003). Climate, soil, and vegetation controls upon the variability of water balance in temperate and semi-arid landscapes: Downward approach to water balance analysis. *Water Resources Research,* **39**(2), 1–21.

FAO. (2003). *The state of food insecurity in the world,* Food and Agriculture Organization of the United Nations, Rome.

Gleick, P. H. (1996). Water resources. In S. H. Schneider, Ed., *Encyclopedia of Climate and Weather.* Oxford University Press, New York.

Helsel, D. R., and Hirsch, R. M. (2002). *Statistical methods in water resources.* US Geological Survey, Washington, DC.

IWMI (2001). *Water for rural development: draft background paper on water for rural development prepared for the World Bank.* International Water Management Institute, Colombo, Sri Lanka.

Kent, K. (1972). Section 4: Hydrology. In *National Engineering Handbook.* United States Department of Agriculture, Washington, DC.

Leavesley, G. H., Lichty, R. W., Troutman, B. M., *et al.* (1983). *Precipitation-runoff modeling system: user's manual.* US Geological Survey, Washington, DC.

Lee, K. N. (1999). Appraising adaptive management. *Conservation Ecology,* **3**(2), article 3.

Loucks, D. P., and Van Beek, E. (2005). *Water resources systems planning and management: an introduction to methods, models and applications.* Studies and reports in hydrology. UNESCO, Paris.

Lyimo, B. M. (2005). *Energy and sustainable development in Tanzania,* Helio International, Paris.

Lyon, S. W., Gerard-Marcant, P., Walter, M. T., *et al.* (2004). Using a topographic index to distribute variable source area runoff predicted with the SCS–Curve Number equation. *Hydrology Proceedings.* **18**(15), 2757–71.

McMahon, T. A., and Mein, R. G. (1986). *River and reservoir yield.* Water Resources Publications, Littleton, CO.

McMahon, T. A., Pegram, G. G. S., Vogel, R. M., *et al.* (2007). Review of Gould-Dincer reservoir storage-yield-reliability estimates. *Advances in Water Resources,* **30**, 1873–82.

Milly, P. C. D. (1994). Climate, soil water storage, and the average annual water balance. *Water Resources Research,* **3**(7), 2143–56.

Nearing, M. A., Ascough, L. D., and Chaves, H. M. L. (1989). WEPP model sensitivity analysis. In L. J. Lane and M. A. Nearing, Eds., *USDA-Water Erosion Prediction Project: Hillside profile model documentation, NSERL Report No. 2.* USDA-ARS-NSERL, West Lafayette, IN.

Oyebande, L. (1998). Effects of tropical forest on water yield. In E. R. C. Reynolds and F. Thompson, Eds., *Forests, climate, and hydrology: regional impacts,* pp. 16–50. United Nations University; Kefford Press, Singapore.

Pagiola, S., Bishop, J. and Landell-Mills, N. (2002). Selling forest environmental services: Market-based mechanisms for conservation and development. Earthscan, London.

Porporato, A., D'Odorico, P., Laio, F., *et al.* (2002). Ecohydrology of water-controlled ecosystems. *Advances in Water Resources,* **25**, 1335–48.

Potter, N. J., Zhang, L., Milly, P. C. D., *et al.* (2005). Effects of rainfall seasonality and soil moisture capacity on mean annual water balance for Australian catchments. *Water Resources Research,* **41**(6).

Rodriguez-Iturbe, I. (2000). Ecohydrology: A hydrologic perspective of climate-soil-vegetation dynamics. *Water Resources Research,* **36**(1), 3–9.

Siebert, J., and Purkey, D. (2007). *WEAP: water evaluation and planning system—users guide.* Stockholm Environment Institute, Sommerville, MA.

Slack, J. R., Lumb, A. M., and Landwehr, J. M. (1993). *Hydro-Climatic Data Network (HCDN): Streamflow data set, 1874–1988.* US Geological Survey Water-Resources Investigation Report 93-4076 (CD).

Steenhuis, T. S., Winchell, M., Rossing, J., *et al.* (1995). SCS runoff equation revisited for variable-source runoff areas. *Journal of Irrigation and Drainage Engineering,* **121**(3), 234–8.

Tallis, H., Kareiva, P., Marvier, M., *et al.* (2008). An ecosystem services framework to support both practical conservation and economic development. *Proceedings of the National Academy of Sciences,* **105**(28), 9457–64.

Winrock International (2004). *Financial incentives to communities for stewardship of environmental resources.* Feasibility study: LAG-A-00-99-00037-00. USAID, Washington, DC.

Winter, T. C., Harvey, J. W., Frankey, O. L., *et al.* (1998). Ground water and surface water a single resource. *US Geological Survey Circular 1139.* US Government Printing Office, Denver, CO.

Yew, D. T., Dlamini, E. M., and Biftu, G. F. (1997). Effects of model complexity and structure, data quality, and objective functions on hydrologic modeling. *Journal of Hydrology,* **192**, 81–103.

Zektser, I. S., and Loaiciga, H. A. (1993). Groundwater fluxes in the global hydrologic cycle: past, present and future. *Journal of Hydrology,* **144**, 405–27.

Zhang, L., Dawes, W. R., and Walker, G. R. (2001). Response of mean annual evapotranspiration to vegetation changes at catchment scale. *Water Resources Research,* **37**, 701–8.

CHAPTER 5

Valuing land cover impact on storm peak mitigation

Driss Ennaanay, Marc Conte, Kenneth Brooks, John Nieber, Manu Sharma, Stacie Wolny, and Guillermo Mendoza

5.1 Introduction

Images of floods displacing or even killing people provide a constant reminder of the power of nature and human vulnerability to natural disasters. Although storms and storm events are highly unpredictable, it is possible to use hydrological models to predict the magnitude of a particular flood, given information on the local geology, soil properties, vegetation, and management practices. We have developed approaches for quantifying the link between changes in land use and land cover (LULC), and flood risk. In flood management, risk has three ingredients: the hydrological response to a storm, the possible failure of flood protection infrastructure (such as a levee breaking), and the value of what might be destroyed by a flood. We focus on the hydrology and economic value of what may be destroyed, leaving structural integrity to be addressed by civil engineers. Given a well-defined storm, we estimate the severity of flooding in terms of water volumes and flow rates, and corresponding damages from the storm.

In general, the combination of meteorological (e.g., rainfall intensity, extent and duration of the event) and geophysical (e.g., basin size, basin geomorphology, soil characteristics, and land use) characteristics are the main factors influencing major flooding following large rainfall events (Hamilton and King 1983; Kattelmann 1987; Bruijnzeel 1990, 2004). In some situations, natural landscapes and vegetation can offer storm peak mitigation. For example, forests and deep permeable soils often reduce runoff as the result of enhanced soil infiltration and soil water storage capacity. Conversion of forests and wetlands to agricultural or developed land covers will tend to increase the volume of runoff and the flooding associated with storm events for medium and small return period events (Ennaanay 2006). However, forests have limited ability to mitigate flooding associated with large return period storm events because enhanced soil infiltration only captures a small fraction of total precipitation depth for such storms (FAO and CIFOR 2005). We develop models in this chapter that can help decision-makers take advantage of nature-based mitigation of floods and storm damage from medium and small return period events to avoid unnecessary flood risk due to poor land management.

This chapter presents two different types of models for quantifying the impact of LULC on storm outcomes. The data-sparse tier 1 model quantifies the reduction in storm peak volume due to LULC relative to bare soil on a parcel-by-parcel basis and values this reduction based on each parcel's relative contribution to mitigation. In the present formulation the tier 1 model is not set up to predict the extent of downstream flooded area associated with a storm peak. The more robust, data-intensive tier 2 model provides probabilistic output for flood magnitudes as affected by incremental changes in the landscape mosaic and quantifies the incremental changes in risks associated with a specific flood volume, where risk is associated with a cost. In tier 2, the extent of flooded area is determined using the

Hydrologic Engineering Centers River Analysis System software (HEC-RAS) and streamflow time series from the Precipitation Runoff Modeling System (PRMS) model.

5.1.1 Storm peak mitigation modeling theory

For small to medium storms vegetation may retain water as it falls and flows through the landscape (through canopy interception, enhanced infiltration, soil water storage) and thus reduce peak flow. In a modeling study using the Hydrologic Simulation Program Fortran (HSPF) in the Cottonwood River watershed within the Minnesota River Basin, Ennaanay (2006) showed that conversion of different percent acreage (60, 75, and 86% of watershed area) of annual cropping systems (soy and corn) to perennial vegetation over a 50-year simulation period showed a decrease in annual instantaneous peak flows for small event storms, but not for larger event storms. Similarly, studies from small paired experimental basins showed that clear-cutting and road building increased only some peak storm discharges (Wright *et al.* 1990). Indeed, in the Pacific Northwest, increases in the peaks were greater for small early wet season storms but there was no significant increase in peak flows for the largest storms (Rothacher 1970).

Wetlands in both up- and downstream areas, and floodplains have a significant role in mitigating floods and storm peaks. Both land-use types have storage capacities higher than many other LULC types. Ennaanay (2006) showed that conversion of 27% of the annual cropping systems to wetlands could significantly reduce peak flows for small and moderate storm events. Wetlands not only reduce the peak flows but also significantly delay time to peak flows, and alter the inflow–discharge relationship and roughness. Floodplains have impacts similar to those of wetlands on flood mitigation and storm peak attenuation.

The spatial resolution of our models differs between tier 1 and tier 2. In tier 1, the analysis takes place at the parcel level, where parcel size is defined by the spatial resolution of the input data. In tier 2, the analysis is based on hydrologic response units

(HRUs), which are homogeneous with regard to LULC, soil, and slope. It should be noted that while the LULC category is a key determinant of an area's ability to intercept rainfall, the equations presented below will not include direct references to LULC categories, as they are captured by the parcel and HRU indices. It should also be noted that while the analyses in tiers 1 and 2 occur at the parcel and HRU level, respectively, the model results can be reported at other scales more relevant to management decisions such as individually owned parcels or counties.

The key parameters linking LULC to storm peak mitigation are canopy interception, soil infiltration, LULC type, soil water storage, and land-use positioning on the landscape. Peak flows will increase as soil infiltration, interception by canopy, or soil water storage is reduced. However, if the magnitude of storm depth becomes large, the impact that soil and plants have on storm flow peaks is small relative to water inputs (Bruijnzeel 1990; Brooks *et al.* 2003; Ennaanay 2006) and thus of reduced value in terms of flood risk reduction. The impacts of land use on flooding also depend on the size of the area being examined. In particular, land-use impacts on flooding are most evident in watersheds less than 1000 km^2 (Kiersch 2001) because the sheer length of stream channels and extent of floodplains in larger basins provide storm peak mitigation, swamping the signal of land use. For example, forest harvesting has produced detectable changes in peak discharges in basins up to 600 km^2 in size. Increases are a result of changes in flow routing (due to roads) rather than to mere changes in water storage due to vegetation removal in small basins.

5.2. Tier 1 biophysical model

5.2.1 Modeling storm mitigation properties

The tier 1 model for storm peak mitigation focuses on a storm event of a specific size defined by the user. Our approach estimates the impact of land use on flood mitigation at the parcel level by determining the amount of on-parcel storm runoff retained by each parcel following a rainfall event. We use GIS capabilities to generate a synthetic hydrograph for the defined size of storm (Martinez *et al.* 2002; Melesse *et al.* 2003) (Figure 5.1a). To do this, we estimate storm runoff volumes for each parcel given a LULC using the SCS–curve number (CN) method (Mockus 1972) and keep track of potential travel times from parcel to the watershed point of drainage. The synthetic hydrograph is formed by aggregating runoff of parcels with similar travel time class. Factors affecting travel time in the model include LULC surface roughness and slope. By identifying landscape units with equal travel times to the watershed outlet and summing the storm water that reaches the stream from these units, we create a hydrograph of a uniform storm depth, and associated duration. For model output to be useful for land managers, we generate a map of landscape contributions to these peak flows.

The tier 1 storm peak model runs for a single specified storm event and predicts the magnitude and the timing of the peak flow. The tier 1 storm peak is mainly based on the SCS-CN method, which is known to work well for the design of culverts and civil works storm flow infrastructure. Admittedly this is a very simple approach, but rainfall-runoff modeling always demands balancing model complexity versus available data. Interestingly, it has been found that more complex models are not necessarily more accurate than their simpler alternatives (Branson *et al.* 1962, 1981; Loague and Freeze 1985).

In order to evaluate the damage-mitigating properties of a landscape we apply the tier 1 model for storm peak mitigation for a known flood return period, and the expected damages of such a storm. Prior to applying the model in the watershed of interest, one must use regional rainfall data or rainfall-runoff models coupled to flood routing models to characterize the storm event P_s related to flood event s of probability π_s. This tier 1 model assumes that the storm rainfall depth is constant in time during the storm event and spatially uniform over the watershed of interest.

The tier 1 storm peak mitigation model calculates the direct runoff generated by each parcel on the landscape using the SCS-CN equations for the user-defined storm event (Kent 1972). The SCS-CN method is a simple, widely used, and efficient method for determining the approximate amount of direct runoff from a rainfall event within any particular parcel. We use this method to compute the rainfall excess as the remainder of precipitation after on-parcel infiltration loss. Direct runoff is generated by a wide variety of surface and sub-surface flow processes, of which the most relevant ones are the Hortonian overland flow, saturation overland flow, shallow sub-surface flow, and through-flow (Ponce and Hawkins 1996). The Hortonian overland flow occurs when rainfall exceeds infiltration capacity. It is a characteristic of dry to semi-dry regions and areas where vegeta-

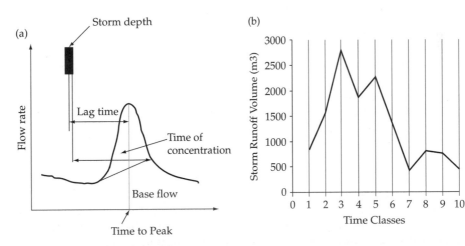

Figure 5.1 (a) Storm hydrograph terminology and (b) synthetic storm hydrograph generated by the storm peak mitigation model.

tion is sparse and the soil surface is highly disturbed. The saturation overland flow occurs after the soil profile has become saturated because of either high antecedent soil moisture or high rainfall depth that fills in the soil profile; this mechanism is referred to as the Dunne mechanism. Shallow sub-surface flow describes the process that takes place when water flows downslope in the shallow soil profile quickly and contributes to storm flow. Kirkby and Chorley (1967) show that the shallow sub-surface flow can be similar to the through-flow that occurs in heavily vegetated landscapes with thick soil covers and less permeable soil profiles atop impermeable bedrock (Kirkby and Chorley 1967). In our approach we use the curve number to quantify runoff under the assumption that the storm event occurs uniformly throughout the watershed and that Hortonian runoff is the dominant process that generates the storm peak. Nonetheless, the CN method can be used in landscapes dominated by saturation overland flow as demonstrated by Boughton (1987), Steenhuis *et al.* (1995), and Lyon *et al.* (2004), but would require some adjustments in how we develop the synthetic hydrograph. Although initially designed for watershed catchment runoff estimates, a distributed CN method has been applied effectively to large parcels (from 900 m² up to several hectares) defined as HRUs in the Soil and Water Assessment Tool (SWAT) model (Arnold *et al.* 1998, and see Hawkins *et al.* 2009). There is no published guidance on an acceptable lower limit to parcel size for application of the CN approach.

The parcel size in our model is defined by the user and can be as small as is commensurate with input data. A common input data resolution is 900 m², corresponding with the resolution of widely available digital elevation models. Our approach of applying the CN model on a parcel-by-parcel basis that is then aggregated (distributed approach) rather than on a watershed catchment (lumped approach) is not standard. We argue that runoff differences between distributed and lumped approaches are minimized in evaluations of mid- to larger sized storm events since the curve number becomes more linear at greater volumes. In addition, since differences are smallest in homogeneous landscapes, such as natural landscapes, the user

should be most wary of heterogeneous landscapes, such as heavily urbanizing encroachments into forests.

The SCS-CN equation determines storm direct runoff depth at parcel x for a flood event s that occurs with probability π_s generated by rainfall depth P_s as

$$S_{xj} = \frac{25400}{CN''_{xj}} - 254$$

$$\begin{cases} Q_{sxj} = 0 & \text{if } P_s \leq 0.2S_{xj}, \\ Q_{sxj} = \frac{\left(P_s - 0.2S_{xj}\right)^2}{P_s + 0.8S_{xj}} & \text{if } P_s > 0.2S_{xj} \end{cases} \quad (5.1)$$

where CN''_{xj} is the medium soil moisture condition curve number adjusted for slope, $0.2\,S_x$ is the initial abstraction, which accounts for the amount of precipitation occurring before runoff, or the rainfall interception by vegetation. The value has been set to $0.2\,S_x$ through developmental history and documentation; however, Hawkins (1979) showed that using $0.2\,S_x$ did not result in good runoff prediction unless S_x was dependent on rainfall amounts. Q_{sx} is the direct runoff or quick-flow that is potentially generated by P_s at x.

The SCS-CN method uses the CN values that were developed and assumed to be appropriate for slopes of 5%. Our model uses the CN adjustment recommended by Williams (1995) to slopes different than 5%,

$$CN''_{xj} = \left(\frac{100 - CN_{xj}}{3}\right) * (1 - 2 \cdot e^{-13.86\theta_{xx}}) + CN_{xj}, \quad (5.2)$$

where CN_{xj} is the CN value associated with LULC j in x applied for soil conditions of medium wetness type II (USDA 1986) with a 5% slope. However, the user can specify one of three wetness conditions (dry, medium, or wet): in temperate conditions, generally forested lands are drier than annual crops after long dry periods, and θ_x is the parcel's slope.

Once direct runoff is calculated, we estimate travel time of that water to the point of interest. When defining the drainage area to evaluate the storm-mitigating properties of the landscape, it is important to exclude parcels that drain into intermediate flood control reservoirs. Thus, we define X_D as the set of all parcels that flow into point of interest D that are not routed through a flood control reservoir. The following equations provide a total time of travel for excess rainfall originating on a parcel in the landscape to a drainage point of interest, D,

$$v_x = \frac{1}{c_x} \cdot \sqrt{\theta_x^\%}$$

$$T_{X_D} = \sum_{\forall x \in X_D} \frac{y_x}{v_x}, \tag{5.3}$$

where x indexes parcels in the landscape; v_x is an estimate for overland flow velocity; $\theta_x^\%$ is the mean percent slope of parcel x; $c1/c_x$ is a roughness coefficient for each LULC type based on the *National Engineering Handbook* (Kent 1972; ASCE 1996) that relates slope and surface vegetation to velocities; y_x is the distance travelled on parcel x equal to the width of the parcel if flow direction is north, south, east, or west and equal to the hypotenuse of the parcel if flow direction is otherwise; and T_{X_D} represents the potential travel time that storm runoff from parcel x is routed to point of interest D. There may be several drainage points of interest that might denote an area of high importance for flood mitigation. If so, the model can be applied iteratively for each point of interest.

Once travel times are calculated for each parcel, isochrones (sets of parcels with the same travel time class to the point of interest) are defined. Let us introduce τ_{iD} to represent isochrone i, where τ_{iD} is defined as follows for demand point D,

$$i = 1,..,I$$

$$\tau_{iD} \in \left[\frac{(i-1) \cdot T_{x \to D \max}}{I}, \frac{i \cdot T_{x \to D \max}}{I} \right], \tag{5.4}$$

where τ_{iD} is a set of parcels x with a similar class of travel time to point D; D is the flood risk reduction demand point and is the reference for estimates of flow travel times; i is an isochrone identifier; I is the number of isochrones (time classes) in the analysis that is input by the user; and $T_{x \to D \max}$ is the maximum time of travel to the outlet. For example, Figure 5.1b depicts how the tier 1 storm peak mitigation model aggregates storm runoff of parcels within the same time class. In this example, most of

the storm runoff generated by the storm event that reaches flood risk point D arises from parcels in time class 3. In effect, our tier 1 model develops a lumped synthetic storm runoff hydrograph to identify the sources of storm peaks in the watershed.

The synthetic hydrograph allows the model to select parcels that contribute to mitigating the storm peak. The parcels, x', most likely to contribute to mitigating the storm peak are those that lie between the peak and the demand point D.

The parcels x' are found by determining the peak $Q_{i^*D}^*$ of the synthetic hydrograph. Q_{iD}, defined as aggregated runoff at each isochrones i, is defined as

$$x' \in \tau_{i'D} \quad i' \le i^*$$

$$Q_{i^*D}^* \ge Q_{iD} \quad \forall i, \tag{5.5}$$

$$Q_{iD} = \sum_{x \in \tau_{iD}} Q_{sxj}$$

where, i^* is the isochrone number that corresponds to the peak of the hydrograph and $Q_{i^*D}^*$ contains the highest volume of runoff among all isochrones. Q_{iD} is the parcel runoff summed for each isochrone i. x' are all the parcels that lie on the flow path between the point of demand D and potential parcels that are likely in synchronicity with the storm peak.

5.2.2 Modeling the landscape benefits of storm peak mitigation

The service we want to represent on each parcel x is the reduction in storm flow volume provided by vegetation. The tier 1 model does not account for infiltration or storage of storm flow reaching a given parcel x from upslope. This means we give a conservative estimate of the minimum amount of storm peak flow mitigation provided by each parcel. The model calculates mitigated runoff by each parcel on the landscape and the contribution of each parcel to the storm peak at the point of interest.

The amount of storm peak mitigation associated with extant vegetation, here called direct mitigation, DM_{sx}, is determined by the runoff depth at each parcel x retained by vegetation on that parcel,

$$DM_{sx} = \begin{cases} P_s - Q_{sx} & \text{when } x = x' \\ 0 & \text{otherwise} \end{cases}, \tag{5.6}$$

where P_s is the storm depth and Q_{sx} is the direct storm runoff at parcel x for storm s, as defined in Eq. (5.1). It is important to note that as a storm event, P_s, gets higher such as in 50-year return period storms or greater, the amount of potential storm mitigation at x, DM_{sx}, tends to become smaller than the storm depth. In other words, large storms saturate the soil and make infiltration and storm peak mitigation negligible relative to the total storm depth. This is consistent with earlier assertions and observations that LULC properties to mitigate storm peak is reduced for larger storms.

We estimate the marginal storm runoff mitigation provided by land cover, Δ_{xsD}, as a function of the marginal change in runoff with respect to total runoff for peak-contributing parcels,

$$\Delta_{xjsD} = \begin{cases} \dfrac{DM_{xjs} - DM'_{xs}}{\sum_{\forall x \in X}(DM_{xjs} - DM'_{xs})} \; ; & x = x' \\ 0 \; ; & \text{otherwise} \end{cases} , \quad (5.7)$$

where Δ_{xjsD} is the parcel x's contribution in the overall storm peak attenuation at the point of interest D under storm s, and DM_{xjs} and DM'_{xs} are the mitigation by parcel x for the storm s with the current LULC j under the current scenario and a bare soil scenario, respectively. One can use Eq. (5.7) to map relative scores of storm peak mitigation as a means of identifying those watersheds with the highest priority for management. However, this method only counts flood mitigation by parcels that contribute to the peak flow. We do this because flooding damage occurs before and during the peak flood. Waters arriving after peak flood seldom cause additional damage. Given this timing–damage relationship, we do not want to assign social value to mitigation of flows that do not cause damage to humans. Again, this is a conservative estimate of the service provided.

Example 1: Determining cell runoff, sources of the storm peak, and storm peak mitigation

For illustrative purposes, we model a watershed located in South-West Tanzania that flows into the town of Ifakara. The watershed has a drainage area of 32 km² with a diverse land cover. We applied the tier 1 model for a 30-year return period storm event of 150 mm. The results show areas where storm flow is being generated (Figure 5.2a). One can also identify separate travel time classes, and see which class contributes the most flow to the peak flood (Figure 5.2b). The darker zones contain parcels that contribute to the peak volume while the lighter are arriving either early or later to the watershed outlet, which is the point of interest. The tier 1 model generates output at the parcel level; however, a user can aggregate these outputs at larger scale to respond to specific needs (e.g., values could be aggregated to a map of individually owned parcels). Finally, we show the expected pattern of storm peak mitigation that occurred under this specific storm and this specific LULC scenario (Figure 5.3). These maps can help managers interested in stabilizing or improving natural flood control to identify two important parts of the landscape; (1) areas to protect because of their current high contribution to the reduction of storm peak flows, and (2) areas that currently contribute high flow to the storm peak itself. Managers may focus restoration or other improved management practices in the latter parts of the landscape to help reduce flood risk and damage (see also Box 5.1).

Since antecedent soil moisture is one crucial element that defines how much runoff is generated following a rainfall event, the user may need to modify the condition of expected antecedent soil moisture. As default we use medium antecedent soil moisture CN for given LULC. However, this can be adjusted upward or downward according to what is known about soil moisture conditions when the storms of interest arrive.

5.3   Tier 1 valuation

Storm peaks with longer return periods (i.e., lower probabilities of occurrence) are associated with larger expected areas of flooding. To mitigate the risk of flooding, society can (1) better manage the natural landscape to reduce the volume of water coming out of each parcel and to delay this water as much as possible so to spread the volume over a longer time period, (2) invest in man-made infrastructure such as levees and reservoirs to stop and store flood waters, and (3) manage people's behav-

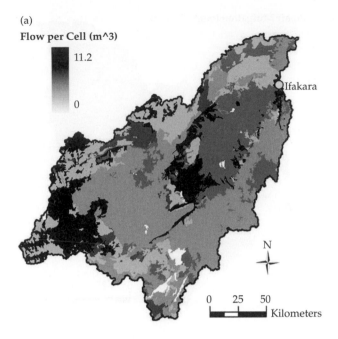

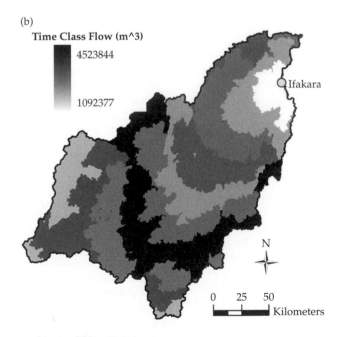

Figure 5.2   Storm volume upstream of the city of Ifakara, Tanzania.
Areas with high CN, bare soil, and urban areas, presented by dark color, generate high volumes after the modeled storm event (a). Time classes of the storm hydrograph (b) show areas of the landscape that yield water that arrives in Ifakara during the storm peak flow (dark zones) or before or after the peak flow (light zones).

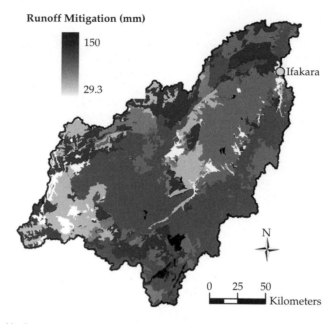

Figure 5.3 Storm peak flow mitigation map.
The land uses in this watershed have the capability to reduce and attenuate the storm volume from 29.3 to 150 mm for a storm event of 150 mm and antecedent moisture content (AMC) II.

Box 5.1 Integrated flood risk management: gaining ecosystem services and increasing revenue

David Harrison

It has become commonplace around the world to plan and construct dams for multiple purposes. If a dam is primarily conceived for hydroelectric generation, for example, it has seemed opportune to include other purposes, most commonly flood control. After all, if a large public works project is being built, it is only logical that it should provide as many benefits as possible. However, this simple proposition has serious downsides and has often led to perverse effects.

Flood control operations of a dam are achieved by lowering the reservoir water level during seasons of higher flood risk to maintain space to receive and hold flood waters for subsequent release at lower flow rates. The goal is to reduce the peak of the flood, or reduce the frequency of a flood of a given peak. Reducing the frequency of a flood may indeed produce calculable economic benefits. Water damage to property and disruption of economic activity is spread over more years and thus the annual cost is reduced. However, small increments of flood control can produce unintended consequences. For example, if the risk of a certain area of floodplain is reduced from a one in 10-year frequency to one in 40, the annual costs may be reduced by one-fourth. However, the exposure to serious human disaster—loss of life—is potentially increased as society relaxes vigilance with a false sense of security and neglects emergency preparedness.

Moreover, flood control imposed on a hydropower reservoir will generally cause substantial reduction of its revenue-generating potential. Full reservoirs produce more energy than reservoirs held partly empty, which is especially important in today's energy situation. Making things worse, the trade-off of hydropower for flood control often occurs at inopportune times. In many monsoonal systems, the flood control season coincides with the high energy demand season. The time that reservoir levels are lowered in anticipation of monsoonal floods is the very time of highest energy demand in the system—the hot summer season with high demand for air conditioning and industrial cooling.

There is an alternative perspective in providing for multiple benefits from hydropower projects. Suppose the flood control reservoir function were shifted out of the reservoirs, thereby allowing increased energy generation.

This is feasible if flood risk management were to be accomplished by management in the flood plain itself. Increased hydropower revenues deriving from fuller reservoirs could be directed to a specific fund to provide revenues for this management.

The management of flood risks in the flood plain can be accomplished through the following measures:

1. Develop and maintain a comprehensive and effective emergency preparedness plan—early warning, orderly evacuation, equipped refuge locations, and orderly system for reoccupation and recuperation of property.
2. Restore flood plain ecosystems that provide not only ecological values but also flood attenuation values.

3. Use funds for regular assessment of the condition of flood plain infrastructure—inspection and maintenance of those levees and detention facilities—to reduce risk of infrastructure failure.
4. Develop new financial instruments for flood risk coverage. Reinvent flood insurance applications to be based on incremental hydropower revenues. Recognize that inevitably water will occupy the land at some times, but that productive uses of the land will otherwise continue.

The fundamental idea is that integrated planning for flood risk management, hydropower production, and ecosystem protection has much to offer over simply imposing multiple purposes on planned infrastructure.

ior in flood prone areas—i.e., flood plain management, through Federal Flood insurance in the USA—limiting development in flood prone areas is often the most feasible and economically viable option. None of these approaches is foolproof, and both have to be designed in reference to the particular severity of storm events. When severe storms occur, flooding is likely to ensue, particularly if the mitigation efforts are focused on less extreme events.

5.3.1 Calculating flood damage as a function of storm peak and LULC pattern

Our models are not designed to predict the flooded area. This means that if one wants to calculate damages they need to obtain flood area maps. In the USA, flood footprint maps can be obtained from Federal Emergency Management Agency (FEMA) or flood insurance companies. In the developing world local knowledge (Tran *et al.* 2009) can help piece together scant information. It may often be necessary to develop flood prints directly using river channel hydraulic software such as HEC-RAS (USACE, 2002; ESRI-HEC 2004) applied to each storm of interest. Once floodmaps have been obtained, using economic valuation techniques, we attach a damage value to each of the mapped flood events. To isolate the value of storm peak mitigation provided by LULC, we must compare the damages associated with events on bare soil with events on extant vegetation.

5.3.2 Determining D_{ks}

Flooding can affect crops, infrastructure, and the production of valuable ecosystem services. Under certain conditions, the magnitude of flood events is impacted by the pattern of land use in a watershed. The value of storm peak mitigation increases with the ability of a LULC scheme to reduce the peak flow after a significant rainfall event in the watershed.

In this model, we determine each parcel's contribution to a flooding event at the watershed's base. Next, we identify each parcel's contribution to the economic damages in the flood area. Finally, we determine the potential savings or additional losses in total flood damage when a parcel changes LULC.

Let V_{xt} indicate the total economic value of parcel x in year t. If possible, V_{xt} should include all market values and monetized ecosystem service values. The total present value of damage due to a flood in area k with storm peak s is given by

$$D_{ks} = \sum_{x \in ks} \sum_{h=t}^{T} \frac{\alpha_{xh} V_{xh}}{(1+r)^{h-t}} + \frac{C_{xT+1}}{(1+r)^{T+1}}, \quad (5.8)$$

where $x \in ks$ indicates the parcels in k that are flooded given storm peak s. T indicates the number of years for which the value of parcel x cannot be fully realized due to storm peak s. α_{xt} is an approximation of x's portion of the value damaged by the flood in each year (i.e., if a flood covers land upon which a

20-story skyscraper is built, the entire value of that building may not be lost as a result of the storm event, although the ability of workers to go to their office might be). C represents the costs that must be incurred in order to return parcel x to its productive capacity (e.g., construction costs to rebuild structures). r represents the annual discount rate. While the linear damage function (area-discharge) might seem arbitrary, Dutta *et al.* (2003) specify depth–damage curves for urban damage estimation that are linear until 3 m of depth above the floor level, the depth at which the damage function plateaus at 60% of total structure value.

As noted in Merz *et al.* (2004), the majority of flood damage estimation methodologies are based on direct tangible damage caused to structures in the flood zone. However, the cost of inundating land may be greater than the structural damage caused by the storm event. The true costs imposed by the flooding are the stream of profit and service values foregone during the period of inundation and drainage as well as any expenses that must be incurred to return the parcel to productive use.

Information on the stream of expected profit and service values at the parcel level may be difficult to gather. In many cases, especially if the flood event in question affects urban or rural-residential areas, the value of real estate can proxy for the discounted stream of all future profits emanating from a parcel (Polasky *et al.* 2008). Note that such an approximation is only reasonable in areas with fully functioning land markets. Because the value of real estate in functioning markets reflects the infinite profit stream associated with optimal use of a parcel, but a parcel will only be impaired for a period of time (t to T), V_{xt} could be set equal to

$$\sum_{h=t}^{T} \frac{RV_{xt}}{(1+r)^{h-t}}, \qquad (5.9)$$

where RV_{xt} is the annual rental value of real estate in parcel x in year t.

The value of storm peak mitigation is captured by the difference in expected damages due to the presence of vegetation in the landscape. In the tier 1

model, the area of the landscape inundated by storm s is not impacted by the LULC pattern. As a result, we are unable to identify how the presence of native vegetation mitigates the damages caused by flooding. We provide an estimate of the true value of this impact as

$$B_{ks} = D_{ks}(1 - \frac{\omega_{fs}}{\omega_{bs}}), \qquad (5.10)$$

where ω_{fs} represents the volume of post-peak water delivered to the damage point during storm s when the landscape is covered by the current LULC pattern and ω_{bs} represents the volume of post-peak water delivered to the damage point when the landscape is covered by bare soil.

The model then aggregates sources of runoff from parcels into isochrones. Finally, we distribute B_{ks} across the parcels that are in isochrones that precede the peak. These groups include all parcels that contribute water to the storm flow that impacts area k. B_{ks} is distributed among these parcels according to each parcel's relative contribution of water volume. Let B_{xks} represent parcel x's share of B_{ks}, where $B_{ks} = \sum_{x=1}^{X_{ks}} B_{xks}$, x indexes the parcels contributing to the storm volume, and X_{ks} is the set of all such parcels. Then B_{xks} can be defined as

$$B_{xks} = B_{ks} * \Delta_{xjsD}, \qquad (5.11)$$

where $D_{xks} = 0$ if x lies within a time class that precedes the storm peak, or the storm return period is greater than 100 years, assuming that at 100 years, LULC has no significant impact on reducing storm peak.

5.3.4 Accounting for the distribution of storm peak return periods

Whether or not a particular return period storm will occur on a landscape in any given year (i.e., the probability that a particular storm peak will occur on the landscape in any given year) is uncertain. The return period indicates on average whether a flood of a particular size will occur at least once during that time period.

Accurate valuation of storm peak mitigation requires calculation of the expected damages avoided due to extant vegetation. To calculate expected damages, the user must determine the damages associated with several different return periods. To determine damages and flooding extent for each it may be necessary to run a flood routing model, such as HEC-RAS, for different storm events because flood hazard maps are often tied to a few specific probability events—often targeting the longer return periods (50–100) not affected considerably by land use.

Let π_{ks} indicate the probability that a storm that will produce storm peak s at k will occur on the landscape in any given year. Then,

$$D_k \equiv E[D_{ks}] = \sum_s \pi_{ks} D_{ks}, \qquad (5.12)$$

where $D_{ks} = 0$ for any s that has a return period of 100 years or more. In order to identify the expected value of storm peak mitigation by vegetative cover on parcel x, we can modify Eq. (5.12) above to develop the expected value of mitigation on cell x with LULC j as

$$B_{xk} = E[\ B_{xks}] = D_k * \Delta_{xjsD}. \qquad (5.13)$$

Example 2: Economic valuation of the landscape for its storm peak mitigation

In our example catchment in Tanzania (Figure 5.4), we applied the tier 1 storm peak mitigation model for the watershed flowing into the city of Ifakara for a 150-mm storm depth and hypothetical US$10 000 000.00 flood damage. Damage values are distributed on the upstream parcels that have travel time within the time classes that are less than or equal to time to peak. The assignment of value is proportional to the amount of storm peak reduction caused by the parcel's vegetation, relative to bare, saturated soil. The value of a parcel is a function of its capability to reduce the excess-rain volume since the travel time is mainly a function of the distance to the outlet, slope, and a small impact of the parcel roughness associated with the land-use type. This map shows values ranging from US$0 to US$1022 per hectare, representing the wide range of flood mitigation provided by the wide range of land-use categories present in this watershed. However, these results do not reflect the full natural system characteristics such as storm depth variability within the watershed, and the storage capacity of different land-use categories.

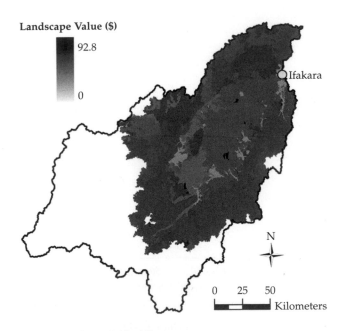

Landscape Value ($)

92.8

0

Ifakara

N

0   25   50
Kilometers

Figure 5.4 Economic valuation of the landscape for avoided flood damages.

5.4   Tier 2 supply and use model

5.4.1   Peak flow model and flood analysis

We are in the process of developing a tier 2 storm peak flow mitigation model, which will be based on assessing incremental changes in risk given changes in land use. This model under development combines several components of existing models in order to bring attention to the role of extant vegetation in regulating storm peak flows. The tier 2 approach will use PRMS (Leavesley *et al.* 1983) or a similar model to evaluate specific characteristics of landscape vegetation in mitigating storm peaks. Currently, the CN approach lumps all known LULC infiltration and storage functions into one number. In contrast, our tier 2 approach is being designed to

independently evaluate canopy interception, rooting effects, soil litter, and many other functions attributed to mitigating runoff. Moreover, the model will incorporate hydraulic routing impacts on mitigating storm peaks, such that the user is informed about the relative impact of vegetation versus landscape geomorphology and scale.

As in tier 1, the tier 2 model will generate maps of a parcel's contribution to each peak flow of interest and the parcel's mitigation of the storm that caused that peak. Specifically, we will assess the incremental changes in risk given a change in landscape or climate using the flood frequency analysis methods by running different LULC scenarios with incremental storm intensities (Figure 5.5). This figure is known as the flood frequency analysis curve. We

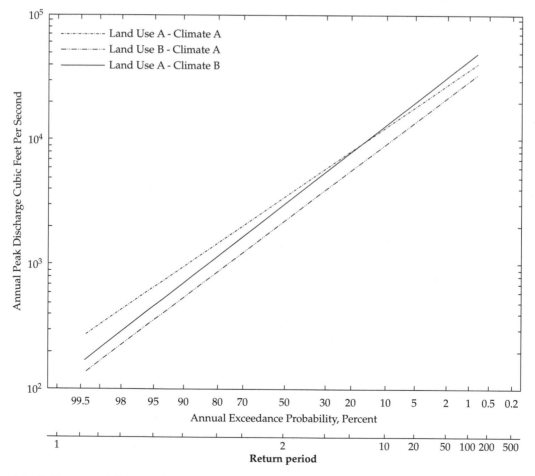

Figure 5.5   Flood frequency analysis for several scenarios of land use and climate.

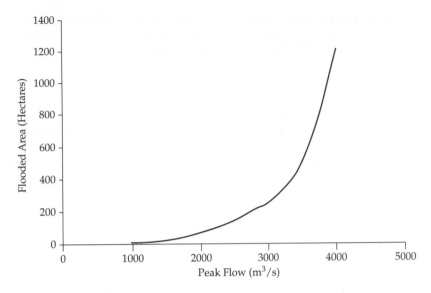

Figure 5.6 Peak flow–flooded area relationship.
The HEC-GeoRAS model runs for different peak flows at steady state modeling or the whole time series at unsteady-dynamic state modeling to draw a relationship between peak flows and flooded area at each point of interest. This relationship will be used to determine the flood damage every time there is a flooding. However, studies show that the channel geomorphology is constantly changing such that the river cross sections will be different between separate flood events. Changes in river cross section will change this relationship; therefore the modeler should include the new cross section or modify and adjust the existing one to match the stream reality.

will develop such curves using the simulated or observed annual peak flows in a statistical method, such as that used in USGS-PeakFQ (Flynn *et al.* 2006) software, to construct frequency distributions for different recurrence intervals. Different land-use scenarios and/or different climate scenarios will be used to generate different annual peak time series that reveal the impacts of land-use changes and climate changes on the peak flow and extent of flooded area.

Using the annual peak time series generated above, one can run flood frequency analyses to determine the peak flow associated with each return period. Then, for each return period, HEC-RAS can be used to generate flooded area profiles at river cross sections of interest. Valuation will require the input of an area-value index map that, when overlaid with the surface area profiles maps, will allow estimation of approximate damages in the flooded area around the stream (Figure 5.6).

Area-value index maps usually focus on damages associated with loss of property and infrastructure. We will provide the additional capacity in our tier 2 model to estimate agricultural crop

damage from drowning or soil saturation. The modular flow estimation component of tier 2 will include a module that tracks the soil moisture in every HRU. The module will contain a component to define the duration of soil moisture greater than the field capacity and create a time series identifying the duration of excess moisture conditions. This information will be coupled with the tier 2 agricultural production model's information on crop growth stage and crop flood to estimate the agricultural crop damage occurring after storm events.

5.5 Tier 2 valuation

The economic valuation for tier 2 will be similar to that in tier 1, except that increased spatial and temporal resolution will allow for the extent of flooding to vary across LULC scenarios. Given this enhanced realism, we replace our proxy for the value of native vegetation on the landscape as described in Eq. (5.12) above, with the true value,

$$B_{xks} = D_{kbs} - D_{kfs}, \qquad (5.14)$$

where D_{kbs} represents the damages when the landscape is covered by bare soil, and D_{kfs} represents the damages with the current LULC pattern.

5.6 Limitations and next steps

The tier 1 storm peak mitigation model uses the CN model to only evaluate landscape parcels for their potential to store water from precipitation and keep it from becoming direct runoff. Runoff from snowmelt, ice, sleet, or rain on frozen ground is not estimated with this model because under these conditions vegetation plays a negligible role in mitigating storm peaks. Our current tier 1 approach also does not value the benefits from vegetation to extend flow routing times (i.e., expand the base of the synthetic hydrograph resulting in lower storm peak). This can be resolved in the next steps. Additionally, our model does not include the in-stream channel and flood plain attenuation due to friction and expansion–contraction energy losses in routing. This means that floodplain management, which is one of the most effective strategies for reducing damages associated with storm peaks, is not accounted for by our tier 1 model (it is addressable with a tier 2 model). When running a tier 1 analysis that does not include the value of floodplains, one is assessing a complementary strategy to flood plain management or restoration—a strategy that places value on the benefits that a forested landscape might have to mitigate storm peaks.

The main limitation of our approach is that we are applying a tool created for lumped watershed analysis at the parcel scale. Curve numbers are dependent on land cover, hydrologic position, soil type, and moisture content, each of which vary considerably spatially. In most models, an area-weighted average CN approach is used to assign a single value for a region or for a small subgrouping of areas considered homogenous (SWAT, HEC-1). Recently, however, work has been completed to assess whether improvements can be attained from using a more distributed CN approach. In comparing the effects of composite versus distributed CNs on estimates of storm runoff depths, Grove et al. (1998) showed that distributed CNs provide closer estimates particularly for wide CN ranges, low CN values, and low precipitation depths. However, the authors noted that for larger design storms (> 50- to 100-year return period) the difference in the runoff computed using composite and distributed CNs is minimal.

There is no explicit provision for the appropriate spatial extent at which to apply the CN approach. By design the method is assumed to apply to small and mid-sized catchments (Ponce and Hawkins 1996). Simanton et al. (1973, 1996) found that CN varied inversely with drainage area and noted a CN decrease of 2.2 units/100 ha of drainage area reflecting the substantial role of transmission losses. White (1986) showed the CN approach effectively predicted stream flow for a large (421 km²) watershed in eastern Pennsylvania. Several studies have also shown that rainfall heterogeneity within larger watersheds is an issue when applying the CN method (Van Mullem et al. 2002). Our model requires one uniform storm depth where in reality storms can hit watersheds with different rainfall depths. Furthermore, with the CN approach rainfall intensity and duration are not considered, only total storm depth. However, Hawkins (1975) noted that for a considerable range of precipitation values, accurate CNs are more important than accurate rainfall estimates.

Given these limitations, one should not use outcomes from these tier 1 models to argue for the replacement of a flood reservoir. However, they can be used to value portions of the landscape that are often neglected and that contribute to flood mitigation. We will continue to develop these models, and are working to incorporate a new approach where the CN is used to compute runoff from variably saturated source areas (i.e., saturation overland flow process of runoff generation). This new approach began with Boughton (1987) and has been followed by Steenhuis et al. (1995), Lyon et al. (2004), and Nachabe (2006). The latter studies have shown that by incorporating surface topographic characteristics into quantifying CNs for an area the accuracy of runoff prediction improves significantly. Further development of this new approach should improve the accuracy of the CN method for prediction of runoff from agricultural, forested, and rangeland areas.

We are also exploring the addition of a method for including infiltration of runon from upslope parcels. This would provide a more full accounting of the role parcels play in drawing down flood peak flows and give a more realistic (rather than conservative) estimate of the storm flow peak mitigation benefits provided by vegetation and soils.

Lastly, it should be noted that although flooding has many adverse impacts, some beneficial impacts do exist. Flood waters inundate floodplains, leaving the soil moisture content and soil fertility high, which can prove beneficial for agriculture, depending on the crop cycle (World Bank 1990). In many parts of the world, floodwaters replenish groundwater aquifers, allowing them to fully recover. These represent additional ecosystem services that will need to be represented with other models. It may be important to consider these positive features of floods when comparing the cost-effectiveness of nature-based as opposed to concrete-based approaches to mitigating flood risk.

References

ASCE. (1996). *Hydrology handbook.* American Society of Civil Engineers, New York.

Arnold, J. G., Srinivasan, R., Muttiah, R. S., *et al.* (1998). Large-area hydrologic modeling and assessment: Part I. Model development. *Journal of the American Water Resources Association,* **34**, 73–89.

Boughton, W. C. (1987). Evaluating partial areas of watershed runoff. *Journal of Irrigation and Drainage Engineering,* **113**, 356–66.

Branson, F. A., Gifford, G. F., Renard, K. G., *et al.* (1981). *Rangeland hydrology.* Society for Range Management, Range Science Series No. 1. Kendall/Hunt, Dubuque, IA.

Branson, F. A., Miller, R. F., and Queen, I. S. (1962). Effects of contour furrowing, grazing intensities, and soils on infiltration rates, soil moisture and vegetation near Fort Peck, Montana. *Journal of Range Management,* **15**, 151–8.

Brooks, K. N., Folliott, P. F., Gregersen, H. M., *et al.* (2003). *Hydrology and the management of watersheds.* 3rd edn. Iowa State Press, Ames.

Bruijnzeel, L. A. (1990). *Hydrology of moist tropical forests and effects of conversion: a state of knowledge review.* Humid Tropics Programme, UNESCO International Hydrological Programme, UNESCO, Paris.

Bruijnzeel, L. A. (2004). Hydrological functions of tropical forests: not seeing the soil for the trees? *Agriculture Ecosystems and Environment,* **104**, 185–228.

Dutta, D., Herath, S., and Musiake, K. (2003). A mathematical model for flood loss estimation. *Journal of Hydrology,* **277**, 24–49.

Ennaanay, D. (2006). *Impacts of land use changes on the hydrologic regime in the Minnesota River Basin.* PhD Thesis, Graduate School, University of Minnesota.

Environmental Systems Research Institute Hydrologic Engineering Center (ESRI-HEC). (2004). *HEC GeoRAS Tools Overview Manual.* US Army Corps of Engineers, Davis, CA.

FAO and CIFOR. (2005). Forests and floods: Downing in fiction or thriving on facts? *RAP Publication* 2005/03. United Nations Food and Agricultural Organization, Regional Office for Asia and the Pacific, Bangkok.

Flynn, K. M., Kirby, W. H., and Hummel, P. R. (2006). *User's manual for program peak FQ, annual flood frequency analysis using bulletin 17B guidelines.* US Geological Survey Techniques and Methods Book 4, Chapter B4.

Grove, M., Harbor, J., and Engel, B. (1998). Composite versus distributed curve numbers: Effects on estimates of storm runoff depths. *Journal of the American Water Resources Association,* **34**, 1015–23.

Hamilton, L. S. and King, P. N. (1983). *Tropical forested watersheds: hydrologic and soils response to major uses or conversions.* Westview Press, Boulder, CO.

Hawkins, R. H. (1975). "The importance of accurate curve numbers in the evaluation of storm runoff." *Water Resources Bulletin,* **11**(5), 887–91.

Hawkins, R. H. (1979). Runoff curve numbers from partial area watersheds. *Journal of the Irrigation and Drainage Division,* **105**, 375–89.

Hawkins, R. H., Ward, T. J., Woodward, D. E., *et al.* (2009). *Curve number hydrology: state of the practice.* American Society of Civil Engineers, Reston, VA.

Kattelmann, R. (1987). Uncertainty in assessing Himalayan water resources. *Mountain Research and Development,* **7**, 279–86.

Kent, K. (1972). Travel time, time of concentration, and lag. In *National Engineering Handbook; Section 4: Hydrology.* US Department of Agriculture, Washington, DC.

Kiersch, B. (2001). *Land use impacts on water resources: a literature review.* Food and Agriculture Organization of the United Nations, Rome.

Kirkby, M. J. and Chorley, R. J. (1967). Throughflow, overland flow and erosion. *Hydrological Sciences Journal,* **12**, 5–21.

Leavesley, G. H., Lichty, R. W., Troutman, B. M., *et al.* (1983). *Precipitation-Runoff Modeling System: User's Manual.* US Geological Survey Water Resource Investigations Report 83-4238.

Loague, K. M., and Freeze, R. A. (1985). A comparison of rainfall runoff modeling techniques on small upland catchments. *Water Resources Research,* **21**, 229–48.

Lyon, S. W., Walter, M. T., Garard-Marchant, P., *et al.* (2004). Using a topographic index to distribute variable source area runoff predicted with the SCS–curve number equation, *Hydrological Processes*, **18**, 2757–71.

Martinez, V., Garcia, A. I., and Ayuga, F. (2002). Distributed routing techniques developed on GIS for generating synthetic hydrographs. *Transactions of the American Society of Agricultural Engineers*, **45**, 1825–34.

Melesse, A. M., Graham, W. D., and Jordan, J. D. (2003). Spatially distributed watershed mapping and modeling: GIS based storm runoff response and hydrograph analysis. Part 2. *Journal of Spatial Hydrology*, **3**, 2–28.

Merz, B., Kreibich, H., Thieken, A., *et al.* (2004). Estimation uncertainty of direct monetary flood damage to buildings. *Natural Hazards and Earth System Sciences*, **4**, 153–63.

Mockus, V. (1972). Estimation of direct runoff from storm rainfall. In *National Engineering Handbook; Section 4: Hydrology*. US Department of Agriculture, Washington, DC.

Nachabe, M. H. (2006). Equivalence between topmodel and the NRCS curve number method in predicting variable runoff source areas. *Journal of the American Water Resources Association*, **42**(1), 225–35.

Polasky, S., Nelson, E., Camm, J., *et al.* (2008). Where to put things? Spatial land management to sustain biodiversity and economic returns. *Biological Conservation*, **141**, 1505–24.

Ponce, V. M., and Hawkins, R. H. (1996). Runoff curve number: Has it reached maturity? *Journal of Hydrologic Engineering*, **1**, 11–18.

Rothacher, J. (1970). Increases in water yield following clear-cut logging in the Pacific Northwest. *Water Resources Research*, **6**, 653–8.

Simanton, J. R., Renard, K. G., and Sutter, N. G. (1973). Procedures for identifying parameters affecting storm runoff volumes in a semiarid environment. USDA-ARS Agricultural Reviews and Manuals ARM-W-1. USDA-ARS, Washington, DC.

Simanton, J. R., Hawkins, R. H., Mohseni-Saravi, M., *et al.* (1996). Runoff curve number variation with drainage area, Walnut Gulch, Arizona. *Transactions of the American Society of Agricultural Engineers*, **39**, 1391–4.

Steenhuis, T. S., Winchell, M., Rossing, J., *et al.* (1995). SCS runoff equation revisited for variable-source runoff areas. *Journal of Irrigation and Drainage Engineering*, **121**, 234–8.

Tran, P., Shaw, R., Chantry, G., and Norton, J. (2009). GIS and local knowledge in disaster management: a case study of flood risk mapping in Viet Nam. *Disasters*, **33**(1), 152–69.

United States Army Corps of Engineers (USACE). (2002). *HEC-RAS River Analysis System*. US Army Corps of Engineers, Davis, CA. Available at http://furat.eng.uci.edu/wsmodeling/Software/USACE_HEC/HEC-RAS/V4Beta/docs/HEC-RAS_Reference_Manual.pdf

United States Department of Agriculture (USDA). (1986). *Urban Hydrology for Small Watersheds*. Soil Conservation Service, Engineering Division, Technical Release 55 (TR-55).

Van Mullem, J. A., Woodward, D. E., Hawkins, R. H., *et al.* (2002). *Runoff Curve Number Method: Beyond the Handbook*. US Geological Survey Advisory Committee on Water Information—Second Federal Interagency Hydrologic Modeling Conference. July 28—August 1, Las Vegas, NV.

White, D. (1986). Synoptic-Scale Assessment of Surface Runoff: An Analysis of the Soil Conservation Service Runoff Curve Number. *Proceedings of the Annual Pittsburgh Conference*, **17**, 159–63.

Williams, J. R. (1995). The EPIC model. In V. P. Singh, Ed., *Computer Models of Watershed Hydrology*, pp. 909–1000. Water Resources Publications, Highlands Ranch, CO.

World Bank. (1990). *Flood Control in Bangladesh: A Plan for Action*. World Bank, Washington, DC.

Wright, K.A., Sendek, K.H., Rice, R.M., and Thomas, R.B. (1990). Logging effects on streamflow: Storm runoff at Casper Creek in Northwestern California. *Water Resources Research*, **26**, 1657–67.

CHAPTER 6

Retention of nutrients and sediment by vegetation

Marc Conte, Driss Ennaanay, Guillermo Mendoza, Michael Todd Walter, Stacie Wolny, David Freyberg, Erik Nelson, and Luis Solorzano

6.1 Introduction

As water flows across the land, its physical and biochemical characteristics are shaped by human activities and the vegetative cover on the landscape. Of particular importance to society is the transport of nutrients and sediment from upstream locations to downstream water bodies. For example, the transport of nitrogen from upstream agricultural lands is partially responsible for the extreme eutrophication of coastal waters that has led to the dead zone in the Gulf of Mexico. From the perspective of sediment pollution, more than 0.5% of global reservoir storage is lost annually due to sedimentation (White 2001) at possible costs of US$13 billion per year (Palmieri *et al.* 2003). The degree to which surface flows remove and deliver nutrients and sediment from their sources to downstream locations is highly dependent on the pattern of land use and land cover (LULC) on the landscape. By retaining portions of the nutrients and sediment released and transported by surface flows, vegetation can help to mitigate damages downstream.

The quantities of nutrients and sediment retained by vegetation are largely functions of slope, vegetation type, volume of water flowing across the land, and the vegetation's location in the watershed. Lands with intact natural vegetation will tend to be net retainers of both nutrients and sediment, whereas lands used intensively for agricultural production will tend to be sources of both nutrients and sediment. The likelihood of retaining nutrients and sediments increases as proximity to source areas increases.

Also, vegetated areas with low slopes tend to provide higher levels of nutrient and sediment retention than steeply-sloped areas. Pulling these ideas together, we present two methods of landscape analysis regarding sediment and nutrient retention. The simple, tier 1 models are based on average annual precipitation levels as well as data related to landscape topography and nutrient loading. The outputs from these models include annual average maps of nutrient export and retention, sediment export and retention, and the value of different parcels of land based on their nutrient and sediment retention.

Because the rate of water flowing across the landscape influences nutrient and sediment retention, and can vary enormously from day to day, annual water volumes may not capture the differential retention properties of assorted parcels of land. In particular, analysis conducted at a fine temporal resolution (e.g., daily) will be able to provide more accurate estimates of nutrient and sediment retention. The tier 2 models capture these temporal dynamics by using the Agriculture Non-Point Source (AnnGNPS) model to estimate nutrient export and retention and the Precipitation Runoff Modeling System (PRMS) to predict the amount of sediment eroded and delivered to focal water bodies. Both models can provide daily estimates. While models in tier 2 provide more realistic estimates of nutrient and sediment retention in the landscape, they require data that are costly to collect and a detailed understanding of local hydrology.

The models here focus solely on the nutrients and sediment that move with surface water flows across the landscape from non-point sources. There are several other damaging non-point source particulates that move with surface water, including heavy metals and pesticides; our modeling approach could be adapted for these other pollutants as long as the reactions of these pollutants with soil and the ambient environment could be described.

6.1.1 Modeling flows relevant to nutrient and sediment retention

The tier 1 and tier 2 models that we develop rely on the principle of saturation excess runoff to generate nutrient and sediment runoff estimates. These models use a runoff index obtained from a topographic index that contains a permeability function. By combining the concept of hydraulic connectivity with pixel-specific runoff indices, these models allow us to estimate the amount of nutrient and sediment retention on individual pixels based on their location in the watershed and their position along the flow path to the downstream body of water.

6.1.2 Implementing theory to model ecosystem-service provision

Many processes drive the impairment of water bodies by nutrient and sediment loading. The magnitude and timing of precipitation events, irrigation scheduling, stream and soil cycling, soil characteristics, and landscape geomorphology and topography are some of the variables that affect the impact of nutrient pollutant loading and sediment transport to a water body. Detailed information about all of the above processes may not be available in all regions struggling to cope with the effects of nutrient and sediment deposition in water bodies. Our tier 1 model of nutrient retention uses as key inputs a water yield index (as a proxy for contributing area, precipitation, soil type, and slope), LULC, export coefficients, and downslope retention. Our tier 1 model of sediment retention uses as key inputs LULC, downslope retention, slope, rainfall erosivity, and soil erodibility. The processes chosen as inputs for our models tend to be associated with more easily-accessible data, making them more practical.

The economic valuation for nutrient retention is based on the cost of water treatment, which is meant to represent the lower bound of the social value of nutrient retention. For sediment retention, the social value is approximated using the dredging costs that are avoided by having sediment-retaining vegetation on the landscape, which should be interpreted as a lower bound of the value of this ecosystem service.

In areas with more-detailed information, planners can apply our tier 2 models that extend the hydrologic principles of tier 1 to include the temporal dynamics of hydrology and nutrient and sediment loading on the landscape. The setup of tier 2 models requires additional data and expertise for calibration, verification, and validation. Our tier 2 approach to nutrient retention has the capability of distinguishing between infiltration and saturation excess runoff transportation pathways. It simulates the timing of nutrient application activities with respect to the generation of runoff.

The tier 2 framework for sediment retention uses methods from PRMS to simulate erosion and sediment processes. PRMS simulates sediment detachment using a revised form of the universal soil loss evaporation (USLE) method (Leavesley *et al.* 1983). The detachment rate of sediment in PRMS is dependent on rainfall intensity, geomorphology, and runoff volumes (Hjelmfelt *et al.* 1975), and the movement of sediment in the stream channel is coupled with the energy in simulated flows. PRMS accounts for both sheetwash and streambank erosion, by incorporating stream scouring and deposition processes, which improves the model's ability to assign value to sediment retention on vegetated parcels.

6.2 Tier 1 biophysical models

The models presented in this chapter use LULC category as a principle driver of the processes being modeled, including water yield, pollutant loading, and sediment retention. The unit of analysis in these models is the parcel, whose size is based on the spatial resolution of the available data in the study area. The model assumes that each parcel is homogeneous with respect to LULC, meaning that there is a single LULC category associated with each parcel. For this reason, the parcel index captures the impact of LULC on each of the processes being modeled, and we do not include LULC indices in the chapter equations unless this assumption is relaxed.

6.2.1   Nutrient retention

There are two key components to nutrient retention: (1) the ability of vegetation and soils to avoid initial nutrient loss (on-parcel retention) and (2) the ability of vegetation and soils to take up nutrients exported to the parcel from upstream parcels (corss-parcel retention). To capture these processes, we first need to know how much nutrient is exported from each parcel on the landscape. We do this using export coefficients derived from previously published field measures of how much nutrient leaves a farmfield under average conditions of nutrient (fertilizer, plant-driven nitrogen fixation) inputs, field management, slope, rainfall, and soil type. We use export coefficients in our tier 1 approach because they are widely available and relatively easily applied around the globe. However, we cannot represent on-parcel retention because export coefficients bundle the effects of vegetation and soils on nutrient retention with the effects of other factors. As such, our tier 1 model only accounts for the role of vegetation and soils in cross-parcel nutrient retention.

6.2.1.1   Modeling non-point source nutrient loads

The nutrient retention model is based on the theory that the damages caused by excess nutrient application on runoff-generating surfaces (i.e., non-point sources) can be mitigated by intercepting vegetated filters (Baker *et al.* 2006; Schneiderman *et al.* 2007).

The first step in the tier 1 model is to characterize the landscape based on nutrient pollutant loadings. Pollutant loading at parcel x with a given LULC category, pol_x, is based on export coefficients directly derived by associating a specific LULC category with a lookup table of corresponding export values. The US Environmental Protection Agency (USEPA) and other environmental agencies in the USA and abroad publish nutrient pollutant export coefficient tables (Gotaas 1956; Reckhow *et al.* 1980; Athayde *et al.* 1983), which can be applied to different regions and replaced by local estimation and local knowledge when possible. Other pollutant proxies, such as manure or fertilizer application, can be used with plant nutrient uptake to estimate the potential export values per LULC if export coefficients are not readily available. Although the goal of this

Table 6.1   Example phosphorus and nitrogen export coefficients (Reckhow *et al.* 1980)

LULC	Nitrogen export coefficient (kg/ha/yr)	Phosphorus export coefficient (kg/ha/yr)
Feedlot or dairy	2900	220
Business	13.8	3
Soybeans	12.5	4.6
Corn	11.1	2
Cotton	10	4.3
Residential	7.5	1.2
Small grain	5.3	1.5
Industrial	4.4	3.8
Idle	3.4	0.1
Pasture	3.1	0.1
Forest	1.8	0.011

model is the management of non-point sources of nutrient pollutants, it is still critical to map point sources using relevant lookup tables to help managers determine the impact of non-point source management with respect to effluent sources. Table 6.1 displays the USEPA export coefficient values for nitrogen and phosphorous for different LULC categories. The table illustrates that intensive management is associated with dramatically more nitrogen and phosphorous loading than native forests; however, loading of these nutrients from forests is non-zero. Both nitrogen and phosphorous are naturally present in aquatic systems.

The export coefficients, developed by Reckhow *et al.* (1980), are annual averages of pollutant loadings derived from field studies that measured export from representative, or average, agricultural parcels in the USA. Since we want to apply this model in other regions, we include a proxy factor that helps offset differences between the fields where the measures were developed and the locations at which the user is applying the model. Factors likely to vary from site to site and influence how much nutrient is exported from a given parcel include slope, soil type, and precipitation. These same factors are key drivers of water yield, so we use a relative surface yield index to correct for non-average conditions in application parcels. We do this by first estimating a runoff index based on the amount of water yield that will accrue to parcel x,

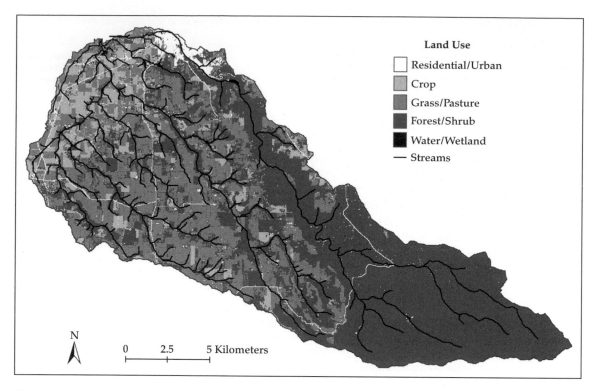

Figure 6.1   The Willamette Valley, Oregon (USA).
This catchment is characterized by upstream forested areas and more intensive agricultural, residential, and urban land uses downstream. Streams run from the southeast to the northwest. Most of the cropland is concentrated in the northwest portion of the catchment, which is an area without many forests or shrublands.

$$\lambda_x = \text{Log}\left(\sum_U Y_u\right), \qquad (6.1)$$

where $\sum_U Y_u$ is the sum of the water yield of parcels along the flow path above parcel x (including the water yield of parcel x). This value can be derived from outputs of the surface water yield model described in Chapter 4, or from any other water yield model. Then, because we want to represent increased nutrient export in areas with above-average factors associated with nutrient export and water yield, and decreased nutrient export in areas with below-average values, we develop the export adjustment factor,

$$EAF_x = \frac{\lambda_x}{\overline{\lambda}_w}, \qquad (6.2)$$

where $\overline{\lambda}_w$ is the mean runoff index in the watershed of interest. Finally, we combine the export adjustment factor with the export coefficient to estimate an adjusted loading value ($ALVx$),

$$ALV_x = EAF_x \cdot pol_x, \qquad (6.3)$$

where pol_x is the export coefficient at parcel x.

To demonstrate the predictive capacity of the methodology described above, we modeled phosphorous retention in a catchment in the Willamette Valley of Oregon (USA) (Figure 6.1). In this example, which we will use to illustrate the complete set of outputs from the nutrient retention model, parcels are 900 m².

The tier 1 model shows that the majority of the sources of phosphorous in this catchment are croplands located in the northwest portion of the catchment (Figure 6.2a). Note that the rivers in this sub-catchment flow from the southeast to the northwest. Figures 6.2a and 6.2b provide maps of ALV at different spatial resolutions.

6.2.1.2   Modeling vegetative filtration and its effects on non-point source pollution

To model cross-parcel nutrient retention on the landscape, we must know the filtering capacity of

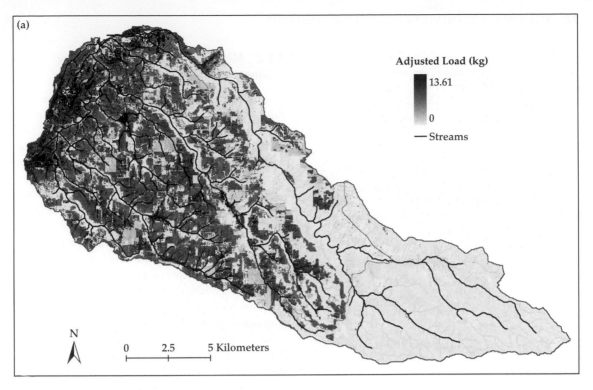

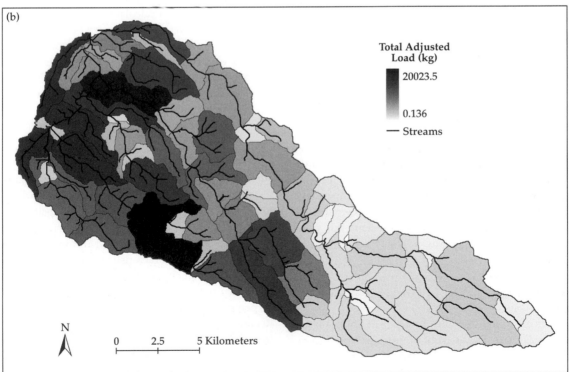

Figure 6.2 Phosphorous adjusted loading values at the parcel (a) and sub-catchment (b) scale in the Willamette Valley, Oregon (USA).
Note that, due to the positions of the sub-catchments, the sub-catchment with the highest total loading in the catchment is not one that borders the focal body of water. This example illustrates the importance of spatial scale in identifying priority areas for the regulation of nutrient application.

different LULC categories as well as the flow path along which nutrient-carrying surface water will flow. We first identify all landscape parcels that include filtering vegetation. Next, we route the ALV_x through the downslope vegetated parcels, which will each retain some of this loading based on their filtration efficiencies and export the remainder of the nutrient load to the next downslope parcel. Let E_x represent the filtering efficiency of parcel x's LULC, where E_x represents the percentage of nutrients reaching parcel x that will be retained on the parcel. This means that $(1 - E_x)$ of the nutrients reaching parcel x will be exported to the next downslope parcel. We assume that the uptake mechanisms acting on the pollutants are never saturated.

To calculate the nutrient retention provided by parcel x, we index all parcels along a given flow path based on their distance to the destination water body, with $x = 1, 2, \ldots, X$. Parcel 1 is the parcel furthest from the water body along the flow path, while parcel X is adjacent to the body of water. In this system, the nutrient retention on parcel x in a given flow path is given by

$$Nuret_x = E_x \cdot \sum_{y=1}^{x-1} ALV_y \prod_{z=y+1}^{x-1} (1 - E_z), \qquad (6.4)$$

where $Nuret_x$ is the nutrient retention in kilograms on parcel x based on its filtration efficiency (E_x), ALV_y is the adjusted loading value of parcels higher up the flow path than parcel x, and E_z is the filtration efficiency of parcels higher up the flow path than parcel x.

We also estimate the amount of nutrients originating on parcel x that reach the downstream water body. Let Exp_x, which represents this portion of the nutrients originating on parcel x, be given by

$$Exp_x = ALV_x \prod_{y=x+1}^{X} (1 - E_y), \qquad (6.5)$$

where all of the variables in Eq. (6.5) are as previously defined. The amount of nutrient export from a parcel will increase as loading on that parcel increases and as the number of downstream filtering parcels, or their filtering efficiencies, decrease. Figure 6.3 illustrates the retention provided by parcels downstream of nutrient loading sources.

6.2.2 Modeling soil erosion to quantify sediment retention by vegetation

For sediment retention, the tier 1 approach starts by calculating the potential sheetwash erosion on the landscape of interest. This model uses the USLE method to incorporate the geomorphological, climatic, and land management characteristics of the landscape (Wischmeier and Smith 1978). Modified forms of the USLE continue to be used, at a minimum, to provide the relative potential of a landscape parcel for sheetwash erosion (excluding gully and streambank erosion) (Reid and Dunne 1996). Hydraulic connectivity is used to account for the location of sediment generation, retention, and transport in the landscape. Outputs of the model include maps of estimated sediment retention and the cumulative amount of sediment exported to downstream bodies of water.

Unlike the nutrient retention model, the sediment retention model allows parcels to retain some of the sediment for which it is a source. In other words, this model accounts for both on-parcel and cross-parcel retention. As such, to calculate the total amount of sediment retention taking place on a parcel, we must calculate both the amount of erosion avoided from parcel x and the sediment reaching parcel x from upslope parcels that is retained by parcel x. First, we let $USLE_x$ represent the amount of sediment originating from parcel x, where $USLE_x$ is defined as

$$USLE_x = R_x \cdot K_x \cdot LS_x \cdot C_x \cdot P_x, \qquad (6.6)$$

where R_x is the rainfall erosivity, which represents the ability of rainfall to move and erode soil and is a function of average regional rainfall intensity and duration; K_x is the soil erodibility, which represents the soil's susceptibility to erosion and is a function of soil texture and characteristics; and LS_x is a slope-length index that characterizes the potential energy associated with the uninterrupted slope leading up to parcel x. Breaks in slope length are based on Renard et al. (1997) and the algorithm for LS_x comes from Stone and Hilborn (2000). C_x is a dimensionless ground cover variable that varies from 1 on bare soil to 0.001 for forest. Finally, P_x is a management factor that accounts for specific erosion control practices such as contour tilling or mounding, or contour ridging. P_x varies from 1 on bare soil

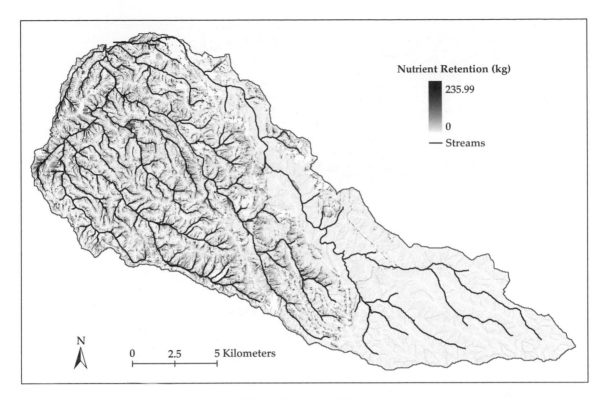

Figure 6.3 Average phosphorous retention values in the Willamette Valley, Oregon (USA).
This map underscores the point that our model formulations make it such that native vegetation must be downstream of loading sources to provide nutrient retention of value to society. As the large area of contiguous forest in the eastern part of the catchment lies upstream of anthropogenic phosphorous sources, it does not provide much, if any, cross-parcel nutrient retention in the eastern part (light zones). However, in buffer areas, forested lands around streams provide much retention. Loadings in the western part of the watershed are retained and removed by forested lands abutting streams (dark zones close to streams).

with no erosion control to about 0.1 with tiered ridging on a gentle slope (Roose 1996).

The tier 1 model of sediment retention determines the potential sediment release on each parcel on the landscape of interest. These soil particles are detached and move through the watershed with runoff along flow paths. As in the nutrient retention model, we index all parcels along a given flow path based on their distance to the destination water body, with $x = 1, 2, \ldots, X$. Parcel 1 is the parcel furthest from the water body along the flow path, while parcel X is adjacent to the body of water. The model routes the sediment originating on parcel x, $USLE_x$, along the flow path, with vegetated parcels retaining some of this sediment based on their sediment retention efficiency and exporting the remaining loading to the next parcel in the flow path. $SEDR_x$, the retention by parcel x's LULC of

sediment originating on parcels higher up the flow path, is given by

$$SEDR_x = SE_x \sum_{y=1}^{x-1} USLE_y \prod_{z=y+1}^{x-1} (1 - SE_z), \qquad (6.7)$$

where SE_x is the sediment retention efficiency of parcel x, $USLE_y$ is the sediment generated on upstream parcel y, and SE_z is the sediment retention efficiency of upstream parcel z.

The potential amount of sheetwash sediment trapped by landscape vegetation or best management practices in soil conservation upstream of reservoir D, $SEDRET_{xD}$, can be estimated by the difference between the geomorphological characteristics of x that might promote soil loss, and the retention properties of the parcel's LULC that help contain sediment released by on-site erosion (C_x and P_x) and upstream transport ($SEDR_x$):

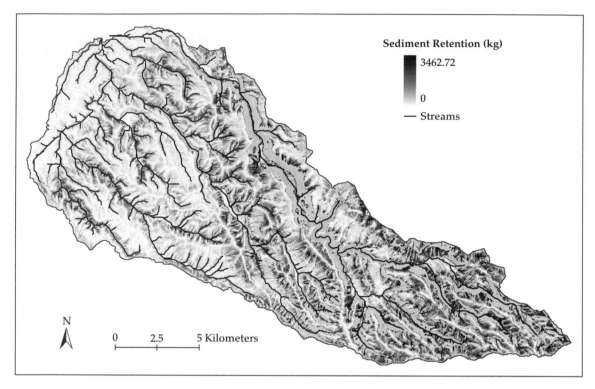

Figure 6.4 The map of annual average sediment retention emphasizes the retention capacity of forested pixels.
Streams throughout the sub-catchment have forested pixels nearby, which provide substantial sediment retention that should serve to extend the productive life of downstream reservoirs. Interestingly, the forested pixels in the southeastern section exhibit significant retention, which illustrates the influence of slope in the sediment retention model.

$$SEDRET_{xD} = R_x \cdot K_x \cdot SL_x \cdot (1 - C_x \cdot P_x) + SEDR_x. \quad (6.8)$$

The first term on the right-hand side of Eq (6.8) is the amount of sediment originating on parcel x retained by that parcel, while the second term represents the amount of sediment originating on upslope parcels retained by parcel x. $SEDRET_{xD}$, can be mapped to display the supply of sheetwash sediment retention by vegetated parcels on the landscape. Figure 6.4 illustrates sediment retention at the parcel level in the Willamette basin sub-catchment where we previously considered nutrient retention.

6.3 Tier 1 economic valuation

6.3.1 Valuation of nutrient retention

It is possible to use the avoided cost of cleanup at water treatment plants as a proxy for the value of

filtration on the landscape in achieving a certain nutrient load. It should be noted that this value will represent the lower bound of the social value of maintaining water below the specified concentration level: if the community is willing to incur these costs, then the value of achieving this threshold must be at least as great as the treatment costs. While we include the avoided treatment cost methodology in the model, the dollar value assigned to achieving a certain standard is an input to the model, so the user is free to choose any of alternative valuation methodology such as contingent valuation or one of the reaveled-preference approaches (Desvousges *et al.* 1987; Englin and Cameron 1996; Leggett and Bockstael 2000; Hanley *et al.* 2003).

Let s_{Dh} represent a standard regarding the concentration of nutrient pollutant h in focal water body D. We assume that s_{Dh} represents the minimum annual loading of nutrient h below which there are no damages associated with the presence of h in water body D. Let L_{Dh} represent the actual total pollutant load of h in D. The total cost of cleaning the water to the desired threshold concentration level, $C(L_{Dh})$, will be a function of the desired threshold, the incoming concentration, and the technology employed in treatment:

$$C(L_{Dh}) = f(L_{Dh}, s_{Dh}, tech_D). \qquad (6.9)$$

Utilities are generally considered natural monopolies because their operations exhibit decreasing average and marginal costs across their range of output. Water treatment utilities are no exception. In this situation, there is no producer surplus, and the welfare impacts of increased environmental quality accrue solely to consumers. Furthermore, marginal cost pricing would lead to financial losses for the utility, requiring subsidies for ongoing operation, meaning that the supply curve is the average cost curve (Foster and Beattie 1981). Assuming perfectly inelastic demand allows us to define the social value of pollutant regulation as the rectangle representing the change in price and the quantity produced (Holmes 1988).

The average treatment cost, $AC(L_{Dh})$, can be defined as

$$AC(L_{Dh}) = \frac{C(L_{Dh})}{Volume} = \frac{f(L_{Dh}, s_{Dh}, tech_h)}{Volume}, \qquad (6.10)$$

where *Volume* represents the total volume of water treated. The estimated average cost of treatment can be obtained through discussion with operators of the appropriate treatment plant. The average cost can then be used to assign value to water of any quality reaching the focal water body given certain assumptions. These assumptions are that there is a constant average cost of treatment and that the treatment technology will remain constant through time. Thus, the proxy for the cost of treatment is

$$\hat{C}_t(L_{Dh}) = AC(L_{Dh}) \times (L_{Dh} - s_{Dh}). \qquad (6.11)$$

It is important to note that in Eq. (6.11), we only assign costs to pollutant loading that exists in excess of the target water quality criterion. The above equation provides the cost of achieving the target criterion in a given year, t. In order to accurately capture the costs of treatment, we need to identify

the present value of the stream of costs incurred over a pre-specified length of time. The present value of a stream of treatment costs is given by

$$\hat{C}(L_{Dh}) = \sum_{t=1}^{T} \hat{C}_t(L_{Dh})$$
$$= \sum_{t=1}^{T} \frac{AC_t(L_{Dh}) \times (L_{Dh} - s_{Dh})}{(1+r)^t}. \qquad (6.12)$$

Note that in the above equation, we allow the average treatment cost, the pollutant load, and the water quality standard criterion to vary across years, though it is possible for them to be constant over time. r represents the market interest rate used to discount future payments into present value.

This model focuses on a single contaminant that impacts the value of water for downstream use. The valuation methodology described above is predicated on the user's ability to identify the cost of treating water for removal of a focal contaminant (nitrogen, sediment, fecal coliform, mercury, etc.). If isolation of individual treatment costs is possible, the above methodology can be used to identify the value of contaminant load reduction for all contaminants present in a given watershed. However, the isolation of water cleanup costs on a per-contaminant basis is often not possible. In these cases, we suggest two alternative methodologies. One approach would be to use the average cost of treating water for all its contained contaminants as the basis for valuing avoided treatment costs for a single contaminant. Note that this approach will necessarily result in double-counting if applied to multiple contaminants. An alternative approach is to rely on the development of a weighting matrix to allocate the total cost of treatment across the vector of contaminants removed during treatment. The allocation of treatment costs across pollutants might be based on existing expert opinion or estimates of such per-pollutant costs in the literature. Applying this approach to estimate contaminant-specific costs of water treatment relies on the assumption that the average treatment cost is constant across pollutants. Each contaminant's contribution to total treatment costs could be estimated by its contribution to turbidity, a measure of the sediment load in the water, which is frequently used to determine treatment costs (Dearmont *et al.* 1998).

We can think of a parcel's value in the context of nutrient loading in two ways. First, we can consider how land that exists in native vegetation is able to restrain the flow of pollutant h into the focal body of water, D. Filtration by vegetation represents an ecosystem service that provides value to society. We can also think of the conversion of land that is currently an anthropogenic source of pollutant h into an alternative land use that would no longer be an anthropogenic source of pollution or of how the filtration currently offered by a parcel covered in extant vegetation would change if the parcel were converted to development. These two approaches come at the issue of a parcel's impact on nutrient loading from different angles, and together they offer the complete picture of how a planner might think of the regulation of nutrient pollution in a watershed.

To determine the treatment costs that might be avoided if parcel x were converted from pollutant-contributing land use j to land use j' not associated with pollutant h, we must identify the portion of loading from each parcel that is costly. We attribute costs to a constant portion of the loading coming from each parcel, based on the difference between actual loading and the desired maximum threshold in the water body. Given this assumption, the cost of water quality impairment induced by source parcel x is

$$c_x = \sum_{t=0}^{T-1} \frac{AC_t(L_{Dh}) \times \dfrac{\sum_x Y_x}{L_{Dh}} \times \dfrac{(L_{Dh} - s_{Dh})}{s_{Dh}} \times Exp_x}{(1+r)^t}, \quad (6.13)$$

where c_x is the contributing cost at water body D of pollutant h from parcel x with LULC j given pollutant loading at the parcel and downstream filtration by natural vegetation, ΣY_x represents the cumulative water yield in the watershed, and all other variables are as previously defined. We include the ratio of water yield to pollutant load to allow our biophysical measure of pollutant loading, measured in kilograms, and our cost of treatment, which is measured in units of currency per volume of water treated, to give us a result measured in currency terms.

We assign value to each parcel based on the treatment costs avoided due to the presence of filtering vegetation. The benefits provided by parcel x for nutrient retention by filtering vegetation, b_x, are given by

$$b_x = \sum_{t=0}^{T-1} \frac{AC_t(L_{Dh}) \times \dfrac{\sum_x Y_x}{L_{Dh}} \times \dfrac{(L_{Dh} - s_{Dh})}{s_{Dh}} \times Nu\_ret_x}{(1+r)^t}, \quad (6.14)$$

where all variables in Eq. (6.14) are as previously defined. Note that a parcel with filtration capacity will be assigned zero value if there is no pollutant to filter—namely, vegetation with filtering properties in a pristine watershed has no value. This is a result of our inability to capture on-parcel filtration, and the fact that vegetation in landscapes with higher nutrient inputs do filter out more nutrients because more nutrients are moving across the landscape and available for filtration.

We applied this valuation approach to phosphorous retention in the Willamette Basin. We assumed that the basin's land use will stay stable for 15 years, and that average treatment costs are US$68 per one million gallons of water. We used a discount rate of 5%. Figure 6.5 illustrates the value of cross-parcel nutrient retention by vegetation based on these values.

6.3.2   Valuation of sediment retention

The more sediment a landscape upstream of a reservoir can retain, the longer the life expectancy of the reservoir, or the less a reservoir manager has to spend on sediment removal. In this model, we approximate the value of this benefit using the avoided cost of sediment removal. It should be noted that not all sediment that reaches a reservoir is removed (e.g., dredged): a certain rate of sedimentation in a reservoir is usually tolerated (Palmieri *et al.* 2003). For those reservoirs whose sedimentation has caused the reservoir to reach its dead volume, which is the point at which reservoir function is impacted by sedimentation, we assume that all sediment originating from upstream parcels will be removed. In all other reservoirs, we assume that none of the sediment reaching the reservoir will be removed. The present value of retained sediment on parcel x, $PVSR_{xD}$, is given by

$$PVSR_{xD} = (SEDRET_x \times R_D \times MC_D) \times \sum_{t=0}^{T} \frac{1}{(1+r)^t} \quad (6.15)$$

where the index D indicates that x is upstream of reservoir D, $SEDRET_x$ is the amount of sediment removed by the LULC type on parcel x annually (see Eq. (6.8)), and T indicates the number of years we expect present landscape conditions to persist or the expected lifetime of reservoir D (set T to the smallest

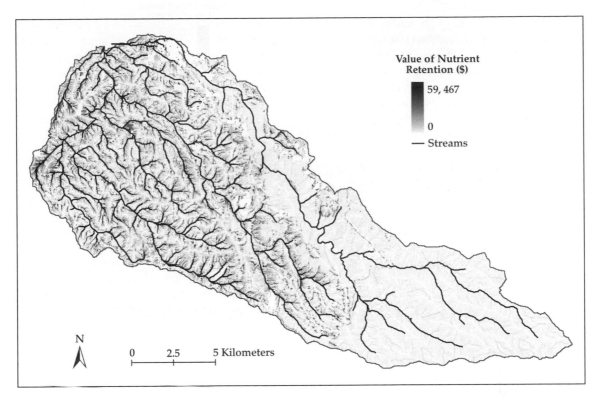

Figure 6.5   Avoided water treatment costs in the Willamette Valley, Oregon (USA).
The most valuable pixels providing nutrient retention in this catchment are those located closest to the sources of pollution rather than those that have the highest filtration capacity (a). When results were aggregated to the sub-catchment level (b), the maximum value is assigned to those that do not include the pixels with the highest value, again emphasizing the importance of spatial scale regarding management decisions.

value if the two time values differ). R_D, the retention factor of reservoir D, describes the fraction of delivered sediment that is retained in reservoir. MC_D is the marginal cost of sediment removal from reservoir D and is based on the appropriate regional technologies available, and the size and purpose of the reservoir. For simplifying purposes, we assume a constant marginal cost of sediment removal.

Sediment removal technologies have widely different costs and include flushing, sluicing density current venting, dredging, dry excavation, and hydrosuction (Palmieri *et al.* 2003). Note that the specification of Eq. (6.15) treats sediment retention and the marginal cost of removal as constant across time and assumes a constant marginal cost of removal.

We applied this approach to a sub-catchment of the Willamette River, Oregon (USA) (Figure 6.6). We

assumed that sediment dredging costs US\$1.37 ton^{-1} (based on costs for removing river sediment in 1983; Moore and McCarl 1987). We assumed a productive lifetime for the reservoir of 100 years and a discount rate of 5%.

6.4   Tier 2 biophysical models

The tier 1 models are created with the goal of providing a credible depiction of nutrient and sediment retention on the landscape, while imposing minimal data requirements on model users. Achieving this goal required the development of novel hydrological models. The tier 2 models aim to provide more realistic depictions of these processes, which is partially achieved by relying on existing hydrologic models that provide finer scale temporal resolution.

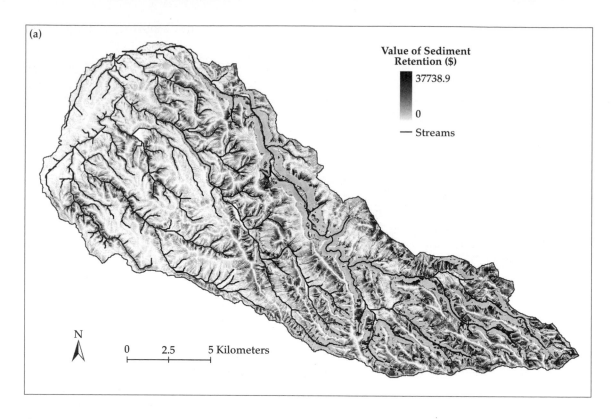

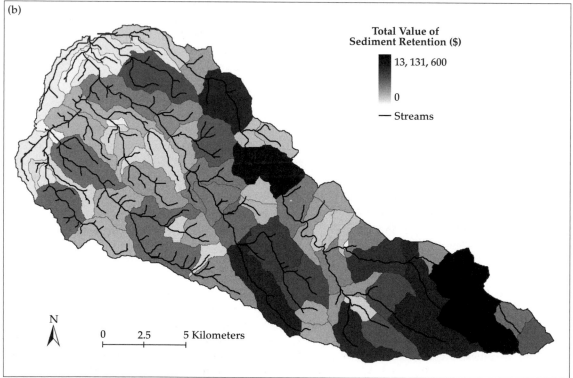

Figure 6.6 Avoided reservoir dredge costs in the Willamette Valley, Oregon (USA).

The highest values, up to $1667 per 30 × 30 m pixel (~$18 500 per ha), occur near the outlets into the reservoirs because sediment has few potential opportunities for re-deposition (a). Aggregation of the value to the sub-catchment level changes the view of the basin somewhat (b).

6.4.1 Tier 2 nutrient retention

In tier 2 we simulate non-point source nutrient run-off using the Annualized Agricultural Non-Point Source (AnnAGNPS) model (Bingner and Theurer 2007) to simulate non-point source loadings. AnnAGNPS is a joint USDA–Agricultural Research Service (ARS) and Natural Resource Conservation Service (NRCS) system of computer models developed to predict non-point source pollutant loadings within agricultural watersheds. The AnnAGNPS model consists of several different programs for realistically capturing pollutant loading, the flow of pollution across the landscape, and the in-stream impacts of this pollution (Bingner and Theurer 2007). The model contains a continuous simulation surface runoff model designed to assist in identifying best management practices, setting Total Maximum Daily Loads (TMDLs), and incorporating risk and cost–benefit analyses (NRCS 2008). We structure the tier 2 nutrient retention model with the needs of watershed managers in mind. Linking AnnAGNPS to a process of economic valuation allows landscape managers to incorporate ecosystem services into their management practices.

There are several fundamental differences between the tier 1 and tier 2 nutrient retention models. First, the tier 2 model allows the user to incorporate management decisions made on sub-annual timeframes. Second, the Tier 2 model recognizes that nutrient pollution export is highly dependent on dynamic hydrology. Finally, the tier 2 model incorporates nutrient-specific characteristics when tracking the transport of the pollutant across the landscape. The challenge of using the more sophisticated tier 2 model is that model setup and interpretation require the user to possess relatively sophisticated knowledge of hydrology. Useful simulations will, at a minimum, require extensive calibration, thoughtful definition of modeling units, and an understanding of the driving processes of the landscape in question.

The tier 2 nutrient retention model provides estimates of non-point source pollutant loading at fine temporal scales that are linked to hydrology via climate and land use. Runoff hydrology in the AnnAGNPS model is based on a modified SCS-Curve Number (CN) approach and is taken to represent variable source area hydrology in wet (saturation excess) landscapes (Steenhuis *et al.* 1995; Schneiderman *et al.* 2007; Easton *et al.* 2008b) and Hortonian flows (Horton 1940) in dry (arid rainfall excess) landscapes. In order to apply this approach one must define Hydrologic Response Units (HRUs) that incorporate, at a minimum, parcel hydraulic and topographic index characteristics but that might also consider management boundaries, such as tenure or zoning units (Fig. 6.7).

In most water quality models (e.g., Soil Water Assessment Tool (SWAT)) HRUs are defined by the coincidence of land use and soil infiltration capacity. In landscapes where variable source areas control runoff generation, it is more appropriate to define HRUs by land use and a soil wetness index (e.g. SWAT-VSA; Easton *et al.* 2008a). Wetness index classes are determined by dividing a watershed into several (at least ten) equal areas delineated by lines of equal topographic index. Any additional information needed to estimate pollutant loading other than LULC, such as topography (e.g., slope position and length) and soil chemical properties, are averaged within each HRU. This is generally acceptable because there is evidence that soil variability roughly correlates with topographic features, which are captured by the topographic index (Page *et al.* 2005; Sharma *et al.* 2006; Thompson *et al.* 2006).

The tier 2 nutrient retention model can characterize each loading and retaining parcel. Furthermore, the model can evaluate to what extent parcels impact overall loads and how incremental changes at each of these parcels affect nutrient loading at the demand point. In tier 2, we keep track of nutrient application, pollution runoff, and vegetative filtering on a daily time step. This arrangement allows the user to evaluate not only changes in magnitude but also to see changes in nutrient export and retention as a function of seasonal hydrology, vegetation growth stages, and seasonal management changes.

Tier 2 outputs are aggregated within the modeling HRUs at any desired temporal period with less than daily frequency. There are two key tier 2 nutrient retention model outputs. First, there is the expected non-point source contribution of nutrient pollutant h from HRU_x to D during time period I, $NPSP_{DhI}(HRUx)$. Next, there is the expected vegetative filtration from HRU_x of nutrient pollutant h upstream of D during time period I, $FILT_{DhI}(HRU_x)$.

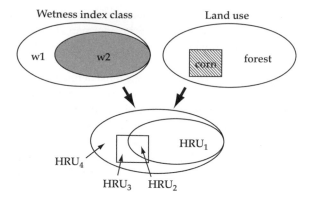

Figure 6.7 Schematic showing how digital maps of wetness index and land use are combined to delineate HRUs.
In this case, two wetness index classes, w1 and w2, are combined with two land uses, corn and forest, to produce a watershed with four HRUs. The HRU is an homogeneous land segment with same wetness index and land use.

Both of these outputs are modeled using the routines and functions of AnnAGNPS.

The largest period, I, is one year and the model is run using hydrology forcings (rainfall events) of several years, such that the natural variability in precipitation can be accounted for as best as possible. Pollutants are aggregated in the form of average daily load averages (mass) within each period that can be combined with water yield simulations per HRU to produce effective concentrations. The biophysical service processes for pollutant removal for each modeling HRU_x is simply

$$FILTSERV_{DhI}(HRU_x) = FILT_{DhI}(HRU_x.$$ (6.16)

The relative impairment of contributing pollutant is thus expressed as follows for each modeling HRU_x:

$$NPSPIMP_{DhI}(HRU_x) = \frac{NPSP_{DhI}(HRU_x)}{\sum_{A_{xj} \in w} NPSP_{DhI}(HRU_x)}.$$ (6.17)

6.4.3 Tier 2 sediment retention

The tier 2 model simulates sediment detachment and transport for storm events using the modules in the PRMS developed by the US Geological Survey (Leavesley *et al.* 1983). In order to identify retention by LULC j on HRU x, we must run PRMS twice, first with bare soil on each HRU and then again with current LULC on each HRU, with the difference representing retention on each HRU attributable to its LULC.

As noted earlier, the detachment rate of sediment is dependent on rainfall intensity, geomorphology, and runoff volumes (see Hjelmfelt *et al.* 1975), and its movement in the stream channel is coupled with the energy in simulated flows. We define the stock of sediment delivered to reservoir or impoundment D at time t, z_{Dt}, as

$$z_{Dt} = z\left(\sum_V \sum_{T'} h_{xt'}, sdr(x, D)\right) + \chi_D.$$ (6.18)

where z_{Dt} is a function of $h_{xT'}$, the cumulative sediment detached from V, the set of upstream HRUs indexed by $v = 1, 2, \ldots, V$ across all times before t, given by T', where T' is indexed by $t' = 1, 2, \ldots, t-1$ (i.e., $t' < t$). The likelihood of this sediment reaching D is mitigated by a sediment delivery ratio, sdr, which is a function of the spatial configuration between x and D. Finally, χ_D accounts for all the sediment that arises from other sources or processes not modeled, such as landslides or stream bank erosion, which can be assumed to be a constant value based on site- or region-specific sediment budget studies.

6.5 Tier 2 economic valuation models

The tier 2 economic valuation models do not differ fundamentally from the tier 1 models. The value of nutrient and sediment retention is still estimated using the costs avoided by vegetative retention of nutrients and sediment on the landscape. The main advance in the tier 2 models is that management decisions can be incorporated into the valuation of the focal ecosystem services.

6.5.1 Tier 2 valuation of nutrient retention

Simple maps of valuation are informative but are by no means an end unto themselves. For many land-use or resource management decisions, the real question is what mixes of land use and activities achieve some goals at minimum costs. For example, there are two ways to achieve reduced nutrient loading in a focal body of water, namely reduced nutrient pollutant loading or increased vegetative filtration in the landscape. The question then is: how might the targeted level of nutrient loading be achieved in the least-cost manner? The tier 2 model provides users with several frameworks to answer this question.

Let us assume a specific load threshold, s_{Dh}, and a single entity, a benevolent social planner, that has the opportunity to make land-use decisions across the landscape. There are several different objectives such a land-use planner might pursue regarding development and nutrient loading. One possibility is that the social planner will attempt to maximize the development value of the landscape (i.e., the sum of the development value across developed parcels), while accounting for the social costs of nutrient loading associated with different LULC categories. Typically, we think of minimizing the cost of regulation, but in the case of pollution from land use, the costs of regulation are opportunity costs rather than costs of technology adoption. Under certain assumed conditions, this objective is represented in the expression

$$\max_{I_x} \sum_{x=1}^{X} (d_{xj^*} - c_{xj^*})I_x,$$
(6.19)

where c_{xj^*} is as defined in Eq. (6.13), d_{xj^*} represents the net present value of parcel x in LULC j^*, where j^* represents the highest value LULC on parcel x and is taken as exogenous, and I_x is a binary random variable, where 0 indicates that the land is not in use (i.e., not acting as an anthropogenic source of pollutant h) and 1 indicates that the parcel is in LULC j^*. In landscapes with functioning land markets, d_{xj^*} represents the net present value of future

revenue streams per hectare of LULC j^*, meaning that

$$d_{xj^*} = \sum_{t=1}^{T} \frac{V_{xj^*t}}{(1+r)^{t-1}},$$
(6.20)

where V_{xj^*t} represents the private per-hectare value accruing on parcel x in year t from LULC j^*. Solution of this problem leads to the outcome that land will only be placed in nutrient-loading LULC j^* so long as

$$\frac{d_{xj^*}}{Exp_{xj^*}} \geq \sum_{t=1}^{T} \frac{1}{(1+r)^t} AC_t(L_{Dh}) \frac{(L_{Dh} - s_{Dh})}{s_{Dh}}.$$
(6.21)

Equation (6.21) shows that the net benefit of LULC j^* per unit of nutrient exported must be greater than or equal to the net cost of nutrient loading in water body D for the planner to allow nutrient-loading LULC j^* on parcel x.

In areas where estimates of the cost of water treatment are not available, we might restate the objective function presented in Eq. (6.19) as

$$\max_{I_x} \sum_{x=1}^{X} d_{xj^*} I_x + \lambda(s_{Dh} - \sum_{x=1}^{X} Exp_x I_x),$$
(6.22)

where d_{xj^*}, Exp_x, s_{Dh}, and I_x are as defined above, and λ represents the additional development value that might be realized by increasing the amount of loading in water body D. This specification allows users to evaluate the trade-offs between intensive land use and pollutant loading in water body D without necessitating the identification of the damages caused by such loading.

In Eqs. (6.19) and (6.22), the unit of management is assumed to be the parcel. In reality, decision units may have different spatial resolution, with boundaries based on property lines or sub-catchments. In this case, there may be several LULC categories on a single decision unit. To accommodate this possibility, we can rewrite Eq. (6.22) as

$$\max_{A_{zj}} \sum_{z=1}^{Z} \sum_{j=1}^{J} d_{zj} A_{zj} +$$
$$\lambda(s_{Dh} - \sum_{z=1}^{Z} \sum_{j=1}^{J} Exp_{zj} A_{zj}),$$
(6.23)

where A_{zj} represents the area of decision unit z that is in LULC j, and all other variables are as previously defined. In order to obtain the efficient pattern of land use on a decision unit, the planner must have precise information about how revenues associated with LULC j vary with the area of land engaged in the given land use. Note that in addition to identifying the least-cost means of achieving a targeted threshold loading, the frameworks presented above can also be used to target cost-effective restoration sites, which would entail identifying those parcels whose value of nutrient retention is greatest per lost value of being converted from production to native vegetation.

6.5.2   Tier 2 valuation of sediment retention

Let g_{dt} represent a mass of sediment stock in impoundment system d at time t. Next, let y_{dt} represent the mass of sediment removed from d in time period t, while z_{dt} represents the rate at which sediment is delivered to d in period t via landscape runoff. The calculation of z_{dt} is a function of the LULC pattern and rainfall during period t on the landscape. Therefore,

$$g_{dt+1} = g_{dt} - y_{dt} + z_{dt} \qquad (6.24)$$

gives the stock of sediment in d at time $t+1$.

When deciding how much sediment to remove from the reservoir in time period t, a manager is interested in minimizing the sum of all future costs of sediment removal and the dis-benefits associated with any future remaining sediment stocks. The decision-rule on how much sediment to remove from d in time period t (i.e., what level to set y_{dt} for each year t) is given by

$$\min_{y_{dt}} \sum_{t=1}^{T} \frac{-B_d(g_{dt}) - c_{yt}y_{dt}}{(1+r)^t}, \qquad (6.25)$$

subject to $g_{dt+1} = g_{dt} - y_{dt} + z_{dt}$ where $B_d(g_{dt})$ represents the dis-benefit associated with a sediment stock of g_{dt}, c_{yt} is the per unit cost of sediment removal in time period t and is a function of the volume of the reservoir and regional technical capabilities (see Palmieri *et al.* (2003) for costs and appropriate technologies for sediment removal), and r is the discount rate. $B_d(g_{dt})$ is estimated by applying the tier 2 water allocation model and determining the difference in benefits associated with reservoirs with lower storage capacity. However, to do this requires field data, such as bathymetry observations, to establish general relationships between reservoir

storage loss and influx of sediment stock. See the supplementary online materials (SOM) for technical details on finding the path of y_{dt} for $t = 1$ to T that minimizes the negative impact of sediment loading on the reservoir. Let this path by given by y^{*}_{dt} and let the net present value of sedimentation at system d until time T be given by

$$NPVS_d = \sum_{t=0}^{T-1} \frac{-B_d(y^{*}_{dt}) - c_{yt}y^{*}_{dt}}{(1+\gamma)^t}, \qquad (6.26)$$

where T indicates the number of years we expect present landscape conditions to persist or the expected lifetime of reservoir d (set T to the smallest value if the two time values differ). $NPVS_d \leq 0$ for all solutions. By comparing $NPVS_d$ scores from a landscape covered by bare soil with those from a landscape with a different LULC pattern, positive ecosystem service values will be generated. See Box 6.1 for a discussion of efforts in China to provide incentives for land management associated with increased nutrient and sediment retention.

6.6   Constraints and limitations

One major simplification made in these models is that we ignore in-stream processes in quantifying the amounts of nutrient and sediment delivery, retention, and value. In doing so, we attach artificially high significance to the contribution of upstream parcels regarding total nutrient and sediment loads in the focal downstream body of water. The nutrient retention model is based on surface and subsurface flows in saturation excess regions. In watersheds where the interaction between surface water and ground water is significant (i.e., shallow aquifers), nutrient modeling using tier 1 will likely misrepresent export and retention, again overestimating these values in most cases. Application to such systems should be undertaken cautiously, and other factors such as travel time between the source parcel and point of interest should be analyzed.

Not surprisingly, the scales of the watershed and the river basin being modeled play a significant role in determining the extent to which LULC changes impact water quality. In large basins, landscape complexity can become important in mitigating the impacts of land use on the hydrologic regime and water quality by providing storage capacity in the watershed and in stream beds. For the same percentage of LULC change, impacts will be much greater in smaller watersheds.

In addition to the scale impact on overall water quality processes, the tier 1 nutrient retention model assumes that saturation excess hydrology and proximity to water bodies dominate water impairment. Also, this model measures all factors at an annual time step including rainfall-runoff dynamics and processes, fertilizers or pollutant applications, and plant growth dynamics. In the real world, water quality components and processes depend upon a variety and complexity of sub-annual processes not simulated in this tier 1 model.

The tier 2 nutrient retention model does not simulate in-stream processes and/or any biochemical transformations. However, temporal dynamics are more fully incorporated as all calculations are done at the HRU level on daily time step.

The tier 1 and tier 2 sediment retention models are both based on the USLE, which is best suited for agricultural land and moderate slopes. Furthermore, these models only incorporate sediment from sheet-wash sources, thus conceptually reflecting a lower limit of the true relationship of total sediment yield and landscape practices. The tier 2 model, PRMS, is based on USLE and revised to include the temporal dynamics of precipitation and their impacts on erosivity.

6.7 Testing tier 1 models

One risk associated with simple models that have limited data requirements is the possibility that they will misrepresent the processes being modeled. To ensure that our models provide reasonable outputs, we have started testing them in watersheds around the world. Here we report the results of these efforts for the nutrient retention model.

6.7.1 Sensitivity

The application of a pollutant on the landscape is a direct and linear function—uncertainty in these values will be directly proportional to errors in pollutant loading modeling. However, stream threshold (the aggregate number of analysis drainage parcels that define the formation of an ephemeral stream) and the efficiency of pollutant removal per unit of downhill vegetation are uncertain variables that can affect outputs in nonlinear ways.

To understand the impact of uncertainty in these two variables, we performed a preliminary sensitivity analysis of these two variables in the Baoxing County watershed within the Upper Yangtze River Basin in China. The selected catchment exhibits highly varied soil, LULC, and management types in a humid monsoon landscape. It is of interest because it suffers from a significant nutrient loading problem due to livestock and intensive agricultural production. Furthermore, tens of millions of people downstream of this watershed rely on this water for consumptive uses, so ensuring adequate water quality is essential.

Prior to evaluating the sensitivity of modeled loading with regard to various parameters, we compared modeled water yield totals (used as one of the input parameters) with annual average observed water yield from a ten-year time series and, following adjustments for groundwater sources outside of the watershed, found that modeled water yield represented 92% of the observed annual average yield. We performed the sensitivity analysis for both nitrogen and phosphorous pollutant loadings. We find that, for both nitrogen and phosphorous, the model is more sensitive to changes in export coefficients than retention efficiency.

6.7.2 Validation testing

The theoretical foundations of our contaminant model have been successfully applied in previous studies dealing with saturation excess runoff watersheds (Endreny 2002; Walter et al. 2003; Schneiderman et al. 2007). Nevertheless, it is important to confirm that our model predictions align with observed outcomes in as many different systems as possible. However, testing our predicted outcomes relative to observed nutrient loading and sediment deposition, particularly at the sub-catchment level, is quite difficult due to the lack of such spatially explicit information in most watersheds. As a proxy for observed data, we compare our predicted outcomes with those of a well-respected model that has been calibrated to a specific region.

Here we present a comparison of our tier 1 predictions to the predictions of the much more data-intensive SWAT model (Arnold and Fohrer 2005). In particular, we compare the total phosphorus load scores from our tier 1 nutrient retention model with those of simulated soluble phosphorus by the SWAT model in 111 sub-catchments in the Willamette River Basin as displayed in Figure 6.8 (see Plate 2). Before describing the details of the comparison, we offer a brief description of the method applied by SWAT to describe nutrient loading and retention.

The nitrogen and phosphorus cycles are dynamic systems influenced by atmospheric, hydrologic, plant, and soil conditions. These two nutrients are modeled in SWAT separately in very specific methods and algorithms. They are added to the soil by fertilizers, manures or residue applications, fixation (N), and rain (N). They are removed from the soil by plant uptake, leaching, volatilization, de-nitrification, and erosion. SWAT monitors different pools of nutrients in the soil, generally categorized in mineral and organic forms. It calculates nitrogen leaching as it calculates surface runoff and lateral flow when it uses an exponential decay weighting function in groundwater response. Organic N

attached to soil particles may be transported by surface runoff. Unlike nitrogen, phosphorus solubility is very limited. Soluble P is leached only from the top 10 mm of the soil. Surface runoff is the primary mechanism by which phosphorus is removed from the catchment; organic and mineral P attached to soil particles may be carried off land by erosion processes. The SWAT model is satisfactorily calibrated to observed water yield at several different sub-catchments within the Willamette River Basin, with an R^2 value of 0.17 at the daily unit of observation. Given its correlation with observed data, which is not available at the desired spatial resolution, we use the SWAT output as a proxy for observed conditions.

SWAT and our model differ significantly with regard to methodology. SWAT models specific processes individually while making restrictive assumptions about uniform land use within HRUs summed to the sub-catchment scale. Our model allows for fine LULC detail across the landscape with a coarser model of the hydrology (annual average water yield) behind the focal processes. That said, we might expect our modeled outputs to correlate closely with observed conditions, as proxied for by calibrated SWAT outputs (Hernandez *et al.* 2008),

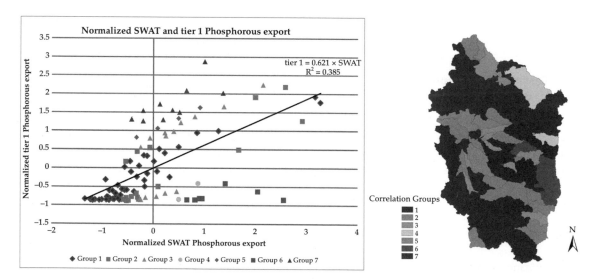

Figure 6.8 Aggregated sub-catchment phosphorous export comparison between our model and SWAT (graph) and agreement of spatial phosphorous export patterns predicted by the two models (map) in the Williamette Valley, Oregon (USA). The graph on the left depicts a correlation between the normalized tier 1 model outputs and the normalized SWAT outputs. The groupings in the graph depict sub-catchments whose tier 1 outputs lie within a threshold distance of SWAT outputs given the correlation between normalized tier 1 and SWAT outputs. Note that Groups 4 and 6 represent sub-catchments in which tier 1 outputs are unexpectedly low and Groups 5 and 7 represent sub-catchments in which tier 1 outputs are unexpectedly high. The map illustrates the sub-catchment groups identified in the graph. (See Plate 2.)

Box 6.1 China forestry programs take aim at more than floods

Christine Tam

In the wake of the devastating Yangtze River floods of 1998 and unprecedented extent of the drying of the Yellow River in 1997, the Government of China (GOC) initiated a number of forestry programs aimed at water and soil conservation to mitigate these impacts in the future. Since these initial forestry programs began, provincial- and national-level government investment has continued and even expanded to protect watershed services more widely. Often with little quantification of actual service delivery, these programs experienced relatively rapid approval and initiation and generally sustain widespread support and compliance, in part because of a strong belief in the overall benefits of forests and in part for their ability to address a variety of associated objectives.

For the past two decades, China has experienced a consistent level of flood damage despite heavy investment in structural solutions to flood control. These damages culminated in 1998 with the Yangtze floods. A follow-up UNEP study (1999) attributed the effects of upper watershed deforestation and overgrazing in reducing water storage capacity as a key factor that aggravated the impacts of prolonged and substantial rainfall. Not only were the degraded soils less able to retain water, but eroded soils washed downstream filling river channels, lakes, and reservoirs, reducing their ability to mitigate flood waters (Zong and Chen 2000).

A direct response to this disaster was consideration and development of a more comprehensive system of flood management that addressed both structural and non-structural components. Among the non-structural components that included flood forecasting, resettlement from high flood risk areas, and flood detention basin compensation, the Chinese government also enthusiastically initiated the Natural Forest Protection Program (NFPP) and Sloping Lands Conversion Program (SLCP) to address upper watershed impacts. These programs, first piloted then approved for the period 2000 to 2010, banned logging of natural forests on 30 million ha and supported reforestation of 15 million ha of sloping cropland, respectively, along the upper Yangtze and upper and middle Yellow rivers. The original investment of the SLCP reached nearly US$40 billion through annual grain and cash payments to farmers for retiring and reforesting cropland, while the GOC has allocated over $11 billion for the NFPP over the same timeframe, numbering these

among the world's largest payments for ecological services schemes (Liu *et al.* 2008).

These two forestry programs were initiated with substantial funding from the central government for soil and water retention benefits but rapidly gained further political favor for addressing an additional broader suite of government priorities and objectives. Besides reducing soil erosion, the SLCP also aims to alleviate poverty, especially in remote, mountainous areas with low productivity lands, reducing urban–rural development imbalances. In addition, the SLCP promotes local economic development (Xu and Cao 2002), facilitates shifts from farming to more sustainable production structures (SFA 2003), and bolstered a then-troubled State Grain Bureau (Bennett 2008). Subscription to the program was highly successful given the relatively high rates of payments such that local and regional implementation almost always met or exceeded targets.

The NFPP additionally supported restructuring of rural economies, retraining and relocating over 700 000 former forestry workers (ADB 2006). Economies once solely dependent on timber production were now more reliant on multiple industries, including forest management, plantation farming, and tourism (Liu *et al.* 2008). Compliance with the NFPP has been relatively successful, with little illegal commercial logging. Villagers, even for subsistence use, must apply to their local forestry bureaus for fuelwood or domestic construction timber allocations. State-run logging companies were eliminated virtually overnight, and new economies, such as that of Jiuzhaigou in Sichuan or Zhongdian, renamed "Shangri-La," in Yunnan have successfully embraced tourism as their alternative income source.

The success of these two programs in compliance and participation coupled with belief in their ecological benefits has led the GOC to expand its forestry programs to conserve further watershed services. More recently, the Forest Ecological Benefit Compensation Fund (FEBCF) was approved. First conceived of in the late 1980s, the FEBCF achieved legal backing in April 1998, initiated pilot work in 2001, and was expanded across China in 2004, providing payments of 5 rmb (7 rmb = US$) per mu to maintain natural forest land (Sun and Chen 2000). Land demarcated as "state ecological forests" based on a suite of mostly ecological criteria are subject to use restrictions and associated payment, spanning both state- and collective-owned forests. Provincial governments are encouraged to

continues

Box 6.1   *continued*

supplement the basic subsidy payments, and wealthy provinces such as Guangdong have structured payment systems above and beyond the central government supported amounts.

In Jinxiu County, Guanxi, which spans the biologically diverse Dayaoshan region in the upper watershed of the Pearl River basin and eventually drains to the economic powerhouses of Guangzhou, Shenzhen, Macao, and Hong Kong, seven surrounding counties paid Jinxiu County a total of over 2 280 000 rmb in 2009 for water and soil conservation supplied by its forests with a potential 10% increase per year. With little scientific information of quantity or quality of water provisioning based on land use or management actions, these counties have continued payments for over 20 years. The payments, combined with Central Government FEBCF investment, provide important funds for local poverty alleviation and nature reserve

management (i.e., enforcement patrols and management bureau administration). Indeed, Jinxiu Count y obtains roughly 50% of its annual fiscal revenue from these watershed payments (Lu, pers. comm.).

Thus, the GOC has reacted swiftly to address the flood mitigation issue within critical basins largely because of a fundamental belief in the myriad of ecosystem services forest systems provide. By expanding the benefits beyond the ecological, these programs have been able to get a strong toehold, enabling them to play an increasing role in China's current socio-economic transformation. However, with limited information of service provisioning and an unclear understanding of trade-offs especially under complex hydrological conditions, it is increasingly difficult to ensure the continuation or expansion of payments and programs. This, in fact, may ultimately impact achievement of China's broader goals and objectives.

when considering ranked or normalized sub-catchment nutrient export. This, in fact, is what we see.

Several conclusions emerge from the map of these ranking groups. First, there is general agreement between the calibrated SWAT results and our model outputs in much of the Willamette River Basin, which is characterized by relative flat terrain on the valley floor and forest or shrubland. Next, the differences between the model outputs are most dramatic along the banks of the Willamette River (in the middle of the map) and on the slopes of the Coastal Mountains (on the left edge of the map). Along the Willamette River (groups 5 and 7), which includes sub-catchments with significant cropping systems and agricultural land, our model predicts much higher phosphorus loading than that predicted by calibrated SWAT. This dissemblance could be due to a number of factors, including SWAT's assumption of homogeneous LULC within HRUs, which could deviate significantly from the actual LULC pattern in the sub-catchment, which is well represented in our model. Indeed, our model uses the LULC pattern on the landscape without any lumping and routes the pollutant loading from each LULC to the downslope parcels. This latter point is not represented in SWAT, which directly links all HRUs to streams. SWAT simulates in-stream

nutrient processes as a one-dimensional, well-mixed stream at the daily time step. This could be a major reason for this dissemblance since soluble phosphorus could be used by riparian vegetation, algae, and other biota in the stream, which could reduce the amount of soluble phosphorus reaching the outlet of the watershed. The source of the deviation between the calibrated SWAT outputs and our outputs in the Cascades is less obvious, though it is not a result of the different assumed LULC patterns across the models as these areas are dominated by forest.

6.8   Next Steps

We are taking several steps to strengthen the models described in this chapter to ensure that they can adequately serve users' need. First, we continue to validate the outputs of our nutrient and sediment retention models through comparison with observed data. We are running the model in different watersheds in different eco-regions to test the patterns of nutrient and sediment transport with different LULC types, distributions, and patterns under different slopes. This will allow us to characterize model strengths and weaknesses vis-à-vis these different conditions and factors. We are also adjusting the model to allow for the inclusion of biological pathogens (e.g., fecal coliform). To achieve this goal, we have included the option of adding the pathogen source map, pathogen concentration (intensity) and vegetation filtration efficiency for this pathogen to the nutrient pollutant model.

Finally, the models described above are functions of several parameters whose values are drawn from a distribution in reality although they have been described as deterministic throughout the chapter. We are currently taking steps to incorporate uncertainty into our models, which will allow the users to understand the impacts of this uncertainty on model outputs. For a limited set of input parameters, we will allow the user to explore the impacts of parameter uncertainty on model outputs in one of two ways. The first methodology involves treating each parcel in the landscape as a realization of an underlying distribution. The second involves providing a confidence interval as well as a point estimate for use in supply estimation.

References

Agnew, L. J., Lyon, S., Gérard-Marchant, P., et al. (2006). Identifying hydrologically sensitive areas: bridging the gap between science and application. *Journal of Environmental Management*, **78**(1), 63–76.

Arnold, J. G., and Fohrer, N. (2005). SWAT2000: Current capabilities and research opportunities in applied watershed modeling. *Hydrological Processes*, **19**(3), 563–72.

Asia Development Bank (ADB). May 2006. Poverty Revolution in Key Forestry Conservation Programs. Final Draft Report.

Athayde, D. N., Shelley, P. E., Driscol, E. D., et al. (1983). *Results of the nationwide urban runoff program: Final report.* US Environmental Protection Agency, Water Plannning Division, Washington, DC.

Baker, M., Weller, D., and Jordan, T. (2006). Improved methods for quantifying potential nutrient interception by riparian buffers. *Landscape Ecology*, **21**(8), 1327–45.

Bennett, Michael. (2008). China's sloping land conversion program: institutional innovation or business as usual. Ecological Economics 65: 699–711.

Bingner, R. L., and Theurer, F. D. (2007). Research: AGNPS. *US Department of Agriculture–Agricultural Research Service.* [Online]. Available: http://www.ars.usda.gov/Research/docs.htm?docid=5199 Accessed 10 October 2008.

Dearmont, D., McCarl, B. A., and Tolman, D. A. (1998). Costs of water treatment due to diminished water quality: A case study in Texas. *Water Resources Research*, **34**(4), 849–54.

Desvousges, W. H., Smith, V. K., and Fischer, A. (1987). Option price estimates for water quality improvements: A contingent valuation study for the Monongahela River. *Journal of Environmental Economics and Management*, **14**(3), 248–67.

Easton, Z. M., Fuka, D. R., Walter, M. T., et al. (2008a). Re-conceptualizing the Soil and Water Assessment Tool (SWAT) model to predict saturation excess runoff from variable source areas. *Journal of Hydrology*, **348**(3–4), 279–91.

Easton, Z. M., Walter, M. T., and Steenhuis, T. S. (2008b). Combined monitoring and modeling indicate the most effective agricultural best management practices. *Journal of Environmental Quality*, **37**, 1798–1809.

Endreny, T. A. (2002). Forest buffer strips: Mapping the water quality benefits. *Journal of Forestry*, (January/February), 35–40.

Englin, J., and Cameron, T. A. (1996). Augmenting travel cost models with contingent behavior data. *Environmental and Resource Economics*, **7**(2), 133–47.

Foster, H. S. J., and Beattie, B. R. (1981). On the specification of price in studies of consumer demand under block price scheduling. *Land Economics*, **57**(2), 624–9.

Gotaas, H. B. (1956). *Composting: Sanitary disposal and reclamation of organic wastes*, Colombia Ubiversity Press, New York.

Hanley, N., Bell, D., and Alvarez-Farizo, B. (2003). Valuing the benefits of coastal water quality improvements using contingent and real behavior. *Environmental and Resource Economics*, **24**(3), 273–85.

Hernandez, M., Kepnerb, W. G., Goodrich, D. G., et al. (2008). *The use of scenario analysis to assess water ecosystem services in response to future land use change in the Willamette River Basin, Oregon.* IOS Press, Amsterdam, The Netherlands.

Hjelmfelt, A. T., Piest, R. P., and Saxon, K. E. (1975). Mathematical modeling of erosion on upland areas. In *Congress of the 16th International Association for Hydraulic Research*, pp. 40.

Holmes, T. (1988). Soil erosion and water treatment. *Land Economics*, **64**(3), 356–66.

Horton, R. E. (1940). An approach toward a physical interpretation of infiltration-capacity. *Soil Science Society of America Proceedings*, **5**, 399–417.

Leavesley, G. H., Lichty, R. W., Troutman, B. M., et al. (1983). *Precipitation-runoff modeling system: user's manual.* US Geological Survey, Washington, DC.

Leggett, C. G., and Bockstael, N. E. (2000). Evidence of the effects of water quality on residential land price. *Journal of Environmental Economics and Management*, **39**(2), 121–44.

Liu, Jianguo, Shuxin Li, Zhiyun Ouyang, Christine Tam, and Xiaodong Chen. (2008). Ecological and Socio-economic effects of China's policies for ecosystem services. Proceedings of the National Academy of Sciences. 105(28): 9477–9482.

Moore, W. B., and McCarl, B. A. (1987). Off-site costs of soil erosion: a case study in the Willamette valley. *Western Journal of Agricultural Economics*, **12**(1), 42–9.

Page, T., Haygarth, P. M., Beven, K. J., *et al.* (2005). Spatial variability of soil phosphorus in relation to the topographic index and critical source areas: Sampling for assessing risk to water quality. *Journal of Environmental Quality*, **34**, 2263–77.

Palmieri, A., Shah, F., Annandale, G., *et al.* (2003). *Reservoir conservation*, vol. 1: *The RESCON approach*. World Bank, Washington, DC.

Reckhow, K. H., Beaulac, M. N., and Simpson, J. T. (1980). *Modeling phosphorus loading and lake response under uncertainty: a manual and compilation of export coefficients*. US Environmental Protection Agency, Washington, DC.

Reid, L. M., and Dunne, T. (1996). *Rapid evaluation of sediment budgets*. Catena Verlag GMBH, Reiskirchen, Germany.

Renard, K. G., Foster, G. R., Weesies, G. A., *et al.* (1997). Predicting soil erosion by water: a guide to conservation planning with the revised universal soil loss equation (RUSLE). USDA Agriculture Handbook 703. USDA, Washington, DC.

Roose, E. (1996). *Land husbandry—components and strategy*, 70 FAO Soils Bulletin. Food and Agriculture Organization of the UN, Rome, Italy.

Schneiderman, E. M., Steenhuis, T. S., Thongs, D. J., *et al.* (2007). Incorporating variable source area hydrology into a curve-number-based watershed model. *Hydrological Processes*, **21**, 3420–30.

Sharma, S. K., Mohanty, B. P., and Zhu, J. T. (2006). Including topography and vegetation attributes for developing pedotransfer functions. *Soil Science Society of America Journal*, **70**, 1430–40.

State Forestry Administration (SFA). (2003). Sloping Land Conversion Program Plan (2001–2010). (In Chinese.)

Steenhuis, T. S., Winchell, M., Rossing, J., *et al.* (1995). SCS runoff equation revisited for variable-source runoff areas. *Journal of Irrigation and Drainage Engineering*, **121**(3), 234–8.

Stone, R. P., and Hilborn, D. (2000). Universal soil loss equation (USLE): Factsheet. *Government of Ontario, Ministry of Agriculture, Food and Rural Affairs*. [Online]. Available at: http://www.omafra.gov.on.ca/english/engineer/facts/00-001.htm#tab3a Accessed May 2008.

Thompson, J. A., Pena-Yewtukhiw, E. M. and Grove, J. H. (2006). Soil-landscape modeling across a physiographic region: topographic patterns and model transportability. *Geoderma*, **133**, 57–70. UNEP Assessment of 1998 Yangtze Floods. 1999.

White, W. R. (2001). *Evacuation of sediments from reservoirs*. Thomas Telford, London.

Wischmeier, W. H., and Smith, D. (1978). *Predicting rainfall erosion losses: a guide to conservation planning*. USDA-ARS Agriculture Handbook, Washington, DC.

Xu, J. and Cao Y. (2002). On sustainability a of converting farmland to forests/grasslands. International Economics Review 22: 56–60 (in Chinese), Zong, Yongqiang and Xiqing Chen. 2000. The 1998 Flood on the Yangtze, China. Natural Hazards 22: 165–184.

CHAPTER 7

Terrestrial carbon sequestration and storage

Marc Conte, Erik Nelson, Karen Carney, Cinzia Fissore, Nasser Olwero, Andrew J. Plantinga, Bill Stanley, and Taylor Ricketts

7.1 Introduction

Ecosystems help regulate Earth's climate by adding and removing greenhouse gases (GHGs) such as carbon dioxide (CO_2) from the atmosphere (IPCC 2006). Terrestrial ecosystems currently store four times more carbon than is found in the atmosphere (3060 versus 760 gigatons (Gt); see Lal 2004). Changes in land use and land cover (LULC) due to timber harvesting, land-clearing for agriculture, and fire can release substantial amounts of terrestrially stored carbon. For example, tropical deforestation was responsible for 15–25% of the globe's total GHG emissions in the 1990s (including fossil-fuel use and other land-use change; see Gibbs *et al.* (2007) and the Technical Summary in IPCC (2007a)). In Africa, deforestation accounted for nearly 70% of the continent's total GHG emissions at the end of the twentieth century (Gibbs *et al.* 2007).

We can mitigate the expected economic damages due to climate change by slowing down GHG accumulation in the atmosphere (Stern 2007). This fact has intensified global interest in enlarging or at least maintaining the size of the terrestrial carbon pool (e.g., IPCC 2007a, Lehmann 2007). A variety of land management techniques can be used to achieve this end, including lengthening harvest rotation time in plantation forests (e.g., Sohngen and Brown 2008), planting trees to restore forests (reforestation; e.g., Canadell and Raupach 2008), planting trees in abandoned croplands (afforestation; e.g., Nilsson and Schopfhauser 1995), improving soil management (e.g., Schuman *et al.* 2002, Lal 2004), reducing forest fire and forest disease risk (e.g., Brown *et al.* 2002), reducing deforestation and forest degradation (REDD; Ebeling and Yasue 2008), and replacing annual crops with perennials (Fargione *et al.* 2008).

Concerns about climate change have led to both regulated and voluntary emissions reduction. Regulated markets such as the European Union Emissions Trading System (EU ETS; Victor *et al.* 2005) and the Regional Greenhouse Gas Initiative (Burtraw *et al.* 2006), and voluntary offset markets like the Chicago Climate Exchange provide a forum in which landowners can generate carbon credits, through increased carbon sequestration on their lands, that can be bought by entities looking to offset their own emissions (Marechal and Hecq 2006). Further, the Kyoto Protocol allows signatory nations to claim GHG emission reductions by funding afforestation and reforestation projects in developing nations (e.g., Pfaff *et al.* 2000). Finally, discussion at the 2007 United Nations Framework Convention on Climate Change Conference focused on developing financial incentives to reward developing nations for reducing emissions by avoiding deforestation and forest degradation (e.g., Mollicone *et al.* 2007; Ebeling and Yasue 2008).

Given the growing interest in carbon markets and offset programs, decision-makers need a way to simply and quickly understand how much carbon is held in landscapes today, and how storage and sequestration will change under different management options. Such information will help managers find and create opportunities for additional carbon sequestration and associated payments in their landscapes. We present two tiers of relatively simple models that address these needs by using minimal data to estimate carbon sequestration and storage on a landscape.

Unlike other ecosystem services, there is no difference between the supply and use of carbon sequestration and storage (see Chapter 3 for definitions of "supply" and "use" used throughout the book); every unit of sequestered GHG emission will allow us to avoid some economic damage that would have occured otherwise (although current and proposed sequestration and storage markets only pay for a portion of the avoided damage; see below). Further, because GHGs uniformly mix in the atmosphere the whole world benefits from a unit of sequestration regardless of where it occurs. After presenting our ecosystem service supply model, we present an approach for estimating the economic value of this supply. We also discuss how we might approximate carbon offsets and avoided emission credits with our model.

Gathering the data needed to estimate total carbon sequestration and storage on a landscape, including species-specific growth rates, associated carbon storage, and species compositions, can be a time-intensive effort. The IPCC has developed an approach that assigns carbon stocks to different LULC categories based on a meta-analysis of field storage studies (IPCC 2006). The models presented in this chapter are grounded in this approach. In our simplest storage and sequestration model, tier 1, we associate carbon storage with different LULC categories and each LULC category is assumed to be in storage equilibrium at any point in time (Section 7.2.1). In tier 1, the amount of carbon sequestered in a parcel is the difference in steady-state storage at two points in time (Section 7.2.2). In this simple model, carbon sequestration is not registered in a parcel unless the parcel's LULC changes or it is subject to harvest of wood.

7.2 Tier 1 supply model

7.2.1 Carbon stored on a landscape

We disaggregate terrestrial carbon storage into five pools: (1) aboveground biomass, (2) belowground biomass, (3) soil, (4) other organic matter, and (5) harvested wood products (HWPs). Aboveground biomass is composed of all living plant material above the soil (e.g., grass, herbaceous material, bark, tree trunks, branches, leaves, and other woody understory). Belowground biomass is the root system of the aboveground biomass. Sequestration and storage accounting in the soil pool is generally concerned with soil organic carbon (SOC) in mineral soils; in certain land cover types, however, SOC in organic soils (e.g., wetlands, peatlands, rice paddies) is the dominant soil carbon (Post and Kwon 2000). The other organic matter pool includes plant litter and dead wood. The HWPs pool includes all carbon stored in products made with wood removed from the landscape (e.g., furniture, paper, charcoal; Harmon *et al.* 1990; Smith *et al.* 2006).

Let C_{aj}, C_{bj}, C_{sj}, and C_{oj} indicate the metric tons of carbon stored per hectare (Mg of C ha^{-1}), in the aboveground, belowground, soil, and other organic matter pools of LULC j respectively, where $j = 1, 2,\ldots, J$ indexes all LULC found on the landscape. LULC types can simply indicate land use and or cover (e.g., conifer forest, cropland) or include other landscape details that affect pool-storage values such as time since disturbance (e.g., conifer forest 120 years or more since a clear-cut, cropland that is plowed annually), extent of disturbance (e.g., heavily disturbed conifer forest due to illegal timber harvest), soil properties (cropland on clay soil; e.g., Torn *et al.* 1997), and climatic conditions (conifer forest below 1000 m, where elevation bands proxy for precipitation gradients; see McGuire *et al.* 2001; Raich *et al.* 2006). Adding specialized LULC classes such as these can help add more reality to the estimates of this steady-state model. For example, using age-specific forest classes with fine enough temporal resolution can help represent additional carbon sequestration as forests mature and change LULC classes.

The carbon stored in a parcel's HWPs pool, C_{p}, is the sum of the carbon in the woody material removed from the parcel in the past (even if the removed material eventually leaves the parcel) less the decay over time in the products that were made from this woody material (which releases carbon in the form of CO_2) and the carbon-equivalent emissions from manufacturing and transporting the final products. Let

$$C_{px} = \sum_{q=0}^{W_x-1} \theta_{hqx} C_{hqx} f\left(\theta_{qx}, q\right) - \sigma_{qx}, \qquad (7.1)$$

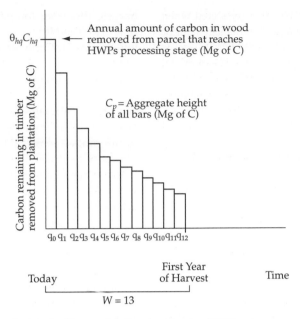

Figure 7.1 Calculating the amount of carbon stored in a parcel's harvested wood products (HWPs) pool.

Wood harvested from a parcel is converted into various products (e.g., furniture, firewood), which burn or decay at a given rate over time, returning CO_2 to the atmosphere. In the graph, the carbon remaining in the HWPs made from wood harvested 13 years ago is given by the height of q_{12}'s bar, and so on. The sum of the bars' heights indicates the amount of carbon still stored in HWPs made from that parcel's wood removed over the past 13 years. The rate at which the height of the bars decays over time is a function of the decay rates associated with the product mix made with the harvested wood.

where C_{px} is measured in metric tons of C in parcel x, where $x = 1, 2, \ldots, X$ indexes all unique parcels on a landscape, C_{hqx} gives the metric tons of C in the woody biomass when it was removed from parcel x q years ago, θ_{hqx} is the portion of the harvested woody biomass removed q years ago from x that makes it to the initial wood processing stage, $f(\omega_{qx}, q)$ gives the fraction of $\theta_{hqx}C_{hqx}$ still stored in wood products q years after removal from x, ω_{qx} is the decay rate of the products made with wood removed from x q years ago, and σ_{qx} is the carbon-equivalent emissions in metric tons associated with the production and distribution of the product made from C_{hqx}. The variable W_x indicates how many years in the past we want to account for harvest on parcel x. An illustration of Eq. (7.1) is given in Figure 7.1 and Section 7.2.5. Appropriate inclusion of the HWP pool requires thoughtful accounting, given its significant storage potential and the emissions associated with a product's life cycle (Niles and Schwarze 2001; USEPA 2009).

The carbon stored on a parcel at time t, given by C_{xt} and measured in metric tons of C, is equal to the sum of the carbon stored in each pool in the parcel at time t,

$$C_{xt} = C_{pxt} + \sum_{j=1}^{J} A_{xjt} \left(C_{aj} + C_{bj} + C_{sj} + C_{oj} \right), \qquad (7.2)$$

where C_{pxt} is parcel x's HWPs pool-storage level at time t, A_{xjt} is the area of LULC j in parcel x at time t, and $A_x = \sum_{j=1}^{J} A_{xjt}$ (parcel area does not change over time). If data or interest in some pools is lacking, the model can be used with any subset of the five carbon pools. To determine the metric tons of C stored across the whole landscape at time t, symbolized by C_t, we sum all parcel-level carbon storage values,

$$C_t = \sum_{x=1}^{X} C_{xt}. \qquad (7.3)$$

See Glenday (2006) and Ruesch and Gibbs (2008) for applications of Eq. (7.2).

7.2.2 Carbon sequestered by a landscape

Carbon sequestration represents an increase in carbon storage over time. In the tier 1 model, carbon sequestration or loss in a parcel is registered when the parcel's LULC mix changes, its management or production of HWPs changes, or its HWPs pool is not in equilibrium. Otherwise, for the sake of modeling simplicity, in tier 1 we assume storage in a parcel is in equilibrium or steady state and will not change over time despite any evidence to the contrary.

To determine the amount of carbon sequestered in a parcel from year t to year T ($t < T$), given by ΔC_{xtT}, we use Eq. (7.2) to calculate the amount of carbon stored in a parcel in year t, C_{xt}, and in year T, C_{xT}, which accounts for any changes in the parcel's LULC or HWP management or production between t and T, and then subtract C_{xt} from C_{xT},

$$\Delta C_{xtT} = C_{xT} - C_{xt} , \qquad (7.4)$$

where

$$\Delta C_{tT} = \sum_{x=1}^{X} \Delta C_{xtT} \qquad (7.5)$$

gives the change in carbon storage from t to T over the entire landscape.

In this case, C_{xt} is the point of reference or baseline for determining whether net sequestration has occurred from t to T. If ΔC_{xtT} or ΔC_{tT} is positive, then sequestration has occurred from t to T in the parcel or landscape, respectively. If they are negative, carbon has been lost between t and T. See Cairns *et al.* (2000) and Glenday (2006) for applications of Eq. (7.4).

7.2.3   Additional sequestration in offset markets

Markets and programs that give carbon credits to landowners in exchange for enhancing terrestrial carbon sequestration on their lands currently only recognize additional sequestration in market-eligible pools (see Box 7.1 for an illustration). Additional sequestration is the amount of sequestration above and beyond the sequestration that would have occurred in the absence of a carbon offset market or program (i.e., baseline sequestration). Additional sequestration in a parcel's eligible pools from t to T, ΔC_{xtT}^{Off}, can be approximated in our model by calculating a parcel's sequestration from time t to T under its baseline LULC mix as of time T, $\Delta C_{xtT}^{Off,B}$, and sub-

tracting this from the parcel's sequestration from time t to T under an alternative, offset program-influenced LULC mix as of time T, $\Delta C_{xtT}^{Off,A}$:

$$\Delta C_{xtT}^{Off} = \max \left\{ 0, \Delta C_{xtT}^{Off,A} - \Delta C_{xtT}^{Off,B} \right\}. \qquad (7.6)$$

Such new, program-influenced scenarios are not generated by our models, but rather can be assessed by our models to reveal the additional carbon benefits associated with the offset program.

Two other carbon market issues are permanence and leakage, and both are topics of intense policy debate (e.g., Brown 2002; Vohringer *et al.* 2006; Murray *et al.* 2007) (Box 7.1). Permanence is a concern because trees and other biomass eventually decay; fires and illegal logging can occur regardless of management efforts; soil can be disturbed, releasing trapped carbon back into the atmosphere; and offset providers may decide to convert their forest to another use even with an existing offset contract. Therefore, when a landowner is compensated for additional sequestration, we might say that they are renting a temporary benefit to society. How much society should pay for this temporary benefit and at what point temporary sequestration becomes permanent sequestration in a climatic sense is contested (e.g., Chomitz 2002; Marechal and Hecq 2006). Leakage is an additional concern with offset, rather than regulatory, programs because when a landowner decides to manage her land for additional sequestration instead of clearing it for agriculture or urban development the economic pressure to clear the land for other uses will not disappear. Instead it may be "leaked" or displaced to other parts of the landscape or globe where subsequent clearing will release stored carbon, decreasing the amount of additional sequestration. Leakage is accounted for in this model if it occurs within the study landscape and between the years t and T, so broadening both spatial and temporal scales can reduce the issue.

7.2.4   Preventing emissions from deforestation and degradation

REDD is a policy mechanism, currently under debate, that rewards efforts to reduce deforestation, forest degradation, and their associated carbon emissions in developing countries (e.g., Ebeling and

Box 7.1 Noel Kempff case study: capturing carbon finance

Bill Stanley and Nicole Virgilio

Policies that constrain carbon dioxide (CO_2) emissions, and include provisions for trading carbon sequestered in forests and other ecosystems, are anticipated to be the largest driver of payments for carbon capture and storage and are a major topic of discussion in the development of national and international climate change policies and legislative language. If successful, these programs could generate billions of dollars of annual funding for forest carbon projects.

The Noel Kempff Mercado Climate Action Project (NKCAP) in Bolivia, borne of a partnership between The Nature Conservancy and Fundación Amigos de la Naturaleza with funding from industry and contributions by the Bolivian Government, is one example of such a project, designed to simultaneously address climate change, conserve biodiversity, and bring sustainable development benefits to local communities by avoiding logging and agricultural land conversion. The 1.5 million-acre project, which began in 1997 and expanded the Noel Kempff Mercado National Park, is the largest effort of its kind. It is expected to prevent the release of up to 5.8 million tons of CO_2 into the atmosphere over 30 years. As with any carbon sequestration project, there were a number of issues that had to be addressed to ensure high-quality carbon offsets.

Key factor 1: additionality

Nearly all voluntary frameworks, as well as the few compliance programs in place or proposed, require that project-based emissions reductions and removals must be beyond "business as usual" to be credited (e.g., Alig and Butler 2004, UNFCCC 1995, UNCCCS 1997, and several US Senate bills including the Forest Resources for the Environment and the Economy Act (S. 1547), the International Carbon Sequestration Incentive Act (S 2540), and the Lieberman–Warner Climate Security Act (S.3036)). The guidelines suggest that a scenario of what would have happened without the project must be developed for comparison against project management. The difference in carbon storage and other greenhouse gas emissions (GHGs) between the two represents the GHG impacts that are truly additional rather than simply the result of incidental or non-project factors such as recent market or

environmental changes (IPCC 2000). In the case of NKCAP, the project provided carbon financing to stop logging in the park and deforestation around communities. Without this funding, these activities would have continued, leading to the loss of forest cover and release of carbon dioxide.

Key factor 2: baseline

The business-as-usual, or "without-project" scenario, is also called the *baseline*, and its development should be among the first steps to assessing the carbon benefits of a project. Baselines are essentially predictions, or future projections based empirically upon historical information or a performance benchmark, of what may have happened had the project not been put into place. The success of any forest carbon emissions reduction project will depend on the estimated carbon storage in the baseline or benchmark, subtracted from the performance of the project itself. Most recently, in the context of programs being considered within the context of the United Nations Framework Convention on Climate Change for activities to reduce emissions from deforestation and degradation (REDD), several national-scale baseline emission scenarios have been proposed. These proposals vary widely, and the baseline approach established as policy will have a significant impact on the emissions reductions that could be claimed. To determine the total emission reductions that were additional to business-as-usual activities at Noel Kempff, baseline rates of deforestation were developed by GIS analysis of a time series of satellite photos from 1986, 1992, and 1996–7 and then were applied in a spatially explicit land-use change model to project future deforestation. The rates and models are to be reassessed periodically. Also, a national-scale economic model of Bolivian timber markets was created to develop baseline rates of logging and field work was undertaken to determine the amount of emissions associated with the logging.

Key factor 3: leakage

Leakage has been defined as "the unanticipated decrease or increase in GHG benefits outside of the project's accounting boundary…as a result of project activities" (Chapter 5 of IPCC 2000). A straightforward example of leakage from conservation is where a farmer who is seeking to clear a plot of uninhabited forest for conversion

continues

Box 7.1 *continued*

to agricultural fields is told that the land has been made part of a carbon project and is now protected. Leakage occurs if the farmer moves to the next available land and clears that forest instead. The project would simply displace the farmer's activity and emissions, and not result in any real reductions of atmospheric CO_2. The risk of leakage can be minimized through thoughtful project design, taking local conditions into consideration and addressing the underlying drivers of deforestation (Schwarze *et al.* 2002). For example, NKCAP attempted to limit leakage just outside the project area by working with the bordering community to develop a sustainable management plan, with which they applied for legal land title through the Bolivian government. This reduced the risk of uncontrolled forest conversion (Aukland *et al.* 2002). Market leakage that was not avoided was estimated using an economic model (Sohngen and Brown 2004) and deducted from the claimed carbon savings.

Key factor 4: permanence

One of the first issues that critics of forest carbon crediting will cite is the possibility for the reversal of the carbon sequestration. Project success is dependent upon keeping the forest healthy and standing. Illegal logging, invasive

pests, fire, and political turnover all have the potential to release carbon sequestered in the forest (IPCC 2000). Thus, permanence must be taken into consideration during the project planning stage, targeting areas where land is likely to remain intact indefinitely and using approaches like permanent conservation easements and sustainable forestry, which will reduce likelihood that the carbon storage will be lost, or that those losses will be sustained, over time. Carbon offset policies generally establish liability for losses. To ensure against these losses one tool is to set aside "buffer" credits that can be drawn upon should the stored carbon be unexpectedly lost. To ensure the permanence of NKCAP, the national park was expanded to include the project area and a permanent endowment was established to fund protection activities into the future.

Conclusion

Obviously, there are many factors that must be considered in the successful employment of forest carbon emissions reductions programs. Additionality, baseline, leakage, and permanence are challenges, but pilot projects such as NKCAP have shown that it is possible to address and overcome them.

Yasue 2008). In this case, if deforestation and degradation that would have occurred under a business-as-usual LULC trajectory is prevented, then ecosystems retain additional carbon, thereby avoiding climate change-related economic damages that were originally expected.

Here we illustrate how we can use our models to approximate the carbon credits that could be generated under REDD or similar programs to compensate avoided forest loss and degradation. This illustration relies on several assumptions. First, we assume that forest type j''s carbon stock in parcel x is in steady state at time t. Next, we assume that if forest type j' is cleared from x in order to establish LULC j, the carbon storage at time T is lower than it was at time t (i.e., $C_{xt} > C_{xT}$). Let the change in storage on x be given by ΔC_{xtT}, where $\Delta C_{xtT} = 0$ if forest type j' remains on the parcel as of time T and $\Delta C_{xtT} < 0$ if the forest is cleared. (We assume that there are only two possible future states for parcel x at time T:

it can remain as is or be converted to a LULC with lower carbon storage.) Let ΔC_{xtT}^{Loss} indicate the case where $\Delta C_{xtT} < 0$.

The expected loss in a parcel's carbon stock due to conversion from LULC j' to some other LULC by time T is the product of the difference in carbon stock ΔC_{xtT} and the overall probability that the forest in parcel x will be converted at some point between t and T, π_{xtT}, under the business-as-usual baseline. Let there be $i = 1, 2, \ldots, I$ events that can cause forest type j' in parcel x to be converted by T. If we assume that the probability of each event occurring, given by $\pi_{xtT, i'}$ under the baseline is independent and additive, then

$$\pi_{xtT} = \min\left\{1, \sum_{i=1}^{I} \pi_{xtT, i}\right\}. \tag{7.7}$$

The expected amount of aboveground biomass carbon storage in parcel x at time T under the baseline is given by

$$E[C_{xT}] = C_{xt} + \pi_{xtT}\Delta C_{xtT}^{Loss}. \qquad (7.8)$$

In order to achieve a net reduction in expected carbon emissions at time T, the parcel's manager must intervene such that expected storage in x at time T is greater than $E[C_{xT}]$. Let $\hat{\pi}_{xtT}$ and $\hat{\pi}_{xtT,i}$ represent the overall probability and specific event probability, respectively, that parcel x will be converted from j' by time T under alternative management where $\hat{\pi}_{xtT}$ is calculated similarly to π_{xtT} (Eq. (7.7)). Finally, the net reduction in expected emissions in parcel x is given by

$$\Delta C_{xtT}^{Avoid} = E[\hat{C}_{xT}] - E[C_{xT}]$$
$$= \min\{0, (\pi_{xtT} - \hat{\pi}_{xtT}) \times \Delta C_{xtT}^{Loss}\}, \qquad (7.9)$$

where

$$E[\hat{C}_{xT}] = C_{xt} + \hat{\pi}_{xtT}\Delta C_{xtT}^{Loss}. \qquad (7.10)$$

In the exposition above we assume that data on LULC conversion probabilities at the parcel level are available. In practice, only historical conversion rates at the country level are available globally. For example, national estimates of annual deforestation rates are available for the periods 1990–2000 and 2000–5 for most nations (FAO 2005). However, we may be able to predict conversion rates at the parcel level using basic economic principles. Under such a framework whether forested parcel x will be converted to another LULC in the future will be a function of the net value of conversion, the availability of substitutable parcels for conversion, the enforcement of property rights, labor and capital constraints, and other socio-economic variables. In this chapter's supplementary online material (SOM), we introduce one method for estimating conversion probabilities π_{xtT} and $\hat{\pi}_{xtT}$ as a function of predicted gross value of production following deforestation in x and its surrounding parcels, x's distance to the nearest road, and x's distance to the nearest source of labor and capital (e.g., a population center).

7.2.5 Tier 1 example: terrestrial carbon storage in the Eastern Arcs Mountain Watershed, Tanzania

We illustrate our tier 1 carbon storage model as described in Eq. (7.2) on a landscape defined by the Eastern Arc Mountains of Tanzania and their watersheds. The mountains contain over 1000 endemic species and are widely considered a global conservation priority (Burgess et al. 2007). They also provide a range of ecosystem services, including water for drinking, agriculture, and hydropower, carbon sequestration, and non-timber forest products (Ndangalasi et al. 2007; Mwakalila et al. 2009). The map used for this analysis estimates land cover as of 1995 (Valuing the Arc 2008) and consists of 590 LULC categories (each LULC type is a unique combination of land cover, elevation range, and terrestrial ecoregion (Olson et al. 2001)).

For mean aboveground (C_a), belowground (C_b), and soil (C_s) pool values, we average across reported values for each LULC type from various sources (see the chapter's SOM for values and their sources and details on the 1995 land cover map). Low and high pool values for each LULC are set equal to the lowest and highest values observed in the literature. We calculated the HWP pool only for the forest plantation LULC type. We used data from Makundi (2001) and IPCC (2006) and an assumption of even-age rotation forestry to calculate mean, high, and low C_p, C_a, C_b, and C_s values for the plantation LULC (see the chapter's SOM for details).

The tier 1 models produce maps of per-hectare mean, lower, and upper bound estimates of aggregate storage in the biomass, soil, and HWPs pools as of 1995 (we ignore the other organic material pool) (Figure 7.2; Plate 3). The highest densities of stored carbon occur in the tropical montane forests found at the highest elevations of the Arc Mountains. These areas remain forested today, amid continued agricultural conversion in the lower elevations. Many of the areas of high carbon storage are already protected by national parks or forest reserves, but unprotected forests on the edges of protected areas remain. Several aggressive reforestation projects— undertaken mostly to garner carbon offset payments—have begun, which could increase stored carbon values in parts of this landscape over time.

Our assumption that forest plantations are the only LULC types that produce HWPs is the greatest shortcoming of this illustrative example. Non-plantation forests in this watershed provide fuel wood, charcoal, and timber for many households on the landscape (e.g., Luoga et al. 2000; ECCM

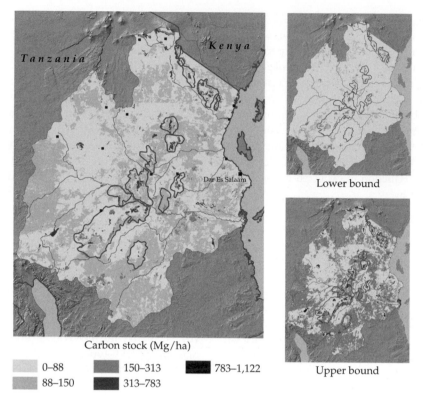

Figure 7.2   Tier 1 carbon storage estimates for 1995 in Tanzania's Eastern Arcs Mountains and their watersheds.
The polygons formed with the dark lines represent Eastern Arc Mountain blocks, which rise from the surrounding woodlands and savannas. These blocks were once largely forested, but now consist of a mixture of agriculture, forest, and woodlands. Gray lines are major rivers. Black squares represent major cities. Timber plantations cover approximately 0.3% of the study landscape. Spatially explicit land cover and other landscape data are from the Valuing the Arc project (2008; Mwakalila 2009). See the chapter's SOM for details on data used in the maps. (See Plate 3.)

2007; Ndangalasi *et al.* 2007). If we could obtain better estimates on these harvests we could include the C_p pool for more forest types and adjust their C_a and C_b values accordingly. In addition, plot level measures of standing carbon stocks being taken in this region will be used to groundtruth our model estimates and improve input values for the aboveground carbon pool (see Box 7.2).

7.3   Tier 1 valuation model: an avoided economic damage approach

By slowing GHG emissions, terrestrial carbon sequestration and storage decrease the severity of future climate change and its associated damages. Integrated assessment models (IAMs) estimate these potential damages by relating rising GHG concentrations in the atmosphere and expected

temperature increases to expected global or regional net economic damage over time (e.g., Nordhaus 1992; Mastrandrea and Schneider 2004). Damages in these models include property damage due to rising sea levels and storm occurrence, the net changes in crop yields due to drier and hotter climates, fishery damage due to ocean acidification, and the net gain in human mortality and morbidity due to greater disease prevalence in a warmer world (Stern 2007).

The social cost of carbon (SCC) is the marginal cost, manifested in the economic damages due to the resultant higher atmospheric GHG concentration, of emitting an additional metric ton of C, all else equal. Global IAMs provide SCC estimates.

Mathematically, the societal value of sequestration on parcel x across the years t to T is given by multiplying x's supply of sequestration (positive or

Box 7.2 Valuing the Arc: measuring and monitoring forest carbon for offsetting

Andrew R. Marshall and P. K. T. Munishi

Carbon offsetting programs and associated policy tools (REDD, MAC, CERs, etc.) require accurate quantification of terrestrial carbon stocks. While there are continuing efforts to determine vegetation structure remotely, permanent sample plots (PSPs) of woody vegetation continue to give the most reliable estimates (Brown 2002; Gibbs *et al*. 2007). These estimates can both calibrate remotely sensed measures and provide inputs to models like those described in this book. The many environmental and human impacts on carbon storage levels in forests must be considered when interpreting PSP data. Outputs of PSPs can also be combined with data on other ecosystem services produced by the forest to determine the overall economic value of a forest ecosystem.

The Valuing the Arc project, funded by the Leverhulme Trust, is doing just that in the Eastern Arc Mountains of Tanzania, one of the world's most important areas for species richness and endemism (http://www.valuingthearc.org). The combined importance of carbon and biodiversity production in a forest is likely to be far more appealing to policy-makers than biodiversity alone.

Plot location and size

There are many considerations for planning the location of PSPs and the methods to use for estimating its carbon content. First, PSP locations need to be representative of the area. This will require prior knowledge of the variation in environmental and human influences. In montane habitats such as those found in the Eastern Arc Mountains, elevation is the primary predictor of vegetation composition and tree allometry, particularly the height to diameter ratio (Marshall *et al*. in prep.). For this reason a stratified random design for locating plots is advisable. The size and number of plots should provide a balance between time and resources to sample the maximum range of environmental conditions. One-hectare plots have been adopted by many international projects, but may be impractical for short-term projects or for sampling broad environmental gradients.

The majority of aboveground carbon in forest habitats is contained within the large trees and lianas (as much as 95% in some tropical forests; Chave *et al*. 2003). A huge amount of measurement and taxonomic identification time can be saved in limiting surveys to these stems.

Tree volume

The most important measurement for estimating tree volume is the diameter at breast height (dbh, typically 1.3 m) (e.g., Kuebler 2003). Where environmental conditions are thought to cause the dbh-to-height ratio to deviate from observed norms, it is also important to measure tree height. This can be time-consuming and it is advisable to simply measure a representative sample (e.g., 10 from each size class; 10–20, 20–30, 30–40, 40–50, and > 50 cm dbh). The dbh-to-height relationship in the sample can then be determined from regression analysis and extrapolated to the remaining stems in each PSP.

The above measures are used to estimate the volume of a tree bole, but other components of the tree are more complicated. Calculating the volume of branches and roots requires destruction of the trees, so that they can then be placed in water to observe the volume displaced. Estimating the volume of foliage and dead wood requires litterfall traps, but litterfall is highly variable among species, elevations, and seasons. Because of these complications, most studies either focus on the bole carbon alone or apply a simple expansion factor to adjust the final carbon estimate (see below).

Calculating tree biomass and carbon content

The next step is to calculate the biomass of a tree, given its estimated volume. Many equations have been employed for this purpose, and these vary according to habitat and location, due to variations in environmental conditions and tree allometry (Chave *et al*. 2005). All approaches, however, involve multiplying tree volume by its wood density (g cm ^{3}). The standard measure of wood density is wood specific gravity (WSG), which is the oven-dry mass divided by green volume. WSG is typically determined by taking a core from a live tree, or a section of a freshly dead tree. WSG for many species is unknown. Studies therefore often use estimates from the most closely related species, genus, or family.

Around 50% of the biomass in woody vegetation is composed of carbon; therefore, the final amount of elemental carbon is calculated by simply multiplying biomass by 0.5. If this calculation is limited to the tree bole, IPCC recommends expansion factors of 0.24, 0.25, and 0.05 for the carbon in branches, roots, and foliage, respectively (IPCC 2003). These have mostly been derived

continues

Box 7.2 *continued*

from studies carried out in association with commercial logging, as felled trees can be physically measured.

There are many sources of error from the various stages of calculation. In terms of allometry some studies have incorporated measurements from the bole to buttress edge (e.g., Glenday 2006). Lianas and stranglers also pose problems due to their unpredictable shape. Estimates of wood density are also highly dependent on the availability of data, and the reliability of using estimates from closely related species is debated. Furthermore the impact of human disturbance on forest ecosystems is particularly hard to quantify. Despite the many sources of error, published estimates for intact tropical forests are usually relatively consistent between studies (typically 100–250 t ha^{-1}) and precision within 10% of the mean is usually possible (Brown 2002). Remote methods including optical, radar, or laser sensors are becoming more widely used, but do not have the accuracy and consistency of inventories using vegetation plots (e.g., Gibbs *et al.* 2007).

Implementation of offset programs

Accurate calculation of carbon emissions is paramount to determining national terrestrial carbon budgets, including sinks by woody biomass, sinks by soil, and releases by human activity. Accurate calculation is also important for the development of a market that pays landowners for reduced emissions from deforestation and degradation (REDD). Interest in REDD markets is driven by the fact that 20% of global CO_2 emissions come from forest destruction and degradation, and that much of this is occurring in developing countries that have limited funds and capacity to address it. A key question is whether the developing world is already, or can be assisted to become, ready for REDD. In some countries this answer is likely to be no due to inadequate policy and environmental law enforcement. More positively, Tanzania may be among the developing countries with highest potential for carbon trading and entering the REDD market in coming years. Tanzania is politically stable, with an advanced forest policy and legislation, has huge forest resources, and has for many years been attractive for international development funding. It also has one of the world's most extensive protected area networks, with nearly 40% of land area covered by National Parks and various other forms of reserved land (Chape *et al.* 2008). These combined factors have led the Norwegian government to approve a US$100 million grant for the implementation of REDD in Tanzania. There is also considerable interest from other nations, the United Nations, and the World Bank.

negative) over that time period by the SCC, discounting all annual values to year t values, and then summing across all discounted values:

$$VAD_{xtT} = \sum_{z=t}^{T-1} \frac{\Delta C_{x,z,z+1} SCC_{z+1}}{(1+r)^{z-t}}, \qquad (7.11)$$

where VAD_{xtT} (value of avoided damage) is the present value of all economic damage avoided (or additional damage caused if negative) due to carbon sequestration on x (emitted from x) from time t to T. In Eq. (7.11) $\Delta C_{x,z,z+1}$ measures the metric tons of C sequestered (emitted) on x between year z and $z + 1$ (if we only have estimated ΔC_{xtT} or sequestration for several time steps in between t and T, we can approximate annual sequestration with linear interpolation), SCC_{z+1} is the SCC in year $z + 1$, and r is the real (inflation-adjusted) discount rate. The landscape-level analog of VAD_{xtT} is given by

$$VAD_{tT} = \sum_{x=1}^{X} V_{xtT}. \qquad (7.12)$$

In Eqs. (7.11) and (7.12) we assume every unit of sequestration after time z, even if it is just temporary, is valuable to society because it reduces the stock of carbon in the atmosphere, thereby (marginally) mitigating or delaying climate change and related damages associated with the continuation of C_{xz} levels into the future (i.e., C_{xz} is the storage baseline in year z). However, if we only want to value the supply of additional sequestration or avoided emissions then $\Delta C_{x,z,z+1}$ in Eq. (7.11) could be replaced by an annualized ΔC_x^{Off} (see Eq. (7.6)) or ΔC_x^{Avoid} (see Eq. (7.9)). Finally, by indexing the SCC with z in Eq. (7.11), we allow it to change over time. There is some expectation that the SCC will grow over time as the marginal impact of avoided emissions becomes even more valuable in a world dealing with a rapidly changing climate (e.g., the IPCC assumes that the SCC will grow at a rate of 2.4% per annum; see Chapter 20 of IPCC 2007b).

It is not clear whether the real (inflation-adjusted) discount rate in the denominator of Eq. (7.11), r, should match the discount rate embedded in our chosen SCC (see below). The denominator of Eq. (7.11) discounts all future costs and benefits from $\Delta C_{x,z,z+1}$ over $z \in [t, T]$ to t's value of money, allowing for the aggregation of costs and benefits incurred over time. The real discount rate used in Eq. (7.11) measures our preference for more immediate consumption over investment and our expectations regarding economic growth. For short time spans, and when comparing VAD_{tT} to other benefits and costs realized in the near term, we may want to use a real discount rate closer to an observed market discount rate.

There remains much uncertainty and debate over the appropriate value of the SCC (e.g., Nordhaus 2007; Stern 2007; Weitzman 2007). Any SCC estimate is a function of the assumptions in its source IAM, including the trajectory of global GHG emissions over time and the impact that climate change will have on the social welfare of people around the world today and in the future. As a result, estimates of SCC vary widely. By using an average or median SCC estimate, however, we can avoid assuming some of the more idiosyncratic predictions of global emission, economic, and demographic trends. Tol (2009) surveyed the peer-reviewed SCC literature and found representative current estimates that range from $46 to $91 Mg^{-1} of C but with a large variance (values are given in 1995 dollars).

Another cause of divergence in SCC estimates is disagreement over the appropriate discount rate to use when determining the SCC in an IAM. The discount rate, which combines expectations about future economic growth and our preferences for present consumption over future consumption, equates future climate change-related damages to more immediate damages. A lower discount rate will lead to higher estimates of the SCC, as future events, including the risk of catastrophic change, are more heavily weighted with lower interest rates (Chapter 20 of IPCC 2007b). Some economists argue that the discount rates used in climate change analysis should be lower than what is typically used for cost–benefit analyses (e.g., a discount rate of 1.4%

per annum in Stern (2007) and 2–4% in Weitzman (2007) versus a "typical" cost–benefit analysis rate of 7%) whether due to intergenerational equity concerns (Stern 2007) or concerns over the highly unlikely but particularly disastrous scenarios of catastrophic change (Weitzman 2007). Other economists argue that in order to avoid a potentially massive misallocation of monetary resources, climate change mitigation activities should be judged with the discount rate we use to judge all other policies (i.e., 5 to 10% per annum; e.g., Nordhaus 2007).

Further, the SCC is a marginal cost—that is, it measures the economic damage avoided for very small changes in carbon emissions relative to total global stock. While significant amounts of carbon are at stake at landscape scales, most applications of this model will only involve small changes in carbon stocks relative to any emission baseline. For our purposes, then, SCC is a legitimate and useful estimate of social cost.

Finally, the SCC will invariably differ from the market price for carbon offsets as the market-clearing price of an offset (the price that sellers and buyers agree on) has no functional relationship to the social value of carbon sequestration. In regulated markets, buyers and sellers will settle on a price that is a function of the offset provider's cost of participating in the offset program and the buyer's willingness to pay for an offset (a function of GHG abatement costs in the industrial, electrical, and transportation sectors and emission caps). In voluntary markets, several other factors may impact price, as the motivation of offset buyers is not solely to achieve least-cost emission reduction (see Conte and Kotchen (2009) for further discussion of determinants of voluntary offset prices).

7.4 Tier 2 supply model

In the tier 1 supply model, sequestration between time t and T only registers in parcel x if x's LULC mix changes between t and T, x's wood harvest rates or harvest management changes between t and T, or x's HWP carbon pool is not in a steady state at time t. However, terrestrial carbon storage levels, especially on recently disturbed land, tend to change continuously due to vegetation growth and decay and organic matter accumulation in the soil. In the tier 2 terrestrial carbon supply model we account for any continual changes with carbon sequestration functions.

Similar to tier 1, carbon storage in a parcel in the tier 2 supply model is a function of the carbon stored in the aboveground biomass, belowground biomass, soil, other organic matter, and HWPs pools. Unlike tier 1, the tier 2 supply model incorporates a series of terms that account for the subsequent change in terrestrially stored carbon in x after a disturbance or land-use change. Formally, the metric tons of C stored on parcel x in year $m \in [t, T]$ is

$$C_{xm} = C_{pxm}$$
$$+ \sum_{j'=1}^{J} \sum_{j=1}^{J} A_{xj'jm} \left(\begin{array}{l} \alpha_{j'i'ji} C_{aj} + \beta_{j'i'ji} C_{bj} + \\ \gamma_{j'i'ji} C_{sj} + \eta_{j'i'ji} C_{oj} \end{array} \right), \quad (7.13)$$

where each pool-specific coefficient (the Greek letters) gives the fraction of the pool's maximum storage capacity achieved as of time m in an area that transitioned to LULC j i years ago from LULC j' that was i' years old at the time of transition to j (a LULC is i' years old if it has been i' years since the last major disturbance in the area occupied by the LULC). Further, $A_{xj'jm}$ is the area of parcel x that is in LULC j as of year m but was previously in LULC j' and the HWPs pool variable C_{pxm} is the same as the tier 1 HWPs pool variable (see Eq. (7.1)). Assuming that a pool-storage variable in Eq. (7.13) gives the pool's maximum storage capacity, then a value of 1 for its associated sequestration coefficient means that the pool has reached its maximum storage value as of year m. If for some reason the pool-storage variables in Eq. (7.13) give another reference value (e.g., the average storage value), then the value of the associated sequestration coefficient will need to be recalibrated such that the maximum value of the coefficient multiplied by the pool's reference storage value equals the pool's maximum storage capacity.

In our approach a series of pool-specific coefficients approximates the pool's sequestration function after a change to LULC j. For example, $\alpha_{j'i'j1}, \alpha_{j'i'j6}, \alpha_{j'i'j11}, \dots$ describes the relative change in aboveground carbon storage levels every 5 years beginning the year after LULC j' of age i' transitioned to LULC j. If parcel x is completely cov-

ered by LULC type j that was established 11 years prior to time m on land formally in LULC j' of age i' at the time of conversion then $\alpha_{j'i'j11} C_{aj}$ gives the aboveground carbon storage levels as of time m. See Figure 7.3 and the SOM for an illustration on the use of carbon sequestration coefficients (in general all coefficients behave in the same manner). See USEPA (2009) for an example of a tier 2 approach.

If we are to use tier 2 modeling in conjunction with an avoided emission analysis (Section 7.2.4) then the exact timing of deforestation will matter. In other words, we will have to define the probability of deforestation in each parcel that is forested as of time t for each time period m (i.e., π_{xtm} and $\hat{\pi}_{xtm}$ will need to be defined for each $m \in [t, T]$) as well as biomass carbon storage at each time period m in case deforestation is avoided.

7.5 Tier 2 valuation: an application of the avoided economic damage approach

The valuation approach used for tier 1 is also applied to tier 2 carbon sequestration estimates. We provide an example of tier 2 carbon sequestration and valuation modeling in a 22 × 20 km landscape located in northwest Minnesota, USA (Figure 7.4; Plate 4). The eastern half of the modeled landscape is dominated by early succession tree stands of aspen, white birch, maple, basswood, and oak. The western half is primarily in row crops and pasture, with a smattering of Conservation Reserve Program (CRP) perennial grasslands. The developed area in the northwest corner of the landscape is the town of Mahnomen.

In Figure 7.4 we give the year 2000 LULC pattern on the landscape (USDA-FSA 2000; Minnesota DNR—Division of Forestry 2000; USDA/NRCS 2008). We also generate two visions of land use by 2050 (not pictured). The Carbon Sequestration Scenario includes the restoration of 6 km² of prairie potholes on cropland, 6 km² of afforestation on cropland, 17 km² of cropland conversion to pasture, and 39 km² of cropland to perennial grassland by 2050 (we convert the least valuable croplands to these new uses and assume all other parcels retained their year 2000 LULC). Conversely, the CRP Loss Scenario

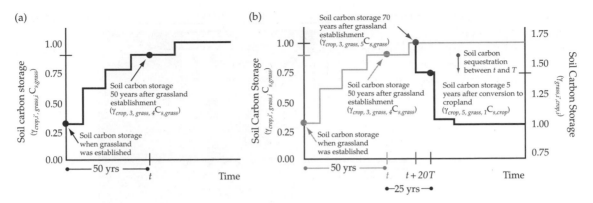

Figure 7.3 A tier 2 model illustration: carbon sequestration in soil from years t to T.

Assume a parcel is in grassland at time t. Assume the grassland was established 50 years ago on land that had been in row crops. Over time the carbon in the parcel's soil accumulates according to the sequestration curve in (a) (in this illustration soil sequestration rates are given for discrete steps in time). By time t the carbon stored in the soil of each hectare of the parcel has reached $\gamma_{crop, 3, grass, 4} C_{s, grass} = 0.9 \times C_{s, grass}$, where $C_{s, grass}$ is grassland's per hectare maximum storage potential, and the carbon sequestration coefficient $\gamma_{crop, 3, grass, 4}$ indicates the portion of the storage maximum that has been reached in a grassland that is in its fourth age-class bin since conversion from cropland in its third age-class bin at the time of conversion. Now suppose the parcel converts back to cropland 20 years after time t, or 5 years before T (i.e., $T = t + 25$). Between t and $t + 20$ soil carbon levels continue to increase according to the grassland sequestration curve (the light gray curve in (b)). The conversion to cropland at time $t + 20$ causes an immediate and significant loss of soil carbon (the initial vertical portion of cropland's sequestration curve given by the black curve in (b)). At time T the soil carbon level in the parcel is given by $\gamma_{pg, 5, crop, 1} C_{s, crop} = 1.45 \times C_{s, crop}$, where $C_{s, crop}$ is cropland's per hectare maximum storage capacity and $\gamma_{pg, 5, crop, 1}$ is the soil carbon sequestration coefficient for cropland in its first age bin at time T that was in grassland's fifth age bin at the time of conversion from grassland to cropland. Finally, sequestration in a hectare of this parcel from time t to T is given by $1.45 \times C_{s, crop} - 0.9 \times C_{s, grass}$ (a negative value).

assumed 877 ha of the perennial grassland (CRP land) that existed in 2000 converts to either row crops or hay production by 2050 (we assume all other parcels retained their year 2000 LULC).

We use tier 2 models to evaluate the consequences of the two land-use change scenarios on the soil carbon pool (C_s). Because estimates of C_{sj} and $\gamma_{j'i'ji}$ are uncertain, we estimate distributions for each of these model inputs using data from Smith *et al.* (2006), Anderson *et al.* (2008), and Nelson *et al.* (2009). We also use distributions (instead of point estimates) for the year of LULC conversion on the parcels that changed LULC, the discount rate (r), and the SCC (Tol 2009). For each scenario, we simulate VAD_{xtT} for all x on the landscape and associated VAD_{tT} 1000 times, each time drawing a unique value from our distributions for model parameters that are uncertain.

The mean present value of economic damage avoided because of soil carbon sequestration under the Carbon Sequestration Scenario is \$25 114 953 (SD\$23 244 250). If we only consider the parcels that experience LULC change under this scenario (15% of the landscape), the mean monetary value of soil

sequestration is \$6 036 494 (SD \$5 606 984) or \$892 ha⁻¹. In contrast, mean present value under the CRP Loss Scenario is \$19 043 572 (SD\$17 634 696). Including only the parcels that experienced LULC change (1.94% of the landscape), the mean monetary value of soil sequestration is \$210 952 (SD\$219 300) or \$241 ha⁻¹. Either way it is summed, the Carbon Sequestration Scenario results in more carbon sequestration, worth roughly \$6 million in avoided economic damages, than the CRP Loss Scenario. See this chapter's SOM for more details on scenario creation and tier 2 model variables.

In this example, we use variable distributions to account for some of the uncertainty in storage values, sequestration rates, SCC, the market discount rate, and dates of LULC transitions or disturbances in a scenario. However, we do not address several potential biases in our modeling. First, the dynamics of carbon storage and sequestration are complex and greatly simplified, even in the tier 2 models; whether or not output produced with a simplified model is systematically biased when compared to output from more detailed carbon sequestration models is an issue that warrants further investigation

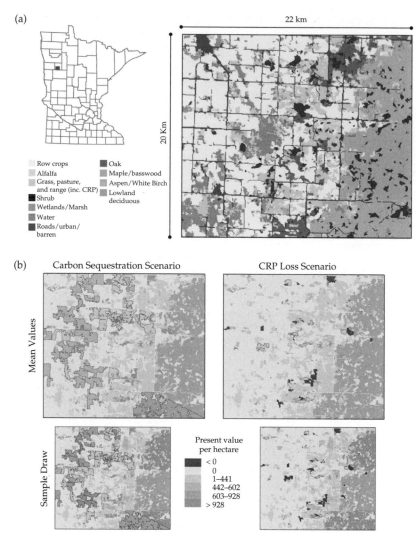

Figure 7.4 The value of carbon sequestered in soil across two alternative LULC scenarios.

(a) The year 2000 landscape. (b) The per-hectare monetary value of carbon sequestration in soil from 2000 to 2050 for each LULC scenario. The top row of maps gives mean results across all model simulations. The bottom rows of maps give the results from one particular run of the model. The black outlines on the parcels indicate parcels that experience LULC change in some portion of its area at some point between 2000 and 2050. The Carbon Sequestration Scenario map reflects a program of afforestation, restoring prairie pothole, and converting row crops to pasture and perennial grassland. In the CRP Loss Scenario any parcel that was primarily in CRP in 2000 was converted to row crops or a hayfield by 2050. (See Plate 4.)

(see Chapter 15). Second, measurement error and non-standardized sampling methods of storage and sequestration rates in the field may introduce systematic bias in storage and sequestration rate data used in our models (Brown 2002). Third, because the map used in this analysis and others like it represents a simplification of the actual landscape, additional error and potentially systematic bias are introduced into model results.

7.6 Limitations and next steps

7.6.1 Limitations

Our analysis of carbon sequestration and storage across a landscape is limited by several constraints. First, our models are driven by patterns of LULC and change in these patterns over time. While LULC change is likely the dominant factor in determining terrestrial carbon sequestration and storage (IPCC

2006), the nature and rate of disturbance events are also important (e.g., see IPCC 2006; Bond-Lamberty *et al.* 2007; Kurz *et al.* 2008). Prairie and forest fires, forest disease outbreak, and exotic species invasion can have significant impacts on storage and sequestration, but may not be reflected in typical LULC maps. Our models do not explicitly consider such events, but using more nuanced LULC classifications may begin to capture them (e.g., diseased conifer forest, disease-free conifer forest).

Second, our models largely ignore local variations in climate, which also have a significant impact on carbon storage and sequestration rates (e.g., McGuire *et al.* 2001). Rainfall and temperature patterns determine such ecosystem processes as net primary productivity (NPP) and soil erosion, the primary inputs in terrestrial carbon storage capacity and sequestration rates. The best strategy for minimizing this limitation is to use "local" carbon sequestration and storage data when possible and to limit the extent of the modeled landscape in order to minimize over-extrapolation of carbon estimates. Another way to deal with this limitation is to stratify LULC categories by landscape features that may reflect microclimate. This requires, however, observations of model parameters for each of the climate-related strata created within each broad LULC type. Further, as a climate changes, vegetation patterns may shift on the landscape. To account for this we can modify LULC types on future scenario maps according to climate–vegetation models. See Chapter 17 for such an exercise.

Finally, land-use management practices that do not have a large impact on carbon storage in plant biomass and soils can still be significant sources of other GHG emissions. For example, crop residue burning (e.g., IPCC 2006), livestock use (e.g., IPCC 2006), and various soil management practices on farms (e.g., Bouwman *et al.* 2002) can significantly increase the rate of methane and nitrous oxide emissions, two powerful GHGs (see Chapter 9). In addition, all of the GHGs emitted by fossil fuel-burning machinery used to support farms and other LULC are not reported except for the production and delivery of HWPs (see Eq. (7.1)).

In part to understand the consequences of our simple model and its corresponding constraints, we have begun to validate its results against more sophisticated models. For scenarios of LULC change in California that consider climate change (Shaw *et al.* 2009), we are comparing our results to those derived from a more detailed, process-based model of carbon storage used in Shaw *et al.* (2009). We will compare our simple results in Tanzania (see Section 7.2.5) to more detailed models developed by the Valuing the Arc Programme (Mwakalila *et al.* 2009). These comparisons will allow us to estimate the accuracy of our simple approach and the limitations and biases it presents.

7.6.2   Next steps

The models presented in this chapter can be used for many purposes. We can layer maps of sequestration and storage and their value with other ecosystem service maps to identify areas of ecosystem service synergies and trade-offs on the landscape (see Chapter 14). We can also compare sequestration supplies and values to those of other ecosystem service across a landscape to determine the opportunity costs of increasing carbon sequestration in both biophysical and economic terms (e.g., Jackson *et al.* 2005; Nelson *et al.* 2008).

Further, we can estimate the additional carbon sequestration on a landscape by comparing a baseline scenario of LULC with an alternative projection that reflects efforts to sequester carbon or reduce emissions (see the SOM for further details). By overlaying these maps with an opportunity cost map we can begin to predict market prices for offsets and avoided emissions credits (Kindermann *et al.* 2008), and to construct additional sequestration supply curves for different policies (e.g., Lubowski *et al.* 2006). Please see the SOM for descriptions of our efforts to validate the biophysical and deforestation risk models.

Finally, these models can serve as the foundation for a tier 3 approach to carbon storage and sequestration modeling. A tier 3 sequestration and storage model would not only consider how land use and conversion decisions effects storage, but it would also simulate the affect of climate and landscape-level disturbance stochasticity on biomass growth and carbon formation in and release from soils. Models such as CENTURY (Parton *et al.* 1992) and the vegetation model LPJ (Sitch *et al.* 2003) are examples of tier 3 models.

References

Alig, R. J., and Butler, B. J. (2004). Projecting large-scale area changes in land use and land cover for terrestrial carbon analyses. *Environmental Management*, 33, 443–56.

Anderson, J., Beduhn, R., Current, D., *et al.* (2008). *The potential for terrestrial carbon sequestration in Minnesota: a report to the Department of Natural Resources from the Minnesota Terrestrial Carbon Sequestration Initiative.* University of Minnesota, St. Paul.

Aukland, L., Sohngen, B., Hall, M., *et al.* (2002). *2001 Analysis of leakage, baselines, and carbon benefits for the Noel Kempff Climate Action Project.* Winrock International, Arlington, VA.

Bond-Lamberty, B., Peckham, S. D., Ahl, D. E., *et al.* (2007). Fire as the dominant driver of central Canadian boreal forest carbon balance. *Nature*, 450, 89–92.

Bouwman, A. F., Boumans, L. J. M. and Batjes, N. H. (2002). Modeling global annual N_2O and NO emissions from fertilized fields, *Global Biogeochemical Cycles*, 16, 1080.

Brown, S. (2002). Measuring, monitoring, and verification of carbon benefits for forest-based projects. *Philosophical Transactions of the Royal Society of London, Series A-Mathematical Physical and Engineering Sciences*, 360, 1669–83.

Brown, S., Swingland, I. R., Hanbury-Tenison, R., *et al.* (2002). Changes in the use and management of forests for abating carbon emissions: issues and challenges under the Kyoto Protocol. *Philosophical Transactions of the Royal Society of London Series A-Mathematical Physical and Engineering Sciences*, 360, 1593–605.

Burgess, N. D., Butynski, T. M., Cordeiro, N. J., *et al.* (2007). The biological importance of the Eastern Arc Mountains of Tanzania and Kenya. *Biological Conservation*, 134, 209–31.

Burtraw, D., Kahn, D., and Palmer, K. (2006). CO2 allowance allocation in the regional greenhouse gas initiative and the effect on electricity investors. *Electricity Journal*, 19, 79–90.

Cairns, M. A., Haggerty, P. K., Alvarez, R., *et al.* (2000). Tropical Mexico's recent land-use change: A region's contribution to the global carbon cycle. *Ecological Applications*, 10, 1426–41.

Canadell, J. G., and Raupach, M. R. (2008). Managing forests for climate change mitigation. *Science*, 320(5882), 1456–7.

Chape, S., Spalding, M., and Jenkins, M., Eds. (2008). *The worlds protected areas: status, values and prospects in the 21st century.* University of California Press, Berkeley.

Chave, J., Andalo, C., Brown, S., *et al.* (2005). Tree allometry and improved estimation of carbon stocks and balance in tropical forests. *Oecologia*, 145, no. 1, 87–99.

Chave, J., Condit, R., Lao, S., *et al.* (2003). Spatial and temporal variation of biomass in a tropical forest: results from a large census plot in Panama. *Journal of Ecology*, 91, 240–52.

Chomitz, K. M. (2002). Baseline, leakage and measurement issues: how do forestry and energy projects compare? *Climate Policy*, 2, 35–49.

Conte, M. N., and Kotchen, M. J. (2009). Explaining the price of voluntary carbon offsets. NBER Working Paper 15294.

Ebeling, J., and Yasue, M. (2008). Generating carbon finance through avoided deforestation and its potential to create climatic, conservation and human development benefits. *Philosophical Transactions of the Royal Society B: Biological Sciences*, 363, 1917–24.

Edinburgh Centre for Carbon Management (ECCM). (2007). *Establishing Mechanisms for Payments for Carbon Environmental Services in the Eastern Arc Mountains, Tanzania.*

FAO. 2005. *State of the World's Forests 2005.* FAO, Rome. Also available at http://www.fao.org/docrep/011/i0350e/i0350e00.HTM.

Fargione, J., Hill, J., Tilman, D., *et al.* (2008). Land clearing and the biofuel carbon debt. *Science*, 319, 1235–8.

Gibbs, H. K., Brown, S., Niles, J. O., *et al.* (2007). Monitoring and estimating tropical forest carbon stocks: making REDD a reality. *Environmental Research Letters*, 2, 045023.

Glenday, J. (2006). Carbon storage and emissions offset potential in an East African tropical rainforest. *Forest Ecology and Management*, 235, 72–83.

Harmon, M. E., Ferrell, W. K., and Franklin, J. F. (1990). Effects on carbon storage of conversion of old-growth forests to young forests. *Science*, 247, 699–702.

Intergovernmental Panel on Climate Change (IPCC). (2000). *IPCC special report on land use land-use change, and forestry.* Cambridge University Press, Cambridge, UK.

Intergovernmental Panel on Climate Change (IPCC). (2003). *Good practice guidance for land use, land-use change and forestry.* Institute for Global Environmental Strategies (IGES), Hayama, Kanagawa, Japan.

Intergovernmental Panel on Climate Change (IPCC). (2006). *2006 IPCC guidelines for national greenhouse gas inventories.* Institute for Global Environmental Strategies, Hayama, Kanagawa, Japan.

Intergovernmental Panel on Climate Change (IPCC). (2007a). *Climate change 2007: mitigation of climate change: Contribution of Working Group III to the Fourth Assessment Report of the Intergovernmental Panel on Climate Change*, Cambridge University Press, New York.

Intergovernmental Panel on Climate Change (IPCC). (2007b). *Climate change 2007: impacts, adaptation and vulnerability: contribution of Working Group II to the Fourth*

Assessment Report of the Intergovernmental Panel on Climate Change, Cambridge University Press, New York.

Jackson, R. B., Jobbagy, E. G., Avissar, R., *et al*. (2005). Trading water for carbon with biological sequestration. *Science*, **310**, 1944–7.

Kindermann, G., Obersteiner, M., Sohngen, B., *et al*. (2008). Global cost estimates of reducing carbon emissions through avoided deforestation. *Proceedings of the National Academy of Sciences*, **105**, 10302–7.

Kuebler, C. (2003). *Standardized vegetation monitoring protocol*. Centre for Applied Biodiversity Science, Conservation International, Washington, DC.

Kurz, W. A., Dymond, C. C., Stinson, G., *et al*. (2008). Mountain pine beetle and forest carbon feedback to climate change. *Nature*, **452**, 987–90 .

Lal, R. (2004). Soil carbon sequestration impacts on global climate change and food security. *Science*, **304**, 1623–7.

Lehmann, J. (2007). A handful of carbon. *Nature*, **447**, 143–4.

Lubowski, R. N., Plantinga, A. J., and Stavins, R. N. (2006). Land-use change and carbon sinks: econometric estimation of the carbon sequestration supply function. *Journal of Environmental Economics and Management*, **51**, 135–52.

Luoga, E. J., Witkowski, E. T., F., and Balkwill, K. (2000). Economics of charcoal production in miombo woodlands of eastern Tanzania: some hidden costs associated with commercialization of the resources. *Ecological Economics* **35**, 243–57.

McGuire, A. D., Sitch, S., Clein, J. S., *et al*. (2001). Carbon balance of the terrestrial biosphere in the twentieth century: Analyses of CO2, climate and land use effects with four process-based ecosystem models. *Global Biogeochemical Cycles*, **15**, 183–206.

Makundi, W. R. (2001). Carbon mitigation potential and costs in the forest sector in Tanzania. *Mitigation and Adaptation Strategies for Global Change*, **6**, 335–53.

Maréchal, K., and Hecq, W. (2006). Temporary credits: A solution to the potential non-permanence of carbon sequestration in forests? *Ecological Economics*, **58**, 699–716.

Marshall, A. R., Lewis, S., Lovett, J. C., Burgess, N., *et al*. (In preparation). Variation in carbon storage and tree allometry with elevation in the eastern arc.

Mastrandrea, M. D., and Schneider, S. H. (2004). Probabilistic integrated assessment of dangerous climate change. *Science*, **304**, 571–5.

Minnesota Department of Natural Resources (DNR)— Division of Forestry (2000). *GAP Land Cover Map of Minnesota*. Minnesota Department of Natural Resources, St. Paul, MN.

Mollicone, D., Achard, F., Federici, S., *et al*. (2007). An incentive mechanism for reducing emissions from conversion of intact and non-intact forests. *Climatic Change*, **83**, 477–93.

Murray, B. C., Sohngen, B., and Ross, M. (2007). Economic consequences of consideration of permanence, leakage and additionality for soil carbon sequestration projects. *Climatic Change*, **80**(1), 127–43.

Mwakalila, S., Burgess, N. D., Ricketts, T., *et al*. (2009). Valuing the Arc: linking science with stakeholders to sustain natural capital. *Arc Journal*, 23, 25–30.

Nelson, E., Mendoza, G., Regetz, J., *et al*. (2009). Modeling multiple ecosystem services, biodiversity conservation, commodity production, and tradeoffs at landscape scales. *Frontiers in Ecology and the Environment*, **7**, 4–11.

Nelson, E., Polasky, S., Lewis, D. J., *et al*. (2008). Efficiency of incentives to jointly increase carbon sequestration and species conservation on a landscape. *Proceedings of the National Academy of Sciences*, **105**, 9471–6.

Ndangalasi, H. J., Bitariho, R. and Dovie, D. B. K. (2007). Harvesting of non-timber forest products and implications for conservation in two montane forests of East Africa. *Biological Conservation*, **134**, 242–50.

Niles, J. O., and Schwarze, R. (2001). The value of careful carbon accounting in wood products. *Climatic Change*, **49**, 371–6.

Nilsson, S. and Schopfhauser, W. (1995). The carbon-sequestration potential of a global afforestation program. *Climatic Change*, **30**, 267–93.

Nordhaus, W. D. (1992). An optimal transition path for controlling greenhouse gases. *Science*, **258**, 1315–19.

Nordhaus, W. D. (2007). Critical assumptions in the Stern review on climate change. *Science*, **317**, 201–2.

Olson, D. M., Dinerstein, E., Wikramanayake, E. D., *et al*. (2001). Terrestrial ecoregions of the worlds: A new map of life on Earth. *Bioscience*, **51**, 933–8.

Parton, W. J., McKeown, B., Kirchner, V., *et al*. (1992). *CENTURY users manual*. NREL, Colorado State University, Fort Collins, Colorado.

Pfaff, A. S. P., Kerr, S., Hughes, R. F., *et al*. (2000). The Kyoto protocol and payments for tropical forest: An interdisciplinary method for estimating carbon-offset supply and increasing the feasibility of a carbon market under the CDM. *Ecological Economics*, **35**, 203–21.

Post, W. M., and Kwon, K. C. (2000). Soil carbon sequestration and land-use change: processes and potential. *Global Change Biology*, **6**, 317–27.

Raich, J. W., Russell, A. E., Kitayama, K., *et al*. (2006). Temperature influences carbon accumulation in moist tropical forests. *Ecology*, **87**, 76–87.

Ruesch, A. S., and Gibbs, H. K. (2008). New IPCC tier-1 global biomass carbon map for the year 2000. Available online from the Carbon Dioxide Information Analysis

Center (http://cdiac.ornl.gov), Oak Ridge National Laboratory, Oak Ridge, Tennessee.

Schuman, G. E., Janzen, H. H., and Herrick, J. E. (2002). Soil carbon dynamics and potential carbon sequestration by rangelands. *Environmental Pollution,* **116,** 391–6.

Schwarze, R., Niles, J. O., and Olander, J. (2002). *Understanding and managing leakage in forest-based greenhouse gas mitigation projects.* Institute for Environmental Economics at the Technische Universität Berlin, Berlin.

Shaw, R., Pendleton, L., Cameron, R., *et al.* (2009). *The impact of climate change on California's ecosystem services.* California Climate Change Center. Draft Paper.

Sitch, S., Smith, B., Prentice, I. C., *et al.* (2003). Evaluation of ecosystem dynamics, plant geography and terrestrial carbon cycling in the LPJ dynamic global vegetation model. *Global Change Biology,* **9,** 161–85.

Smith, J. E., Heath, L. S., Skog, K. E., and *et al.* (2006). *Methods for calculating forest ecosystem and harvested carbon with standard estimates for forest types of the United States.* General Technical Report NE-343, US Department of Agriculture, Forest Service, Northeastern Research Station, Newtown Square, PA.

Sohngen, B., and Brown, S. (2004). Measuring leakage from carbon projects in open economies: a stop timber harvesting project in Bolivia as a case study. *Canadian Journal of Forest Research-Revue Canadienne De Recherche Forestiere,* **34,** 829–39.

Sohngen, B., and Brown, S. (2008). Extending timber rotations: carbon and cost implications. *Climate Policy,* **8,** 435–51.

Stern, N. (2007). *The economics of climate change: the Stern review.* Cambridge University Press, Cambridge, UK.

Tol, Richard S. J. (2009). The economic effects of climate change. *Journal of Economic Perspectives,* **23,** 29–51.

Torn, M. S., Trumbore, S. E., Chadwick, O. A., Vitousek, *et al.* (1997). Mineral control of soil organic carbon storage and turnover. *Nature,* **389,** 170–3.

United Nations Climate Change Convention Secretariat (UNCCCS). (1997). *UNFCCC AIJ Methodological Issues.* UNCCCS, Bonn, Germany.

United Nations Framework Convention on Climate Change (UNFCCC). (1995). *Decision 5/CP.1 from Report of the Conference of the Parties on its first session, held at Berlin from 28 March to 7 April 1995. Addendum. Part two: Action taken by the Conference of the Parties at its first session.* UNFCCC, Berlin, Germany.

United States Department of Agriculture-Natural Resource Conservation Service (USDA/NRCS). (2008). *USDA-NASS Cropland Data Layer. 2006.* USDA/NRCS, Washington, DC.

United States Department of Agriculture-Farm Service Agency (USDA-FSA). (2000). *Conservation reserve program map of Minnesota. 1997.* USDA-FSA, Washington, DC.

United States Environmental Protection Agency (USEPA). (2009). *Inventory of US greenhouse gas emissions and sinks: 1990–2007.* USEPA, Washington, DC.

Valuing the Arc (2008). *Land use and land cover map of the Eastern Arc Mountains and surrounding watersheds.* Valuing the Arc, Cambridge, UK.

Victor, D. G., House, J. C., and Joy, S. (2005). A Madisonian approach to climate policy. *Science,* **309,** 1820–1.

Vohringer, F., Kuosmanen, T., and Dellink, R. (2006). How to attribute market leakage to CDM projects. *Climate Policy,* **5,** 503–16.

Weitzman, M. L. (2007). A review of the Stern review on the economics of climate change. *Journal of Economic Literature,* **45,** 703–24.

CHAPTER 8

The provisioning value of timber and non-timber forest products

Erik Nelson, Claire Montgomery, Marc Conte, and Stephen Polasky

8.1 Introduction

Forests play an iconic role in environmental conservation campaigns and are habitat to much of the world's known terrestrial biodiversity (Repetto and Gillis 1988). Forests also provide and regulate many important ecosystem services (e.g., Williams 2003; Ricketts 2004; Ticktin 2004; Maass *et al.* 2005), including carbon sequestration (e.g., Scholes 1996; Sohngen and Brown 2006), potable water supply (e.g., Núñez *et al.* 2006), and a stock of plants and animals that can be used to meet human needs for food, materials, and medicine (e.g., Milner-Gulland and Clayton 2002; Belcher *et al.* 2005; Damania *et al.* 2005; Ndangalasi *et al.* 2007). In fact, the legal and illegal harvest of timber and non-timber forest products (NTFPs) supports and supplements the livelihoods of millions of families around the world (e.g., Justice *et al.* 2001; Pattanayak and Sills 2001; Vedeld *et al.* 2004; Box 8.1). The ability of relatively intact forests to provide habitat and valuable ecosystem services year in and year out is the main argument given for forest conservation (e.g., Peters *et al.* 1989; Boot and Gullison 1995; Bawa and Seidler 1998; Arnold and Pérez 2001; Silvertown 2004; Sinha and Brault 2005).

In this chapter we present approaches for modeling the quantity and value of timber and NTFP harvest from forested parcels across a landscape (the role that forests play in the regulation and provision of other ecosystem services is covered in other chapters of this book). Timber and NTFP harvest levels and patterns are largely determined by forest ecology and property rights structure. Open-access forests are those in which anyone can harvest forest stocks. At the other extreme, a forest may be regulated with exclusive harvest rights, where only a limited number of individuals or entities have the legal right to harvest (Feder and Feeny 1991). An individual or entity with an exclusive right to harvest forest stocks has an incentive to maximize the net present value of economic returns to harvest over time, whereas open-access harvest tends to be characterized by a race to exploit the resource. Therefore, forests with well-enforced property rights will tend to have lower harvest rates and greater biological stocks at any point in time than open-access forests (e.g., Luoga *et al.* 2005; Birdyshaw and Ellis 2007; for a discussion on exceptions to this general rule see Larson and Bromley 1990).

As with other models presented in this book we present two tiers of analysis. Tier 1 involves approaches that are analytically simpler and require less data than the approaches in tier 2. In tier 1 models, we assume that the rate of harvest equals the harvested stock's natural regeneration rates, leaving the biological stock unchanged or in steady state over a harvest period. Steady-state harvest implies that the forest is being sustainably managed and future harvests of stocks can continue indefinitely. However, in many forested landscapes around the world stock levels are declining (e.g., FAO 2006). To accommodate the reality of forest degradation, we develop a tier 2 dynamic approach that allows for an estimation of harvest volume and value when harvest rates do not necessarily equal stock growth rates and forest stocks can degrade (or improve) over time.

The focus of this chapter is on forest and forest stocks but our approach can be used to model other land-cover types that provide extractive resources

Box 8.1 Wildlife conservation, corridor restoration, and community incentives: a paradigm from the Terai Arc Landscape

Eric Wikramanayake, Rajendra Gurung, and Eric Dinerstein

The Terai Arc Landscape (TAL) extends from Nepal's Chitwan National Park to Rajaji Tiger Reserve, India (Figure 8.A.1), and includes the forests and grasslands along the base and inner valleys of the Himalayas. The landscape was designed to conserve metapopulations of Asia's largest mammals, especially the tiger (*Panthera tigris*), Greater one-horned rhinoceros (*Rhinoceros unicornis*), and Asian elephant (*Elephas maximus*). These grasslands and forests are the most fragmented and converted Himalayan ecosystems (Wikramanayake *et al.* 2001); thus, the endangered species are mostly confined to protected areas where they face an uncertain future from the inevitable consequences of genetic

inbreeding, and breakdown of natural ecological dynamics and behavioral interactions that structure their populations and communities. The conservation challenge in the TAL is to conserve these species as ecologically, demographically, and genetically viable populations. The strategy is to create a conservation landscape that links twelve protected areas with corridors to facilitate dispersal.

We used a GIS-based habitat analysis to identify the potential corridors (Wikramanayake *et al.* 2004). The analysis revealed several bottlenecks in the potential network, including six restoration priorities and three transboundary corridors between protected areas in Nepal and India (Figure 8.A.1).

There are several challenges to restoring and maintaining corridors. Over 7 million people live in the TAL

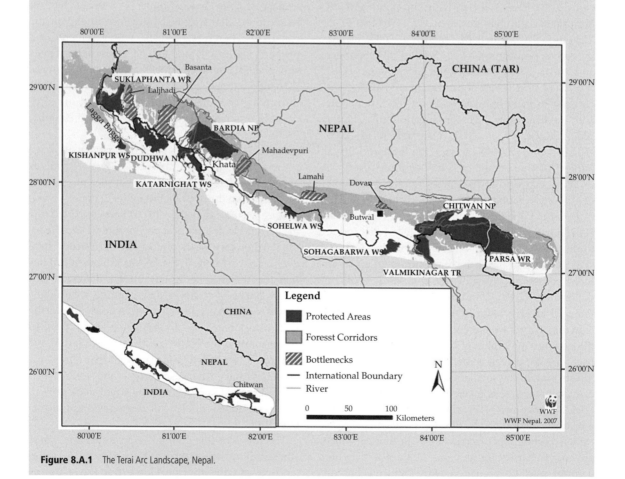

Figure 8.A.1   The Terai Arc Landscape, Nepal.

and tend over 4.5 million head of livestock (MFSC 2004). Immigration to the Terai still continues, and contributes significantly to the 2.86% population growth rate (WWF 2006). Most immigrants encroach into state forests, clear land, and begin to cultivate. After a couple of years, some sell the land and move on to encroach and occupy another forest patch, creating a chronic process of forest degradation.

Most of the livestock are "scrub" cattle which are allowed to free-range in state forests. As a result, these forests are overgrazed. Controlling cattle grazing in state forest land is difficult because the forest department is under-resourced.

To restore bottlenecks the TAL program identified and facilitated the conversion of strategic corridor areas to community forests, a forest management strategy that has worked well to restore forests in Nepal (Nagendra *et al.* 2005). Under community forestry, degraded state forests are assigned to local forest user groups to manage, based on plans approved by the Forest Department (Nagendra 2002). Because user groups receive management and usufruct rights they have a vested interest in sustainably managing the forests.

Even within the context of community forestry, however, user groups still require fuelwood and other forest resources. Therefore, alternatives to reduce this demand were necessary. The fuelwood demand was eased by providing subsidies to the communities for biogas plants, which use cattle dung to produce methane for cooking and lighting. Better cattle breeds promote stall-feeding, instead of free-grazing, with the added incentive of making it easier to collect cattle dung for biogas plants.

Five years after bottleneck restoration began the "big three" species—tigers, rhinoceros, and elephants—have begun to use the corridors. Tigers are present in four corridors, rhinoceros in two, and the frequency of elephant movement has increased in five. In some corridors elephants have returned after a lapse of 50 years, according to local residents.

Since 2002, the TAL program has facilitated the conversion of over 193 km^2 to community forests in strategic areas of six bottlenecks to 196 forest users groups. These forests are being used by over 24 500 households. The biogas plants and fuel-efficient cooking stoves provided from 2001 to 2006 have saved an estimated 21 000 metric tons of fuelwood annually, representing an equivalent of over 1.6 km^2 of clear-cut forests. Biogas as an alternative energy also represents an opportunity for carbon trading in the voluntary market. As of 2005, the biogas program qualified as a Gold Standard CDM-VER project.

The transboundary Khata corridor represents an interesting case. It links Bardia National Park with Katerniaghat Tiger Reserve, and tigers harbored in both parks now use the corridor. Bardia also supports an important rhinoceros population. Several rhinos now use the corridor, and five rhinoceros from Bardia have begun to reside in Katerniaghat. Elephants have also begun to use the corridor. Thus Khata is now a functional wildlife corridor.

In Khata, the TAL program has promoted community-based enterprises that use forest products. For instance, *Bel* fruits (*Aegel marmelos*) are used to produce juice under a community-based project. During the fiscal year 2005–6 over 25 metric tons of fruit were harvested and sold to the cooperative. Over 17 400 bottles of *Bel* juice were produced and marketed, which earned a net profit of about US$6000, a considerable sum for rural communities.

The TAL program has also promoted *mentha* (*Mentha piperita*) farming, and provided a distillery to extract oil. Mentha was originally promoted as a "live fence" to prevent crop damage by elephants. But the financial benefits from mentha oil proved to be so lucrative that more farmers adopted it as a primary cash crop and the output exceeded the carrying capacity of the distillation plant, and three additional distillation plants had to be built to accommodate the demand. In fiscal year 2008, the income from 80 ha of mentha production was US$59 000. The TAL program also established a community-based rattan furniture-making enterprise that earns the user groups over US$3000 annually. The rattan is sustainably harvested from the community forests.

Over 160 men and women from the user groups are also involved in community-led anti-poaching operations in Khata. This group patrols the corridor to safeguard the forests and wildlife from illegal activities such as encroachment, logging and collection of forest products, and poaching. Even though several rhinos have been poached from within the core of Bardia National Park, no poaching incidents have occurred in the corridor, under the vigilance of the community anti-poaching units.

The experiences from the TAL show that restoration and conservation of wildlife habitat in a larger landscape is possible through local stewardship, as long as the local communities benefit financially from their natural resources management.

under a range of property right structures. For example, open-access grasslands can provide food and fuel for local households as well as a food and water source for their livestock (e.g., Swallow and Bromley 1995; Thwaites *et al.* 1998; Adger and Luttrell 2000).

8.2 The supply, use, and value of forests' provisioning service in tier 1

In our ecosystem service taxonomy only the portion of a forest's stock that is harvested represents the forest's provisioning service. We place a value on the use of a forest's provisioning services when we convert the forest's harvest volume into a monetary value. Accurately quantifying timber and NTFP harvest levels and values requires socio-economic and ecological data. Because data availability and quality will vary, we provide several alternative approaches in tier 1 to quantify harvest volume and value. We first introduce a method that uses current or historical harvest volume or harvest effort, harvest costs, and stock price data to measure a current or baseline period's volume and value of harvest in a forest parcel (Section 8.2.1). We then provide a method for estimating steady-state harvest volume and value in a forest parcel when data on harvest volume or effort are not available (Section 8.2.2).

8.2.1 Calculating harvest value when harvest volume or harvest effort is observable

In some cases, such as with commercial timber harvests in developed countries, relatively good data exist on harvest volumes, harvest costs, and product prices. This data can come from one of several sources, including field monitoring of harvests, surveys of harvesters, government statistics, or, if the stocks are sold in the market, market data. We denote harvest volume of some product from a forest over a time period (typically a year) with H. If we cannot find data on H but instead can find data on harvesting "effort" (measured in hours) in the forest over the time period, denoted by E, and the harvest volume collected per hour by a typical harvester, given by V,

we can generate an estimate of H from E and V; $H = E \times V$.

The cost of harvesting a stock is comprised of two components, the wages paid to labor and the capital costs associated with harvesting equipment. Let the total cost of harvesting the stock in the forest parcel over a harvest period be given by C. In rare cases, information on costs will be directly available. More commonly, cost estimates can be generated by multiplying harvest effort by the sum of hourly average wage rate of labor, W, and the amortized cost of capital equipment used per unit of harvest effort, Y (in this cost calculation we ignore the opportunity costs associated with harvesters traveling to and back from forest parcels and any costs associated with transporting harvested stocks to processing centers).

The price or value of the product made from the harvested stock, denoted by p, is the final data element needed to compute harvest value. If the harvested product is traded in a market then market prices can be used. Otherwise, if the product is generally consumed directly by the harvester and not traded in a market, then other methods, such as non-market valuation methods, will need to be used to estimate p.

The net value of a product's harvest from the forest parcel over the given time period is equal to the revenue from that harvest minus the costs incurred in the process,

$$NV = (p \times H) - C = E \times ((p \times V) - W - Y), \qquad (8.1)$$

where harvest volume can be observed, H, or calculated with $E \times V$, and costs can be observed, C, or calculated with $E \times (W + Y)$.

In Eq. (8.1) increasing harvest effort in a time period increases both revenue and costs in a linear fashion (each additional unit of E generates $p \times V$ in revenue but costs $W + Y$). However, large jumps in effort may lead to nonlinear changes in net returns. This may occur because a large pulse in effort over the course of a harvest period could drastically reduce the stock's level, making the stock harder to find as the time period progresses. Such a dynamic will reduce the harvest rate V as the harvester has to use more time to search for the increasingly scarcer stock. If V does fall as effort increases, then net value will fall as well because per unit effort costs remain the same no matter the effort level. Therefore, caution should be exercised in using the same V value across very different effort levels.

We can map all harvest volumes and values on the landscape by expanding the net value equation over all harvested biological stocks and forest parcels on the landscape,

$$NV_{xz} = p_z \times H_{xz} - C_{xz} = E_{xz}\left((p_x \times V_{xz}) - W - Y\right), \quad (8.2)$$

where $x = 1, 2, \ldots, X$ indicates distinct forest parcels on the landscape and $z = 1, 2, \ldots, Z$ indexes biological stocks available in the landscape's forest parcels. The sum of H_{xz} and NV_{xz} over all z represents the use and net value of forest parcel x's provisioning service over the given time period, respectively, and the sum of H_{xz} and NV_{xz} over all z and x represents the use and net value of the landscape's forest provisioning service over the given time period, respectively.

If we do not observe harvest volumes or harvest effort then we cannot calculate the current or baseline period's net value of harvest with Eq. (8.1) and we need to apply alternative methods described below.

8.2.2   Calculating steady-state harvest volume and value

Let $G(S)$ indicate the growth of a biological stock over a time period in a forest parcel where S indicates the level of the stock. For many stocks growth slows down as stock increases due to resource limitations (density-dependent growth rates; see Boot and Gullison 1995). In steady-state harvesting, the per period harvest of a stock from a forest parcel is equal to its biological growth rate in the parcel,

$$H = E \times V(S) = G(S), \quad (8.3)$$

where we now explicitly indicate that a stock's harvest rate, V, is a function of its biological stock level. If we can observe the level of the biological stock from a forest inventory and can specify the stock's growth function $G(S)$ then we can use Eqs. (8.3) and (8.1) to estimate a period's steady-state harvest and its net value (assuming we have cost and price data). We can map all steady-state biological stocks, harvest volumes, and values on the landscape if we

can specify the biological stock levels and their growth functions in every forest parcel.

However, even if we lack observations of biological stock, we can still estimate harvest volumes and values under steady-state conditions if we can specify $V(S)$ and $G(S)$ and ascribe certain behavior to harvesters. In the next sub-sections we describe several methods for finding harvest volumes and values without stock data. The method we use to find harvest volumes and values will depend on the presence and protection of harvest rights and our assumptions about harvest behavior. We begin with a situation of open-access harvesting, followed by exclusive harvest rights, and conclude by discussing a case of intermediate access.

8.2.2.1   Open-access steady-state harvest volume and value

In many parts of the world, especially in developing countries, households harvest stocks from forests to supplement their incomes (e.g., Monela *et al.* 1993). Household harvesting is most active in forest parcels where harvest rights are not established or not enforced, so-called "open-access" parcels (Feder and Feeny 1991; Hyde 2003). When using such open-access forests, households typically do not consider the value of leaving biological stock *in situ* to mature and harvest at a later date because other harvesters are likely to harvest it in the meantime. As long as household labor is not in short supply on the landscape, we can expect households to continue to enter the forest and harvest resources until the last entrant cannot make an economic profit from further harvesting (where foregone wages in the market and any harvesting equipment costs represents household harvesting costs; see Gordon 1954; Conrad and Clark 1987; Clark 1990; Lopez-Feldman and Wilen 2008).

This means that if all households earn the same return from a unit of harvest effort, then in open-access equilibrium no household will make an economic profit from harvesting (i.e., the revenues from harvesting equal the wages a household foregoes to harvest plus any equipment costs). However, when households are differentiated, either because they have different harvesting skills, different travel distances to harvest sites, or have different transport access to harvest sites (e.g., bicycle versus truck), then some households may earn positive net returns in open-access equilibrium. With differentiation, there will be a critical distance or skill level at which a household choosing to harvest will just break even (the value of harvest will just equal the household's opportunity cost of harvesting plus any equipment costs). In open-access equilibrium, those households closer to the forest or with higher skill level will harvest while other households will not.

Here we illustrate the case of open access where households are identical in all ways except for their proximity to the forest. Suppose that a standard harvesting trip to a forest parcel for a certain stock involves e hours of harvesting in the forest as well as the travel time to and from the forest. Let d^h be the hours spent traveling to and from the forest for household h. We can either model the actual locations of all households, or if spatially explicit census data is coarser, we can divide the landscape into regions and treat all households in a region as having the same distance.

The household most distant from the forest that engages in harvest, labeled household n, will earn zero economic profit from a harvesting trip,

$$pH = pV(S)e = (W+Y)(e+d^n).$$ (8.4)

We can use Eq. (8.4) to solve for the distance (measured in hours) that household n travels to harvest,

$$d^n = \frac{[pV(S)-(W+Y)]e}{(W+Y)}.$$ (8.5)

If we order households from closest to furthest from the forest, then all households with a travel time less than or equal to d^n, given by $h = 1, 2, \ldots, n$, will enter the forest to harvest. Household harvest over some time period is

$$H^o = V(S) \times nFe,$$ (8.6)

where H^o indicates aggregate household harvest of the stock from the forest parcel over the time period, F is the average number of trips that the household takes during the time period, and $E^o = nFe$ is the household's aggregate harvest effort in the forest over the given time period (the superscript "o" signifies household behavior versus exclusive harvest-right holder behavior). Hereinafter the lower case "e" indicates sub-time period harvest effort and E indicates the effort over the entire time period of

interest (e.g., e indicates daily effort decisions and E is the annual sum of all daily decisions).

On each harvest trip, a household will earn $(W + Y)(d^n - d^h)$ (see the supplementary online material (SOM) for a proof). This represents the difference in travel time to the forest parcel between household h and household n multiplied by the value of an hour to a household. Because household n earns zero net returns, the value added to household h from harvest is simply given by the value of time that h does not have to spend traveling. All household net returns from harvesting over the time period is given by,

$$NV = \sum_{h=1}^{n} F(W+Y)(d^n - d^h).$$ (8.7)

Implementing this model requires specification of travel time to the harvest site as a function of distance, location of households relative to the site, average productivity of harvest in the site, $V(S)$, wage and equipment costs per unit of time, $W+Y$, average amount of effort per trip for households that harvest from the site, e, and average number of harvest trips in a time period, F. We can map all open-access steady-state harvest volumes and values on the landscape by expanding the harvest volume Eq. (8.6) and the net value Eq. (8.7) over all relevant biological stocks and open-access forest parcels on the landscape.

8.2.2.2 Exclusive harvest-right steady-state harvest volume and value

At the other extreme from open access, harvests in a forest can be completely limited to those with a property right to harvest. Exclusive harvest rights can be granted to private companies, communities, or governments (Feder and Feeny 1991; Engel and Lopez 2008; Guariguata *et al.* 2008). In most cases exclusive harvest rights pertain to timber harvest.

If there are no restrictions on forest structure then we assume the holder of a timber concession will create a monoculture forest where a targeted volume of wood can be harvested at regular periods (rotational forestry; see Tahvonen and Salo 1999; Dauber *et al.* 2005). Further, we would expect rotational forest operators to structure their holdings such that operation-wide annual harvest levels are roughly equal to annual biological growth. One reason for such behavior is that it creates a smoother flow of revenue and consumption over time, something many businesses and communities seek (Browning and Crossley 2001).

Because the holder of an exclusive harvest right can manipulate stock levels in a forest, and thus stock growth rates, the holder can theoretically choose any steady-state harvest level they wish. Here we assume the operator of a rotational forest will choose steady-state harvests that maximize the net present value (NPV) of economic returns from harvest over time. To do this, the operator will have to consider how the current harvest affects harvest potential in the future. For example, by harvesting most of the biological stock in the current period the owner will not be able to capture the additional revenue that could be gained by leaving stock *in situ* to grow larger and be harvested in the next period. Therefore, the solution to the owner's problem involves solving a dynamic optimization model that incorporates trade-offs between the current and future harvest. This type of dynamic harvest model has been extensively analyzed (Clark 1990).

The steady-state solution to the harvest-right holder's net revenue maximization problem is characterized by the following system of equations,

$$(p - \alpha d)V(S) = (W + Y) + \mu V(S) \qquad (8.8)$$

$$(p - \alpha d)E \frac{\partial V}{\partial S} = \mu \left(r - \frac{\partial G}{\partial S} + E \frac{\partial V}{\partial S} \right) \qquad (8.9)$$

$$G(S) = H = E \times V(S), \qquad (8.10)$$

where α represents the cost to haul one unit of harvest volume one kilometer, d is distance in kilometers from the forest parcel to the harvested stock's processing site, and μ is the monetary value of the biological stock (the "shadow" value in economics vernacular; see Clark (1990) or the SOM for more information about the derivation of these conditions). According to Eq. (8.8) optimal harvest is found by equating the net value of one more unit of harvest effort (the left hand side) with the marginal cost of additional unit of harvest effort, where this

marginal cost is the sum of the direct cost of effort plus the loss in future value from depleting the biological stock. Combining Eqs. (8.8) and (8.9) yields an expression for optimal harvest-right holder effort,

$$E \frac{\partial V / \partial S}{\mu} \left[\frac{(W + Y)}{V(S)} \right] + \frac{\partial G}{\partial S} = r. \qquad (8.11)$$

In words this condition means that the harvest-right holder will hold a stock *in situ* up to the point where its marginal growth rate plus the marginal benefit in harvest productivity due to the additional stock equals the economy's interest rate r. Finally, Eq. (8.10) enforces the steady-state assumption.

Assuming $G(S)$ and $V(S)$ are defined and p, α, d, r, W, and Y are observed we can use Eqs. (8.8) through (8.10) to find harvest-right holder optimal effort in a forest, optimal stock levels in the forest, and the optimal value of the stock, or E^c, S^c, and μ^c, respectively. Optimal steady-state harvest by the exclusive harvest-right holder, H^c, is determined by evaluating the harvest function at E^c and S^c while the net value of the optimal steady-state harvest each time period, NV^c, is given by evaluating Eq. (8.1) at H^c, E^c, and S^c (however, we now have to add transportation costs to the net revenue equation). If we determine exclusive harvest-right volumes H^c and net values NV^c across all relevant biological stocks and parcels with exclusive harvest rights then we can create a map of these values.

While this approach is theoretically consistent, it may be difficult to define the functional relationships and find the necessary data. In such cases we can fall back on simpler methods to find H^c, such as looking for published harvest volumes, biological growth rates, and harvest costs in government documents, forestry management journals, or other industry documents (this approach to finding harvest-right holder harvest volumes and costs was used in Polasky *et al.* 2008).

If the holder of an exclusive harvest right is a community or government, their objective may not be to maximize the NPV of harvest but to maximize an objective function that includes spiritual, biodiversity, and regulating ecosystem service values associated with the forest. The approach outlined here can be modified to include these different objectives. Alternatively these considerations can be included as constraints on the type of practices that harvest right-holders are allowed to engage in. If a government or community is interested in maintaining a semblance of a forest's natural state then timber harvest rights typically only allow selective logging (Repetto and Gillis 1988; Pinard and Putz 1996). For example, in New Zealand only certain old-growth forests can be logged and the total biomass removed annually from these forests is limited to 20% of annual biomass growth (see http://www.insights.co.nz/Natural_Forests_r.aspx#a). In such systems the most mature trees are generally selected for logging in order to keep harvest costs per unit of volume removed low. When a forest parcel includes selective logging concessions we can use Eq. (8.1) along with information on the selective logging restrictions and costs (including transportation costs) to estimate the net value of harvest by the harvest-right holders.

8.2.2.3   Intermediate access steady-state harvest volume and value

The open-access and exclusive harvest models represent extreme cases of harvesting activity. In many forests around the world households (illegally) harvest stocks that are subject to exclusive harvest rights (Hyde 2003). In response to the threat of illegal harvest, harvest-right holders often expend some effort to protect their property rights (Feder and Feeny 1991). In addition, the models above do not consider the influence that alternative harvest sites would have on household harvest behavior, that households may harvest multiple stocks in a forest, and the unobserved household attributes that influence its harvest effort choices (e.g., household harvesting skills or preferences).

Here we describe an approach that incorporates these factors in a steady-state harvest model. In this approach we model household harvesting decisions by comparing a household's potential harvesting revenues to their opportunity costs where costs include time spent avoiding any forest guards and potential illegal harvest fines (Barbier and Burgess 2001). In this approach we do not model by harvest right-holders but instead recover or estimate their harvest levels by enforcing steady-state equilibrium and comparing household harvest volume to stock

growth. However, if we observe harvest by harvest right-holders or it does not exist in a forest parcel then we can drop the steady-state assumption in this particular tier 1 model.

The first step in this intermediate access model is to calculate a household's expected net revenues generated from harvest, no matter whether it is illegal or not. Like all other tier 1 models this is a function of effort. Let D be the hours in a household's working day if we are modeling day trips or working hours a harvester expects to devote to a harvesting trip if we choose to include multi-day trips in our model, let d be the household's travel time in hours to and from the forest parcel, and let e_z be the hours that the household spends harvesting stock z in the forest on the trip such that

$$D = d + \sum_{z=1}^{Z} e_z, \tag{8.12}$$

where $\sum_{z=1}^{Z} e_z = 0$ if $d \geq D$. We can calculate $\sum_{z=1}^{Z} e_z$ by using GIS software and travel speed assumptions to determine d and subtracting this from D.

In this intermediate access model a household incurs two costs when harvesting, wages lost by not working in the labor market and expected fines from being caught harvesting illegally (we will ignore equipment costs for now). Let $\rho_z \in [0, \bar{\rho}]$ indicate the intensity of efforts to prevent the illegal harvest of stock z where $\rho_z = \bar{\rho}$ indicates maximum intensity and delta (p_z) is the expected daily or trip monetary fine from harvesting z illegally. We assume delta is increasing in p. Efforts to reduce illegal harvest will affect household harvest in two ways. First, they may reduce hourly harvest rates as households have to spend time avoiding detection. Second, forests with less intensive protection of harvest rights will be subject to more household harvest, all else equal.

The household's daily or trip net revenue from harvest in a forest parcel is given by,

$$nr = \left(\sum_{z=1}^{Z} \left(p_z e_z V_z(\rho_z, S_z) \right) - \delta(\rho_z) \right) - (W + Y)D + u, \tag{8.13}$$

where $V_z(\rho_z, S_z)$ is decreasing in ρ_z and increasing in S_z, $\delta(\rho_z)$ is the expected daily or trip monetary fine from harvesting z, which is increasing in ρ_z, and u is a household-specific random variable that includes unobserved household characteristics that affect the net value calculation, and all other variables are as before. If biological stock data is missing then we estimate $V_z(\rho_z)$ instead. If harvest rights for stock z in parcel x do not exist or are not enforced then $V_z(\rho_z = 0, S_z) = V_z(S_z), \delta(\rho_z = 0) = 0$, and Eq. (8.13) reduces to the open-access net revenue function.

We assume the household will choose to spend the day or trip harvesting in the forest parcel that is expected to generate the highest net return when compared to all other working options (Parker *et al.* 2003). Each forest's probability for generating the highest household net revenues for a day or trip time period is given by the joint probability that harvesting from the parcel generates both positive expected net returns (i.e., the probability that harvesting in the parcel is better then working in paid labor) and generates the highest expected net returns when compared to all other potential harvest sites. Let this joint probability for forest parcel x be given by γ_x. The parameter γ_x will only lie between 0 and 1 (and not be equal to 0 or 1 exclusively) if we include the random variable u in Eq. (8.13) or assume that the price, wage, fine, effort, or harvest-right variables in Eq. (8.13) are random variables.

The value of γ_x decreases as enforcement efforts in parcel x increase (higher ρ_z), as the distance between the household and parcel x increases (higher d and lower $\sum_{z=1}^{Z} e_z$), and as household wage rates increase. The value of γ_x will increase with an increase in biological stocks on parcel x or with the price of the harvested stock. If the sum of γ_x across all forest parcels on the landscape is less than 1 (i.e., $\sum_{x=1}^{X} \gamma_x < 1$) then there is some possibility that a household will not harvest at all on a given day or over a potential trip period and instead work as paid labor in another economic sector.

Next we use γ_x values over all households to calculate expected aggregate household harvest effort in parcel x over the modeled time period. Rather than model each individual household, which could be a computationally difficult task on large landscape, we can use a limited number of representative households to account for all household harvesting behavior on the landscape. The representative households should delineate the greatest differences in household locations, harvesting skill sets, and harvesting preferences. For example, if

we have divided the landscape into 1) three distinct regions, 2) assume two classes of harvesting skill, low and high, and 3) assume two harvesting preferences, gathering timber to sell in the market versus gathering NTFPs for home consumption, then there are 12 representative households. Let $v = 1, 2, \ldots, V$ index all household types on the landscape. If N^v is the number of households of type v and β^v is the number of v's working days in the time period if we are modeling daily decisions or the number of trips considered over the course of a time period if we are considering multi-day trips then the expected aggregate household harvest of stock z in parcel x over the time period is the sum of expected daily or trip household harvests over all households.

$$H_{xz}^o = \sum_{v=1}^{V} \beta^v \gamma_x^v N^v e_{xz}^v V_{xz}^v(\rho_{xz}, S_{xz}) , \qquad (8.14)$$

where we have allocated $\sum_{z=1}^{Z} e_{xz}^v$ across all z in parcel x and $\beta^v \gamma_x^v N^v e_{xz}^v$ is equivalent to aggregate illegal harvest effort for z in x over a harvest period by households of type v, or E_{xz}^v. For example, if we are modeling two stocks, timber and food NTFPs, then the representative households that prefer timber could be assigned an effort value of 0 for food NTFPs and vice-versa.

Finally, the aggregate harvest of stock z in forest parcel x over the time period assuming steady-state harvest is given by H_{xz},

$$H_{xz} = H_{xz}^c + H_{xz}^o = G_{xz}(S_{xz}). \qquad (8.15)$$

where H_{xz}^c is the harvest level of z in forest parcel x by z's harvest-right holder in x and is recovered by subtracting H_{xz}^o from $G_{xz}(S_{xz})$. In this model we do not estimate harvest-right holder behavior. Instead we are reliant on observations of ρ_{xz} and $G_{xz}(S_{xz})$ to estimate H_{xz}^c.

The net value of harvest in parcel x over the time period is given by,

$$\begin{aligned} NV_x = &\sum_{z=1}^{Z} p_z H_{xz}^o + \sum_{z=1}^{Z} (p_z - \alpha_z d_{xz}) H_{xz}^c \\ &- \sum_{v=1}^{V} (W^v + Y^v) \beta^v \gamma_x^v N^v D^v , \\ &- (W + Y) E_{xz}^c - \omega_{xz}(\rho_{xz}) \end{aligned} \qquad (8.16)$$

where d_{xz} is the distance from parcel x to stock z's processing site, the superscript "v" on W and Y indicate that wages can vary across household types (versus average wages levels on the landscape), E_{xz}^c is the effort level needed by z's harvest-right holder in x to achieve H_{xz}^c and $\omega_{xz}(\rho_{xz})$ is the cost to z's harvest-right holder in x to achieve illegal harvest prevention effort ρ_{xz}. We do not include the revenues generated from fines because they are a transfer of household wealth to the enforcement authority (either the harvest-right holder or the government); no additional societal value is generated in these transfers. By indexing harvest volumes and values by biological stock and parcels we can easily transfer harvest volumes H_{xz} and net values NV_x to maps.

While this model incorporates many realistic aspects of harvesting, it requires extensive information to implement. It requires information on illegal harvest prevention effort, expected illegal harvest fines, household locations, distance to forest parcels and speed of travel, wages and equipment costs, product prices, and legal harvest rates or biological growth rates. Obtaining all of the data inputs required for use of this approach will most likely require independent research as well as consultation with local experts and modeling with GIS software.

8.2.3   Tier 1 intermediate access model example

To illustrate portions of the household harvest model described in Section 8.2.2 we focus on a small region in the Eastern Arc Mountains watershed of Tanzania with seven urban areas that vary in size from small villages to major urban areas and 32 forest parcels (see Figure 8.1). In this example we model the harvest of wooden poles (poles are used in small-scale construction and building) and mushrooms. The maps in Figure 8.1 indicate in which forest patches mushroom and pole stocks are either high or very high. In this example we assume that it is only worthwhile for households to harvest poles mushrooms in forests where stocks are high or very high. There are no exclusive right harvests in this landscape ($H^c = 0$ for all parcels). Several areas on the landscape have conservation status and technically forbid harvest of biological stocks. The conserved areas are indicated by the dashed lines on the map in

Figure 8.1. The area within the longer-dashed line is much better protected by government officials than areas within the shorter-dashed lines (data on relative stock levels and protected areas come from a panel of Tanzanian experts consulted at a conference in Morogoro, Tanzania in February, 2008).

In this example, we assume that there are two types of harvesting households in each urban area: high-skilled (HS) and low-skilled (LS; for simplicity we assume all household harvest originates from these seven urban areas). We index the urban area-household type combination on the landscape by v (seven urban areas and two household types means $v = 1$, $2,…, 14$). We assume highly skilled households travel to forested-parcel access points by truck and then walk from the road to the parcel's centroid. We assume that low-skilled households travel to a forested parcel access points by bike and then walk from the road to the parcel centroid. In addition, we assume highly skilled households have higher harvest rates than their low-skilled counterparts for both mushrooms and poles, all else equal (i.e., $V^{HS}(S) > V^{LS}(S)$; see the SOM for more information on the values of $V(S)$ for each representative household type). We assume all harvesting trips are daily trips.

We use a modified version of Eq. (8.13) to determine the daily net revenue value of v's harvest in forest parcel x, given by,

$$nr_x^v = \left((1-\rho_x)\sum_{z=1}^{2}\left(e_{xz}^v p_z V_{xz}^v(S_{xz})\right)\right) - (W^v \times D), \quad (8.17)$$

where efforts to prevent illegal harvest anti-poaching efforts are not distinguished by stock type, $\rho_x = 0$ in forest parcels with not conservation status (open-access parcels), $\rho_x > 0$ in conserved forest parcels, and D measures hours in a work day. Here we set $\delta(\rho_x) = 0$ because there is no fine if a household is caught illegally harvesting; instead, when caught, the harvest is confiscated. Daily household effort in forest parcel x (i.e., e_{xz}^v) is found by calculating the time needed by representative household v to travel to parcel x on the landscape's road and path network given v's transportation mode, multiplying this by two (there and back), and subtracting this total travel time from D.

To account for uncertainty in model variables we treat ρ_x for all conserved parcels, p_z for all z, $V_{xz}^v(S_{xz})$

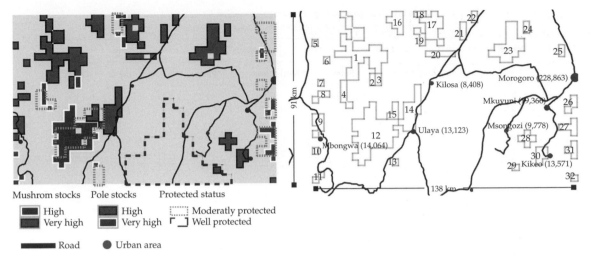

Figure 8.1   Distribution of urban areas, mushroom and pole sources, and protected areas on a portion of the Eastern Arc Mountains watershed in Tanzania.

In map (a) each dark-shaded parcel is a distinct forest stand. Parcels with the same border (either white or black) have similar stock levels of mushrooms and poles. The polygons formed with dashed lines represent areas imperfectly protected by a government agency for conservation. The polygon formed by the longer-dashed lines indicates a very well protected area (although not necessarily strong enough to prevent all illegal household harvest). The black lines are roads and the black dots are urban centers, scaled by population size. Map (b) gives parcel IDs ($x = 1, 2, \ldots, 32$), urban area names, and urban area populations.

for all unique combinations of x, z, and v, W^v for all v, and D as random variables. To account for unobserved harvesting preferences across household types (the random variable u from Eq. (8.13)), we treat v's allocation of daily effort across pole and mushroom harvesting in each forest parcel x as a random variable.

As noted above, the probability that a household of type v would harvest in x in any given day is the joint probability that the daily net revenue of v's harvest in forest parcel x is greater than 0 and that the daily net revenue from v's harvest in forest parcel x is greater than all other parcel-level daily harvests. In this example we assume these two probabilities are statistically independent and that

$$\gamma_x^v = \Pr(nr_x^v > 0)\Pr(nr_x^v > \max_{x \neq j} nr_j^v). \qquad (8.18)$$

We find each probability by calculating nr_x^v, as given by Eq. (8.17), 1,000 times for each x and v combination where we randomly draw variable values from each random variable's assumed distribution for each iteration. $\Pr\left(nr_x^v > 0\right)$ is given by the number of times out of 1,000 that $nr_x^v > 0$ and $\Pr\left(nr_x^v > \max_{x \neq j} nr_j^v\right)$ is given by the number

of times out of 1000 that $nr_x^v > \max_{x \neq j} nr_j^v$ (see the SOM for details).

Next we use Eq. (8.14) to calculate aggregate household effort and harvest levels across a harvest period for each x and z combination 1000 times. In each iteration, values for N^v for each household type v, the number of days in a harvest period, given by β, e_{xz}^v for all unique combinations of x, z, and v, and $V_{xz}^v(S_{xz})$ for all unique combinations of x, z, and v are randomly drawn from their distributions (γ_x^v for all x and v combinations, calculated in the first step of this illustrative example, remain constant in each run). See the SOM for details.

The range in aggregate household effort, harvest, and daily net revenues values for each x and z combination over the harvest period in each parcel x are reported in Tables 8.1 through 8.3. The most aggregate effort is expended in parcel 30, the forest patch right next to the town of Kikeo and within one day's distance of 3 other towns, including Morogoro, by far the largest urban area on the landscape. This parcel attracts a lot of effort because it is right on the highway and some of the other forest patches in the area have protected status. Parcels 10 and 14 also

Table 8.1   Low, mean, and high E^o_{xz} (hours per year) values for each x and z combination (a parcel not represented has a 0 value in every column)

Parcel ID	Low		Mean		High	
	Mushrooms	Poles	Mushrooms	Poles	Mushrooms	Poles
5	1	2	923	953	3 757	4 336
6	0	0	13	13	65	56
7	0	0	3	3	16	16
8	3	4	1 389	1 339	6 203	5 452
9	23	16	2 957	3 040	10 821	12 423
10	33	52	7 251	7 466	32 575	28 131
11	2	3	419	389	1 653	1 333
14	182	606	9 316	9 366	31 101	28 615
20	19	31	1 953	2 027	7 786	6 500
21	114	40	2 588	2 476	9 203	8 633
22	119	150	3 019	2 921	11 627	10 636
23	0	0	1	1	6	5
28	0	0	282	294	1 768	2 094
30	57	131	11 570	11 209	39 340	41 506
32	0	0	0	0	5	5

Table 8.2   Low, mean, and high H^o_{xz} (Mg per year) values for each x and z combination (a parcel not represented has a 0 value in every column)

Parcel ID	Low		Mean		High	
	Mushrooms	Poles	Mushrooms	Poles	Mushrooms	Poles
5	0.0	0.1	9.3	28.5	41.0	124.3
6	0.0	0.0	0.1	0.4	0.7	1.9
7	0.0	0.0	0.0	0.1	0.1	0.5
8	0.0	0.1	13.9	40.2	65.0	181.2
9	0.2	0.4	29.5	90.9	109.5	418.7
10	0.1	1.8	28.8	224.7	130.5	785.9
11	0.0	0.1	2.1	15.4	8.8	59.7
14	1.8	13.8	84.0	236.8	301.5	779.9
20	0.2	0.9	19.7	60.9	75.4	216.6
21	1.1	1.1	25.8	74.5	91.1	262.1
22	1.1	4.5	30.1	87.4	99.8	336.8
23	0.0	0.0	0.0	0.0	0.1	0.2
28	0.0	0.0	2.8	8.8	18.9	64.3
29	0.0	0.0	0.0	0.0	0.0	0.0
30	0.6	3.6	115.4	336.9	431.7	1,239.0
32	0.0	0.0	0.0	0.0	0.1	0.1

attract substantial effort. Not surprisingly, this example illustrates that patches that are right next to highways (traveling speeds off the main roads are very slow), are found between towns, and are not protected attract the most effort (e.g., Luoga et al. 2002). Finally, the mean value of aggregate household harvest in most parcels is approximately 0. This indicates that in general households are indifferent between harvesting in the given parcel and working in paid labor. This calculus would change if wages fell, unemployment in other economic sectors became a problem (here we do not

Table 8.3 Low, mean, and high aggregate household net economic returns from harvesting (*NV*) for each *x* and *z* combination in thousands $US

Parcel ID	Low	Mean	High
5	−25	−5	38
6	0	0	1
8	−26	−7	37
9	−71	−13	74
10	−97	−23	131
11	−6	−2	15
14	−78	−11	253
20	−36	−13	164
21	−77	−39	75
22	−100	−46	21
23	−1	0	0
25	−57	−23	−6
26	−24	−10	−3
27	−7	−2	−1
28	−52	−27	−6
30	−713	−336	−86

formally model unemployment possibilities), or conservation status was given to and enforced in more forest parcels.

This intermediate access model was introduced above as a method for estimating both households and harvest-right holder harvest in a matter that better incorporated harvesting realities. The steady-state restriction is applied in most applications of this method in order to recover harvest-right holder harvest volumes. However, because there is no property right harvest to recover in this example the steady-state assumption is irrelevant in this case.

8.3 The supply, use, and value of forests' provisioning service in tier 2

Unlike tier 1, in tier 2 we do not assume a constant harvest volume per harvest period but instead allow the trajectory of harvest volume and value to change through time. By tracking biological stocks and harvest though time we can model biological degradation or growth dynamics. Just as in tier 1, the solution to these models will depend on the property right structure in the forest and the behavior ascribed to harvesters. In all cases, prices of products made from biological stocks and wages can either be given as part of a scenario or can be determined within the model to be consistent with market equilibrium.

8.3.1 Tier 2 model when harvest volume or harvest effort is observable

In this initial tier 2 model we assume that data on harvest volume of biological stock z or the total effort used to harvest the stock in a forest parcel is available. Let H_{zt} represent harvest volume of stock z in period t in a forest parcel and E_{zt} represent total harvest effort for stock z in period t in a forest parcel. If effort data are observable and harvest volume data are not, then stock z's harvest rate function, given by $V_z(S_z)$, must also be defined. Just as in tier 1, the net value of stock z's harvest in period t from a forest parcel is given by the basic harvest net revenue equation,

$$NV_{zt} = p_{zt} \times H_{zt} - C_{zt} = E_{zt}$$
$$\left(\left(p_{zt} \times V_z \left(S_{zt} \right) \right) - W_t - Y_t \right), \tag{8.19}$$

which is equivalent to Eq. (8.2) except that we have included time subscripts (again, in this cost calculation we ignore the opportunity costs associated with harvesters traveling to and back from forest parcels and any costs associated with transporting harvested stocks to processing centers). The NPV of harvest from periods $t = 1, 2, \ldots, T$ on a forest parcel is equal to

$$NV_z = \frac{\sum_{t=1}^{T} NV_{zt}}{(1+r)^{t-1}}. \qquad (8.20)$$

We can either define prices and wages for each time period t or we can determine them within our model. We can use several methods to define price and wage trajectories. Most simply, we can extrapolate past price and wage trends into the future. Or, if available, we can use published estimates of future timber and NTFP prices and wage rates on the study landscape. Otherwise we can endogenously determine prices by gathering information about demand for the stock and solving for the price that equates demand for timber and NTFP stocks with local supply. Let the function that describes local demand for a stock harvested from the study landscape at time t be given by $F_{zt}(p_{zt})$. The market clearing price for stock z at time t is by solving the following for p_{zt},

$$\sum_{x=1}^{X} H_{xzt} = F_{zt}(p_{zt}) \qquad (8.21)$$

where H_{xzt} indicates z's harvest in parcel x at time t. In Eq. (8.21) we assume no imports or exports of stock z in the study landscape. If the demand function $F_{zt}(p_{zt})$ describes local demand for stock regardless of point of origin we can assume some portion of $F_{zt}(p_{zt})$ is satisfied by imports of z and some portion of z's harvest is exported before we solve for local equilibrium prices.

We can solve for wages in a similar manner if we have data on harvest effort. Let $D_t - \sum_{x=1}^{X} E_{xzt}$ indicate the supply of labor available after accounting for harvest effort where D_t is the aggregate work time of all harvesters in period t and $\sum_{x=1}^{X} E_{xzt}$ is the amount of time all harvesters spend in all forest parcels $x = 1, \ldots, X$ in period t. Further, let local demand for labor at time t be given by $B_t(W_t)$. Finally, we equate $D_t - \sum_{x=1}^{X} E_{xzt}$ with $B_t(W_t)$ and solve for W_t. Because endogenous solutions require defining $F_{zt}(p_{zt})$ for each stock z across all time periods and $B_t(W_t)$ across all time periods, this approach may be difficult to implement.

Finally, we have to verify that the given (or calculated) harvest volumes are possible. In tier 2 we track biological stock in a parcel over time with an equation of motion. Specifically, the stock of z in a forest parcel in harvest period $t + 1$ is equal to period t's stock level plus the growth in stock from t to $t + 1$ less the parcel harvest of the stock in period t.

$$S_{z,t+1} = S_{zt} + G_z(S_{1t}, \ldots, S_{zt}, \ldots, S_{Zt}) - H_{zt}, \qquad (8.22)$$

where $G_z(S_{1t}, \ldots, S_{zt}, \ldots, S_{Zt})$ gives the growth of z's stock in the parcel from time t to $t + 1$. The inclusion of other stock levels in the growth function means that the growth of stock z from one period to the next can be affected by the contemporaneous stock levels of one or more other stocks in the parcel. For example, the growth rate of a bush meat species may be dependent on its own stock *and* the stocks of predators and prey.

If the model ever calculates a harvest of z larger than its stock, $H_{zt} \geq S_{zt}$, then we set $H_{zt} = S_{zt}$ and $H_{z,t+1} = S_{z,t+1} = 0$ for time periods $t + 1$ to T unless the forest parcel is re-colonized later by biological stock z.

8.3.2   Calculating harvest volume and value in tier 2 when such data is missing

In tier 2, when harvest volumes or efforts are not observable, we analyze a unified model of intermediate access that includes the possibilities of exclusive harvest and complete open access as special cases. We begin by describing household harvest, which will be a function of household characteristics, biological stocks, and illegal harvest prevention efforts. Next we describe how the holder of harvest rights should choose harvest and enforcement efforts to maximize the NPV of harvest returns over time. Because households react to efforts to enforce harvest rights, harvest-right holders make their decisions first in each time period. We combine both harvest by households and harvest-right owners to describe total harvest and evolution of the biological stock and harvests through time.

8.3.2.1   Household harvest
Just as in tier 1, in tier 2 a household allocates their labor in a time period between paid labor and harvesting activities in order to maximize their net revenues. Further, just as in tier 1, because an individual household does not consider the affect of their harvest on biological stock levels their optimization procedure is

static; they make harvest decisions without considering the ramifications of such harvest effort on future stock levels. In the tier 2 framework, a household allocates their entire period's labor budget simultaneously (in tier 1 a household made multiple daily or trip-level decisions). We index households on the landscape with $h = 1, 2, \ldots, N$, where household h is located at point h on the landscape. Let D_t^h indicate the total work hours in time period t for household h, E_{xzt}^h indicate the total time (in hours) that h uses to harvest stock z in forest parcel x during time period t, d_{xt}^h indicate the one-way travel time between household h and forest parcel x in hours in period t, and I_{xt}^h indicate the number of trips household h takes to parcel x in period t. Finally, L_t^h indicates household h's time available for work in wage labor during the harvest period.

$$L_t^h = D_t^h - \left(\sum_x \sum_z E_{xzt}^h \right) - \sum_x 2d_{xt}^h I_{xt}^h , \qquad (8.23)$$

where the term $2d_{xt}^h I_{xt}^h$ converts trips to parcel x in period t into the number of hours household h uses to travel to and back from forest parcel x in period t. To simply matters we replace the number of trips in Eq. (8.23) with a function that relates the number of trips to parcel x in period t to total harvest effort in the parcel during period t, $I_{xt}^h = \sigma \left(\sum_{z=1}^Z E_{xzt}^h \right)$, where σ is increasing in effort in parcel x.

Household harvest volume is related to effort with a generic version of the tier 1 production function,

$$H_{xzt}^h = y \left(E_{xzt}^h, S_{xzt}, \sigma_{xzt} \right), \qquad (8.24)$$

where H_{xzt}^h is household h's harvest of stock z from parcel x in period t, S_{xzt} represents the level of biological stock z in parcel x in period t, and θ_{xzt} represents a vector of parcel-level biophysical variables. For example, the vector θ_{xzt} could include information on the elevation or slope of parcel x where greater elevations and slopes reduce the productivity of harvest effort, all else equal. We assume that harvest of z in x increases in effort and that harvest per unit of effort increases in S_{xzt}.

In choosing how much to harvest, the household considers the opportunity cost of harvesting (lost wages), and harvest-related capital costs, plus the costs associated with illegal harvesting. Let $\delta \left(\rho_{xzt}, E_{xzt}^h \right)$ indicate the expected monetary fine for harvesting z from parcel x in period t where δ increases in enforcement effort, ρ_{xzt}, and effort. If the forest parcel x is completely open to households for the harvest of z, $\rho_{xzt} = 0$, and $\delta \left(\rho_{xzt}, E_{xzt}^h \right) = 0$ for all E_{xzt}^h. Let household h's wage in period t be given by W_t^h (different households can command different wages) and the per effort unit cost of harvesting equipment and accessories, such as trucks and saws, in period t be given by Y_t^h.

In the static optimization framework the household allocates their harvest effort over each stock and parcel combination and the number of trips they make to each parcel for each time period t. Specifically, the household maximizes their net revenues in time period t by choosing the effort levels that solve the following for each z and x combination in time period t,

$$
\begin{aligned}
&p_{zt} \left(\frac{\partial y}{\partial E_{xzt}^h} \right) - Y_t^h - \left(\frac{\partial \delta}{\partial E_{xzt}^h} \right) \\
&= W_t^h \left(1 + 2d_{xt}^h \frac{\partial \sigma}{\partial E_{xzt}^h} \right)
\end{aligned}
\qquad (8.25)
$$

If household h spends all of their working time harvesting and never works in the paid labor market then W_t^h in Eq. (8.25) should be replaced by h's marginal value of harvesting. The left hand side of Eq. (8.25) measures the revenue generated by the last unit of effort used to harvest stock z in parcel x less the per effort unit cost of capital costs and the marginal expected fine for that less unit of harvest effort. The right hand side of Eq. (8.25) is the opportunity cost of lost wages due the last unit of effort to harvest z in parcel x in period t (including travel time). If the parcel x is completely open to household harvest then the marginal expected fine term in Eq. (8.25) is dropped.

The solution to Eq. (8.25) gives the optimal harvest effort for stock z in parcel x by household h in period t, given by E_{xzt}^{h*}. Optimal effort levels in period t are a function of p_{zt}, S_{xzt}, d_{xt}^h, θ_{xzt}, ρ_{xzt}, W_t^h, and Y_t^h. Let H_{xzt}^{h*} indicate h's optimal harvest of z from x in period t when evaluated at E_{xzt}^{h*} (see Eq. (8.24)) and let the optimal aggregate household harvest of stock z in parcel x in period t is given by $H_{xzt}^{o*} = \sum_h H_{xzt}^{h*}$. By solving E_{xzt}^{h*} over all h, x, z, and t we specify the complete solution to the tier 2 household model problem.

8.3.2.2 Exclusive harvest-right harvest volume and value

Just as in tier 1, we assume the holder of a harvest right for stock z in parcel x chooses harvest and enforcement effort over time such that the NPV of harvest returns are maximized. In tier 2, unlike tier 1, we do not assume that biological stocks and harvest are constant through time.

Let the harvest of biological stock z from parcel x during harvest period t by its harvest-right holder be given by H^c_{xzt}. Harvest of a stock by harvest-right holders is a function of harvest labor, E^c_{xzt}, capital equipment used, Q_{xzt}, stock in period t, and parcel-level biophysical variables, represented by vector θ_{xzt}. The relationship between H^c_{xzt} and these input variables is defined by production function f, a generic version of the tier 1 production function,

$$H^c_{xzt} = f(E^c_{xzt}, Q_{xzt}, S_{xzt}, \theta_{xzt}). \tag{8.26}$$

We assume harvest increases in labor and capital equipment and that the rate of harvest per unit of effort increases in S_{xzt}.

In tier 2, the harvest-right holder considers three types of costs: the cost of harvest effort, which is a function of the wage labor rate W_t and capital wage rate M_t; the net cost of illegal harvest prevention efforts, $\varphi(\rho_{xzt})$ (the costs of prevention efforts less the revenue from fines); and the cost of transporting harvest to a processing site or market.

The harvest-right owner's NPV of revenues from harvesting stock z in parcel x is maximized by choosing harvest and enforcement over time such that satisfy,

$$\max_{E^c_{xzt}, S_{xzt}, Q_{xzt}, \rho_{xzt}} \sum_{t=1}^{T} \left((p_{zt} - \alpha_{zt}d^c_{xzt}) \, f(E^c_{xzt}, Q_{xzt}, S_{xzt}, \theta_{xzt}) \Big/ (1+r)^{t-1} + \right.$$
$$\left. \varphi(\rho_{xzt}) - W_t E^c_{xzt} - M_t Q_{xzt} \Big/ (1+r)^{t-1} \right), \tag{8.27}$$

subject to the stock constraint

$$S_{xz,t+1} = S_{xzt} + G_{xz}(S_{xzt}) - f(E^c_{xzt}, Q_{xzt}, S_{xzt}, \theta_{xzt}) - H^o_{xzt} \tag{8.28}$$

where α_{zt} represent the cost to haul one unit of stock z one kilometer in period t and d^c_{xzt} is distance in kilometers from parcel x to z's processing site in period t.

Here we assume that harvest-right holders make their harvest and property right protection choices first and then households react accordingly as described in optimality condition (8.25). Therefore the harvest-right holder will replace H^o_{xzt} in equation of motion (8.28) with their expectation for aggregate household harvest as a function of their choice of ρ_{xzt}. According to the previous section, the harvest-right holder can expect H^o_{xzt} to be given by the function $H^{o*}_{xzt}(S_{xzt}, \rho_{xzt}, \beta_{xzt})$ where β_{xzt} is the vector of all other variables and parameters that define the optimal household harvest of stock z in parcel x in period t (e.g., prices, wage). If the tier 2 version of $H^{o*}_{xzt}(S_{xzt}, \rho_{xzt}, \beta_{xzt})$ is too difficult to solve because we cannot define the relationship between effort in parcel x and the number of trips to parcel x or between effort in parcel x and expected trespassing fines we can use a tier 1 approach to find household harvest in the parcel. All that is required of the household harvest expectation function is that it is explained by biological stock and effort to prevent illegal harvest.

The conditions that satisfy the optimization problem for the entity that has right to harvest z in x include

$$(p_{zt} - \alpha_{zt}d^c_{xzt}) \frac{\partial f}{\partial E^c_{xzt}} = W_t + \mu_{xzt} \frac{\partial f}{\partial E^c_{xzt}} \tag{8.29}$$

$$(p_{zt} - \alpha_{zt}d^c_{xzt}) \frac{\partial f}{\partial Q^c_{xzt}} = M_t + \mu_{xzt} \frac{\partial f}{\partial Q^c_{xzt}} \tag{8.30}$$

$$\frac{\partial \varphi}{\partial \rho_{xzt}} = \mu_{xzt} \frac{\partial H^o_{xzt}}{\partial \rho_{zxt}} \tag{8.31}$$

$$(p_{zt} - \alpha_{zt}d^c_{xzt}) \frac{\partial f}{\partial S_{xzt}} = \mu_{xz,t-1}(1+r) - \mu_{xzt} -$$
$$\mu_{xzt} \frac{\partial G_z}{\partial S_{xzt}} + \mu_{xzt} \frac{\partial f}{\partial S_{xzt}} + \mu_{xzt} \frac{\partial H^o_{xzt}}{\partial S_{xzt}} \tag{8.32}$$

$$S_{xz,t+1} = S_{xzt} + G_{xz}(S_{xzt}) - f(E^c_{xzt}, Q_{xzt}, S_{xzt}, \theta_{xzt}) - H^o_{xzt}(S_{xzt}, r_{xzt}, \beta_{xzt}) \tag{8.33}$$

where μ_{xzt} is the value of the biological stock z in parcel x in period t. The first two conditions immediately above are no different from optimality condition (8.8) from tier 1. Further, conditions (8.29) and (8.32) can be combined as they were in tier 1 to solve for optimal effort by the harvest-right holder. Condition (8.31) means that illegal harvest prevention efforts should be increased until the net cost of one more unit of effort equals the value of the last unit of stock z saved from illegal harvest. Equation (8.33) keeps track of the evolution of biological stock (stock in the next period equals stock this period plus growth minus harvest).

Let E_{xzt}^{c*}, Q_{xzt}^{*}, S_{xzt}^{*}, ρ_{xzt}^{*}, and μ_{xzt}^{*} for all t indicate the variable values that solve optimality conditions (8.29) through (8.33). These optimal actions are functions of the model parameters p_{zt}, α_{zt}, d_{xzt}^{c}, θ_{xzt}, W_t, M_t, and r across all t beginning at time $t = 1$. Let H_{xzt}^{c*} indicate the value of harvest production function f when evaluated at E_{xzt}^{c*}, Q_{xzt}^{*}, and S_{xzt}^{*} (see Eq. (8.26)). We can find the NPV of timber harvest for the harvest-right holder of stock z in parcel x by plugging in their optimal choices into the objective function (8.27).

As in tier 1, a harvest-right holder's objective may not be to maximize the NPV of harvest returns but to maximize an objective function that includes spiritual, biodiversity, and regulating ecosystem service values associated with the forest. In these cases we can change the objective function (8.27) accordingly. For example, assume a harvest-right holder's utility is maximized when anthropogenic disturbance in a forest parcel, including the extraction of timber and NTFPs by households, is minimized over time. In this case, the harvest-right holder's problem is to allocate her illegal harvest prevention budget over time such that aggregate household harvests of stocks from the forest parcel over time are minimized.

8.3.2.3 Total harvest volume and value

Once we have solved the harvest-right holder of stock z's problem and each household's harvest effort problem over stock z in forest parcel x the aggregate level of stock z's harvest in parcel x in period t is given by H_{xzt}^{*},

$$H_{xzt}^{*} = H_{xzt}^{c*} + H_{xzt}^{o*} \qquad (8.34)$$

subject to the constraint that $H_{xzt}^{*} \leq S_{xzt}$. If the solutions to the household and harvest-right holder problems generate a collective harvest level that is greater than or equal to a parcel's stock (i.e., $H_{xzt}^{*} \geq S_{xzt}$) then we set $H_{xzt}^{*} = S_{xzt}$ and all subsequent collective harvests of z from x are equal to 0 (i.e., $H_{xz,t+1}^{*} = H_{xz,t+2}^{*} = \ldots = 0$) unless the forest parcel is recolonized by individuals of stock z at a later date.

The net value of z's total harvest from parcel x at time t is given by,

$$
\begin{aligned}
NR_{xzt} = {} & p_{zt}(H_{xzt}^{c*} + H_{xzt}^{o*}) - \alpha_{zt}d_{xt}^{c}H_{xzt}^{c*} \\
& - W_t E_{xzt}^{c*} - M_t Q_{xzt}^{*} - \omega_{xzt}(\rho_{xzt}^{*}) \qquad (8.35) \\
& - \sum_{h=1}^{H}(W_t^h + Y_t^h)E_{xzt}^{h*} - W_t^h \sum_x 2d_{xt}^h\sigma\left(\sum_z E_{xzt}^{h*}\right)
\end{aligned}
$$

where $\omega_{xzt}(\rho_{xzt})$, as in tier 1, is the cost to z's harvest-right holder in x to achieve the anti-poaching effort ρ_{xzt}. As in tier 1 we do not include the revenues generated from fines because they are a transfer of household wealth to the enforcement authority (either the harvest-right holder or the government); no additional societal value is generated in these transfers. We sum across stocks, parcels, and time periods $t = 1$ to T (in a discounted fashion) to generate the NPV of revenues generated by the landscape's forests from $t = 1$ to T.

As discussed in Section 8.3.1, we can exogenously define prices and wages in Eq. (8.33) over the time period span of $t = 1$ to T. Otherwise, if we solve for them endogenously assuming market clearing then we would set local supply of stock z in period t equal to local demand for stock z in period t,

$$\sum_x H_{xzt}^{c*}(p_{zt}) + H_{xzt}^{o*}(p_{zt}) = F_{zt}(p_{zt}), \qquad (8.36)$$

where we explicitly note that optimal harvest volumes are a function of stock prices (this assumes that harvest product z is not exported from the local area; if it is the market demand curve for z would have to include appropriate external demand). Finally, we would set local household labor supply in time period t equal to a demand function for it to solve for market clearing wage levels,

$$\sum_{h=1}^{H}D_t^h - E_{xzt}^{h*}(W_t^h) - \sum_x 2d_{xt}^h\sigma\left(E_{xzt}^{h*}(W_t^h)\right) = B_t(W_t), \quad (8.37)$$

where we explicitly note that optimal harvest effort levels are a function of wages.

8.3.3 Solving the tier 2 model

To solve the tier 2 timber and NTFP model given by equations 8.23 to 8.37 we need at a minimum the following data: a map of forest parcels as of period $t = 1$, stock levels of all relevant stocks in all forest parcels as of period $t = 1$, growth relations for each biological stock, harvest production functions, a spatial database of harvest rights as of period $t = 1$, net cost of efforts to prevent illegal harvesting in each forest with property rights, information on the effectiveness of efforts to prevent illegal harvest, a map of the household locations on the landscape as of period $t = 1$, and a vector of (exogenously or endogenously defined) stock prices and wages, from periods $t = 1$ to T. Household density across the landscape and the spatial distribution of harvest rights for periods $t > 1$ can remain fixed at $t = 1$ levels or can be changed according to some scenario. The location of forest parcels should remain fixed at period $t = 1$ levels unless a scenario of LULC change indicates a parcel will be afforested, reforested, or clear-cut in some period $t > 1$.

Running the tier 2 model will be computationally challenging. We can use several simplifications to make the solution process a bit easier. First, we could divide the landscape into regions and define several household types where each region-household type combination is also indexed by $h = 1, 2, \ldots, N$. In this case, total household harvest of z in x at time t will be given by $\sum_{h=1}^{N} \omega_t^h H_{xzt}^h$ where ω_t^h is the number of households of type h on the landscape at time t. Second, we could create a few very large forest parcels by combining many unique forest parcels. Third, we could ignore capital inputs and costs.

8.4 Limitations and next steps

The tier 1 models assume that households and harvest-right holders make decisions that are net revenue maximizing with full information of options. In reality, most harvesters have limited information and use "bounded rationality" when making decisions. Because of information and computation limitations, agents use "rules of thumb" to make decisions that may not be optimal but are acceptable (Manson and Evans 2007; Brown 2008). In addition, households or harvest-right holders may care about other things, such as environmental stewardship or the provision of other ecosystem services, besides a strict focus on returns from harvesting. Incorporating different preferences into the models is possible by changing the objective function. Incorporating bounded rationality, however, can be

difficult and assuming rational choice is a more tractable method for approximating household decision-making (Manson and Evans 2007).

In tier 1 we can only estimate harvest value over time if we assume the baseline or current period conditions hold into the future. Steady-state solutions will not accurately predict long-run values if the landscape's demographic, socio-economic, or ecological conditions are in flux. For example, increasing unemployment in cities or increasing timber and NTFP prices to wage ratios may encourage greater harvest effort in the future, leading to harvest rates that exceed growth rates and forests with rapidly declining resource stocks. Further a steady-state analysis ignores the various disturbances or shocks to the ecosystem (e.g., fires or pest outbreaks) that could abruptly change stock levels and biomass growth rates. Therefore, the application of a steady-state framework on a landscape with rapidly changing conditions or subject to periodic disturbance should only be used for short-term calculations.

The tier 2 model estimates the rate and value of timber and NTFP harvest over time assuming that households and harvest-right holders are motivated by net revenue maximization. Unlike tier 1, stock harvest and stock levels in forest parcels can change over time. However, this added level of realism also places an increased data collection and modeling complexity burden on the user. A tier 2 approach to valuing the provisioning service provided by forests on a landscape will take considerable time and effort to implement. Further, as in tier 1, in tier 2 we assume households and harvest-right holders make decisions that are net revenue maximizing with full information of options. In reality actors on a landscape make many sub-optimal decisions due to imperfect information and in many cases the inability to actually determine what is optimal.

In this chapter we have presented methods for estimating the volume and value of timber and NTFPs extracted from forested parcels on a landscape. We have neglected uncertainty. The simplest way to add more robustness to modeled tier 1 or tier 2 results is to simulate the response of the model to variation in input values. For example, we can run the tier 1 exclusive harvest-right model with a range of typical biological stock prices, wage rates, and the interest rates to approximate the range in potential steady-state exclusive harvest volumes and values (see Section 8.2.4). We can also explore how sensitive tier 1 and 2 modeled results are to changes in functional relationships, such as harvest rates, growth rates, and expected fines for illegal harvesting. For example, climate change could change stock growth rates in forests over time (Aber *et al.* 2001). We could explore the ramifications of such changes on harvest volumes and values (Irland *et al.* 2001) by varying $G(S)$ for all relevant stocks in a manner that is consistent with climate change estimates.

We can also add structural uncertainty and risk to the model. For example, we could assume a household harvester will not always find wage employment while not harvesting. In this case we could make the prospect of finding wage work probabilistic. And we could add the risk of forest disturbances, such as fire or disease (Dale *et al.* 2001), into stocks' equations of motion (Sohngen and Sedjo 1996). It would be useful then to map out how the inclusion of risk parameters in the model causes it to deviate from deterministic solutions.

In addition to examining uncertainty, it is important to consider the negative impacts that stock harvests can have on other ecosystem processes and services (e.g., Luoga *et al.* 2000). If we can modify LULC definitions according to the harvest volumes and stocks calculated in the models above then the ramifications of predicted harvests can be incorporated into almost every other ecosystem service model described in this book. For example, if the water quality function is a function of forest type (degraded versus intact) then harvest volumes and stocks as calculated in the models above will affect water quality model output. Alternatively, we can modify other ecosystem service model outputs *ex post* with results from this model. For example, suppose we have data on the carbon storage potential of a relatively intact forest type. If modeled harvest volumes in a parcel with this forest type are high then the parcel's predicted carbon storage in the aboveground and belowground biomass pools could be scaled downward to reflect modeled stock loss in the forest.

Finally, a major source of global land-use change is the clearing of forests to open land for agricultural or residential use (Repetto and Gillis 1988). Here we value harvests in forests that begin and, for the most part, remain in forest from time $t = 1$ to T (if it is completely drained of all of its stocks by time T then it is no longer a forest); these are not models of land-use change. To model such changes we rely on LULC change scenarios that are created with land-use change models. These models can provide output that would be useful to such LULC change models. Specifically, estimates of the NPV of forest harvest from forests across a landscape along with estimates of the NPV of returns from alternative uses of forested areas could be used by LULC change models to determine which forest parcels would most likely remain forested.

References

Aber, J., Neilson, R. P., McNulty, S., *et al.* (2001). Forest processes and global environmental change: Predicting the effects of individual and multiple stressors. *Bioscience,* **51**, 735–51.

Adger, W. N., and Luttrell, C. (2000). Property rights and the utilisation of wetlands. *Ecological Economics,* **35**, 75–89.

Arnold, J. E. M., and Pérez, M. R. (2001). Can non-timber forest products match tropical forest conservation and development objectives? *Ecological Economics,* **39**, 437–47.

Barbier, E. B., and Burgess, J. C. (2001). The economics of tropical deforestation. *Journal of Economic Surveys,* **15**, 413–33.

Bawa, K. S., and Seidler, R. (1998). Natural forest management and conservation of biodiversity in tropical forests. *Conservation Biology,* **12**, 46–55.

Belcher, B., Ruíz-Pérez, M., and Achdiawan, R. (2005). Global patterns and trends in the use and management of commercial NTFPs: Implications for livelihoods and conservation. *World Development,* **33**, 1435–52.

Birdyshaw, E., and Ellis, C. (2007). Privatizing an open-access resource and environmental degradation. *Ecological Economics,* **61**, 469–77.

Boot, R. G. A., and Gullison, R. E. (1995). Approaches to developing sustainable extraction systems for tropical forest products. *Ecological Applications,* **5**, 896–903.

Brown, D. R. (2008). A spatiotemporal model of shifting cultivation and forest cover dynamics. *Environment and Development Economics,* **13**, 643–71.

Browning, M., and Crossley, T. F. (2001). The life-cycle model of consumption and saving. *Journal of Environmental Perspectives,* **15**, 3–22.

Clark, C. W. (1990). *Mathematical bioeconomics: the optimal management of renewable resources.* Wiley, New York.

Conrad, J. M., and Clark, C. W. (1987). *Natural resource economics: notes and problems*. Cambridge University Press, Cambridge.

Dale, V. H., Joycel, L. A., McNulty, S., *et al.* (2001). Climate change and forest disturbances. *Bioscience*, **51**, 723–34.

Damania, R., Milner-Gulland, E. J., and Crookes, D. J. (2005). A bioeconomic analysis of bushmeat hunting. *Proceedings of the Royal Society B-Biological Sciences*, **272**, 259–66.

Dauber, E., Fredericksen, T.S. and Peña, M. (2005). Sustainability of timber harvesting in Bolivian tropical forests, *Forest Ecology and Management*, **214**, 294–304.

Engel, S., and Lopez, R. (2008). Exploiting common resources with capital-intensive technologies: The role of external forces. *Environment and Development Economics*, **13**, 565–89.

FAO. (2006). *Global forest resources assessment 2005: progress toward sustainable forest management*. United Nations Food and Agricultural Organization, Rome.

Feder, G., and Feeny, D. (1991). Land tenure and property rights: Theory and implications for development policy. *World Bank Economic Review*, **5**, 135–53.

Gordon, H. S. (1954). The economic theory of a common-property resource: The fishery. *Journal of Political Economy*, **62**, 124.

Guariguata, M. R., Cronkleton, P., Shanley, P., *et al.* (2008). The compatibility of timber and non-timber forest product extraction and management. *Forest Ecology and Management*, **256**, 1477–81.

Hyde, W. F. (2003). Economic considerations on instruments and institutions. In Y. Dube and F. Schmithusen, Eds., *Cross-sectoral policy impacts between forestry and other sectors*, FAO Forestry Paper 142. Rome.

Irland, L.C., Adams, D., Alig, R., *et al.* (2001). Assessing socioeconomic impacts of climate change on US forests, wood-product markets, and forest recreation. *Bioscience*, **51**, 753–64.

Justice, C., Wilkie, D., Zhang, Q., *et al.* (2001). Central African forests, carbon and climate change. *Climate Research*, **17**, 229–46.

Larson, B., and Bromley, D. (1990). Property rights, externalities, and resource degradation: Locating the tragedy. *Journal of Development Economics*, **33**, 235–62.

Lopez-Feldman, A., and Wilen, J. E. (2008). Poverty and spatial dimensions of non-timber forest extraction. *Environment and Development Economics*, **13**, 621–42.

Luoga, E. J., Witkowski, E. T. F., and Balkwill, K. (2000). Economics of charcoal production in miombo woodlands of eastern Tanzania: Some hidden costs associated with commercialization of the resources. *Ecological Economics*, **35**, 243–57.

Luoga, E. J., Witkowski, E. T. F. and Balkwill, K. (2002). Harvested and standing wood stocks in protected and communal miombo woodlands of eastern Tanzania. *Forest Ecology and Management*, **164**, 15–30.

Luoga, E. J., Witkowski, E. T. F., and Balkwill, K. (2005). Land cover and use changes in relation to the institutional framework and tenure of land and resources in Eastern Tanzania miombo woodlands. *Environment, Development and Sustainability*, **7**, 71–93.

Maass, J., Balvanera, P., Castillo, A., *et al.* (2005). Ecosystem services of tropical dry forests: Insights from long-term ecological and social research on the Pacific Coast of Mexico. *Ecology and Society*, **10**, 17.

Manson, S. M., and Evans, T. (2007). Agent-based modeling of deforestation in southern Yucatan, Mexico, and reforestation in the Midwest United States. *Proceedings of the National Academy of Sciences of the United States of America*, **104**, 20678–83.

Milner-Gulland, E. J., and Clayton, L. (2002). The trade in babirusas and wild pigs in North Sulawesi, Indonesia. *Ecological Economics*, **42**, 165–83.

Monela, G. C., O'Kting'ati, A., and Kiwele, P. M. (1993). Socio-economic aspects of charcoal consumption and environmental consequences along the Dar-es-Salaam–Morogoro highway, Tanzania. *Forest Ecology and Management*, **58**, 249–58.

Nagendra, H. (2002). Tenure and forest conditions: Community forestry in the Nepal Terai. *Environmental Conservation*, **29**, 530–9.

Nagendra, H., Karmacharya, M., and Karna, B. (2005). Evaluating forest management in Nepal: Views across space and time. *Ecology and Society*, **10**, 24.

Ndangalasi, H. J., Bitariho, R., and Dovie, D. B. K. (2007). Harvesting of non-timber forest products and implications for conservation in two montane forests of East Africa. *Biological Conservation*, **134**, 242–50.

Núñez, D., Nahuelhual, L., and Oyarzún, C. (2006). Forests and water: The value of native temperate forests in supplying water for human consumption. *Ecological Economics*, **58**, 606–16.

Parker, D. C., Manson, S. M., Janssen, M. A., *et al.* (2003). Multi-agent systems for the simulation of land-use and land-cover change: A review. *Annals of the Association of American Geographers*, **93**, 314–37.

Pattanayak, S. K., and Sills, E. O. (2001). Do tropical forests provide natural insurance? The microeconomics of non-timber forest product collection in the Brazilian Amazon. *Land Economics*, **77**, 595–612.

Peters, C. M., Gentry, A. H., and Mendelsohn, R. O. (1989). Valuation of an Amazonian rainforest. *Nature*, **339**, 655–6.

Pinard, M. A. and Putz, F. E. (1996). Retaining forest biomass by reducing logging damage. *Biotropica*, **28**, 278–95.

Polasky, S., Nelson, E., Camm, J., *et al.* (2008). Where to put things? Spatial land management to sustain biodiversity and economic returns. *Biological Conservation,* **141,** 1505–24.

Repetto, R. C. and Gillis, M., Eds. (1988). *Public policies and the misuse of forest resources.* Cambridge University Press, New York.

Ricketts, T. H. (2004). Tropical forest fragments enhance pollinator activity in nearby coffee crops. *Conservation Biology,* **18,** 1262–71.

Scholes, R. J. (1996). Miombo woodlands and carbon sequestration. In B.M. Campbell, Ed., *The Miombo in transition: woodlands and welfare in Africa.* Centre for International Forestry Research (CIFOR), Bogor, Indonesia.

Silvertown, J. (2004). Sustainability in a nutshell. *Trends in Ecology and Evolution,* **19,** 276–8.

Sinha, A., and Brault, S. (2005). Assessing sustainability of nontimber forest product extractions: How fire affects sustainability. *Biodiversity and Conservation,* **14,** 3537–63.

Sohngen, B. L., and Sedjo, R. (1996). *A comparison of timber models for use in public policy analysis.* Resources for the Future, Washington, DC.

Sohngen, B., and Brown, S. (2006). The influence of conversion of forest types on carbon sequestration and other ecosystem services in the South Central United States. *Ecological Economics,* **57,** 698–708.

Swallow, B. M., and Bromley, D. W. (1995). Institutions, governance and incentives in common property regimes for African rangelands. *Environmental and Resource Economics,* **6,** 99–118.

Tahvonen, O., and Salo, S. (1999). Optimal forest rotation with *in situ* preferences. *Journal of Environmental Economics and Management,* **37,** 106–28.

Thwaites, R., De Lacy, T., Li, Y. H., *et al.* (1998). Property rights, social change, and grassland degradation in Xilingol Biosphere Reserve, Inner Mongolia, China. *Society and Natural Resources,* **11,** 319–38.

Ticktin, T. (2004). The ecological implications of harvesting non-timber forest products. *Journal of Applied Ecology,* **41,** 11–21.

Vedeld, P., Angelsen, A., Bojö, J., *et al.* (2004). Forest environmental incomes and the rural poor. *Forest Policy and Economics,* **9,** 869–79.

Wikramanayake, E., Dinerstein, E., Loucks, C., *et al.* (2001). *Terrestrial ecoregions of the Indo-Pacific: a conservation assessment.* Island Press, Washington, DC.

Wikramanayake, E., McKnight, M., Dinerstein, E., *et al.* (2004). Designing a conservation landscape for tigers in human-dominated environments. *Conservation Biology,* **18,** 839–44.

Williams, M. (2003). *Deforesting the earth: from prehistory to global crisis.* University of Chicago Press, Chicago.

World Wildlife Fund. (2006). *Demographic analysis: Terai Arc Landscape—Nepal.* World Wildlife Fund, Nepal Program, Kathmandu, Nepal.

CHAPTER 9

Provisioning and regulatory ecosystem service values in agriculture

Erik Nelson, Stanley Wood, Jawoo Koo, and Stephen Polasky

9.1 Introduction

Ecological processes combine with human labor and inputs such as fertilizer and irrigation to produce agricultural goods used for food, fodder, fiber, and fuel. The ecosystem processes that influence agricultural production include soil retention, pest control, nutrient recycling in the soil, water capture, and animal pollination (e.g., Wood *et al.* 2005; Swinton *et al.* 2007). By contributing to agricultural productivity, these processes become ecosystem services. The value of these services can be proxied by their contributions to the monetary value of commercial agricultural production or the utility value of subsistence agricultural production. Because of the agricultural sector's importance in regional economic development, especially in developing countries (Byerlee *et al.* 2009), estimating the potential agricultural output on as yet unconverted natural habitat may also be of interest because it represents production value foregone.

There is a range of approaches for estimating the value of agricultural production and the contribution that particular ecosystem services make to it. These approaches are largely differentiated by their geographical scale, the degree to which they incorporate behavioral responses by farmers to changes in biophysical and market variables, and the extent to which they track the impacts of agricultural production on ecological processes.

Landscape-scale, regional, or global agricultural models typically assess expected or potential crop yield as a function of climate, soil type, input use intensity, and, sometimes, basin-wide water resource availability (e.g., Fischer *et al.* 2002; Bruinsma 2003; Rosegrant *et al.* 2005; Nelson *et al.* 2009). These broader-scale models do not provide an explicit framework for examining changes in farmer behavior and profits due to policy, price, or environmental changes. Nor, in general, do they assess the ecological consequences of agricultural production (e.g., Naidoo and Ricketts 2006; Cassman and Wood 2005).

Conversely, in farm-level models, farmers are usually represented as making crop and production method choices that are expected to maximize their economic profits. In farm-level models, the value of a change in an ecosystem service input is equal to the change in profit it induces. For example, the value of an increased supply of irrigation water on a farm can be estimated by the change in farm profits due to the additional water volume. Farm-scale models can be applied to understand how farmers might respond to taxes or subsidies (e.g., Just and Antle 1990; Wu *et al.* 2004; Wossink and Swinton 2007) and—if the models relate agriculture production to environmental impact—how they might respond to payments for ecosystem services (e.g., Holden 2005; Antle and Valdivia 2006; Antle *et al.* 2007). Further, farm-level models can be used in conjunction with commodity demand models to assess the impact of agricultural policies and changes in ecosystem service inputs on commodity supply and food prices (e.g., Gillig *et al.* 2004;

Zilberman *et al.* 2008). When appropriate, farm models can assume alternative optimization strategies among farmers, such as risk minimizing behavior by subsistence households in developing countries. In many cases subsistence farmers adopt production systems that minimize food insecurity risk rather than those that maximize expected profits (e.g., Dercon 1996; Kinsey *et al.* 1998; Kandlikar and Risbey 2000; Luckert *et al.* 2000).

In this chapter we propose an approach to assess and map the expected value of agricultural production and the value added by ecosystem services at the landscape scale. As in other chapters, we present the approach in two tiers that differ by information needs, data availability, model sophistication, output scope, and treatment of time. In tier 1, we use broad-scale crop production, yield, and cost maps along with crop prices to assess the value of agricultural production across coarser time steps. When possible, we also assess the contributions of supporting and regulating services to output value. The tier 1 approach can be implemented with existing maps of agricultural production or productivity (e.g. crop yield) or those we create. In tier 1, our ability to model farmer response to market or environmental change is limited.

In tier 2, we model agricultural production from a more detailed perspective and consider the decisions that farmers make over finer time steps. Unlike tier 1, optimum spatial patterns of agricultural production across the landscape and changes in agricultural output and input prices can be determined within the tier 2 model. We also discuss methods for evaluating landscape-scale ecological consequences of agricultural land use, including the emissions of greenhouse gases (GHGs) as well as water and nutrient uptake.

9.2 Defining agricultural scenarios

Our starting point is to generate alternative scenarios of land use/land cover (LULC) pertinent to agricultural productivity. For each LULC scenario we need to specify the crops grown and the method of production on each agricultural parcel in the landscape, as well as crop prices and crop-production costs. As with all other ecosystem service models in this book, the landscape is divided into parcels, indexed by $x = 1, 2, \ldots, X$, where a parcel can be any geographic unit, including a polygon, grid cell, or hexagon. Let the harvested area (in hectares) of crop c ($c = 1, 2, \ldots, C$) grown under production system k ($k = 1, 2, \ldots, K$) in parcel x over some time period (e.g., a growing season, the whole year) be given by A_{ckx}. In this formulation a crop can represent a single crop (e.g., maize), a crop mix (e.g., an intercrop of maize and beans), a crop rotation (e.g., maize-soybeans), or a livestock type (e.g., cattle). Let p_c be the per-unit output price of crop c. Let s_{ckx} be the ha^{-1} costs incurred by the farmers in parcel x when producing and delivering crop c to market under production system k.

In subsistence agriculture crops may be produced and consumed within a household and not sold in a market, in which case we may not observe a market price. In these cases we will need to use other proxies for value. For example, subsistence agriculture prices could be based on measures of the caloric, protein, or micronutrient value of crops. Costs in subsistence agriculture could be expressed in terms of total wages foregone from working in another economic sector. From an ecosystem service perspective, it is also possible to express the "price" of farm outputs in terms of appropriated ecosystem services (e.g., water consumed or nutrient extracted; see Section 9.5) rather than in monetary terms.

Production systems can be differentiated on the basis of production outputs, technologies and practices utilized, and use of inputs (e.g., flood irrigation, drip irrigation, rainfed commercial, or slash-and-burn subsistence using, say, inorganic fertilizer and genetically modified seeds). Production system definitions can be general or quite specific. Examples of general production systems come from the Global Agro-Ecological Zone (GAEZ) datasets where a combination of a technology/input use category (low, medium, and high; see Fisher *et al.* 2002) and a binary variable that indicates whether a system is irrigated or rainfed defines a production system. Alternatively, a production system can be defined with an exact specification of the fertilizer, pesticide, irrigation water, and other inputs applied over the course of a growing season and a precise definition of production methods and technology used. In tier 2 the timing of input use is also part of a production system definition.

If the agricultural production map being evaluated is from the current or a past year then current or past commodity prices and production costs can be used. If we are using a published agricultural production forecast we can use the forecast's accompanying prices (e.g., the USDA Long-term Projections, the OECD-FAO Agricultural Outlooks); otherwise we will have to use some other method to determine prices. In tier 2 we can exogenously define all scenario components, as in tier 1, or with additional data and analytical effort we can determine crop-production system choices endogenously by assuming profit maximizing or other farmer behavior under a given set of prices. We can extend the scenario creation process even further in tier 2 by determining all prices endogenously under the assumption of market equilibrium.

9.3   Tier 1

9.3.1   The supply and use of agricultural production

First we need to gather information on the potential or expected yield for crop c grown with production system k on parcel x, denoted as $\hat{Y}_{ckx}$. In our ecosystem service taxonomy, $\hat{Y}_{ckx}$ represents the measure of the land's ability to supply the ecosystem service of agricultural production at point x. For agricultural production, ecosystem service supply and use (as defined in Chapter 3) are the same. Expected yield is a function of a set of biophysical characteristics such as soil quality, weather, and animal pollinator abundance, and a set of managed inputs such as fertilizer and irrigation water, as well as the production system itself. Formally, $\hat{Y}_{ckx}$ can be represented as,

$$\hat{Y}_{ckx} = g_c\left(\boldsymbol{\theta}_x, \boldsymbol{\varphi}_k\right). \tag{9.1}$$

where g_c is crop c's yield function and is explained by $\boldsymbol{\theta}_x$, the vector of biophysical characteristics or ecological processes in parcel x, and $\boldsymbol{\varphi}_k$, the vector of managed input levels used in production system k.

Ideally we will be able to find completely specified functions $g_c(\boldsymbol{\theta}_x, \boldsymbol{\varphi}_k)$ for all potential values of $\boldsymbol{\theta}_x, \boldsymbol{\varphi}_k$. This would allow us to estimate the change in yield with respect to any change in biophysical characteristics or managed inputs, including many ecosystem services. However, if such yield functions cannot be found we suggest two methods for deriving information about yield across the landscape: first, using existing yield "lookup" tables or maps and, second, estimating a yield function with observed data.

Prepared maps of $\hat{Y}_{ckx}$ can be found for major crops for many areas in the world. For example, the USDA-NRCS (2001) has published observed yields of various crops in each US county as a function of soil capability class (a parameter in $\boldsymbol{\theta}$) under typical management practices. These county-level lookup tables of yield as a function of soil class have been used in conjunction with the behavioral assumption of profit maximization to predict land-use decisions across the USA in response to a hypothetical carbon sequestration payment policy (Lubowski et al. 2006) and to measure the opportunity costs of biodiversity conservation in the Willamette Basin, Oregon, U.S (Polasky et al. 2008). Similarly, Vera-Diaz (2008) has calculated expected soybean yields (and net economic returns) across a gridded map of Brazil. For the GAEZ project (Fischer et al. 2002), global maps of $\hat{Y}_{ckx}$ were generated with yield models explained by general production management categories, ecological characteristics and processes, and climate. Naidoo and Iwamura (2007) used expected yield maps from GAEZ to plot the expected opportunity costs of conservation in forest ecosystems and across the globe.

While these ready-made yield maps may be sufficient for the purposes of some applications, they will not allow us to answer several questions of interest. For one, static maps of yield are not sensitive to changes in ecological processes or climate. Second, these yield maps may ignore the impacts of atypical conditions on yields, such as droughts, floods, major pest and pathogen infestations, and future climate change. Third, some existing yield maps are composed of large spatial units, masking the heterogeneity in local farming systems and ecological characteristics and processes (e.g., a parcel in a GAEZ map covers approximately 10000 hectares).

As an alternative, we can use statistical techniques to estimate a relationship between expected yields and input management, ecological processes, and environmental conditions. This estimation process requires contemporaneous data on crop yield, Y_c, agricultural input use, φ, and biophysical data, θ, from enough sites on a landscape that statistically robust yield functions can be estimated with regression techniques (see Pender 2005 for a review of regression techniques used to build crop yield functions). It is also useful to have data from a number of growing seasons so variations in climate and atypical events such as drought or pest outbreaks can be reflected in estimated yield functions.

The advantage of the statistical approach is that it allows us to build yield functions that are explained by input and biophysical data we can collect. If these inputs include ecological processes then we can predict changes in agriculture output due to a change in such inputs. For example, if we build yield functions that are explained by crop access to water during the growing season then we could estimate the impact of a crop-production system only receiving half of its prescribed irrigation water needs. Drawbacks associated with the statistical technique include the need for a large data set and the necessary time and expertise to assemble and analyze the data. In addition, the confidence in predicted yields decreases when estimated yield functions are applied with explanatory variable levels that differ significantly from those used to estimate the yield function (e.g., predicting yield on a landscape significantly affected by future climate change with functions estimated with data from historic climate patterns).

9.3.2   The value of agriculture's provisioning services

Maps of crop-production areas, expected yields, and expected costs can be combined with crop price databases to produce a map of expected net agricultural production value. In standard parlance used in this book (see Chapter 3), crop production in a parcel represents the *supply* and *use* of agriculture's provisioning service in that parcel, while the net value of this production represents the use *value* (Bockstael *et al.* 2000).

Expected net production value in parcel x for a growing season, AV_x, is found by multiplying the harvested area of each crop-production system in the parcel over the course of the growing season by its expected ha^{-1} revenue less its ha^{-1} production and transportation costs,

$$AV_x = \sum_{k=1}^{K} \sum_{c=1}^{C} A_{ckx}(p_c \hat{Y}_{ckx} - s_{ckx}). \qquad (9.2)$$

If there is a price premium for crops grown under certain conditions (e.g., organic apples versus conventionally grown apples) then such price effects can be included in Eq. (9.2) by stratifying crop price by production system, p_{ck}. Prices and costs should not include production subsidies or cost share payments from a government even if they are payments for ecosystem services. We are interested in assessing the net social value of agricultural production in this model; taxes and subsidy payments are transfers of wealth from one group in society to another, and not a creation of value (the societal losses created by taxation inefficiencies are beyond the scope of this chapter). The benefits of subsidized environmentally friendly production choices, for example, cleaner water, less soil carbon emissions, are accounted for when we model these specific services.

Easily accessible and reliable estimates of production costs for most crop-production systems grown around the world do not exist (Naidoo and Iwamura 2007). USDA-ERS (2009) estimates typical ha^{-1} production costs for most major commodities grown under best management practices in the USA. Cost estimates can also be generated using published budget sheets that itemize the cost of most or all inputs used in a production process (e.g., Polasky *et al.* 2008). The cost of transporting produce to a market can be calculated with road network data and information on per-unit transportation costs. However, in many cases cost data will need to be gathered from local experts. If the cost term is dropped from Eq. (9.2), then AV_x will measure the gross agricultural production value in parcel x rather than agriculture's provisioning value. Only if production costs are fairly uniform over every combination of c, k and x on the landscape will gross agricultural production value give a reliable picture of the spatial variation in the net value of agriculture across the landscape.

9.3.3   Examples of tier 1 modeling

9.3.3.1   *Using ready-made yield maps*

We used Eq. (9.2) to map gross agricultural production values in Tanzania for the year 2000. We focused on 12 crop-production system combinations, formed by combining 4 crops: maize, sorghum, sweet potatoes, and groundnuts, with each of 3 production systems: high inputs and irrigated (*HI*), high inputs and rainfed (*HR*), and low inputs and rainfed (*LR*). The yield and year 2000 harvested area maps used to create the gross agricultural map in Figure 9.1 are from You and Wood (2006) and IFPRI (2008). Commodity prices for all four crops were calculated using 2000 market prices from several east African countries (Tanzanian prices were not reported; FAO 2008a). See the chapter's supplementary online material (SOM) for more details on this illustrative example.

The map in Figure 9.1 does not fully indicate the use value of agricultural production in Tanzania in 2000 for two reasons. First, we have not included production cost data. Second, the four modeled crops comprised only 54% of Tanzania's total cropped area in 2000 (FAO 2008a) and the map does not include livestock, egg, or dairy production. Futhermore, the size of the individual parcels (roughly 10 000 ha) limits this map's relevance for informing land-use decisions at more local scales.

9.3.3.2   *Using yield functions to generate yield maps*

In general, we need to use yield functions versus yield lookup tables or maps if we want to assess the impact of input management and ecological processes on yield. Here we demonstrate a statistical method for estimating a maize yield function for a landscape in east-central Africa. The modeled

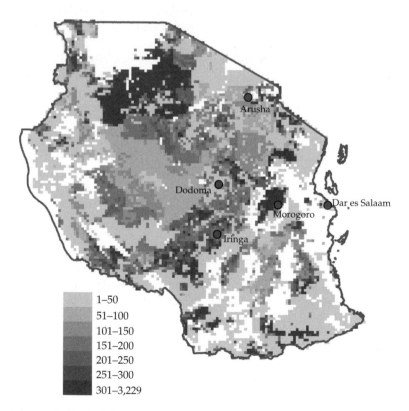

Figure 9.1   Gross agriculture production value in Tanzania.
Parcel (5 minute grid cells) values are measured in thousands of year 2000 US dollars. Data on yield and area devoted to each of the 12 modeled crop-production system combinations come from IFPRI (2008). Year 2000 crop prices are the average market prices observed in ten east African counties in 2000 (Tanzanian prices were not available; FAO 2008a). A white parcel indicates that none of the 12 modeled production systems were produced in that parcel in 2000. See the SOM for more modeling details.

landscape is a 300 × 450 km region made up of 1 551 parcels (5 arc minute, roughly 10 km² grid cells) that straddles five countries (Fig. 9.2a).

Ideally, we would use observed yields, input use, and ecological process levels to estimate yield functions. However, because maize yield data were not readily available for this region, we first used a crop growth model to simulate maize yields for the few locations on the landscape where detailed input data were available. We then estimated a linear regression model to explain simulated yields as a function of fertilizer use and a set of biophysical explanatory variables. Finally, we used the estimated yield function to predict maize yields on all 1 551 parcels on the landscape.

We simulated maize yields with data from 31 locations within the landscape that coincided with the availability of detailed soil profile information (Batjes 2002). Daily weather data for the select locations from 1997 to 2003 were obtained from the NASA Langley Research Center Atmospheric Sciences Data Center POWER Project (http://power.larc.nasa.gov).

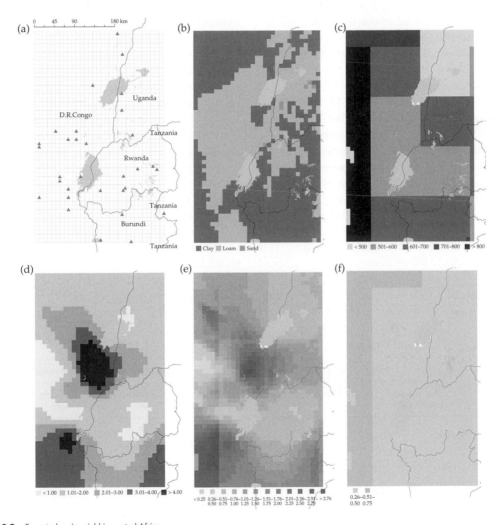

Figure 9.2 Expected maize yield in central Africa.
The modeled landscape straddles five countries and includes 1 551 parcels ((a) 5 minute grid cells). The triangles indicate the locations of maize yield simulations. Explanatory variables used to explain yields include (b) soil texture (*TXTC*), (c) rainfall data in mm season⁻¹ in 2000, and (d) soil organic carbon (*SOC*) percentage in the top 100 cm of soil. Expected yield of maize in Mg ha⁻¹ using the estimated yield function (Eq. (9.3)) is given in (e). We also show the expected maize yield increase in Mg ha⁻¹ if *SOC* percentage in the top 100 cm of soil increased by 1% (f). See Table 9.1 and the SOM for more details on the estimated yield function used in this example.

For each of the 31 locations we assumed that maize was cultivated once a year in six separate but identical plots differentiated only by application rates of inorganic nitrogen (N) fertilizer (0, 20, 40, 60, 80 100 kg N ha⁻¹). Using this set of input data and the Decision Support System for Agrotechnology Transfer (DSSAT) Crop Systems Maize Model v4.02 (Jones *et al.* 2003), growth of a widely adopted, long maturity maize variety (measured in Mg ha⁻¹) was simulated annually at all 31 locations under the 6 different N application for the years 1997 through 2003, creating 1 302 simulated maize yield observations (31 locations × 6 plots × 7 years).

Next we used simple regression techniques to estimate the functional relationship between the 1 302 yield observations and several biophysical variables that are available everywhere on the landscape. These variables describe soil texture (*TXTC*, clay = 1 and non-clay = 0; see FAO/IIASA/ISRIC/ISS-CAS/JRC 2008), growing season precipitation (*RAIN*, mm; NASA POWER Project), soil organic carbon percentage in the top 100 cm of soil (*SOC*; Batjes 2002), and the managed input of inorganic nitrogen (N) fertilizer application over the growing season (*NFRT*, kg N ha⁻¹). The ordinary least squares estimate of a linear specification of the yield function is

$$\hat{Y}_{maize} = 0.02 + (0.034 NFRT) + (0.012 NFRT \times TXTC) + (0.001 RAIN \times SOC) \quad (9.3)$$

where $\hat{Y}_{maize}$ is maize yield in Mg ha⁻¹. See Table 9.1 and the chapter's SOM for regression equation details.

To predict yield in each of the landscape's 1 551 parcels, Eq. (9.3) was applied using a parcel's data on *TXTC*, *RAIN* in the year 2000, *SOC*, and the assumption that 5 kg of N ha⁻¹ would be applied (Morris *et al.* 2007; see Figure 9.2a–d). The resulting map depicts the expected maize yield on each parcel on the landscape (Figure 9.2e) and can be used in conjunction with a maize harvested area map, maize prices, maize production costs, and Eq. (9.2) to determine the expected net value or use value of maize production across the landscape.

The positive coefficients on the N variables in the linear yield Eq. (9.3) imply that each increment of N fertilizer application will generate a proportionate increase in yield. At higher levels of N fertilizer application (or any other managed input use), the yield response declines (Tilman *et al.* 2002). Such diminishing returns to inputs are rarely encountered in Africa because of limited use of inputs like fertilizer (e.g., 5 kg of N ha⁻¹ is very low) and irrigation water (Crawford *et al.* 2003). However, if we are predicting yield with a linear function in an area where managed input use can be high then the value range of the explanatory variables should be limited to the range used to estimate the function. Alternatively, we can use a nonlinear model to estimate yield in such systems.

Besides using the map of estimated maize yield to determine net values of agricultural production on the landscape, we can also use Eq. (9.3) to determine the change in estimated maize yield, $\Delta \hat{Y}_{maize}$, due to small changes in *TXTC*, *RAIN*, *SOC*, or *NRFT*. For example, the organic carbon content of soil,

Table 9.1 Regression results for maize yield in an African landscape

Variables	Estimated Eq. (9.3)			Descriptive statistics of variables			
	Regression coefficient	Standard error	*p* value	Variable mean	Variable std. dev.	Variable min.	Variable max.
Independent							
Maize yield (Mg ha⁻¹)				2.8	1.9	0	8.0
Dependent							
Intercept	0.020	0.068	0.767				
NFRT	0.034	0.001	0.000	50	34	0	100
NFRT × *TXTC*	0.012	0.001	0.000	27	35	0	100
RAIN × *SOC*	0.001	4.03 × 10⁻⁵	0.000	1 038	743	163	3 374
N	1,301						
R²	0.67						
Adjusted *R²*	0.67						

SOC, reflects an ecological process that influences yield and is controlled to some degree by LULC and land management choices on the landscape. Therefore, if the land could be managed such that *SOC* increased by 1 percentage point in a parcel then the expected change in maize yield in the parcel would be equal to 0.001 × *Rain* (i.e., $\partial\hat{Y}_{maize}/\partial SOC$; see Figure 9.2f) In the next section we discuss in more depth a method for finding the service value provided by changes in ecological processes and conditions mediated by land-use and management choices.

9.3.4   Measuring the value of regulatory and supporting ecosystem service inputs to agriculture

A change in the supply of one or more ecological processes that is an input into agricultural production (i.e., an ecosystem service) can change the expected value of crop production. For example, the net value generated by pollinator-dependent crop can decline if the number of animal pollinators in the parcel declines. We can measure the value of the change in the regulating or supporting ecosystem service by comparing the expected net agricultural value before and after the change in the service. Let θ_{ix}^0 and θ_{jx}^0 be the initial level of ecosystem services *i* and *j* in parcel *x* and let θ_{ix}^1 and θ_{jx}^1 be the levels after the change. Using the tier 1 yield function, we can define the pre- and post-change expected yields as $\hat{Y}_{ckx}^0 = g_c(\varphi_k, \theta_{ix}^0, \theta_{jx}^0, \boldsymbol{\theta}_{-x})$ and $\hat{Y}_{ckx}^1 = g_c(\varphi_k, \theta_{ix}^1, \theta_{jx}^1, \boldsymbol{\theta}_{-x})$, respectively, where $\boldsymbol{\theta}_{-x}$ defines all ecological process inputs other than *i* and *j*. If appropriate, $\hat{Y}_{ckx}^0$ and $\hat{Y}_{ckx}^1$ can include even more changes in ecological processes and/or changes in managed inputs, such as water use for irrigation, defined with φ_{ix}^0, φ_{ix}^1, and φ_{-x}.

The change in the net value of agricultural output due to the change in the supply of ecosystem services *i* and *j* is given by ΔAV_x,

$$\Delta AV_x = \sum_{k=1}^K \sum_{c=1}^C \left[A_{ckx}^1 (p_c^1 \hat{Y}_{ckx}^1 - s_{ckx}^1) - A_{ckx}^0 (p_c^0 \hat{Y}_{ckx}^0 - s_{ckx}^0) \right]. \quad (9.4)$$

where ΔAV_x can also be interpreted as the value of the joint change in ecosystem services *i* and *j* as expressed through agricultural production (Shiferaw

et al. 2005; Swinton 2005). In this analysis we allow for the possibility that changes in ecosystem service inputs can lead to changes in other input use, crop-production system combination choices (A_{ckx}^1), production costs (s_{ckx}^1), and crop prices (p_c^1). For example, in response to honeybee declines, farmers of animal-pollinated crops are experimenting with ways to attract native pollinators to their fields (Kremen *et al.* 2008). Such experimentation will change input costs and may change cropping practices. In tier 1 we do not include a formal method for determining changes in A_{ckx}, p_c, or s_{ckx} in response to changes in ecological processes (we do in tier 2).

As presented here, we gain the greatest flexibility in measuring the incremental value of regulatory or supporting services when we use yield functions that are continuous in their levels of provision. However, in some cases we can still approximate the impact of a change in the provision of a service on production values with expected yield maps, as we illustrate next.

9.3.4.1   An example of supporting service valuation: the case of surface water supply in Tanzania

Surface water that flows across a landscape can contribute to agricultural production if it is used for irrigation. We used the map and datasets discussed in Section 9.3.3 and a version of Eq. (9.4) to estimate the decline in gross agricultural production value from a small reduction in surface water availability. Specifically, we estimated the percentage decline in year 2000 aggregate gross production values across Tanzania if 5% of the area of high input irrigated (*HI*) maize and sorghum found in each parcel could no longer be irrigated due to reduced surface water flow (see Figure 9.3). In this analysis we ignored irrigation via groundwater (such practice is rare in Tanzania, FAO 2008b) and we ignored sweet potato and groundnut systems because they are not typically irrigated.

The maize and sorghum yield maps do not explicitly include surface water supply for irrigation as an explanatory variable. Instead, we assumed that farmers of the hectares formerly in *HI* maize or sorghum production, in reaction to the loss in irrigation water, grew the same crops but under rainfed conditions with less intensive input use (*LR* production). In many cases farmers without access to

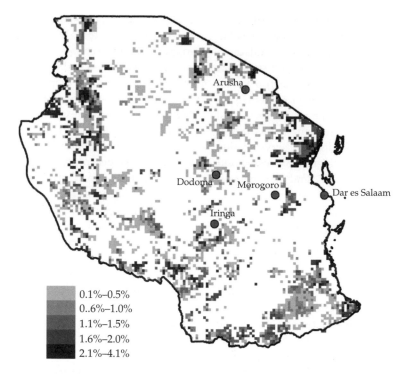

0.1%–0.5%
0..6%–1.0%
1.1%–1.5%
1.6%–2.0%
2.1%–4.1%

Figure 9.3   Decline in gross agricultural production value due to reduced irrigation in Tanzania.
The map gives the percentage decline in a parcel's 2000 agriculture gross production value (Figure 9.1) when 5% of each parcel's area in irrigated production (*HI*) of maize and sorghum were replaced with low input and rain-fed (*LR*) production. Parcels shown in white do not contain any of the 12 modeled production systems in 2000. Price data are the same as those used in Figure 9.1. See the SOM for more modeling details.

irrigation water will reduce the use of costly inputs due to the lower yields and greater crop failure risk associated with rainfed versus irrigated agriculture (Feder *et al.* 1985).

Assuming prices remain at their 2000 levels our analysis indicates that small losses in irrigation water would have caused a range in parcel-level gross value losses of 0 to 4.1%. In aggregate, this decline in irrigation capacity would have caused a 0.02% decrease in the country-wide gross production value represented in Figure 9.1. These results, while simple and illustrative, would indicate that the ecosystem service of irrigation may currently add little to the net value of maize and sorghum production in Tanzania.

9.4   Tier 2

Tier 2 extends the tier 1 approach in several ways. First, we account for the timing of input use during the growing season. Second, we can track agriculture production value over multiple growing sea-

sons. Third, we propose a framework for modeling farmer behavior given government policy, prices, technology and environmental conditions. Finally, we can use the tier 2 model to solve for product output and input prices by assuming that product output and input markets clear (i.e., aggregate quantities supplied balance aggregate demand).

9.4.1   The supply, use, and value of agriculture's provisioning ecosystem services

The tier 2 yield function considers the temporal pattern of input use across a growing season. Let z index growing seasons and let $b = 1, 2, \ldots, B$ index the sub-periods within season z. The net value of agricultural production in parcel x in time period z, AV_{xz}, is given by

$$AV_{xz} = \left(\sum_{k=1}^{K} \left[\begin{array}{c} \displaystyle\sum_{c=1}^{C} p_{cz} A_{ckxz}\, \hat{g}_{cz}\left(\varphi_{1kxz}, \ldots, \varphi_{Bkxz}; \theta_{1xz}, \ldots, \theta_{Bxz}\right) \\ -A_{kxz}\left(\mathbf{q}_{1z}\varphi_{1kxz} + \ldots + \mathbf{q}_{bz}\varphi_{bkxz} + \ldots + \mathbf{q}_{Bz}\varphi_{Bkxz}\right) \end{array} \right] \right) \quad (9.5)$$

where p_{cz} is the market price of crop c in season z (or p_{ckz} if there is a price premium for crop c grown under production system k), A_{ckxz} is the area of x in crop c under production system k in season z, $\hat{g}_{cz}$ is the estimated yield function of crop c in season z and is explained by an intraseasonal series of inputs associated with production system k ($\varphi_{1kxz}, \dots, \varphi_{Bkxz}$) and biophysical properties and processes in x ($\theta_{1xz}, \dots, \theta_{Bxz}$), A_{kxz} is the area of production system k in x in season z, and $\mathbf{q}_{bz}$ is a vector of per-unit input prices during sub-period b in season z, including any harvested crop transportation costs (the vector product $\mathbf{q}_{bz}\varphi_{bkxz}$ gives the total ha^{-1} input use costs incurred in sub-period b of season z under production system k). For simplicity's sake, we assume that the production system determines production costs and not the crop-production system (i.e., costs in equation (9.5) are a function of k and not c). However, the modeler could differentiate costs across each crop-production system choice (i.e., A_{kxz} in equation (9.5) could become A_{ckxz} and the managed input vectors in equation (9.5) could be indexed by c as well). Because a crop's response to inputs over time can change due to technological growth (e.g., Rosegrant *et al.* 2001; Alston *et al.* 2009) or climatic shifts, we index the yield function $\hat{g}_{cz}$ by seasons. For example, from 1966–2005, maize yield in the USA improved 100% largely due to farmer adoption of improved maize production technology (Cassman and Liska 2007).

As in tier 1, we can link tier 2 yield equations with models of ecosystem service input supply. For example, we can determine the amount of surface water flow available for the irrigation of crop c under production system k in sub-period b of season z using the tier 2 surface water flow model (Chapter 4). In addition the vectors $\theta_{1xz}, \dots, \theta_{Bxz}$ can include several other tier 2 outputs from models noted in this book, including pollinator abundance (Chapter 10), soil erosion (Crosson *et al.* 1995), and temporary flooding (Chapter 5). As in tier 1, the change in the net value of agricultural output on parcel x due to the change in the supply of an ecosystem service input is given by evaluating Eq. (9.5) pre- and post-change assuming all input use and prices not affected by the change in the ecosystem service input remain constant. For example, if vegetation changes are made on the landscape such that temporary flooding in parcel x during growing season z is much less than it

was in the previous season—and all other input use and prices remain the same across growing seasons—then $\Delta AV_x = AV_{xz+1} - AV_{xz}$ represents the value of the vegetation change in parcel x as expressed through agricultural production.

9.4.2   Tier 2 scenarios

We propose two approaches to determining agricultural production scenarios in tier 2. One approach is similar to that used in tier 1 in which we define all relevant model parameters and variables exogenously, including crop-production system combinations, managed input use levels, and prices for all sub-period and parcel combinations for growing seasons $z = t, t+1, \dots, T$.

Alternatively, we can assume that farmer behavior will be guided by objectives such as profit maximization. To make this approach more tractable we first define those parcels that will be used for agriculture in each growing period. Then farmers will choose amounts of different crops to cultivate (A_{ckxz}), inputs ($\varphi_{1kxz}, \dots, \varphi_{Bkxz}$), and to the extent possible biophysical properties ($\theta_{1xz}, \dots, \theta_{Bxz}$), over each growing season z such that the expected net present value (NPV) of their agricultural production returns over time are maximized,

$$\max\left(\sum_{z=t}^{T} \frac{\sum_{k=1}^{K}\left[\sum_{c=1}^{C} p_{cz} A_{ckxz}\hat{g}_{cz}(\varphi_{1kxz}, \dots, \varphi_{Bkxz}; \theta_{1xz}, \dots, \theta_{Bxz}) + e_{ckxz} A_{ckxz}\right]}{(1+r)^{z-t}} - \sum_{z=1}^{T} \frac{\sum_{k=1}^{K} A_{kxz}(\mathbf{q}_{1z}\varphi_{1kxz} + \dots + \mathbf{q}_{Bz}\varphi_{Bkxz})}{(1+r)^{z-t}} \right)$$

(9.6)

subject to $\sum_{k=1}^{K} A_{kxz}(\mathbf{q}_{1z}\varphi_{1kxz} + \dots + \mathbf{q}_{Bz}\varphi_{Bkxz}) \le B_{xz}$ for all z where e_{ckxz} is the ha^{-1} subsidy (or tax if negative) for producing crop c under production system k, r is the relevant interest rate, and B_{xz} is the budget constraint for parcel x during season z. In Eq. (9.6) we assume parcel x is in agricultural land use from growing season t to T. If the farmers in parcel x have access to capital (e.g., smallholder financial services) then the budget constraint can be loosened or even dropped. We explicitly include any subsidies (or taxes) in Eq. (9.6) because, while they do not affect the net value of agricultural production explicitly, they do affect farmers' private decisions (and the provision of other ecosystem services; e.g., Antle and Valdivia 2006). If the subsidy is a function of the amount of an ecological process or good produced then the A_{ckxz} after e_{ckxz} can be replaced by an appropriate ecological production function.

Let $A_{ckxz}(\mathbf{p}_z, \mathbf{e}_z, \mathbf{q}_{1z}, \ldots, \mathbf{q}_{Bz}), \varphi_{1kxz}(\mathbf{p}_z, \mathbf{e}_z, \mathbf{q}_{1z}), \ldots, \varphi_{Bkxz}(\mathbf{p}_z, \mathbf{e}_z, \mathbf{q}_{Bz})$, and to the extent they are manageable, $\theta_{1xz}(\mathbf{p}_z, \mathbf{e}_z, \mathbf{q}_{1z}), \ldots, \theta_{Bxz}(\mathbf{p}_z, \mathbf{e}_z, \mathbf{q}_{Bz})$, for all z from growing seasons t to T indicate the choices made in parcel x that are expected to solve the maximization problem described by Eq. (9.6). All solutions are functions that are explained by prices $\mathbf{p}_z, \mathbf{e}_z,$ and $\mathbf{q}_{bz}$ ($\mathbf{p}_z$ and $\mathbf{e}_z$ are vectors of all p_{cz} and e_{ckxz} at time z). In many cases, choices for growing season z will need to be made before some biophysical values for that season, such as rainfall or irrigation water availability (e.g., Perry and Narayanamurthy 1998), are known. In these cases, we hypothesize that farmers will maximize over expectations for biophysical values. Evaluating Eq. (9.5) with the choices that solve Eq. (9.6) gives $AV^*_{xt}, \ldots, AV^*_{xT}$, the series of maximized profit levels in parcel x from growing season t to T.

As with tier 1 models we can use alternatives to profit maximizing assumptions when considering farmer behavior. Subsistence farmers may be primarily concerned with reducing the chance of crop failure in any given growing season rather than maximizing the expected value of yield or profit (e.g., Dercon 1996). For example, if a landscape is exposed to an extended drought period (rainfall levels in $\varphi_{1kxz}, \ldots, \varphi_{Bkxz}$ are expected to be low or unreliable during the growing season) subsistence farmers are likely to make very different crop-production system choices than a profit maximizing farmer; for example, minimizing the potential severity of food shortages during the drought period. In such cases we can replace the expected net revenue function in Eq. (9.6) with a utility function described by a high degree of risk aversion (see Bardhan and Udry 1999 for more details).

To fully specify farmer behavior in a scenario, whether modeled as maximizing the NPV of revenue streams or, more generally, of utility streams, we need to establish output and input prices. We can simply assume some $\mathbf{p}_z$ and $\mathbf{q}_{bz}$ for $z = t >, \ldots, T$ (see Section 9.2) or, to maintain theoretical consistency with typical economic equilibrium analyses, we can endogenously determine market-clearing prices, i.e.,

those that equate supply and demand for all agriculture products and inputs (e.g., Zilberman *et al.* 2008). To establish the trajectory of output market-clearing prices we need to define functional relationships between the demand for each modeled crop and managed inputs and their prices. Specifically, let $D_{cz}(p_{cz})$ represent the demand in season z for crop c produced on the study landscape (we can further segment by production system if it is relevant). Further, let $\mathbf{F}_{1z}(\mathbf{q}_{1z}), \ldots, \mathbf{F}_{Bz}(\mathbf{q}_{Bz})$ represent the supply of inputs available on the landscape in the subperiods of time z. We estimate market equilibrium prices by equating the optimal supply of commodity c to its demand and, similarly, the demand for inputs equal to their supply and then solve for all prices,

$$\sum_{x=1}^{X}\sum_{k=1}^{K}\left[A_{ckxz}(\mathbf{p}_{cz}, \mathbf{e}_{ckz}, \mathbf{q}_{bz}) \times \hat{g}_{cz}\left(\begin{array}{l}\theta_{1kxz}(\mathbf{p}_{cz}, \mathbf{e}_{ckz}, \mathbf{q}_{1z}), \ldots, \theta_{Bkxz} \\ (\mathbf{p}_{cz}, \mathbf{e}_{ckz}, \mathbf{q}_{Bz}); \theta_{1xz}, \ldots, \theta_{Bxz}\end{array}\right)\right] = D_{cz}(p_{cz}) \quad (9.7)$$

and

$$\sum_{x=1}^{X}\sum_{k=1}^{K} A_{kxz}\varphi_{bkxz}(\mathbf{p}_z, \mathbf{e}_z, \mathbf{q}_{bz}) = \mathbf{F}_{bz}(\mathbf{q}_{bz}) \quad (9.8)$$

for $b = 1, 2, \ldots, B$. In general, as the supply of an agricultural output or input increases, its price tends to fall, all else equal, because it has become less scarce in the market. In many cases demand functions for agricultural output, whether describing demand at regional or global scales, are indifferent to point of output origin. In such cases we can assume a certain portion of demand will be met by agricultural production outside of the study landscape and let $D_{cz}(p_{cz})$ represent the residual demand to be met locally. Further, the supply of inputs available to farmers, and thus relevant supply functions, varies greatly across the world. For example, farmers in the USA can generally buy fertilizer and other inputs from a global supply market. Farmers in Africa may be limited to regional market with less choice and higher prices (e.g., Morris *et al.* 2007).

Solving Eq. (9.6) can be computationally difficult, even if 1) we set prices exogenously, 2) assume that farmers in x have no control over $\theta_{1xz}, \ldots, \theta_{Bxz}$, and 3) assume that $\varphi_{1kxz}, \ldots, \varphi_{Bkxz}$ and $\theta_{1xz}, \ldots, \theta_{Bxz}$ are not dependent on current and past management choices on neighboring parcels (i.e., spatial production externalities). When we include production externalities (e.g., water use upstream affects water availability downstream) and simultaneously determine prices with Eqs. (9.7) and (9.8) then solving Eq. (9.6) can be exceedingly difficult (see Conrad and Clark (1987) and Holden (2005) for a review of modeling methods that have been used to solve Eqs. (9.6) through (9.8)). We can reduce the degree of complexity by modeling a minimum number of crops, production systems, and inputs; keeping the number of time steps between t and T small; and using coarsely defined maps to keep the number of agricultural parcels low.

9.5 Mapping the impacts of agriculture on important ecological processes

Agricultural land use and management practices can have profound effects on the supply and condition of ecosystem services and biodiversity on the landscape (Swinton *et al.* 2007). We can measure many of these effects using the models presented elsewhere in this book. Below we present a simple system for tracking three biophysical processes affected by agriculture: greenhouse gas (GHG) flux other than carbon dioxide (CO_2) and water and nitrogen use on agricultural fields. We can use estimates of GHG release and water and nutrient use to summarize some of the landscape-wide "external" costs of an agricultural scenario.

Agricultural production can be the source of methane (CH_4) and nitrous oxide (N_2O) emissions, two powerful GHG gases (see Chapter 7 for the accounting of CO_2 emissions from agricultural land use). Both livestock and flooded agriculture systems (e.g., rice paddies) produce CH_4 (e.g., Neue 1993; IPCC 2007). A fraction of the N produced in and applied to fields, including the N inorganic fertilizer and manure, contributes to the formation of N_2O (e.g., Bouwman *et al.* 2002; Dalal *et al.* 2003). Further, the portions of the N in soil, manure, and biomass residue that do not contribute to the direct formation of N_2O may eventually contribute to the formation of N_2O via other ecological processes (IPCC 2006).

We can model agriculture-derived CH_4 and N_2O flux by tabulating data on per-area or per-yield fluxes, converting them to CO_2-equivalents, and then to elemental carbon.

$$FluxGHG_{xz} = 0.2727 \left(\sum_{k=1}^{K} \sum_{c=1}^{C} A_{ckxz} \hat{Y}_{ckx} \left(\frac{25YCH4_{ckx} + 298}{(YDN2O_{ckx} + YIN2O_{ckx})} \right) \right.$$
$$\left. + \sum_{k=1}^{K} \sum_{c=1}^{C} A_{ckxz} \left(\frac{25CH4_{ckx} + 298}{(DN2O_{ckx} + IN2O_{ckx})} \right) \right)$$

(9.9)

where $FLuxGHG_{xz}$ is the flux of CH_4 and N_2O on parcel x in growing season z in Mg of carbon equivalents, $YCH4_{ckx}$, $YDN2O_{ckx}$, and $YIN2O_{ckx}$ represent the direct emissions of CH_4, direct emissions of N_2O, and the indirect emissions of N_2O, respectively, per unit yield of crop c on parcel x grown under production system k, and $CH4_{ckx}$, $DN2O_{ckx}$, and $IN2O_{ckx}$ represent the direct emissions of CH_4, the direct emissions of N_2O, and the indirect emissions of N_2O, respectively, per hectare of crop c on parcel x grown under production system k. The constants 25 and 298 in Eq. (9.9) convert CH_4 and N_2O, respectively, to CO_2-equivalents (IPCC 2007). The constant 0.2727 in Eq. (9.9) converts CO_2 measurements to elemental carbon equivalents. By indexing each coefficient by x we allow for the possibility that emission rates can also be explained by features that vary across the landscape (soil types, slope, etc).

Values for the coefficients in Eq. (9.9) can come from IPCC (2006), can be estimated using field data, or come from other relevant sources. For example, if we have CH_4 emissions data for multiple fields of c under production system k we could regress the emissions data against corresponding yield estimates to derive coefficients $YCH4_{ck}$ (the estimated slope and) and $CH4_{ck}$ (the estimated intercept). If some coefficients are not relevant we can set their values to zero. For example, if CH_4 and N_2O emission data are only given per hectare of crop c grown under production system k then $CH4_{ckx}$, $DN2O_{ckx}$, and $IN2O_{ckx}$ will be greater than 0 and all other emission coefficients will be equal to 0. Finally, if we multiply $FLuxGHG_{xz}$ by the social cost of carbon (SCC) we estimate the economic damage generated by GHG flux (see Chapter 7 for a discussion of the SCC).

Agricultural systems also modify water and nutrient cycles on the landscape (Lesschen *et al.* 2007). For example, by quantifying water use by all crops on a landscape (i.e., total crop transpiration) under different scenarios of agricultural production we can determine which scenarios result in lower consumptive use by crops. Or we may be interested in determining if agricultural practices on a landscape in period z remove more water or nutrients than the amount replenished (from natural or human mediated sources), i.e., the *net* impact. If the water volumes or nutrient amounts removed are greater than those replenished over a growing season then landscapes run a temporary water or nutrient deficit that might, given successive growing season deficits, lead to a water or nutrient supply crisis.

We can modify Eq. (9.9) to model a water or nutrient cycle within parcel x in growing season z,

$$\Omega_{xz} = -\omega_{xz} + \sum_{k=1}^{K}\sum_{c=1}^{C} A_{ckxz}(\hat{Y}_{ckxz}YU_{ckx} + U_{ckx}),\quad (9.10)$$

where ω_{xz} is the volume of water or mass of nutrient added to parcel x during growing season z and the double summation term represents the amount of water or nutrient consumed by crops in parcel x in period z. For example, suppose we are tracking water use by crops. In this case YU_{ckx} is the volume of water used per unit output of crop c grown under production system k on parcel x while U_{ckx} is the per hectare volume of water used when producing crop c with production system k on parcel x and is independent of c's yield. As with the GHG flux accounting, the uptake coefficient YU is greater than 0 if the use data are a function of yield and U is greater than 0 if some or the entire use is explained by area. If we set $\omega_{xz} = 0$ then Ω_{xz} measures the water used in parcel x by its crops or, in other words, the amount of water no longer available to the rest of the landscape for use in period z. If we set ω_{xt} equal to water production in parcel x during period z (i.e., rainfall and irrigation) then Ω_{xz} measures the net import (if positive) or excess water produced on x (if negative). Crop water models and databases such as *CropWat* (FAO 1992), *AquaCrop* (Steduto *et al.* 2009), and *AquaStat* (FAO 2008b) can provide data on YW_{ckx} and W_{ckx} for many crop-production systems across the globe.

If we are tracking N use instead of water use, YU_{ckx} is the use of N per unit output of crop c under production system k in x and U_{ckx} is N use per unit of area. If we set $\omega_{xz} = 0$ then Ω_{xz} measures the gross use of N in x during time z in (measured in kg). If we set ω_{xt} equal to N (natural and man-made) production and application in parcel x during period

z then Ω_{xz} measures the net decrease in N levels (if positive) or excess N (if negative).

9.6   Uncertainty

In general the models described above ignore the issues of stochasticity in agricultural production and value. For example, yields may differ from expectations due to variability in weather (e.g., droughts, storm damage, and floods) and pest and pathogen outbreaks. Further, problems in transporting crops between farm and processing centers or markets, post harvest losses, and market related instability (e.g., volatile prices for either crops or inputs) means revenues and costs can deviate, perhaps significantly, from expectations.

One approach to dealing with biophysical and market uncertainty in our models is to generate bounds on the range of potential production values by analyzing farmer behavior and production results under a range of climatic and market conditions, including various output and input price scenarios. For example, what agricultural production values would be generated on a landscape with high levels and a timely pattern of rainfall over a growing season, making typical irrigation practices unnecessary, versus the values generated during a drought that causes many rainfed crops to fail and creates high demand for irrigation? Or we could look for published relationships between climate and crop-production system patterns on the landscape and probabilities of pest or pathogen outbreak. Then we could adjust maps of expected net agricultural production and value appropriately *ex post*, reducing output and value in areas where disease and pathogen risks are judged to be high.

Another approach to incorporating uncertainty is to perform a Monte Carlo analysis of production value where we introduce random variation around expected values of environmental conditions, prices, input availability, and farmer behavior to generate histograms of model outputs. Or, if we are using the tier 2 framework to model farmer behavior we might compare profit maximizing to risk minimizing (or utility maximizing) assessments of production values.

9.7 Limitations and next steps

9.7.1 Limitations

In this chapter we propose two tiers of agricultural modeling, aiming for a range of conceptually sound and tractable approaches to estimating and mapping values of agricultural production and the ecosystem services that support agricultural production. However, the approaches outlined do have several limitations that are important to acknowledge. First, we did not explicitly model livestock operations. We propose modeling livestock production systems as if they are a crop with expected yield (e.g., kg of meat ha^{-1}, animal units supported ha^{-1} yr^{-1}) and input use defined by a production system k. However, the physical size of intensive and confined livestock production enterprises may be too small to be represented on the LULC maps that form the basis of the agricultural production models described in this chapter. In such cases we may need to add point data to the LULC maps to account for these systems, or account for them outside the mapping framework. A better understanding of the links between livestock production and water and air pollution, ecosystem processes, and biodiversity conservation is becoming more important as meat demand increases throughout the world (Delgado *et al.* 1999; Steinfeld *et al.* 2006).

Second, many agricultural landscapes include a diversity of crops and systems of input use. Most applications of the approaches described here focus on major crops, both due to data limitations and time constraints. This restriction may underestimate production and ecosystem service value and distort the spatial distribution of that value—especially if the modeled systems are likely to be allocated across the landscape differently than non-modeled systems. Further, by ignoring minor crops we might miss important niche products on the landscape and overlook important crop alternatives that could improve environmental conditions on the land while maintaining the value of agricultural output.

Climate change is likely to have a large impact on agricultural yield functions, crop-production system patterns, crops prices, production costs and management decisions in the future (e.g., Mendelsohn *et al.* 1994; Schlenker *et al.* 2005; Luedeling *et al.* 2009; Mendelsohn and Dinar 2009). If we have yield functions that include climate variables then we may be able to predict some of the changes in the agricultural sector from climate change (see Chapter 18). In addition, we can model some of the expected effects of climate change on agriculture production via changes in ecosystem services that are inputs to agricultural production (see Chapters 4 and 10). However, major climate change might cause unanticipated changes in crop and animal growth patterns that cannot be predicted with statistical models based on current or historic relationships between yields and agricultural inputs (crop simulation models would appear better able to model yields in response to climate change; e.g., Parry *et al.* 1999). Further, predicting commodity supply and demand in a climate change-affected world will also be very difficult, making the endogenous price determination method discussed in the tier 2 approach particularly challenging to implement for future scenarios under climate change.

9.7.2 Next steps

The output from the agriculture models described here can be used for several types of analyses. We can use expected net production value maps to determine the opportunity costs of setting aside agriculture land for conservation, or conversely, the value added to society from bringing land into agricultural production. We can use maps of agricultural production values, the value added to agriculture production by regulating and supporting ecosystem services, and agriculture-related GHG flux maps along with maps of other ecosystem service use to create a series of service trade-off analyses (see Chapter 14).

Other models presented in this book are able to use outputs from the agriculture models described here. For example, when we assign fertilizer and pesticide use to each crop-production system and then map these systems we have also mapped fertilizer and pesticide use. The water pollution regulation model (see Chapter 6) can use these data to estimate agriculture's impact on water quality across the landscape. Surface water withdrawals for irrigation will affect water use in other sectors of the economy (see Chapter 4). We could also use the outputs from this model to make predictions about the number of rural households in developing countries that are involved in timber and non-timber forest product harvesting. In general, the lower the predicted net values from agriculture the more that rural households will look to supplement their incomes with products generated by nearby forests (see Chapter 8).

Moving beyond tier 2 to gain even more modeling realism implies the use of more elaborate yield, biophysical, and economic simulation models. For example, we can better model crop yields by using one of more specialized crop growth simulation models such as DSSAT, APSIM, or EPIC (e.g., Bryant *et al.* 1992). These models, as partly demonstrated in Section 9.3.3 where we used DSSAT to generate "observed" maize yields, are data intensive and usually require detailed datasets and specialized training to use. Complex models that track nutrient cycles in soils (e.g., CENTURY, RothC-26.3, or SCUAF), water balance and quality, and other environmental impacts of farming (e.g., WATBAL, PERFECT, or SWAT; see Wani *et al.* 2005 for a review of all of these and other simulation models) are also available. However, these models often require significant calibration effort and data not readily available in developing countries. Finally, far more elaborate agricultural production models that embed microeconomic principles within macroeconomic and general equilibrium analytical components are available to evaluate trajectories of land-use decisions and ecosystem service provision over time (e.g., Eickhout *et al.* 2006; Bouwman *et al.* 2006; Nelson *et al.* 2009). Once again, however, these models are complex and are often only accessible to the research groups that developed and maintain the models.

References

Alston, J. M., Beddow, J. M., and Pardey, P. G. (2009). Agricultural research, productivity, and food prices in the long run. *Science*, **325**, 1209–10.

Antle, J. M., and Valdivia, R. (2006). Modelling the supply of ecosystem services from agriculture: a minimum-data approach. *Australian Journal of Agricultural and Resource Economics*, **50**, 1–15.

Antle, J. M., Capalbo, S., Paustian, K., *et al.* (2007). Estimating the economic potential for agricultural soil carbon sequestration in the Central United States using an aggregate econometric-process simulation model. *Climatic Change*, **80**, 145–71.

Bardhan, P. K., and Udry, C. (1999). *Development Microeconomics*, Oxford University Press, New York.

Batjes, N. H. (2002). *A homogenized soil profile data set for global and regional environmental research (WISE, version 1.1).* International Soil Reference and Information Centre, Wageningen.

Bockstael, N. E., Freeman, A. M., Kopp, R. J., *et al.* (2000). On measuring economic values for nature. *Journal of Environmental Science and Technology*, **34**, 1384–9.

Bouwman, A. F., Boumans, L. J. M., and Batjes, N. H. (2002). Modeling global annual N_2O and NO emissions from fertilized fields. *Global Biogeochemical Cycles*, **16**, 1080.

Bouwman, A. F., Van Der Hoek, K. W., and Van Drecht, G. (2006). Modelling livestock-crop-land use interactions in global agricultural production systems. In A. F. Bouwman, T. Kram, and K. K. Goldewijk, Eds., *Integrated modelling of global environmental change: An overview of IMAGE 2.4.* Netherlands Environmental Assessment Agency (MNP), Bilthoven.

Bruinsma, J. (2003). *World agriculture: towards 2015/2030.* In: J. Bruinsma, ed., *An FAO perspective.* Earthscan, London.

Bryant, K. J., Benson, V. W., Kiniry, J. R., *et al.* (1992). Simulating corn yield response to irrigation timings: Validation of the EPIC model. *Journal of Production Agriculture*, **5**, 237–42.

Byerlee, D., de Janvry, A., and Sadoulet, E. (2009). Agriculture for development: Toward a new paradigm. *Annual Review of Resource Economics*, **1**, 15–31.

Cassman, K. G., and Liska, A. J. (2007). Food and fuel for all: realistic or foolish? *Biofuels, Bioproducts and Biorefining*, **1**, 23.

Conrad, J. M., and Clark, C. W. (1987). *Natural resource economics: Notes and problems.* Cambridge University Press, New York.

Crawford, E., Kelly, V., Jayne, T. S., *et al.* (2003). Input use and market development in Sub-Saharan Africa: an overview. *Food Policy*, **28**, 277–92.

Crosson, P., Pimentel, D., Harvey, C., *et al.* (1995). Soil erosion estimates and costs. *Science*, **269**, 461–5.

Dalal, R. C., Wang, W., Robertson, G. P., *et al.* (2003). Nitrous oxide emission from Australian agricultural lands and mitigation options: a review. *Australian Journal of Soil Research*, **41**, 165–95.

Delgado, C., Rosegrant, M., Steinfeld, H., *et al.* (1999). *Livestock to 2020. The Next Food Revolution*. International Food Policy Research Institute (IFPRI), Washington, DC.

Dercon, S. (1996). Risk, crop choice, and savings: Evidence from Tanzania. *Economic Development and Cultural Change*, **44**, 485–513.

Eickhout, B., van Meijl, H., and Tabeau, A. (2006). Modelling agricultural trade and food production under different trade policies. In A. F. Bouwman, T. Kram, and K. K. Goldewijk, eds., *Integrated modeling of global environmental change: An overview of IMAGE 2.4*. Netherlands Environmental Assessment Agency (MNP), Bilthoven.

Feder, G. Just, R. E., and Zilberman, D. (1985). Adoption of agricultural innovations in developing countries: A survey. *Economic Development and Cultural Change*, **33**, 255–98.

Fischer, G., van Velthuizen, H., Medow, S. and Nachtergaele, F. (2002). *Global agro-ecological assessment for agriculture in the 21st century*, Food and Agricultural Organization/International Institute for Applied Systems Analysis (FAO/IIASA), Laxenburg.

Food and Agriculture Organization of the United Nations (FAO) (1992). *CROPWAT, a computer program for irrigation planning and management by M. Smith*, FAO Irrigation and Drainage Paper 26, Food and Agriculture Organization of the United Nations (FAO), Rome.

Food and Agriculture Organization of the United Nations (FAO). (2008a). [Online]. Available: http://faostat.fao.org/default.aspx.

Food and Agriculture Organization of the United Nations (FAO). (2008b). *Aquastat*, [Online]. Available: http://www.fao.org/nr/water/aquastat/main/index.stm.

Food and Agriculture Organization of the United Nations/International Institute for Applied Systems Analysis/International Soil Reference and Information Centre/Institute of Soil Science—Chinese Academy of Sciences/Joint Research Centre of the European Commission (FAO/IIASA/ISRIC/ISS-CAS/JRC). (2008). *Harmonized world soil database (version 1.1)*, Food and Agriculture Organization of the United Nations (FAO) and International Institute for Applied Systems Analysis (IIASA), Rome.

Gillig, D., McCarl, B. A., and Sands, R. D. (2004). Integrating agricultural and forestry GHG mitigation response into general economy frameworks: Developing a family of response functions. *Mitigation and Adaptation Strategies for Global Change*, **9**, 241–59.

Holden, S. T. (2005). Bioeconomic modelling for natural resource management impact assessment. In B. Shiferaw, H. A. Freeman, and S. M. Swinton, Eds., *Natural Resource Management in Agriculture: Methods for Assessing Economic and Environmental Impacts*. CABI Publishing, Cambridge.

Intergovernmental Panel on Climate Change (IPCC). (2007). *Climate change 2007: the physical science basis. Contribution of Working Group I to the Fourth Assessment Report of the Intergovernmental Panel on Climate Change*. Cambridge University Press, New York.

Intergovernmental Panel on Climate Change (IPCC). (2006). 2006 IPCC Guidelines for National Greenhouse Gas Inventories. In H. S. Eggleston, L. Buendia, K. Miwa, *et al.*, Eds., *Prepared by the National Greenhouse Gas Inventories Programme*. Institute for Global Environmental Strategies, Hayama, Kanagawa.

International Food Policy Research Institute, The (IFPRI). (2008). *Spatial allocation model (SPAM)*. International Food Policy Research Institute (IFPRI), Washington, DC. Accessed at http://www.mapspam.info

Jones, J. W., Hoogenboom, G., Porter, C. H., *et al.* (2003). The DSSAT cropping system model. *European Journal of Agronomy*, **18**, 235–65.

Just, R. E., and Antle, J. M. (1990). Interactions between agricultural and environmental policies: A conceptual framework. *American Economic Review*, **80**, 197–202.

Kandlikar, M., and Risbey, J. (2000). Agricultural impacts of climate change: If adaptation is the answer, what is the question? *Climatic Change*, **45**, 529–39.

Kinsey, B., Burger, K., and Gunning, J. W. (1998). Coping with drought in Zimbabwe: Survey evidence on responses of rural households to risk. *World Development*, **26**, 89–110.

Kremen, C., Daily, G. C., Klein, A., *et al.* (2008). Inadequate assessment of the ecosystem service rationale for conservation: Reply to Ghazoul. *Conservation Biology*, **22**, 795–8.

Lesschen, J., Stoorvogel, J., Smaling, E., *et al.* (2007). A spatially explicit methodology to quantify soil nutrient balances and their uncertainties at the national level. *Nutrient Cycling in Agroecosystems*, **78**, 111–31.

Lubowski, R. N., Plantinga, A. J., and Stavins, R. N. (2006). Land-use change and carbon sinks: Econometric estimation of the carbon sequestration supply function. *Journal of Environmental Economics and Management*, **51**, 135–52.

Luckert, M. K., Wilson, J., Adamowicz, V., *et al.* (2000). Household resource allocations in response to risks and returns in a communal area of western Zimbabwe. *Ecological Economics*, **33**, 383–94.

Luedeling, E., Zhang, M., and Girvetz, E. H. (2009). Climatic changes lead to declining winter chill for fruit and nut trees in California during 1950–2099. *PLoS ONE*, **4**, e6166.

Mendelsohn, R., Nordhaus, W. D., and Shaw, D. (1994). The impact of global warming on agriculture: A Ricardian analysis. *American Economic Review,* **84**, 753–71.

Mendelsohn, R., and Dinar, A. (2009). Land use and climate change interactions. *Annual Review of Resource Economics,* **1**, 309–32.

Cassman, K. G., Wood S. (2005). Cultivated Systems. In *Millennium Ecosystem Assessment: Global Ecosystem Assessment Report on Conditions and Trends.* Island Press, Washington, DC.

Morris, M., Kelly, V., Kopicki, R., *et al.* (2007). *Fertilizer use in African agriculture: lessons learned and good practice guidelines.* World Bank, Washington, DC.

Naidoo, R., and Iwamura, T. (2007). Global-scale mapping of economic benefits from agricultural lands: Implications for conservation priorities. *Biological Conservation,* **140**, 40–9.

Naidoo, R., and Ricketts, T. H. (2006). Mapping the economic costs and benefits of conservation. *PLoS Biology,* **4**, 2153–64.

Nelson, G. C., Rosegrant, M. W., Koo, J., *et al.* (2009). *Climate change. impact on agriculture and costs of adaptation.* International Food Policy Research Institute (IFPRI), Washington, DC.

Neue, H. (1993). Methane emission from rice fields. *Bioscience,* **43**, 466–74.

Parry, M., Rosenzweig, C., Iglesias, A., *et al.* (1999). Climate change and world food security: a new assessment. *Global Environmental Change,* **9**, S51–S67.

Pender, J. (2005). Econometric methods for measuring natural resource management impacts: Theoretical issues and illustrations from Uganda. In: B. Shiferaw, H. A. Freeman, and S.M. Swinton, Eds., *Natural resource management in agriculture: methods for assessing economic and environmental impacts.* CABI Publishing, Cambridge.

Perry, C. J., and Narayanamurthy, S. G. (1998). *Farmer response to rationed and uncertain irrigation supplies.* International Water Management Institue, Colombo.

Polasky, S., Nelson, E., Camm, J., *et al.* (2008). Where to put things? Spatial land management to sustain biodiversity and economic returns. *Biological Conservation,* **141**, 1505–24.

Rosegrant, M. W., Paisner, M. S., Meijer, S., and Witcover, J. (2001). *Global food projections to 2020: emerging trends and alternative futures.* International Food Policy Research Institute, Washington, DC.

Rosegrant, M. W., Ringler, C., Msangi, S., *et al.* (2005). *International Model for Policy Analysis of Agricultural Commodities and Trade (IMPACT-WATER): Model Description.* International Food Policy Research Institute, Washington, DC.

Schlenker, W., Hanemann, W. M., and Fisher, A. C. (2005). Will US agriculture really benefit from global warming? Accounting for irrigation in the hedonic approach. *American Economic Review,* **95**, 395–406.

Shiferaw, H., Freeman, H. A., and Navrud, S. (2005). Valuation methods and approaches for assessing natural resource management impacts. In B. Shiferaw, H. A. Freeman, and S. M. Swinton, Eds., *Natural resource management in agriculture: methods for assessing economic and environmental impacts.* CABI Publishing, Cambridge.

Steduto, P., Hsiao, T. C., Raes, D., *et al.* (2009). AquaCrop— the FAO crop model to simulate yield response to water: I. Concepts and underlying principles. *Agronomy Journal,* **101**, 426–37.

Steinfeld, H., Gerber, P., Wassenaar, T., *et al.* (2006). *Livestock's long shadow—environmental issues and options.* Food and Agriculture Organization of the United Nations (FAO), Rome.

Swinton, S. M. (2005). Assessing economic impacts of natural resource management using economic surplus. In B. Shiferaw, H. A. Freeman, and S. M. Swinton, Eds., *Natural resource management in agriculture: methods for assessing economic and environmental impacts.* CABI Publishing, Cambridge.

Swinton, S. M., Lupi, F., Robertson, G. P., *et al.* (2007). Ecosystem services and agriculture: Cultivating agricultural ecosystems for diverse benefits. *Ecological Economics,* **64**, 245–52.

Tilman, D., Cassman, K. G., Matson, P. A., *et al.* (2002). Agricultural sustainability and intensive production practices. *Nature,* **418**, 671–7.

United States Department of Agriculture-Economic Research Service (USDA-ERS). (2009). *Commodity Costs and Returns.* Accessed at http://www.ers.usda.gov/Data/CostsAndReturns/.

United States Department of Agriculture-Natural Resources Conservation Service (USDA-NRCS). (2001). *National SSURGO (Soil Survey Geographic) Database.*

Vera-Diaz, M. D. C., Kaufmann, R. K., Nepstad, D. C., *et al.* (2008). An interdisciplinary model of soybean yield in the Amazon Basin: The climatic, edaphic, and economic determinants. *Ecological Economics,* **65**, 420–31.

Wani, S. P., Singh, P., Dwivedi, R. S., *et al.* (2005). Biophysical indicators of agro-ecosystem services and methods for monitoring the impacts of NRM technologies at different scales. In B. Shiferaw, H. A. Freeman, and S. M. Swinton, eds., *Natural resource management in agriculture: methods for assessing economic and environmental impacts.* CABI Publishing, Cambridge.

Wood, S., Ehui, S., Alder, J., *et al.* (2005). Food. In *Ecosystems and human well-being,* vol. 1: *Current state and trends:*

Millennium Ecosystem Assessment. Island Press, Washington, DC.

Wossink, A., and Swinton, S. M. (2007). Jointness in production and farmers' willingness to supply non-marketed ecosystem services. *Ecological Economics*, **64**, 297–304.

Wu, J. J., Adams, R. M., Kling, C. L., *et al.* (2004). From microlevel decisions to landscape changes: An assessment of agricultural conservation policies. *American Journal of Agricultural Economics*, **86**, 26–41.

You, L., and Wood, S. (2006). An entropy approach to spatial disaggregation of agricultural production. *Agricultural Systems*, **90**, 329–47.

Zilberman, D., Lipper, L., and McCarthy, N. (2008). When could payments for environmental services benefit the poor? *Environment and Development Economics*, **13**, 255–78.

CHAPTER 10

Crop pollination services

Eric Lonsdorf, Taylor Ricketts, Claire Kremen, Rachel Winfree, Sarah Greenleaf, and Neal Williams

10.1 Introduction

10.1.1 Importance to agriculture and policy

Crop pollination by bees and other animals is an ecosystem service of enormous economic value (Losey andVaughan 2006; Allsopp *et al.* 2008). Pollination can increase the yield, quality, and stability of crops as diverse as almond, cacao, canola, coffee, sunflower, tomato and watermelon. Indeed, Klein *et al.* (in press) found that 75% of globally important crops benefit from animal pollination. The value of this service, while difficult to quantify properly, has been estimated several times over the past decade (Southwick and Southwick 1992; Costanza *et al.* 1997; Losey and Vaughan 2006; Allsopp *et al.* 2008) with a recent estimate of 195 billion Euros (~$200 billion) worldwide (Box 10.1).

While much research and policy attention has focused on managed bees (especially the honey bee, *Apis mellifera*), wild bees and other insect species also contribute importantly to crop pollination. Our models focus on wild bees, because the pollination they deliver represents an ecosystem service from natural systems. In fact, for some crops (e.g., blueberry), wild bees are more efficient and effective pollinators than honey bees (Cane 1997). Diverse bee communities potentially provide more stable pollination services over time, compared to single (managed) species (Greenleaf and Kremen 2006; Hoehn *et al.*. 2008). Finally, if alarming regional declines in honeybee populations continue (National Research Council of the National Academies 2006; Stokstad 2007; Klein *et al.*, in press), wild pollinators may become increasingly important to farmers.

Maintaining pollinator habitats in agricultural landscapes, therefore, can help ensure food production, quality, and security. While other pollinators (e.g., bats and moths) also pollinate crops, bees are the most important crop pollinators (Free 1993) and are thus the focus of our models.

The pollination models aim to quantify and map scores for relative pollinator abundance across an entire landscape, including farms requiring pollination. The models use these results to indicate areas supplying pollinators that increase crop yields. Alternatively, intermediate results can be integrated into our agricultural model (see Chapter 9) to estimate the economic value of pollination services as an input to crop yields in a more sophisticated manner. Either way, these models can inform agricultural and land management policies in several ways. First, land-use planners could predict consequences of different policies on pollination services and income to farmers (Priess *et al.* 2007). Second, farmers could use these tools to locate crops in places where their pollination needs are most likely to be met. Third, conservation organizations that guide land management and restoration could use the tool to optimize conservation investments for both biodiversity and crop productivity. Finally, governments or others proposing payment schemes for ecosystem services could incorporate the results into plans for who should pay whom, and how much.

10.1.2 Scientific foundations and context

Our pollination models are founded on an increasing number of studies that have investigated the impacts of landscape structure and habitat-quality

Box 10.1 Assessing the monetary value of global crop pollination services

**Nicola Gallai, Bernard E. Vaissière,
Simon G. Potts, and Jean-Michel Salles**

Most major crop species are pollinated by bees or other insect groups (Klein *et al.* 2007). As the abundance and diversity of bees are now declining in many parts of the world, there is a growing need for improved methods to: (1) adequately assess the potential loss in terms of economic value that may result from pollination shortfalls, and (2) link this value to the vulnerability of agriculture confronted with pollination shortages.

To evaluate the monetary value of crop pollination services worldwide, we used the FAO global crop production statistics (http://www.fao.org) coupled with the reported degree of dependency of each crop on biotic pollination (Klein *et al.* 2007). FAO statistics are available for *direct* crops (production data available) and *commodity* crops (individual crop production data is aggregated for each commodity). Although these aggregations of crop production may represent a significant part of the agricultural output of a given country or region, and some of these species depend heavily on biotic pollination (Klein *et al.* 2007), we excluded all aggregated crop complexes from our analysis as prices and production figures were not available for each individual crop.

Following Gallai *et al.* (2009), we defined the economic value of pollinators (EVP) as the value of the pollinator contribution to the total economic crop production value. This contribution was calculated based upon the dependency ratio of crop production on pollinators, defined as the proportion of the yield attributable to insect pollinators. The economic value of pollinators was thus calculated as

$$EVP = \sum_{i=1}^{I} \sum_{x=1}^{X} P_{ix} Q_{ix} D_i \, , \qquad (10.A.1)$$

where P is the producer price per production unit, Q is the quantity produced for each crop $i \in [1; I]$ and for each country $x \in [1; X]$, and D is the dependence ratio for each crop $i \in [1; I]$. For Q_{ix} we used 2005 FAO production data expressed in metric tons. Producer prices, P_{ix}, for 2005 were obtained using data from financial markets, USDA (http://www.fas.usda.gov) and Eurostat (http://epp.eurostat.ec.europa.eu), and actualization of FAO data and expressed in US$per metric ton. The dependence ratios D_i were

calculated based upon the five dependency levels defined in the Appendix 2 of Klein *et al.* (2007). For each crop, we calculated an average dependence ratio based on the reported range of dependence on animal-mediated pollination. Based on this, the 2005 worldwide economic value of pollinators was US$190 billion compared to US$2,013 billion for the overall crop production value (Gallai *et al.* 2009).

Vulnerability is a function of three elements: exposure, sensitivity and adaptive capacity. For crops, the agricultural vulnerability to pollinator decline depends upon the crop dependency on pollinators and the capacity of farmers to adapt to pollinator decline. In this context, we used the ratio of the economic value of pollinators (*EVP*) to the total economic crop production value (*EV*) to calculate a level of vulnerability, which provides a measure of the potential relative production loss attributable solely to the lack of insect pollination. We evaluated the vulnerability in term of the proportion of the agricultural production value that depends on insect pollination (Gallai *et al.* 2009):

$$VR = \frac{EVP}{EV} = \frac{\sum_{i=1}^{I} \sum_{x=1}^{X} P_{ix} Q_{ix} D_i}{\sum_{i=1}^{I} \sum_{x=1}^{X} P_{ix} Q_{ix}} \% \qquad (10.A.2)$$

The vulnerability ratio of global agricultural production used for human food in 2005 was 9.5% (Gallai *et al.* 2009). The ratio varied considerably among different geographical areas, for example, at a national level, the vulnerability of European countries varied between 1% in Ireland to 19.5% in Austria (Figure 10.A.1). In Europe, there was a positive correlation between the vulnerability to pollinators of a crop category and its value per production unit ($r = 0.729$, $n = 10$, $P = 0.017$), indicating that the more a crop is dependent on insect pollination, the higher its value per production unit.

However, our approach provides an incomplete picture of the value of insect pollinators to society because we did not take into account agricultural production not used directly for human food (e.g., fodder crops), seeds produced for plant breeding, and perhaps most importantly, natural vegetation and all its associated ecosystem services which would almost certainly be strongly impacted by pollinator decline. Our estimates are therefore conservative.

continues

Box 10.1 *continued*

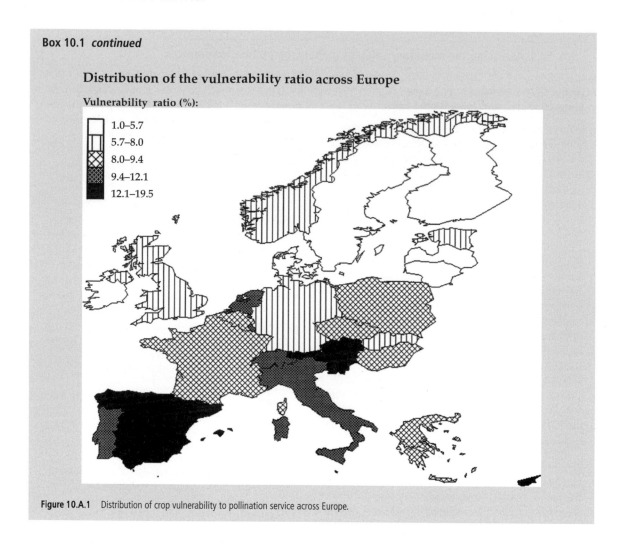

Distribution of the vulnerability ratio across Europe

Vulnerability ratio (%):

- 1.0–5.7
- 5.7–8.0
- 8.0–9.4
- 9.4–12.1
- 12.1–19.5

Figure 10.A.1   Distribution of crop vulnerability to pollination service across Europe.

on pollinator populations (reviewed by Kremen *et al.* 2007). They have found that the availability of nesting substrates (e.g., suitable soils, tree cavities; Potts *et al.* 2005) as well as floral resources (i.e., both nectar and pollen) in both natural and semi-natural habitats can strongly influence the diversity (Hines and Hendrix 2005), abundance (Williams and Kremen 2007), and distributions of pollinators across a landscape (Tepedino and Stanton 1981; Potts *et al.* 2003). In addition, because bees forage from fixed nest sites with limited foraging ranges, their abundance and diversity on a farm, as well as their effect on crop pollination, can be influenced

strongly by proximity to nesting habitats (Morandin and Winston 2006).

For example, Ricketts and colleagues (Ricketts 2004; Ricketts *et al.* 2004) found that bee diversity, visitation rate, pollen deposition rate, and fruit set are all significantly greater in coffee fields near forest than in fields further away. On the other hand, other studies have found little effect of landscape pattern on pollinator visitation, such as Winfree *et al.*'s (2008) study of pollination services to vegetable crops in the northeastern USA. Despite this variation among studies, Ricketts *et al.* (2008) synthesize 23 case studies (including many of those cited above) and find a general "consensus" decline in pollination services with increasing isolation from natural or semi-natural habitat.

Building from these and many other studies, Kremen *et al.* (2007) have proposed a general framework for understanding how pollination services are delivered across landscapes, and how these services are affected by land-use change in agricultural regions (Figure 10.1). Here, we develop a simplified version of this general model (indicated in Figure 10.1), which uses simple landscape indices, governed by a few key parameters that can be estimated from field data or expert opinion, to predict relative pollinator abundances across a landscape and agricultural fields. Moreover, we use the framework that predicts abundance at crop field to attribute the pollinator-dependent gains in yield and crop value to the parcels supplying the pollinators.

10.1.3   Model intuition and difference between tiers

10.1.3.1   *Overview of data requirements*

Pollinators require two basic types of resources to persist on a landscape: nesting substrates and floral resources (Westrich 1996; Kremen *et al.* 2007). The model therefore requires estimates of availability of both of these resource types for each land-use and land cover type (LULC) in the map. These data can be derived from quantitative field estimates or from expert opinion. Pollinators move between nesting habitats and foraging habitats (Westrich 1996; Williams and Kremen 2007), and their foraging distances, in combination with arrangement of different habitats, affects their persistence, their abundance, and the level of service they deliver to farms. Our model therefore also requires a typical foraging distance for pollinators. These data can be supplied, e.g., from quantitative field estimates

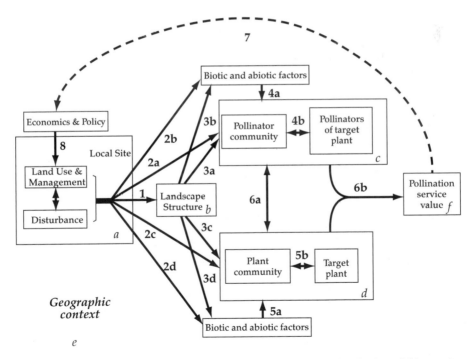

Figure 10.1   General conceptual model describing pollination services and their delivery across an agricultural landscape (full framework, reproduced from Kremen et al. (2007)).

Land-use practices (Box a) determine the pattern of habitats and management on the landscape (Box b). The quality and arrangement of these habitats affect both pollinator and plant communities (Boxes c and d). The value of pollination services (Box f) depends on the interaction between specific plants (e.g., crops) and their specific pollinators. Our pollination model is a simplified version of this full model, capturing the following arrows only: 3a, 3c, 6a, and with economic model 6b.

(Knight *et al.* 2005), from proxies such as body size (Gathmann and Tscharntke 2002; Greenleaf *et al.* 2007), or from expert opinion.

The ultimate level of pollination service provided to a farm depends on the crops grown, the ability of each modeled species to pollinate them effectively, the crop's response to animal pollination and the abundance of pollinators at the crop. The model therefore incorporates data on location of farms of interest, the crops grown there, and how effective each species is as a pollinator for a given crop.

10.1.4 Model intuition

Using these data, the model first estimates a relative abundance score of each pollinator species in each parcel (hereafter, pollinator "supply" to follow the conventions in Chapter 3), based on the available nesting resources in that parcel and the floral resources in surrounding parcels. We define

parcel as the analytical spatial unit on the landscape. In our case, it is a 90 meter by 90 meter grid cell that may have more than one land cover within it. In calculating floral resources, nearby parcels are given more weight than distant parcels, based on the species' average foraging range. The result is a map of relative abundance scores (0–1) for each species in the model (the "supply map").

Given this pattern of pollinator supply, the model then estimates the relative abundance of foraging bees arriving at each farm ("farm abundance"). It sums the relative bee abundances in neighboring parcels, again giving more weight to nearby parcels, based on average foraging ranges. This weighted sum is our relative index (0–1) of abundance for each pollinator in the farm. If the crop type at each parcel and its pollinators are known, the model will limit the weighted sum only to relevant pollinators.

We use a very simple yield function to translate farm abundance into relative crop yields. Alternatively,

Table 10.1 Comparison of model complexity and parameters used in tier 1 versus tier 2

Parameter	Description	Tier 1	Tier 2
HN	Habitat suitability for nesting	X	X
HF	Habitat suitability for foraging	X	X
J	Number of land cover types (each indexed by j)	X	X
N_j	Compatibility of habitat j for nesting	X	X
F_j	Compatibility of habitat j for foraging	X	X
M	Number of parcels in landscape (indexed by m)	X	X
D	Distance between parcels	X	X
α	Expected pollinator foraging distance	X	X
P	Pollinator abundance score	X	X
O	Number of farm parcels (indexed by o)	X	X
X	Pollinator source parcel	X	X
ψ	Farms' average change in normalized scores (used for sensitivity analysis)	X	X
Y	Crop yield (indexed by o)	X	X
V	Crop value based on agricultural production function (indexed by o)	X	X
v	Proportion of a crop's yield attributed only to wild pollination (indexed by c)	X	X
K	Pollinator abundance to achieve ½ of pollinator-dependent yield (indexed by c)	X	X
PS	Pollinator service provided to crops (indexed by m)	X	X
S	Number of species (indexed by s)		X
I	Number of nesting types (indexed by i)		X
W	Weight describing importance of floral season for pollinator		X
K	Number of floral seasons (indexed by k)		X
C	Crops' pollinator requirement		X
ε	Relative abundance of pollinator in landscape		X

The last seven parameters are unique to tier 2. The model can be run with a mixture of tier 1 and tier 2 parameters, allowing a continuum of model complexity to match data availability.

one can use farm abundance as an input in the more sophisticated agricultural production model (Chapter 9) to determine the crop yield and value on each farm parcel. Finally, our model redistributes crop value back onto the landscape to estimate the service value provided from each parcel to surrounding farms (equivalent to the "use" and "value" in the parlance of this book, Chapter 3). It does so using the same foraging ranges, so that parcels that are sources of abundant pollinators and near to farms tend to have relatively high service value.

The pollination model has two tiers permitting use of different amounts of information. The tier 1 models are nested within tier 2 (i.e., tier 1 is a simpler version of tier 2). Both tiers are based on a LULC map, showing both natural and managed land types. Onto this landscape, tier 1 models a single pollinator that represents the overall pollinator community, while tier 2 considers multiple pollinator species or guilds, allowing them to differ in flight season, resource requirements, and foraging distance. Because the models are nested, tier 1 and tier 2 are actually endpoints of a continuum of potential model complexity (Table 10.1). For example, one can recognize multiple species and use a different foraging radius for each (Eq. (10.2)), but model only a single flowering season and nesting guild. This nestedness also allows easy comparisons of model outputs between tiers, as we explore in Chapter 15.

10.2   Tier 1 supply model

Pollinators require habitat for nesting and within some foraging distance they require floral resources for food. The first step of our model is to translate a LULC map (Figure 10.2a; Plate 5a) into a nesting suitability map and a floral resource availability map. Then based on the amount and location of nesting and floral resources, we calculate the pollinator supply map.

The first step in calculating the pollinator source score at each parcel is identifying the proportion of suitable pollinator nesting habitat in a parcel x as a function of LULC j, HN_x:

$$HN_x = \sum_{j=1}^{J} N_j p_{jx},\qquad(10.1)$$

where $N_j \in [0,1]$ represents compatibility of LULC j for nesting and p_{jx} is the proportion of parcel x that is covered by LULC j. This provides a landscape map of nesting suitability where $HN_x \in [0,1]$ (Figure 10.2b; Plate 5b). A score of 1 would indicate that the entire area of the parcel provides habitat suitable for nesting (e.g., forest habitat in Supplemental Online Material (SOM) Table 10.S1) while a score of 0.2 would indicate 20% of the parcel's area provides suitable nesting habitat (e.g., coffee/pasture habitat).

We calculate the proportion of suitable foraging habitat surrounding a parcel x, given by $HF_x \in [0,1]$. We assume that foraging frequency in parcel m declines exponentially with distance (Cresswell et al. 2000), and that pollinators forage in all directions with equal probability. Therefore, parcels farther away from nest parcel x contribute less to total resource availability than parcels nearby, and leads to the following prediction for the potential floral resources available to pollinators nesting in parcel x, HF_x:

$$HF_x = \frac{\sum_{m=1}^{M}\sum_{j=1}^{J} F_j p_{jm} e^{\frac{-D_{mx}}{\alpha}}}{\sum_{m=1}^{M} e^{\frac{-D_{mx}}{\alpha}}}\qquad(10.2)$$

where p_{jm} is the proportion of parcels m in LULC j, D_{mx} is the Euclidean distance between parcels m and x, α is the expected foraging distance for the pollinator (Greenleaf et al. 2007) and $F_j \in [0,1]$ represents relative amount of foraging resource in LULC j. The numerator is the distance-weighted resource summed across all M parcels. The denominator represents the maximum possible amount of forage in the landscape. This equation generates a distance-weighted proportion of habitat providing floral resources within foraging range, normalized by the total forage available within that range (Winfree et al. 2005) (Figure 10.2c; Plate 5c).

Supply map: Since pollinator abundance is limited by both nesting and floral resources, the pollinator abundance score on parcel x is simply the product of foraging and nesting such that $P_x = HF_x HN_x \in [0,1]$. This score represents the location and supply of pollinators *available* for crop pollination from

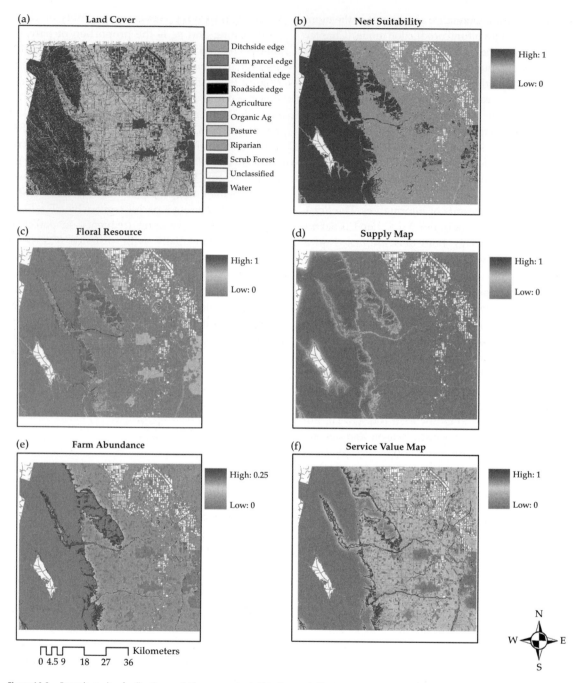

Figure 10.2 Example results of pollination model for watermelon in Yolo County, California.
The model uses (a) land cover data as input and derives maps of (b) nesting habitat and (c) floral resources. From this, it generates (d) a pollinator supply map that describes an index of pollinator abundance on the landscape. Based on the supply map, the model generates (e) a pollinator abundance map on farm parcels (i.e., "farm abundance"). After using a simple yield function to translate farm abundance into relative yield the model distributes yield or economic value back onto the surrounding landscape to generate (f) the value map. All steps are the same for tier 1 and tier 2 models; results here are tier 2, based on data supplied in supplemental online appendix. (See Plate 5.)

parcel x and results in the supply map. This map does not account for the location or type of crops present in the landscape, and as such has not adjusted pollination to show the actual service supplied to people, but rather all potential pollination on the landscape.

10.3 Tier 1 farm abundance map

For pollinators' actions to provide crop pollination benefits to people, pollination must take place on a farm growing a crop that requires insect pollination. In the next modeling step, we identify farms on the landscape, and the relative abundance of wild pollinators on ech farm.

Pollinators leave their nesting sites to forage in surrounding parcels, so farms surrounded by a higher abundance of nesting pollinators should experience higher abundances of pollinating visitors to their crops. We use the foraging framework described in Eq. (10.2) to determine the contribution to pollinator abundance from a single nest parcel m to forage on a crop in farm o:

$$P_{om} = \frac{P_m e^{\frac{-D_{om}}{\alpha}}}{\sum_{m=1}^{M} e^{\frac{-D_{om}}{\alpha}}} , \quad (10.3a)$$

where P_m is the relative supply of pollinators on map unit m, D_{om} is distance between source parcel m and farm o, and α is species' average foraging distance. O can be used to index specific farms of interest or every agricultural parcel on the landscape. The numerator of Eq. (10.3a) represents the distance-weighted proportion of the pollinators supplied by parcel m that forage within farm o and the numerator is a scalar that normalizes this contribution by the total area within foraging distance to farm (Winfree *et al.* 2005). The total pollinator abundance on farm o, P_o, is simply the sum over all M parcels,

$$P_o = \sum_{m=1}^{M} P_{om} . \quad (10.3b)$$

10.4 Tier 1 valuation model

Pollination has economic value as an ecosystem service because it is an input to agriculture, from which people derive food and income. In formal terms, pollination can be an important factor in agricultural production functions, which relate yields of a given crop to the quantity and quality of various inputs (e.g., water, soil fertility, labor, chemicals). Production functions (or "yield functions") are a well-established econometric technique used widely in agriculture and product manufacturing (Polasky *et al.* 2008). The agricultural models described in Chapter 9 take exactly this approach, so we do not repeat it here; instead, we offer an extremely simple alternative. Our "farm abundance" results above can be used as inputs to either one.

Using production functions in this way will result in an estimate of the economic value of pollinators *at each farm*. It is most likely of interest, however, to estimate the value of the habitats in the landscape that support these pollinators. For this we can use the ecological models described here, which model movement of pollinators from source parcels to farms, to attribute economic value realized on farms back to the pollinator-supporting habitats.

10.4.1 Estimating crop yield and value

The calculated pollinator abundance from Eq. (10.3b) will be an input into the agricultural production function to determine the crop yield and crop value on each parcel. In lieu of a more detailed agricultural production model (Chapter 9), we use a simple saturating yield function to translate the abundance of pollinators on farms into an expected yield. Yield should increase as pollinator abundance and diversity increase (Greenleaf and Kremen 2006), but crops vary in their dependence on pollinators, i.e., some crop species are self-compatible and yield is less dependent on pollination while other species obligately require pollination to generate any yield (Allsopp *et al.* 2008; Ricketts *et al.* 2008). We account for both observations, and thus calculate the expected yield of a crop c on farm o, Y_o, as

$$Y_o = 1 - \nu_c + \nu_c \frac{P_o}{P_o + \kappa_c} , \quad (10.4)$$

where v_c represents the proportion of total crop c's yield attributed only to wild pollination (e.g., v_c would be equal to 1 if a crop is an obligately out-crossing species and equal to 0 if the crop species were wind-pollinated). In the denominator of the third term, κ_c is a half-saturation constant and represents the abundance of pollinators required to reach 50% of pollinator-dependent yield. The monetary value of the crop on farm o, V_o, is simply the product of yield per hectare, Y_o, the number of hectares of the crop and the price of the crop (Gallai *et al.* 2009).

10.4.2   Assigning value back to pollination sources: service value

We use the pollinator model here to redistribute each farm o's value back onto the landscape based on the actual level of service supplied by each parcel m. Recall Eqs. (10.3a) and (10.3b) that determined the total abundance on farm o by summing across all M supply parcels the proportion of pollinators foraging from each supply parcel to farm o. Here, we instead attribute the pollinator-generated value from the O farms back to the M supply parcels. For each supply parcel m, we sum across all O parcels, weighting the contribution from each farm o to parcel m by their proximity. Thus, supply parcels close by crops should provide a greater service than parcels far from any crops. Formally, we calculate pollinator service provided to O farms from each m parcel, PS_m, as

$$PS_m = v_c \sum_{o=1}^{O} V_o \frac{P_{om}}{P_o} , \qquad (10.5)$$

where V_o represents the crop value in farm o. This score generates the pollinator service map (Figure 10.2f; Plate 5F) and represents the location and value based on supply of pollinators that *provide* crop pollination to surrounding farms (i.e., equivalent to "value" results in the parlance of Chapter 3).

10.5   Tier 2 supply model

The tier 2 model follows the same logic as tier 1, but each step allows for more detailed, season-specific

or species-specific information to be incorporated. While we refer to species throughout, these same models could also be applied to guilds. Specifically, we model multiple pollinators, incorporate multiple nesting types per habitat and allow for multiple seasons of foraging.

As in tier 1, the first step in calculating the pollinator score is identifying compatible nesting habitat for each pollinator species across the landscape, given by $HN_{sx} \in [0,1]$. In tier 2, we account for species (or guild) differences in habitat suitability so that the proportion of suitable nesting habitat in a parcel x for pollinator species s as a function LULC j, HN_{sx} is

$$HN_{sx} = \sum_{j=1}^{J} N_{js} p_{jx} , \qquad (10.6)$$

where $N_{js} \in [0,1]$ represents compatibility of LULC j for nesting by species s.

Some LULC classes can provide habitat suitable for multiple nesting types. For example, in California, we scored oak woodland habitat as providing good habitat for wood-nesting, ground-nesting and cavity-nesting bees, but scored agricultural habitat as providing poorer habitat for ground-nesting bees, and non-habitat for wood or cavity nesters (see Section 10.4.1).

For bee species or groups that span nesting types (e.g., species that nest in the ground and in hollow stems) we assigned the habitat type according to the nest type that maximized its suitability for that bee species. In other words, if there are I nesting types, then $N_{js} = \max[NS_{s1} N_{ji}, \ldots, NS_{sI} N_{jI}]$, where NS_{si} is the nesting suitability of nesting type i for species s and N_{ji} is the suitability of LULC j for nesting type i. This analysis provides a map of nesting suitability (Figure 10.2b; Plate 5B).

As in Tier 1, we calculate the proportion of suitable foraging habitat for pollinator species s nesting in parcel x given by $HF_{sx} \in [0,1]$. In tier 2, though, we allow for production of floral resources to vary among K seasons. We also use data or expert opinion to assess flight period and account for variation among pollinators in their K flight seasons, e.g., some are present in summer only, while others are present in multiple seasons. We calculate the overall floral resources available as a weighted sum across K seasons where the weight $(w_{sk}) \in [0,1]$ represents the relative importance of floral production in season k for species s. We constrain each w_{sk} value such that $\sum_{k=1}^{K} w_{sk} = 1$. This leads to the following prediction for the potential floral resources available to species s parcel x across K seasons, HF_{xs},

$$HF_{sx} = \sum_{k=1}^{K} w_{sk} \frac{\sum_{m=1}^{M}\sum_{j=1}^{J} F_{js,k}\, p_{jm}\, e^{\frac{-D_{mx}}{\alpha_s}}}{\sum_{m=1}^{M} e^{\frac{-D_{mx}}{\alpha_s}}}, \quad (10.7)$$

where p_{jm} is the proportion of parcel m in LULC j, D_{mx} is the Euclidean distance between parcels m and x, α_s is the typical foraging distance for species s and $F_{js,k} \in [0,1]$ represents suitability for foraging of LULC j for species s during season k. The use of $F_{js,k}$ permits attributing different resource levels to the same LULC type for different bee species or guilds by season—for example, in California, riparian habitat produces important early spring resources but many pollinator species are not yet flying at this time. By contrast, riparian habitat produces almost no summer resources. Using the normalized proportion controls for differences among pollinators that vary in their foraging radii, and allows us to estimate total pollinator abundances in subsequent model steps.

As in tier 1, we calculate a supply score for each species on parcel x as the product of foraging and nesting: $P_{sx} = HF_{sx}HN_{sx}$ (Figure 10.2d; Plate 5D).

10.6   Tier 2 farm abundance map

To calculate the abundance of each pollinator species on a crop in parcel o, we use the framework described in Eqs. (10.3a) and (10.3b). First to calculate pollinator visitation by species s from nest parcel m to farm parcel o, P_{osm},

$$P_{osm} = \frac{P_{sm}\, e^{\frac{-D_{om}}{s}}}{\sum_{m=1}^{M} e^{\frac{-D_{om}}{s}}}, \quad (10.8a)$$

where P_{sm} represents the supply of pollinator s on map unit m, D_{om} is distance between map unit m and farm o and α_s is species s' typical foraging distance. The total

pollinator abundance of species s on farm o, P_{os}, is simply the sum of P_{osm} over all M parcels at each farm o,

$$P_{os} = \sum_{m=1}^{M} P_{osm}. \quad (10.8b)$$

This score represents the relative abundance of pollinators visiting farm and results in the farm abundance map (Figure 10.2e; Plate 5E).

To calculate the total pollinator score for farm o from all pollinators, P_o, we calculate the normalized pollinator score for all pollinator guilds or species, such that

$$P_o = \frac{\sum_{s=1}^{S} C_s P_{os}}{\sum_{s=1}^{S} C_s}, \quad (10.9a)$$

where $C_s \in [0,1]$ if the crop requires pollinator s and 0 otherwise. This unweighted summation assumes that all pollinators are equally abundant. However, if some pollinators have higher background abundance than others, then a weighted average may be more appropriate such that

$$P_o = \frac{\sum_{s=1}^{S} \varepsilon_s C_s P_{os}}{\sum_{s=1}^{S} C_s}, \quad (10.9b)$$

where ε_s represents the abundance of pollinator s in the landscape, relative to other pollinator species or guilds. The weights for each species can be determined by expert opinion or with observed data. Additional species-by-crop weights could be added to Eq. (10.9b) in the same fashion as ε_s to account for differences among pollinators in their effectiveness on a given crop (Greenleaf and Kremen 2006).

10.7   Tier 2 valuation model

As in tier 1, the calculated tier 2 abundance from Eq. (10.9b) will be an input into a simplified agricultural production model to determine the crop yield and crop value on each farm parcel. The description in Section 10.4.1 and Eq. (10.4) are the same for tier 2. It follows we again use the ecological model to redistribute the value from all O farms onto each supply parcel m for each species s. The resulting score represents the available supply weighted by the relevant demand, each species' relative abundance in the landscape, effectiveness (ε_s) and the crop value within foraging distance. Thus we calculate pollinator service value from parcel m to other O parcels, PS_m, as

$$PS_m = v_c \sum_{s=1}^{S} \sum_{o=1}^{O} \varepsilon_s C_s V_o \frac{P_{oms}}{P_o}. \qquad (10.10)$$

This score generates the tier 2 pollinator service value map (Figure 10.2f; Plate 5F) and represents the location and pollination service value based on relative abundance of pollinators that *provide* crop pollination from parcel m to farm o.

10.8 Sensitivity analysis and model validation

We first compare model predictions against field data in two contrasting landscapes in California, USA, and San Isidro, Costa Rica. We then illustrate a sensitivity analysis with the Costa Rican data to determine the extent to which our results depend on the precision and accuracy of our parameter estimates.

10.8.1 Model validation

To validate our model, we compare its predictions of total (community-wide) abundance against total observed abundance in farms of crops in two landscapes: coffee in Costa Rica and watermelon in California. The Californian and Costa Rican examples use different levels of model complexity and differing mixes of field- and expert-derived parameters. In all cases, model parameters were derived independently of field validation data (e.g., estimates of typical foraging ranges (α), were derived from bee body size; floral availability was estimated through expert assessment based on other studies, not from field measurements taken simultaneously with pollinator abundances).

10.8.1.1 Costa Rica
We applied the model to an agricultural landscape in the Valle del General, Costa Rica, one of that country's major agriculture regions. The landscape is dominated by coffee, sugar cane, and cattle pasture, all of which surround hundreds of remnants of

tropical/premontane moist forests (Janzen 1983). Studies were conducted on 12 sites in a large coffee farm (approx. 1100 ha) in the center of this landscape.

High-resolution (1 m) aerial photos, supplied by CATIE (Centro Agronómico Tropical de Investigación y Enseñanza), were classified into six major classes of LULC and resampled to 30m spatial resolution. These classes were then assigned values of nesting and floral resources (assuming a single flowering season) based on expert opinion (see SOM Table 10.S1)), informed by field work in the area (Ricketts 2004; Brosi *et al.* 2008). The most common visitors to coffee in this region are 11 species of native stingless bees (*Meliponini*) and the introduced, feral honey bee, *Apis mellifera*. For the model, these 12 species were assigned to two nesting guilds based on expert opinion (SOM Table 10.S2). All 11 species were observed during the period of study, but sampling did not continue year-round. Lacking this information on seasonality, we assumed a single flight season for all species. To estimate typical foraging ranges for each species (Table 10.2), we used intertegular spans for 10 museum specimens and the statistical relationship presented by Greenleaf *et al.* (2007).

During the flowering seasons of 2001 and 2002, Ricketts and colleagues (Ricketts 2004) measured bee activity, pollen deposition, and pollen limitation in 12 sites, varying from 10 to 1600 m from the nearest major forest patch. We used these observations to compare against our model. Our models predict at least 80% of the variance in observed pollinator abundance (Figure 10.3a).

The model's predictions for farm abundance scores were not as strongly related to field measurements of pollen deposition on coffee stigmas (Figure 10.3b), which is a closer correlate to actual pollination service (Ricketts 2004). Modeled abundance scores do not predict pollen limitation of coffee well (Figure 10.3c). Pollen limitation is the degree to which coffee production (seed number and mass) is reduced due to insufficient pollination, and is a close measure of actual pollination services. Pollen limitation does decline with increased modeled service scores, but the fit is weak. Variation in pollen deposition and pollination limitation also

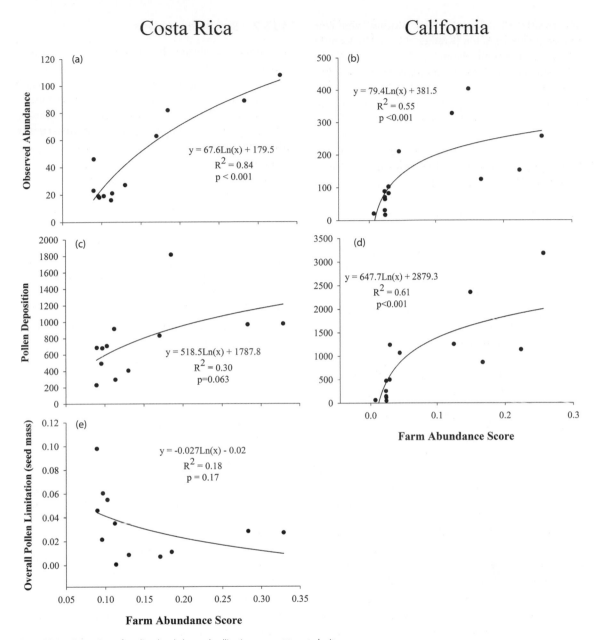

Figure 10.3 Comparison of predicted and observed pollination scores at two study sites.
In each site we compared the model's predicted abundance to pollinator abundance (a: Costa Rica, b: California), pollen deposition (c: Costa Rica, d: California), and in Costa Rica, we also compared the model to pollen limitation as measured by seed mass (e).

depends on pollination efficiency of each bee species, on resource limitation of the coffee plant itself, and other factors not captured in a prediction of pollinator abundance, which likely contributes to the poor fit.

10.8.1.2   California
We applied the model to an agricultural landscape in the Central Valley of California, across a strong gradient in isolation of farms from large tracts of natural habitats (oak woodland, chaparral scrub

and riparian deciduous forest). Studies were conducted on watermelon (Kremen *et al.* 2002b, Kremen *et al.* 2004) across this landscape.

The LULC data were simplified from a 13-class supervised classification of Landsat TM data at 30 × 30 m resolution (described in detail in Kremen *et al.* 2004) into six classes. Four additional cover classes were hand drawn on the landscape using ArcGIS to account for nesting and floral resources that come from edges of roads, agricultural parcels, residential areas and irrigation ditches (Figure 10.2a; Plate 5A). These classes were then assigned values of nesting and floral resources based on expert opinion values (SOM Table 10.S3), informed by studies of bee-plant networks (Kremen *et al.* 2002a, Williams and Kremen 2007, Kremen *et al.*, unpublished; Williams and Kremen, unpublished) and bee-nesting densities (Kim *et al.* 2006) in the same landscape.

During the flowering season of 2001, bee visits were recorded at 12 sites, and median species-specific pollen deposition per visit was estimated (Kremen *et al.* 2002b). Each bee species in the study was characterized by its nesting habit based on expert opinion and the length of its flight period, based on over 12000 bee specimens collected from 1999 to 2004 by pan-trapping and netting at flowers in this landscape (Kremen and Thorp, unpublished; Williams *et al.*, unpublished) (SOM Table 10.S4). Typical foraging distances were calculated from measurements of intertegular span, using the regression in Greenleaf *et al.* (2007). For nearly all bee species, at least five individuals were measured but for a few species, only one measurement was used. Data on *Apis mellifera*, which are managed for pollination in this landscape, were removed prior to analysis.

The model provided a reasonable fit to the observed data on total abundance of native bees on watermelon, although with considerable scatter (Figure 10.3b). Model predictions were strongly related to estimated pollen deposition from native bees (Figure 10.3d), a more direct measure of pollination services that has been used to assess the contributions of wild bees to pollination services (Kremen *et al.* 2002b; Kremen *et al.* 2004; Winfree *et al.* 2007). However, we caution against interpreting the model's ability to predict pollen deposition since pollen deposition is calculated from visitation data, not direct observations (Kremen *et al.* 2002b).

10.8.2 Sensitivity analysis

Sensitivity analysis should identify the model parameters that have the greatest influence on model results. This allows the scientist to focus on improving accuracy and precision of parameters to which the model is most sensitive, and allows managers to determine the major sources of uncertainty affecting model predictions. In our case, we are interested in how estimates of nesting suitability, floral resource availability and bee dispersal distance influence our predicted pollinator abundance scores. If we find them to be quite sensitive, then further research is required to reduce this uncertainty before the model can be used with confidence.

Our model predicts a parcel's pollinator abundance *relative to other parcels on a landscape*, so our sensitivity analysis focuses on these relative scores. We let $\hat{P}_o$ represent a normalized pollinator score on farm o (from Eq. (10.3) or (10.9)) based on the original parameter estimates such that

$$\hat{P}_o = \frac{P_o - P_{min}}{P_{max} - P_{min}} \qquad (10.11)$$

where P_{min} and P_{max} are the minimum and maximum pollinator service scores for all farms on the landscape. We let $\hat{P}_{o,c}$ represent the analogous normalized score on farm o resulting from modified parameter combination c, and let $\hat{\psi}_c$ be the average change in normalized scores from combination c such that

$$\hat{\psi}_c = \frac{\sum_{o=1}^{O} \left| \hat{P}_o - \hat{P}_{o,c} \right|}{O} \qquad (10.12)$$

where O is the number of farms in the analysis.

We use regression analysis to determine sensitivity, similar to McCarthy *et al.*'s (1995) logistic regression approach used in population viability analyses. Our goal is to calculate how variation in each parameter affects estimates of a parcels' pollinator abundance, independent of all other parameters in the model. Given the number of parameters, exploring every combination is impractical. Instead, we create a sample of parameter combinations by selecting parameter values randomly from a uniform distribution, each within its range of uncertainty and then generate an estimated pollinator score $\hat{P}_{o,c}$ for each parcel.

To generate parameter combinations, we set a minimum and maximum for the range of parameter values and drew a random number from a uniform distribution with this range (Table 10.2). For floral and nesting resources we set the range as ±0.1 around the estimate, and we did not allow the max-

Table 10.2 Results of sensitivity analysis for Costa Rican study

	Parameter	Estimate	Max	Min	δ (Slope)	SE_δ	Standardized regression coefficient (t-value)
Forage resource availability (Ff)	**Forest**	**1**	**1**	**0.9**	**4.550**	**1.303**	**3.493***
	Coffee	**0.5**	**0.6**	**0.4**	**7.758**	**0.666**	**11.643***
	Cane	0	0.1	0	0.027	1.356	0.020
	Pasture/grass	0.2	0.3	0.1	0.144	0.652	0.221
	Scrub	0.3	0.4	0.2	0.553	0.657	0.842
	Bare	0.1	0.2	0	0.300	0.663	0.453
	Built-up	0.3	0.4	0.2	0.450	0.676	0.666
Apis nesting suitability (Nf)	Forest	1	1	0.9	0.046	1.300	0.035
	Coffee	0.2	0.3	0.1	0.975	0.648	1.505
	Cane	0	0.1	0	1.950	1.292	1.509
	Pasture/grass	0.2	0.3	0.1	0.535	0.664	0.805
	Scrub	0.3	0.4	0.2	0.784	0.674	1.162
	Bare	**0**	**0.1**	**0**	**2.896**	**1.302**	**2.224***
	Built-up	0.2	0.3	0.1	0.739	0.672	1.101
Native nesting suitability (Nj)	**Forest**	**1**	**1**	**0.9**	**3.381**	**1.287**	**2.626***
	Coffee	0.1	0.2	0	0.067	0.650	0.102
	Cane	0	0.1	0	0.382	1.318	0.290
	Pasture/grass	0.1	0.2	0	0.259	0.683	0.379
	Scrub	0.2	0.3	0.1	0.458	0.658	0.696
	Bare	0.1	0.2	0	0.247	0.656	0.376
	Built-up	0.1	0.2	0	0.640	0.659	0.972
Foraging range(αs) for each species (m)	Apis mellifera	663	776	562	0.001	0.001	1.467
	Huge Black 2002**	**214**	**239**	**191**	**0.007**	**0.003**	**2.376***
	Melipona fasciata	578	634	525	0.001	0.001	0.653
	Nannotrigona mellaria	70	79	61	0.008	0.008	1.037
	Partamona cupira/Trigona fussipennis/Trigona corvina***	**87**	**110**	**69**	**0.007**	**0.003**	**2.134***
	Plebeia jatiformis	28	30	25	0.027	0.024	1.131
	Plebia frontalis	34	36	33	0.005	0.051	0.096
	Trigona (Tetragona) clavipes	55	63	48	0.004	0.009	0.490
	Trigona (tetragonisca) angustula	22	24	20	0.013	0.029	0.453
	Trigona dorsalis	60	66	54	0.006	0.011	0.544
	Trigona fulviventris	**77**	**82**	**73**	**0.046**	**0.015**	**3.158***
	Trigonisca sp.	21	23	20	0.043	0.051	0.829

* $p < 0.05$.

Unidentified species.

*** These species were indistinguishable during field observations and lumped together.

The strength of the model's sensitivity is given by the standardized regression coefficients in the final column. These coefficients result from a multiple regression of the parameter value combinations on the average change in normalized pollination score, $\hat{\psi}c$.

imum to exceed 1 or the minimum to drop below 0. For foraging ranges, we set the range using the minimum and maximum of the 10+ measurements of intertegular span.

By iterating this parameter draw process 1000 times, and then regressing the change in scores, $\hat{\psi}_c$ against randomly varying parameters, we can estimate sensitivity to each parameter while accounting for variation in the others. The sensitivity of each predictor variable is indicated by its standardized regression coefficient (t-value), calculated from the best fit of a multiple linear regression model: $\widehat{\psi}_c = \delta_0 + \delta_1 x_1 + ... + \delta_n x_n$, where x_n are predictor variables (foraging distance, nesting suitability values, etc) and δ_n are the regression coefficients. The standardized regression coefficient is the regression coefficient (slope of a line) divided by its standard error (Cross and Beissinger 2001). This is a unitless quantity that allows one to directly compare the sensitivity among parameters, and because our parameter combinations were created randomly, also accounts for potential interactions among model parameters (Cross and Beissinger 2001). The standard error for one model parameter is caused by the dependence of $\hat{\psi}_c$ on other parameters and the significance of the slope is calculated using a two-tailed t-distribution (a t-value greater than 1.9 or less than -1.9 is significant at $p < 0.05$).

We illustrate the sensitivity analysis using our Costa Rica data set (Table 10.2). The results indicate that a farm's normalized pollinator score, $\hat{P}_o$, is most sensitive to foraging resources present in coffee (t-value = 11.65; $p < 0.05$) and forest (t-value = 3.50; $p < 0.05$) habitats. Interestingly, $\hat{P}_o$ is also sensitive to uncertainty in a group of species' foraging distances, which ranged between 77 and 214 m. Pollinator service scores were not sensitive to species with smaller or greater estimated foraging ranges. These sensitivities are likely due to the variation in forest composition surrounding farms sites at these moderate scales. The implication for conservation is that additional effort to estimate and manage the floral resources within coffee farms, and bee-pollinated crops in general, would be of highest priority for understanding the response of pollination to landscape change.

10.9 Limitations and next steps

10.9.1 Limitations

Despite the promising results, there are several limitations to our model. First, our models estimate the benefits of wild pollinators to agricultural crop production, but pollinators contribute to a much broader set of social benefits that need to be modeled separately (Box 10.2). Second, our models are limited to predicting relative pollinator abundance, which is only one of many potential contributors to crop yield (see Chapter 9). Translating from pollinator abundance to pollinator influence on crop yield will be limited in many cases by gaps in our understanding of pollinator-yield effects. First, we often do not know the functional form of the relationship between increased number or quality of pollen grains deposited and yield, and the functional form may further vary with crop variety as well as water and nutrient availability. In addition to the relationship between pollinator abundance and the amount and quality of pollen delivered, pollination is influenced by pollinator foraging behavior and effectiveness, across scales from within flower, inflorescence, patch and landscape (Klein *et al.* 2007; Kremen *et al.* 2007; Ricketts *et al.* 2008).

The uncertainty in the relationship between the model's output, a relative score, and quantitative pollinator abundance currently limits the models application to land-use planning. Without a quantified relationship between the model score and abundance, it is difficult to determine the precise yield, crop value and subsequent service value of supply parcels. And without these precise values, decisions about land management, often based on a cost-benefit analysis, would be difficult because the benefits are thus uncertain. In other words, the model can determine that one landscape will provide qualitatively more pollinators to a farm, but it cannot determine if the cost of management or habitat restoration is outweighed by the benefits. Parameterizing the model to facilitate this type of cost-benefit analysis is an obvious next research priority.

LULC data are often only available at resolutions coarser than the scale at which they influence pollinator behavior. Thus, while our model predicts the

Box 10.2 Pollination services: beyond agriculture

Berry Brosi

In the middle of the vast Amazon River, a fisherman strains to pull a huge *tambaqui* fish (*Colossoma macropomum*) into his homemade wooden boat. This scene seems about as removed from pollination as you can get—but that couldn't be further from the truth.

This chapter has focused its valuation approach on crop pollination, but pollinators are perhaps even more important in providing a huge range of non-agricultural pollination services, many of which are quite surprising. For example, a sizable proportion of the fish species harvested in the Amazonian freshwater fishery—including the *tambaqui*—eat fruits that drop into the waters of seasonally flooded forests and which have evolved to be fish-dispersed (Correa *et al.* 2007). The bulk of these fruit-producing trees rely on animals to pollinate their flowers as an essential step in producing fruit. Thus, pollination disruptions in flooded forests would have severe economic and nutritional consequences for the people of the Amazon and their multi-million dollar fishery.

Animal-mediated pollination is necessary for the reproduction of the great majority of flowering plant species globally, providing a service that is foundational to the bulk of other ecosystem services and that is essential to life on Earth. The role of pollination interactions in the functioning of ecosystems is particularly important because it is a limiting factor to reproduction in more than two-thirds of plant species (Burd 1994).

The roles that pollination plays in the production of non-agricultural ecosystem services are diverse. In terms of understanding the value of pollination services in this context, a central issue is the *ecological distance* from pollination to the service being considered (Figure 10.B.1). At one end of the spectrum are ecosystem services that are ecologically *proximal* to pollination (left side of Figure 10.B.1), such as the pollination of non-agricultural products derived from fruits or seeds. These services are characterized by:

• A direct dependence on one or a few discrete pollination events (floral visits) to provide a tangible product
• Pollination's role in the value of the service is large relative to that of other ecological interactions over short scales of space and time
• Relatively low resilience of the value of the service to pollination shortfalls over short scales of space and time.

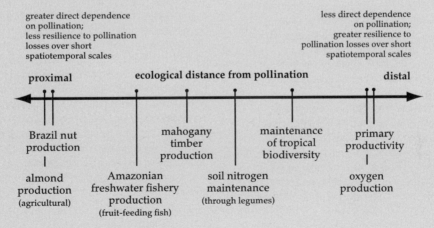

Figure 10.B.1   Spectrum of ecosystem service reliance on pollination.

At the other end of the spectrum, services that are ecologically *distal* to pollination have the opposite set of characteristics (right side of Figure 10.B.1):
• Many pollination events, integrated over large scales of space and time, are needed to support the service

• The role of pollination for the value of the service in any given small scale of space and time is relatively minor
• There is relatively high resilience of the service to pollination losses over short scales of space and time; but if pollination losses were to be sustained over larger

Box 10.2 *continued*

spatiotemporal scales, these functions and services could greatly suffer

Animal-mediated pollination is ultimately derived from the actions of single pollinators moving between a few plants over small spatial scales. Thus, the services that are most proximal to pollination are typically tangible, plant-derived products, while those more distal to pollination are produced by the aggregate actions of countless pollinators at scales larger than that of individual plants (from several square meters to the globe).

There is considerable middle ground in this spectrum. The aforementioned Amazon freshwater fishery is relatively proximal to pollination (disruptions in flooded forest pollination would have major consequences for the fishery over short timescales). The pollination of mahogany (*Swietenia macrophylla*), which provides valuable tropical timber, has fewer ecological linkages than the Amazon fishery example. Yet mahogany timber production could be considered more distal to pollination, because pollination disruptions in any given year would be unlikely to have a strong effect on the value of the mahogany harvest that year. Continued pollination disruptions, however, would eventually damage the mahogany timber industry since the trees could no longer reproduce in the absence of pollination.

Pollination affects ecosystem services in interdependent ways. For example, the weevil-mediated pollination of *Bactris gasipaes*, the peach palm, is a *regulating* service. But that service allows for the production of peach palm fruits (a *provisioning* service), which in turn are a cultural necessity in parts of Latin America—providing important *cultural* services.

Because plants are central to all of the primary supporting services in the terrestrial biosphere—such as primary production, nutrient cycling, and preservation of options (e.g., genetic diversity for future use in pharmaceuticals)—this is perhaps the most important functional role of non-agricultural pollination services. Just as one example, a large proportion of plants in the bean family (legumes) are animal-pollinated; this family is critical for its fixation of atmospheric nitrogen to the soil. If legumes were to suffer pollination reductions, even plant species that are wind-pollinated or self-pollinated would be greatly affected by reductions in available soil nitrogen over timescales as short as a few years.

Flowering plants are also central to a host of climate regulation functions (oxygen production, carbon sequestration, etc.) and hydrological functions (water filtration and flow regulation) that yield vital services. As with supporting services, pollination is important, but ecologically distal, to many of these regulating functions.

The benefits of pollination are most tangible in the production of provisioning services, such as products from non-managed ecosystems, including wild food (e.g., Brazil nuts), fiber (e.g., rattan), and fuelwood resources. Many animals hunted as food for people (not just the tambaqui fish) in turn feed on pollination-dependent fruits and other plant parts. Such products can have a high economic value, particularly when considered in the aggregate (Peters *et al.* 1989). Pollination is key for the population persistence of a number of valuable timber trees as well, not just mahogany.

Pollination interactions are invaluable in the varied roles they play in providing ecosystem services beyond agriculture. Yet we still know little about how ongoing anthropogenic environmental changes will affect communities of pollinators or the pollination functions they perform. For example, there is serious concern in the scientific community that global climate change will lead to changes in the timing of flowering and of pollinator foraging behavior, disrupting pollination interactions worldwide (e.g., Memmott *et al.* 2007). Such disruptions would have major impacts on global ecological functioning and thus on a huge range of non-agricultural pollination services. This makes the need for understanding and ameliorating the effects of environmental change on pollination all the more pressing.

---

likelihood that a pollinator could reach a given 90 meter parcel, pollination delivery may be influenced by plant composition within 90 meters (Morandin and Winston 2006; Kremen *et al.* 2007). This was not as much of a limitation in the landscapes in this chapter but has been in other landscapes (Winfree *et al.* 2008).

Finally, our model, while quantitative, is essentially a statistical evaluation of the landscape so it cannot project pollinator abundance over time. Rather it assumes population stasis given a particular landscape configuration. In other words, our model does not provide an estimate of pollinator population viability or predict pollinator temporal dynamics or interaction of time and space through meta-population dynamics. As such, it does not incorporate stochastic events, which may influence long-term population dynamics and yield.

10.9.2   Next steps

While new ecological data are needed to gain a better understanding of the relationship between crop pollination and yield, we can use this current model framework to advance our understanding in a number of ways. First, we can apply this model to a much larger set of crop studies conducted at the landscape scale (viz, studies in Ricketts *et al.* 2008). Second, using statistical techniques, we can relate the landscape-level outputs of the model (pollinator supply) to the observed measure of pollination services in each study (e.g., pollen deposition, pollen limitation) to attempt to develop a direct relationship between landscape and yield effects via pollinator abundances. Third, by manipulating modeled landscapes (e.g., by increasing floral or nesting resources in different spatial configurations), we can estimate the effects on pollinator abundances and pollination services across a range of changes in resources, and look for generalities across landscapes in the density and arrangement of resources needed to provide adequate pollinators and pollination services. This would inform efforts to preserve existing habitats within degraded landscape and also guide planning of habitat restoration. Similar to our sensitivity analysis of model parameters, we also envision analyses exploring the sensitivity of modeled pollination services to resource patchiness at different grain sizes or to different landscape configurations.

Acknowledgments

This work was facilitated by McDonnell Foundation 21st Century and University of California Chancellor's Partnership awards to C.K., and by two National Center for Ecological Analysis and Synthesis working groups (*Restoring an ecosystem service to degraded landscapes: native bees and crop pollination*; PI's C.K. and N.M.W. *Conservation priorities: Can we have our biodiversity and ecosystem services too?* PI's P. Kareiva, T. Ricketts, G. Daily) supported by NSF grant DEB-00–72909, the University of California at Santa Barbara, and the State of California. Kirsten Almberg helped with figures and Jaime Florez provided measurements of intertegular spans for Costa Rican bees. Berry Brosi provided expert assessment of nesting and floral resources for the Costa Rica validation. Peter Kareiva, Erik Nelson, Berry Brosi, and Kai Chan all provided input in early development of the model. Saul Cunningham and Alexandra Klein provided valuable comments and corrections that improved the chapter.

References

Allsopp, M. H., de Lange, W. J., and Veldtman, R. (2008). Valuing insect pollination services with cost of replacement. *PLoS One* **3**, e3128.

Brosi, B. J., Daily, G. C., Shih, T. M., *et al.* (2008). The effects of forest fragmentation on bee communities in tropical countryside. *Journal of Applied Ecology* **45**, 773–83.

Burd, M. (1994). Bateman's principle and plant reproduction—the role of pollen limitation in fruit and seed set. *Botanical Review* **60**, 83–139.

Cane, J. H. (1997). Lifetime monetary value of individual pollinators: the bee *Habropoda laboriosa* at rabbiteye blueberry (*Vaccinium ashei* Reade). *Acta Horticulturae* **446**, 67–70.

Correa, S. B., Winemiller, K. O., Lopez-Fernandez, H., *et al.* (2007). Evolutionary perspectives on seed consumption and dispersal by fishes. *Bioscience* **57**, 748–56.

Costanza, R., dArge, R., deGroot, R., *et al.* (1997). The value of the world's ecosystem services and natural capital. *Nature* **387**, 253–60.

Cresswell, J. E., Osborne, J. L., and Goulson, D. (2000). An economic model of the limits to foraging range in central place foragers with numerical solutions for bumblebees. *Ecological Entomology* **25**, 249–55.

Cross, P. C., and Beissinger, S. R. (2001). Using logistic regression to analyze the sensitivity of PVA models: A comparison of methods based on African wild dog models. *Conservation Biology* **15**, 1335–46.

Free, J. B. (1993). *Insect pollination of crops*. Academic Press, San Diego.

Gallai, N., Salles, J. M., Settele, J., *et al.* 2009. Economic valuation of the vulnerability of world agriculture confronted with pollinator decline. *Ecological Economics* **68**, 810–21.

Gathmann, A., and Tscharntke, T. (2002). Foraging ranges of solitary bees. *Journal of Animal Ecology* **71**, 757–64.

Greenleaf, S., Williams, N., Winfree, R., *et al.* (2007). Bee foraging ranges and their relationships to body size. *Oecologia* **153**, 589–96.

Greenleaf, S. S., and Kremen, C. (2006). Wild bee species increase tomato production and respond differently to surrounding land use in Northern California. *Biological Conservation* **133**, 81–7.

Hines, H. M., and Hendrix, S. D. (2005). Bumble bee (*Hymenoptera apidae*) diversity and abundance in tall-grass prairie patches: Effects of local and landscape floral resources. *Environmental Entomology* **34**, 1477–84.

Hoehn, P., Tscharntke, T., Tylianakis, J. M., and Steffan-Dewenter, I. 2008. Functional group diversity of bee pollinators increases crop yield. *Proceedings of the Royal Society B: Biological Sciences* **275**, 2283–91.

Janzen, D. H., Ed. (1983). *Costa Rican natural history*. University of Chicago Press, Chicago.

Kim, J., Williams, N., and Kremen, C. 2006. Effects of cultivation and proximity to natural habitat on ground-nesting native bees in California sunflower fields. *Journal of the Kansas Entomological Society* **79**, 309–20.

Klein, A. M., Mueller, C. M., Hoehn, P., *et al.* (in press). Understanding the role of species richness for pollination services. In: D. Bunker, A. Hector, M. Loreau, *et al.*, Eds., *The consequences of changing biodiversity—solutions and scenarios*. Oxford University Press, Oxford.

Klein, A. M., Vaissière, B. E., Cane, J. H., *et al.* (2007). Importance of pollinators in changing landscapes for world crops. *Proceedings of the Royal Society* **274**, 303–13.

Knight, M. E., Martin, A. P., Bishop, S., *et al.* (2005). An interspecific comparison of foraging range and nest density of four bumblebee (*Bombus*) species. *Molecular Ecology* **14**, 1811–20.

Kremen, C., Bugg, R. L., Nicola, N., *et al.* (2002a). Native bees, native plants and crop pollination in California. *Fremontia* **30**, 41–9.

Kremen, C., Williams, N. M., and Thorp, R. W. (2002b). Crop pollination from native bees at risk from agricultural intensification. *Proceedings of the National Academy of Sciences* **99**, 16812–16.

Kremen, C., Williams, N. M., Bugg, R. L., *et al.* (2004). The area requirements of an ecosystem service: crop pollination by native bee communities in California. *Ecology Letters* **7**, 1109–19.

Kremen, C., Williams, N. M., Aizen, M. A., *et al.* (2007). Pollination and other ecosystem services produced by mobile organisms: a conceptual framework for the effects of land-use change. *Ecology Letters* **10**, 299–314.

Losey, J. E., and Vaughan, M. (2006). The economic value of ecological services provided by insects. *Bioscience* **56**, 311–23.

McCarthy, M. A., Burgman, M. A., and Ferson, S. (1995). Sensitivity analysis for models of population viability. *Biological Conservation* **73**, 93–100.

Memmott, J., Craze, P. G., Waser, N. M., *et al.* (2007). Global warming and the disruption of plant-pollinator interactions. *Ecology Letters* **10**, 710–717.

Morandin, L. A., and Winston, M. L. (2006). Pollinators provide economic incentive to preserve natural land in agroecosystems. *Agriculture, Ecosystems & Environment* **116**, 289–92.

National Research Council of the National Academies. 2006. *Status of pollinators in North America*. National Academy Press, Washington, DC.

Peters, C. M., Gentry, A. H., and Mendelsohn, R. O. (1989). Valuation of an Amazonian rainforest. *Nature* **339**, 655–6.

Polasky, S., Nelson, E., Camm, J., *et al.* (2008). Where to put things? Spatial land management to sustain biodiversity and economic returns. *Biological Conservation* **141**, 1505–24.

Potts, S. G., Vulliamy, B., Dafni, A., *et al.* (2003). Linking bees and flowers: how do floral communities structure pollinator communities? *Ecology* **84**, 2628–42.

Potts, S. G., Vulliamy, B., Roberts, S., *et al.* (2005). Role of nesting resources in organising diverse bee communities in a Mediterranean landscape. *Ecological Entomology* **30**, 78–85.

Priess, J. A., Mimler, M., Klein, A. M., *et al.* (2007). Linking deforestation scenarios to pollination services and economic returns in coffee agroforestry systems. *Ecological Applications* **17**, 407–17.

Ricketts, T. H. (2004). Tropical forest fragments enhance pollinator activity in nearby coffee crops. *Conservation Biology* **18**, 1262–71.

Ricketts, T. H., Daily, G. C., Ehrlich, P. R., *et al.* (2004). Economic value of tropical forest to coffee production. *Proceedings of the National Academy of Sciences of the USA* **101**, 12579–82.

Ricketts, T. H., Regetz, J., Steffan-Dewenter, I., *et al.* (2008). Landscape effects on crop pollination services: are there general patterns? *Ecology Letters* **11**, 499–515.

Southwick, E. E., and Southwick, L. (1992). Estimating the economic value of honey-bees (hymenoptera, apidae) as agricultural pollinators in the United States. *Journal of Ecological Entomology* **85**, 621–33.

Stokstad, E. (2007). The case of the empty hives. *Science* **316**, 970–2.

Tepedino, V. J., and Stanton, N. L. (1981). Diversity and competition in bee-plant communities on short-grass prairie. *Oikos* **36**, 35–44.

Westrich, P. (1996). Habitat requirements of central European bees and the problems of partial habitats. In: A. Matheson, S. L. Buchmann, C. O'Toole, *et al.*, Eds., *The conservation of bees*. Academic Press, London, pp. 1–16.

Williams, N., and Kremen, C. (2007). Floral resource distribution among habitats determines productivity of a solitary bee, *Osmia lignaria*, in a mosaic agricultural landscape. *Ecological Applications* **17**, 910–21.

Winfree, R., Dushoff, J., Crone, E., *et al.* (2005). Testing simple indices of habitat proximity. *American Naturalist* **165**, 707–17.

Winfree, R., Williams, N. M., Dushoff, J., *et al.* (2007). Native bees provide insurance against ongoing honey bee losses. *Ecology Letters* **10**, 1105–13.

Winfree, R., Williams, N. M., Gaines, H., *et al.* (2008). Wild bee pollinators provide the majority of crop visitation across land-use gradients in New Jersey and Pennsylvania, USA. *Journal of Applied Ecology* **45**, 793–802.

CHAPTER 11

Nature-based tourism and recreation

**W. L. (Vic) Adamowicz, Robin Naidoo, Erik Nelson,
Stephen Polasky, and Jing Zhang**

11.1 Nature-based tourism and recreation values in context

International tourism and recreation generated over $1 trillion in receipts in 2007 (roughly equivalent to South Korea's 2007 gross domestic product; World Tourism Organization 2008). Environmental attributes of tourism sites are important in determining visitation rates and the value of tourism and recreation. For example, the earliest writings on tourism emphasize the environment quality at seaside resorts, parks, and wilderness areas (Towner and Wall 1991). National parks are often located in areas with photogenic biodiversity (e.g., Serengeti National Park in Tanzania and Krueger National Park in South Africa) or areas of scenic beauty (mountains, coasts, etc.). Many forms of recreation require natural amenities (clean water for swimming, species diversity for birdwatching). By providing the natural features that attract tourists, ecosystems provide a tourism and recreation service.

Tourism generally refers to travel for pleasure and typically involves overnight stays away from home. Recreation typically refers to activities that occur over part or all of a single day (e.g., hiking or fishing) that may take place during a tourism trip or as a day trip from home. For simplicity in this chapter we use "tourism" as the general category for both recreation and tourism.

Economists have developed a variety of techniques for assessing the economic value of tourism (Champ *et al.* 2003; Phaneuf and Smith 2005; Bockstael and McConnell 2007) and how the value of tourism is affected by changes in the supply of environmental attributes (Phaneuf and Smith 2005).

In this chapter a site's environmental attributes include its quantity and quality of ecological processes such as water delivery and cleanliness, biodiversity, quality and diversity of habitat, net primary productivity, etc. Tourism dynamics, however, are not explained solely by environmental attribute supply. Tourism valuation models are essentially models of human behavior applied to the decisions of where, when, and how to engage in tourism. As such, the value that a tourist places on a particular site will depend on his/her personal characteristics (including past behavior and social interactions), the local geography (including distance and cost of accessing sites as well as the presence of substitute tourism sites) and the individual's perception of congestion, environmental quality, and other site-level attributes.

The value of nature-based tourism in various parks, landscapes, or regions has been estimated in a large number of economic studies (Phaneuf and Smith 2005). Most of these studies have involved primary data collection. However, if a tourism site or landscape has not been the focus of a detailed economic analysis, which represents the majority of cases, we have to rely on secondary data (e.g., data collected by government agencies) to assess tourism values. Unfortunately, these secondary datasets often leave out key variables required for understanding the linkages between visitation rates and characteristics of sites such as their environmental attributes.

Here, we outline the conceptual basis for assessing the values people place on engaging in tourism activities across a landscape and develop methods that increase in sophistication with increased data availability. Each potential tourism site has environmental characteristics that influence the attractiveness of tourism at the site. Tourism tends to increase with improvement in environmental characteristics but is also influenced by the distance of sites from tourists' starting location, availability of substitute tourism sites and other factors. To isolate the effect of the environment on tourism use and value, we need to construct models that control for these other influences. (We use the term "tourism use" instead of "tourism demand," the term of choice in the valuation literature, to remain consistent with the taxonomy of ecosystem services presented in this book; see Chapter 3.)

We present three methods for assessing tourism use and values in a landscape. In tier 1 we present a mapping methodology for spatially representing important tourism areas. Overlaying these site maps on maps of environmental attributes displays spatial correlations between tourism use and environmental attributes. The tier 2 model provides a more theoretically appropriate mechanism for measuring the change in tourism use and value given marginal changes in environmental attributes on the landscape. Tier 2 models can be used to approximate the change in tourism values under future scenarios of land use/land cover (LULC) vis-à-vis the current landscape. We conclude with a discussion of the state-of-the-art tourism valuation models (tier 3). In these models individuals examine the attributes of alternative destinations and choose the destination that generates the highest utility (Train 2003; Phaneuf and Smith 2005). In principle this approach can capture linkages among environmental attributes, substitute sites, substitute activities, demographic factors affecting value and other aspects of tourism valuation, though practical complexities and data requirements are both high.

11.1.1 Major social and environmental processes that affect tourism values

In general, the value of a tourism site will increase as the quantity or quality of environmental attributes at the site increases. For example, the value of a site visit for a bird watcher increases in the abundance and diversity of species, for an angler with cleaner water and fish stocking, and for a beachgoer with improved water quality. Figure 11.1 outlines the relationships between individual tourists, tourism destinations, and four factors that form the main components of the linkages between tourism, the environment, and value: environmental attributes, tourism site infrastructure, costs of visiting sites (illustrated by travel distance) and the availability of substitutes. Figure 11.1 provides an example with two cities where individual tourists live, and two tourism sites. At each tourism site environmental attributes and infrastructure affect the attractiveness of the site and thereby influence the number of trips taken and/or the choice to visit site A versus site B. The costs of visiting a site, captured in the travel distances, and the availability and impact of substitute sites complete the characterization.

In assessing the value of a change in environmental attributes, consider a base case in which all of the tourists visit the closest site, site A for city 1 tourists and site B for city 2 tourists. Suppose the quality of the water in the river that flows through site A improves. The improved water quality increases the attractiveness of site A, thereby increasing the total number of trips from residents of city 1. In addition, some residents from city 2 may now be willing to travel further to enjoy the improved supply of water quality at site A. Both the increase in number of visits from residents of city 1 and the change in site choice by residents of city 2 are reflections of the value of the improvement in an environmental attribute at site A.

Figure 11.1 also presents some of the complexities associated with the measurement of tourism values. The assessment of value depends on modeling the choice of sites and/or number of trips by residents of cities 1 and 2. Predicting trips requires information on the characteristics of these residents (income, perceptions of site attributes, etc.), environmental and infrastructure attributes, cost of trips (travel distances, time costs) and substitute sites. In Figure 11.1 the set of substitutes is defined as the two sites A and B. In reality, there may be hundreds of substitute sites. Information about the spatial location and costs associated with travel to each of these sites from the relevant residence zones is necessary for predicting trips. Finally, the model of tourism should be able to translate changes in environmental attributes (e.g., water quality/quantity, species

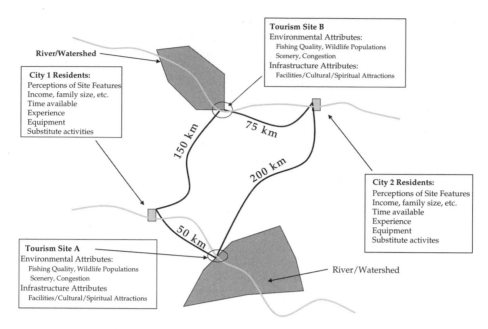

Figure 11.1 Tourism linkages to environmental attributes.

abundance and composition) into impacts on site attractiveness and the subsequent impact on number of trips or site choice.

11.2 Tier 1 tourism supply and use model

Creating a complete model of tourism that incorporates all the dimensions discussed above requires more data than are typically available. In the tier 1 tourism model we start with a more modest aim that requires much less data. We use maps to plot and characterize the current spatial pattern of nature-based tourism across the landscape. We measure use of a site for nature-based tourism with the number of visits to the site. We compare tourism use with several categories of site features and characteristics, including environmental attributes, site infrastructure (e.g., campgrounds), and site accessibility (e.g., proximity to population centers, roads, airports). By overlaying these data layers we can correlate use of a site for tourism activities with its supply of environmental attributes. For example, is

recreational fishing particularly popular at sites with the cleanest water in a landscape?

The main strength of tier 1 is that it gives a picture of the current state of nature-based tourism on the landscape and only requires relatively easily collected data to implement. The tier 1 data can also form the basis of the data collection efforts needed to run a tier 2 analysis. In interpreting tier 1 results, it is important to note that spatial correlation in tourism and environmental attributes does not imply causality. Further investigation with more detailed data and analytical techniques, such as those developed in tier 2 and tier 3 approaches, are needed to fully disentangle the effects of environmental attributes from other factors such as site accessibility and availability of substitute sites.

11.2.1 Developing tier 1 maps

The tier 1 approach involves compiling and overlaying five categories of data to investigate the spatial relationship between use of sites for tourism, environmental attributes, and landscape features.

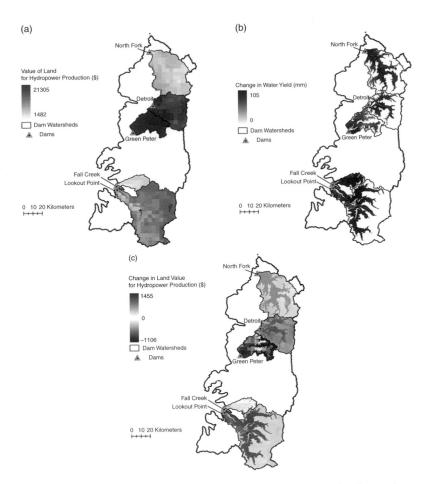

Plate 1 Hypothetical example application of tier 1 model of water provisioning for hydropower generation in the Willamette river watershed. The example evaluates five sub-catchments of hydropower stations at North Fork (41MW), Detroit (115MW), Green Peter (92MW), Fall Creek (6.4MW), and Lookout Point (138MW). (a) The net present value of landscape water provision services for hydropower; (b) changes in water yield as a result of hypothetical deforestation of all land below 1000 m above sea level; and (c) the changes in landscape value for hydropower under the deforestation scenario. (See Figure 4.1.)

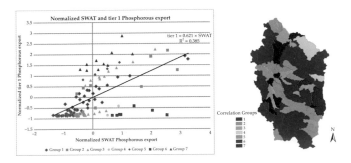

Plate 2 Aggregated sub-catchment phosphorous export comparison between our model and SWAT (graph) and agreement of spatial phosphorous export patterns predicted by the two models (map) in the Williamette Valley, Oregon (USA). The graph on the left depicts a correlation between the normalized tier 1 model outputs and the normalized SWAT outputs. The groupings in the graph depict sub-catchments whose tier 1 outputs lie within a threshold distance of SWAT outputs given the correlation between normalized tier 1 and SWAT outputs. Note that Groups 4 and 6 represent sub-catchments in which tier 1 outputs are unexpectedly low and Groups 5 and 7 represent sub-catchments in which tier 1 outputs are unexpectedly high. The map illustrates the sub-catchment groups identified in the graph. (See Figure 6.8.)

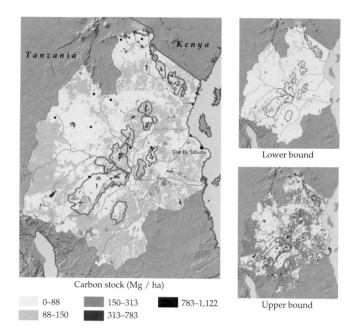

Carbon stock (Mg / ha)

| 0–88 | | 150–313 | | 783–1,122 |
| 88–150 | | 313–783 | | |

Lower bound

Upper bound

Plate 3 Tier 1 carbon storage estimates for 1995 in Tanzania's Eastern Arcs Mountains and their watersheds. The polygons formed with the green lines represent Eastern Arc Mountain blocks, which rise from the surrounding woodlands and savannas. These blocks were once largely forested, but now consist of a mixture of agriculture, forest, and woodlands. Blue lines are major rivers. Black squares represent major cities. Timber plantations cover approximately 0.3% of the study landscape. Spatially explicit land cover and other landscape data are from the Valuing the Arc project (2008; Mwakalila 2009). See the chapter's SOM for details on data used in the maps. (See Figure 7.2.)

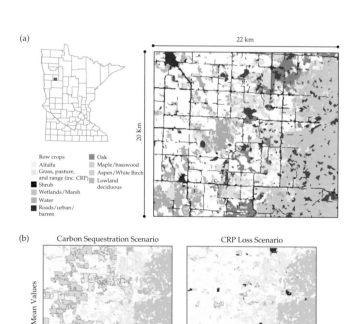

Plate 4 The value of carbon sequestered in soil across two alternative LULC scenarios. (a) The year 2000 landscape. (b) The per-hectare monetary value of carbon sequestration in soil from 2000 to 2050 for each LULC scenario. The top row of maps gives mean results across all model simulations. The bottom rows of maps give the results from one particular run of the model. The black outlines on the parcels indicate parcels that experience LULC change in some portion of its area at some point between 2000 and 2050. The Carbon Sequestration Scenario map reflects a program of afforestation, restoring prairie pothole, and converting row crops to pasture and perennial grassland. In the CRP Loss Scenario any parcel that was primarily in CRP in 2000 was converted to row crops or a hayfield by 2050.
(See Figure 7.4.)

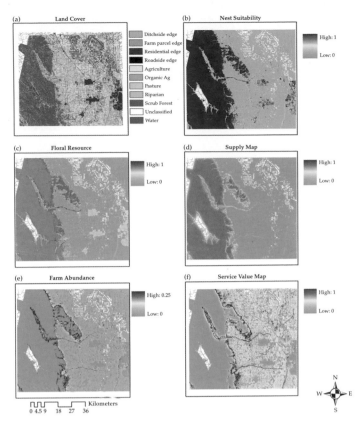

Plate 5 Example results of pollination model for watermelon in Yolo County, California. The model uses (a) land cover data as input and derives maps of (b) nesting habitat and (c) floral resources. From this, it generates (d) a pollinator supply map that describes an index of pollinator abundance on the landscape. Based on the supply map, the model generates (e) a pollinator abundance map on farm parcels (i.e., "farm abundance"). After using a simple yield function to translate farm abundance into relative yield the model distributes yield or economic value back onto the surrounding landscape to generate (f) the value map. All steps are the same for tier 1 and tier 2 models; results here are tier 2, based on data supplied in supplemental online appendix. (See Figure 10.2.)

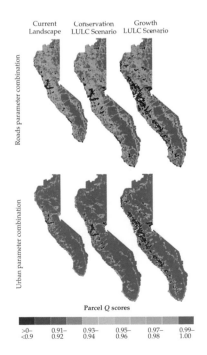

Plate 6 Maps of parcel habitat quality scores when the "Roads" and "Urban" parameter combinations are used in the Sierra Nevada illustrative example.

We ran the tier 1 model on a grid map with a cellular resolution of 400 m × 400 m (16-ha grid cells). In these maps we present the mean habitat quality score (Q) of all grid cells within 500 hectare hexagons, our parcels in this illustrative example. There are 23 042 500-ha hexagons in the Sierra Nevada Conservancy. In both future LULC scenarios the majority of residential development is centered on Sacramento, and generally along the western foothills. In the Growth scenario, montane hardwood is the land cover type that loses the most area to development (158 268 ha). In the Conservation scenario, annual grassland is the land cover type that loses the most area (17 798 ha). See the chapter's SOM for all tier 1 model details. (See Figure 13.2.)

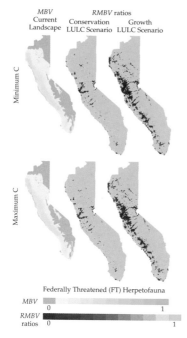

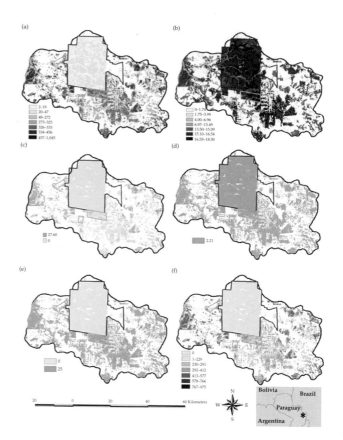

Plate 7 The spatial distribution of MBV and RMBV ratio scores for federally threatened herpetofauna (FT) using minimum and maximum species-habitat suitability scores in the Sierra Nevada illustrative example.

The Growth scenario creates a much greater loss in FT subgroup species effective habitat area in the foothills of the Sierra Nevada than the Conservation scenario does. See the chapter's SOM for all tier 2 model details. (See Figure 13.4.)

Plate 8 Net present values in US$ha⁻¹ for selected ecosystem services in the Mbaracyau Forest Biosphere Reserve, Paraguay.

(a) Sum of all five services; (b) sustainable bushmeat harvest; (c) sustainable timber harvest; (d) bioprospecting; (e) existence value; and (f) carbon storage. (See Figure 14.2.)

Source: Naidoo and Ricketts (2006).

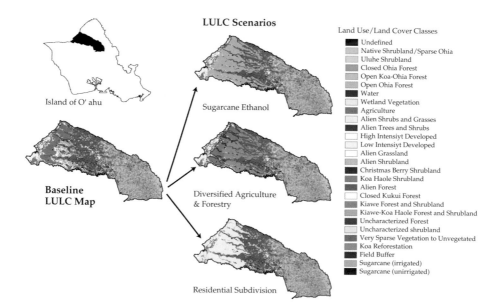

Plate 9 Land use/land cover maps on the north shore of O'ahu.

The area shown here includes all of Kamehameha Schools' north shore land holdings, as well as small adjacent parcels that make for a continuous region. The baseline map is from the Hawai'i Gap Analysis Program's land cover layer for O'ahu (Hawai'i Gap Analysis Program 2006). (See Figure 14.6.)

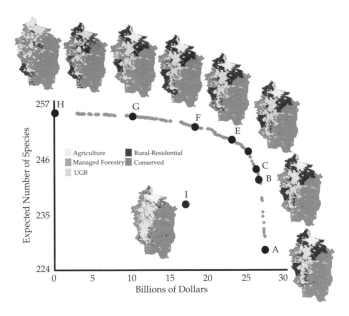

Plate 10   Efficiency frontier showing maximum feasible combinations of economic returns and biodiversity scores.
Land-use patterns associated with specific points along the efficiency frontier (points A–H) and the current landscape (point I). (See Figure 14.8.)

Source: Polasky *et al.* (2008).

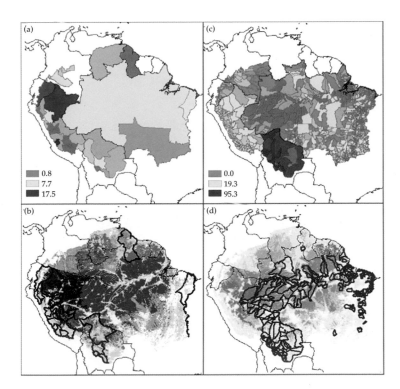

Plate 11   Poverty indicators and representative forest product harvest distributions in the Amazon Basin.
The incidence of underweight children is highest in northern Peru and eastern Ecuador (a) while unsatisfied basic needs are highest in Bolivia (c). High poverty areas defined as those above the 75th percentile for underweight children (outlined in dark black) are shown in a direct pairing, overlaid with the harvest index of fruits and nuts for subsistence (b). High poverty areas defined as those above the 75th percentile for unsatisfied basic needs are shown in an indirect pairing, overlaid with the harvest index of wood for market sale (d). Units for underweight children are percentage of the population under the age of 5 that is underweight. Units for unsatisfied basic needs are the percentage of the population with unsatisfied basic needs. The legends and units for (b) and (d) are the same as those in Figure 16.1c and h. (See Figure 16.2.)

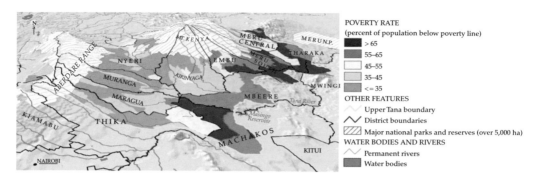

Plate 12 Map of the Tana River headwaters in Kenya, and the distribution of poor communities. (See Figure 16.B.1.)

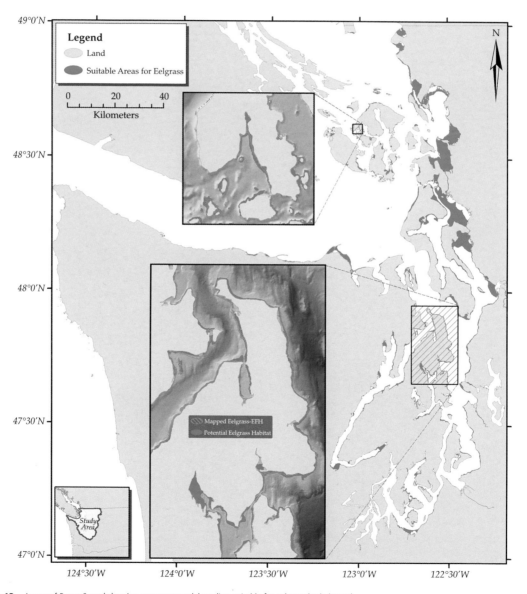

Plate 13 A map of Puget Sound showing areas our model predicts suitable for eelgrass beds (green).
Inset maps show higher detail; orange represents currently mapped eelgrass from the NOAA Essential Fish Habitat data (TerraLogic GIS Inc. 2004). (See Figure 17.1.)

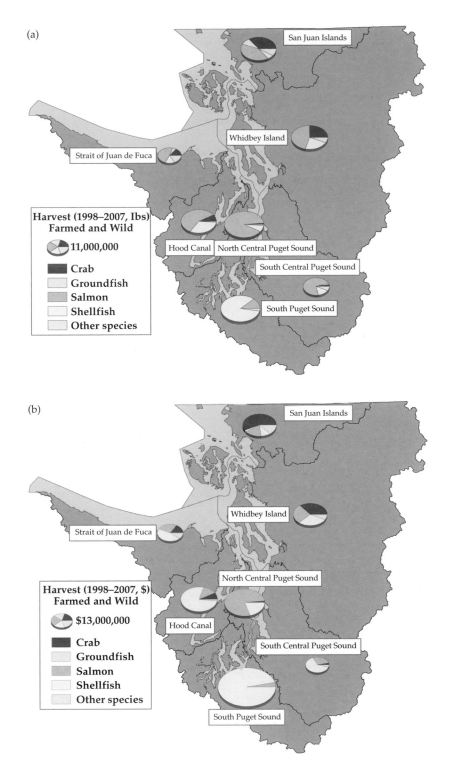

Plate 14 A map of the Puget Sound Partnership's action areas showing the distribution of (a) landings (in UK£) and (b) revenue (in US$) of farmed and wild seafood from 1998 to 2007. (See Figure 17.2.)

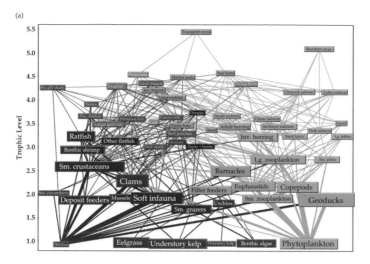

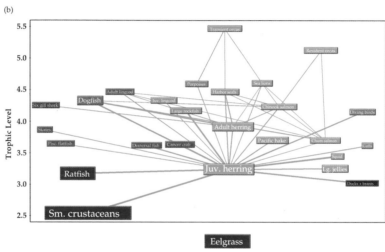

Plate 15 (a) The structure of the EwE food web model of the Central Basin of Puget Sound (without fisheries) and (b) a subset of the EwE food web model focusing on eelgrass and herring.

(a) Box size is proportional to standing stock biomass; line thickness is proportional to the flow of energy/material from the prey to the predator. Red colors represent detritus and the portion of the food web it supports, blues are benthic primary producers and those they support, and greens are phytoplankton and phytoplankton-supported groups. Consumers' colors are a mix proportional to the amount of production that ultimately stems from those sources. In (b) dashed arrows indicate groups whose predation on herring eggs is mediated by the biomass of eelgrass. Colors are as those in (a). (See Figure 17.3.)

Plate 16 Change in annual average water availability (yield) between present day and mid-century for the Willamette Basin. (See Figure 18.2.)

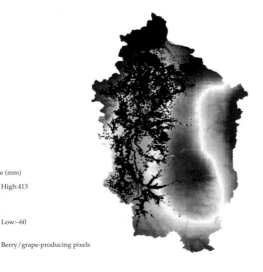

Difference (mm)

High: 413

Low: –60

Berry/grape-producing pixels

The first data layer includes the sites that provide tourism opportunities. Tourism sites can include national, provincial, state, county, or privately held parks and recreation areas. In some cases, tourism activities are not restricted to easily defined places on the landscape. Some tourism activities take place across a broad region or landscape. For example, duck hunting occurs on private and public land throughout the Prairie Pothole Region of the Midwest USA. To account for this type of activity we may subdivide the landscape into zones to identify the portions of the landscape where diffuse tourism activity is more and less popular.

The second data layer maps visitation data. Visitation data can be measured a number of ways: site visits per unit time (where one person can register multiple visits over the course of the time period), number of visitors to the site per unit time, visitor days per unit time (the sum of all visitors' length of stay in days), number of visitors purchasing entrance permits per unit time, etc. Stratifying visit data by visitor place of origin, reason for visit, and time allows more detailed and comprehensive analysis. For example, stratifying visitors by international and domestic origin can show differences in visitation rates by cost (on average, international visitors pay much more to tour than domestic tourists). We discuss a way of placing a monetary value on the annual number of visits in Section 11.4 below.

The third data layer maps information on environmental attributes and other landscape features at sites. For example, does the site include a lake that would be a draw for swimming or fishing, major changes in elevation good for hiking or dramatic views, habitat for charismatic species that are a draw for wildlife viewing? High water quality, sufficient water flow, and abundant game fish are vital for certain stretches of rivers if they are to provide tourism value via recreational fishing.

The fourth data layer maps information on important infrastructure at each site. Infrastructure important for determining tourism site visits includes roads, hiking trails, lodges, camping sites, and interpretative facilities.

The fifth data layer maps major transportation infrastructure and urban areas. With these data we can calculate distance and travel time from urban areas to each tourism site.

11.2.2 Example application of tier 1: Willamette Basin, Oregon, USA

We applied the tier 1 approach to tourism in the state parks of the Willamette Basin, Oregon, USA. The Willamette River and the Basin's major highways and cities are located in the Basin's valley floor (Figure 11.2A). Most state parks lie near cities and are on the main stem of the Willamette River or one of its tributaries; a few parks are in the Cascades mountain range on the eastern side of the Basin (Figure 11.2A).

Figures 11.2b through 11.2d present maps of environmental attributes and other features in the Basin (ODFW 2005; OGEO 2008; PNW-ERC 2008).

The state parks in the Basin with the most day visitors (not deconstructed by activity) are located in the Cascade Mountains and offer outstanding scenic attractions or recreation opportunities (Figure 11.3a). Silver Falls State Park, which contains many large waterfalls, had the most visits in 2004 even though its aggregate distance to the Basin's cities was more than most other state parks.

Detroit Lake State Park, higher up in the Cascades than Silver Falls and of greater distance from cities, was the most popular destination in the Basin for overnight camping (an activity that is generally more costly than a day visit, given the time requirements and the price of camping permits; see Figure 11.3b). Many sites in the northeast corner of the Basin (and just outside the Basin along the Columbia River) were popular tourism destinations both because they are close to the largest urban center in the Basin (Portland) and feature the spectacular scenery around Mount Hood and the Columbia River Gorge.

The greater use of nature-based tourism in the northeast corner of the Basin is also reflected in hunting data. The Santiam hunting region (the region that includes Mount Hood National Forest, Silver Falls State Park, Detroit Lake State Park, and part of the Columbia River Gorge) is the most popular region for big game hunting, with 127 446 hunter days in 2004 (see Figure 11.4a).

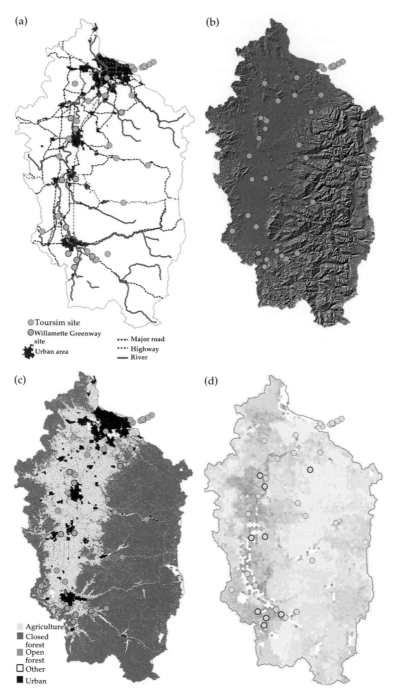

Figure 11.2 Willamette Basin state parks and landscape features, characteristics, and environmental attributes.
Major landscape features and characteristics and state parks in the Basin (a; OGEO 2008, PNW-ERC 2008, site data provided by Terry Bergerson, Oregon Parks and Recreation Department). Some access sites to the Willamette Greenway, a bicycle path, are considered state parks. The map of hillshade (b) in the Basin (PNW-ERC 2008) indicates areas of significant elevation changes. The map of landcover in the Basin circa 2000 (c; PNW-ERC 2008). The gray gradient in (d) represents an area's marginal biodiversity value (*MBV*) score for 24 at-risk vertebrates, a tier 2 measure of biodiversity supply (see Chapter 8): the darker the parcel the greater the share of the 24 species' total habitat on the landscape found in the parcel (see Hulse and Baker 2002 and Nelson *et al.* 2009 for details).

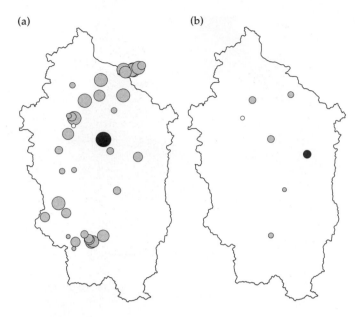

Figure 11.3 State park use in the Willamette Basin.
Total number of day visits in a state park in 2004 (points proportional to number of visits). Silver Falls State Park (the black circle) had the greatest number of day visits in 2004 (981 680). The Lincoln Access of the Willamette Greenway (the white circle just to the northwest of the black circle) had the lowest use with 5 440 day visits. Total number of overnight camping visits in a state park in 2004 (b; the points in (a) and (b) are on the same scale). Detroit Lake State Park (the black circle) had the greatest number of campers in 2004 (84 137). Willamette Mission State Park (the white circle to the northwest of the black circle) had the lowest use with 2 312 campers. Most state parks in the Basin do not have camping facilities. Visit data provided by Terry Bergerson, Oregon Parks and Recreation Department.

All of this suggests that any changes in the supply of environmental attributes or transportation infrastructure in the northeast corner of the Basin may have the greatest impact on tourism use and value in the Basin. Interestingly, this area is expected to experience major forest cover transitions over the next 100 years due to climate change (see Chapter 18).

One way to relate use of a site for tourism to environmental attributes at state parks is to overlay the site map with maps of these attributes. To illustrate this point, we construct a biodiversity supply map using the tier 2 biodiversity model (see Chapter 8), based on the habitat preferences of 24 terrestrial vertebrates that are habitat-limited in the Basin (Nelson *et al.* 2009). Many of the state parks on the Basin floor align spatially with some of the most valuable habitat areas in the Basin (see Figure 11.2d). However, only camping visit rates are correlated with areas that supply the greatest share of habitat for these 24 at-risk species, whereas day visit rates are not (see Figure 11.5). These correlates may

change, however, as the roster of species analyzed changes. Similar analyses can be performed for other environmental attributes of interest.

11.3   Tier 2 tourism supply and use model

The tier 1 tourism methodology has two major shortcomings. First, we cannot quantify how the supply of environmental attributes at a site affects the overall tourism experience and its value at the site. In addition, the tier 1 analysis cannot explicitly estimate how changes on the landscape could change tourism activity or value. For example, does a future LULC and land management scenario reduce environmental attribute supply across portions of the landscape, and would this change impact tourism at particular sites? Or, are new roads or airports being built to facilitate access to a tourism site? An additional challenge comes from evaluating the addition of a new tourism site. A new tourist site is a substitute that could siphon some

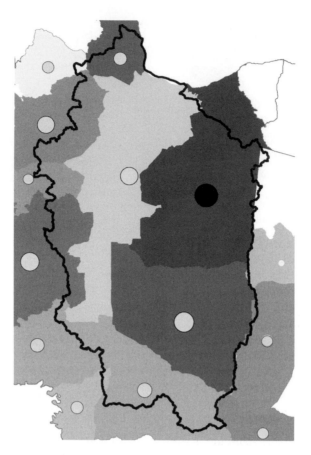

Figure 11.4 Diffuse tourism demand in the Willamette Basin.
The size of a point indicates the total number of big game hunter days in a hunting region (deer and elk with bow or rifle, all seasons) in 2004 (OFWD 2005). Each hunting region is given by a distinct polygon. The Santiam hunting region, the region with the black circle, was the most popular region for big game hunting in 2004 (127 446 hunter days). The Metolius hunting region (to the southeast of the Santiam hunting region), the region with the white circle, was the least popular region for big game hunting in 2004 (8,952 hunter days).

use from existing sites but it could also make tourism on its host landscape more attractive overall and increase the landscape's overall tourism use and value.

In tier 2 we develop a relatively simple model that predicts annual visitor days at each tourism site (or region) as a function of the site's (1) environmental attributes; (2) infrastructure; (3) amenities; (4) distance to relevant population centers; and (5) spatial distribution of potential substitutes. We can use this model to predict the expected changes in visitor days at each tourist site on the landscape due to expected changes in any of these five explanatory landscape variables. In Section 11.4, we discuss a way to place a monetary value on (1) the annual number of visits to a site and (2) the change

in the annual number of visits to a site due to a change on the landscape.

11.3.1 The visits model

The tier 2 model estimates the annual number of visitor days or visits (hereafter, "visits"), to tourism sites as a function of several landscape variables. Environmental attribute and infrastructure variables are site specific and directly affect visitation. The costs of visiting a site depend on the location of the tourists relative to the sites and the activities participated in at the site. To capture costs we examine the proximity of population centers to the site to develop an index of visitation cost for each type of tourism activity. The availability and impact of

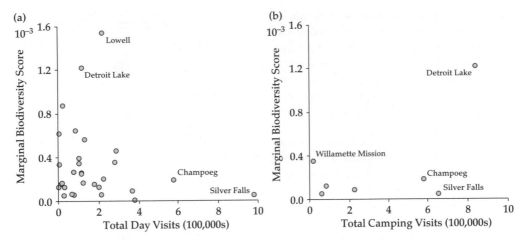

Figure 11.5 State park use versus biodiversity supply.
The *y*-axes of both graphs indicate the marginal biodiversity value (*MBV*) score at a state park's location for 24 at–risk vertebrate terrestrial species (a tier 2 biodiversity supply measure; see Chapter 8 and Nelson *et al.* 2009). The *x*-axes indicate total visits in 2004, either day visits (a) or camping visits (b). Each circle represents a state park; the outliers are labeled. Most state parks do not have overnight camping.

substitutes depends on the spatial location of the site and potential substitute sites, as well as the supply of environmental attributes, infrastructure, and activity opportunities of the competing sites. We illustrate how an index can be constructed to assess the impact of substitutes.

In tier 2 we model the number of annual visits to each site q to participate in a combination of activities a by tourist type ω, represented as $T_{qa\omega}$, as a function of a vector of environmental attributes at q that could impact participation in a ($\mathbf{EA}_{qa}$), a vector of infrastructure variables at q that could impact participation in a ($\mathbf{X}_{qa}$), the relative cost of visiting the site q to participate in a (G_{qa}), and an index of substitute sites that provide at opportunity to participate in a (S_{qa}),

$$T_{qa\omega} = f(\mathbf{EA}_{qa}, \mathbf{X}_{qa}, G_{qa}, S_{qa}), \qquad (11.1)$$

where $a = 1,2,\ldots, A$ indexes any combination of activities, e.g., $a = \{$fishing; camping; hiking; fishing and camping; fishing and hiking;$\ldots$; fishing, camping and hiking$\}$. We deconstruct visits by activities because explanatory variables can affect participation for each activity differently. Further, different types of tourists can react to site attributes and costs differently. Here and throughout, these models can be simplified if we do not have sufficient data to

stratify tourism visits or activities. Doing so involves dropping subscripts a and/or ω and modeling the total number of annual visits to each site q (i.e., $q\omega$, qa, or simply q).

The vector of environmental attributes ($\mathbf{EA}_{qa}$) includes biodiversity, scenic overlooks, boating opportunities, etc., while the vector of infrastructure variables ($\mathbf{X}_{qa}$) can include campgrounds, bathrooms, hiking trails, etc. The costs of visiting site q to participate in activity combination a (G_{qa}) will depend on the distance of the site to population centers and the costs of participating in activity combination a. An index that uses the distance from all relevant population centers to q is one way to determine the relative cost of visiting a site to participate in activity combination a,

$$G_{qa} = (1 - \beta_{qa})g_{qa} + \beta_{qa}\sigma\sum_{i=1}^{I} C_{ia}h(d_{iq}), \qquad (11.2)$$

where g_{qa} is a fixed cost of participating in activity combination a at site q (e.g., an entrance or participation fee), σ is a scalar, C_{ia} is the number of people that might participate in activity combination a at site q that are from population center i, $h(d_{iq})$ is an increasing function of the distance from population center i to site q, d_{iq}, and $\beta_{qa} \in [0, 1]$ is used to weight the relative importance of fixed participation costs versus relative travel costs in the cost index of participating in activity combination a at site q. A commonly used form for $h(d_{iq})$ is $e^{\rho d_{iq}}$ with $0 < \rho \le 1$. In general, as the aggregate distance to site q increases (weighed by C_{ia}) the greater the average cost of visiting site q. In Eq. (11.2), population centers that have a greater number of people that might participate in a are given more weight in the determination of G_{qa}.

The use and value of a tourism site for activity combination a will tend to be lower if there are nearby substitute sites for a. We illustrate a method for measuring the impact of substitute sites on the use and value of a given site q for activity combination a. Our proposed index, S_{qa}, is a function of environmental attributes and infrastructure at nearby sites weighted by the difference in distance that these sites are from population centers,

$$S_{qa} = \sum_{k \in N_{qa}} \sum_{i \in N_{qa}} C_{ia} \times y\left(\mathbf{EA}_{ka}\right) \times z\left(\mathbf{X}_{ka}\right) \times e^{\frac{-(d_{ik} - d_{iq})}{\gamma}}, \quad (11.3)$$

where q is a site suitable for activity combination a, N_{qa} is the set of tourism sites in q's "neighborhood" that provide opportunities to participate in activity combination a (neighborhood sites are indexed by k), population center i is in N_{qa} if population center i could supply people for a in the area formed by N_{qa}, $y(\mathbf{EA}_{ka})$ is an increasing function in k's environmental attributes that affect a, $z(\mathbf{X}_{ka})$ is an increasing function in k's infrastructure that affects a, d_{iq} is as above, d_{ik} is the distance from population center i to site k, and γ is a scalar. The distance to alternate sites is a proxy for the cost of using these sites.

A higher S_{qa} indicates more competition from rival sites for use of site q for activity combination a. All else equal, S_{qa} increases with (1) the number of alternate sites k in site q's neighborhood N_{qa}, (2) the number of population centers i in N_{qa}, (3) the environmental attributes and infrastructure in alternate sites k in N_{qa}, and (4) increasing distance from i in N_{qa} to q vis-à-vis the distance from i in N_{qa} to the set of rivals k.

The extent of site q's neighborhood N_{qa} should be given by the furthest someone would travel to participate in activity a. For example, if a is a wildlife safari, an activity that people will travel around the globe to participate in, then q should be a site that provides safari tourism and N_{qa} should include all

other sites in the world that provide safari tourism and all of the world's major urban centers. On the other hand, if we are modeling recreational stream fishing (typically a one day activity) then N_{qa} should only include alternative stream sites a few hours from q's stream and urban centers that are a few hours from q. Defining N_{qa} for such a local recreation market is illustrated in Figure 11.6.

11.3.2   Estimating a visits model

Assuming we have data on tourist visits to sites and information on the independent variables described above, we can estimate Eq. (11.1) across all potential sites for each combination of activity and tourist type, using standard regression techniques. Or,

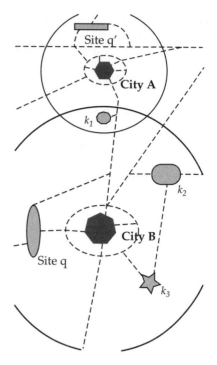

Figure 11.6   Calculating the impact of local tourism substitutes. The neighborhood of site q for stream fishing includes alternate tourist sites k_1, k_2, and k_3 and city B (all sites within the large black circle centered on city B). In other words, people from city B typically choose among q, k_1, k_2, and k_3 when planning a day of fishing. The value of S_q in this case will depend on the supply of environmental and infrastructure attributes at sites k_1, k_2, and k_3 and the distances between city B and the alternate tourism sites (given along road highways, as indicated by dashed lines). The neighborhood of more northern site q' includes alternate tourist site k_1 and city A (both within the black circle centered on city A).

given that all activities in a site will be affected by similar unobserved factors (e.g., the macroeconomic conditions in the demonstration site's country, the macroeconomic conditions in the countries that supply the tourists, country stability), it may be most appropriate to pool all activity and tourist type models. A promising technique for this approach is the seemingly unrelated regression framework (see Greene (2003) for details). In Section 11.4.1 we fit Eq. (11.1) to data for the Willamette Valley in Oregon, USA.

11.4   Tier 1 and 2 use value

A monetary value of tourism at each site can be generated by multiplying the number of observed (tier 1) or estimated (tier 2) visits by an average value per visit (or visitor day). The value of a tourist visit to a site equals the benefit generated by this visit over and above the costs of this visit. This measure of value is known as consumer surplus (Loomis 2005). In general, consumer surplus can vary by site, type of tourist, and tourist activity. By aggregating the consumer surplus generated by each visit to a site over the course of a year we calculate a site's annual value (effectively the value lost if access to the site is removed). In addition, using the tier 2 methodology, an estimate of the monetary effect of a change in environmental attributes can be developed by comparing the predicted value (trips multiplied by value per trip) at the site under the baseline conditions against the predicted value at the site after a change in environmental attribute.

Let the annual value of tourism at site q be given by AVT_q where

$$AVT_q = \sum_{a=1}^{A} \sum_{\omega=1}^{\Omega} V_{qa\omega} T_{qa\omega}, \qquad (11.4)$$

where $T_{qa\omega}$ is trips as above, $V_{qa\omega}$ is the average consumer surplus generated by a visit to site q to participate in activity combination a by tourist type ω (drop all a and/or ω *subscripts* if we do not have data by activity combination or type).

We can measure the change in AVT_q due to a change in environmental attributes at q by using the regression-estimated Eq. (11.1) to predict changes in visitation with a change in environmental attributes, and assuming that consumer surplus per trip

remains constant. If the landscape has changed such that environmental attributes in q have changed, represented by $\Delta \mathbf{EA}_{q'}$ then the expected change in predicted $T_{qa\omega}$ is given by

$$\Delta \hat{T}_{qa\omega} = f_{a\omega}(\Delta \mathbf{EA}_q, \mathbf{X}_q, G_{qa}, \Delta S_{qa}), \qquad (11.5)$$

where $f_{a\omega}$ indicates the regression-estimated Eq. (11.1) for combination a, ω, and ΔS_{qa} indicates any change in the substitute index due to the change in $\Delta \mathbf{EA}_q$ (a change on the landscape that changes $\mathbf{EA}$ at q could also change $\mathbf{EA}$ at some other q). $\Delta \mathbf{EA}_q$ can include just one change (e.g., the change in water quality at q) or multiple changes (e.g., the change in water delivery and quality at q). We can also measure $\Delta \hat{T}_{qa\omega}$ due to changes in $\mathbf{X}_q$ and G_{qa}.

The expected change in monetary value at site q due to the change in $\Delta \hat{T}_{qa\omega}$ is,

$$\Delta AVT_q = \sum_{a=1}^{A} \sum_{\omega=1}^{\Omega} [\hat{V}_{qa\omega}(\Delta \hat{T}_{qa\omega} + T_{qa\omega})] - [V_{qa\omega} T_{qa\omega}], \qquad (11.6)$$

where $\hat{V}_{qa\omega}$ represents the new value per trip after the change in environmental attributes on the landscape.

The landscape-level annual value of tourism and annual change in the value of tourism due to changes in environmental quality is given by summing AVT_q and $\Delta AVT_{q'}$ respectively, across all q on the landscape.

Finally, we can replace $V_{qa\omega}$ in the equations above with other values of interest, for example, expenditures in the region by tourists participating in activity combination a at site q, to generate other tourism-related economic information (e.g., the economic impacts on local businesses). See this chapter's text box (Box 11.1) for a discussion on tourism revenues generated by visits to Tambopata, Peru.

If we do not have consumer surplus estimates for the site we are studying, we may be able to use a consumer surplus estimate from a valuation study of a similar region as a proxy (for example, using an estimate of recreation values per day generated by a study in a nearby state). As noted in Chapter 3 there are limitations to benefits transfer approaches, thus primary studies are preferable. In this case we only transfer the value per unit of recreation activity (days, trips). For behavioral response (visits) to changes on the landscape, we use a locally calibrated

Box 11.1 How the economics of tourism justifies forest protection in Amazonian Peru

Christopher Kirkby, Renzo Giudice, Brett Day, Kerry Turner, Britaldo Silveira Soares-Filho, Hermann Oliveira-Rodrigues, and Douglas W. Yu

The province of Madre de Dios in south-east Peru, an Amazonian frontier region bordering Bolivia and Brazil, is renowned amongst scientists for its biologically and culturally rich landscape. One area of this region in particular, known as Tambopata, is now also firmly entrenched in the minds of international travellers as the quintessential Amazon rainforest destination, attracting 39 565 ecotourists in 2005. Tambopata's growing popularity is largely the result of (i) ease of access, only a 30-minute flight from Cusco (the gateway city to Machu Picchu) to Puerto Maldonado (the gateway town to Tambopata) followed by a few hours of river travel in motorised canoe; (ii) the proximity to two large protected areas, the Tambopata National Reserve (TNR, 274 690 ha) and the Bahuaja-Sonene National Park (BSNP, 1 091 416 ha); (iii) a healthy natural ecosystem showcasing primates, giant otters, macaws, and other charismatic fauna; (iv) intact oxbow lakes and clay-licks that concentrate fauna in predictable ways; (v) a choice of 37 ecotourist establishments (lodges), from 100-bed operations to small research stations and village guesthouses; and (vi) spending on international marketing by the larger lodges. In 2005, ecotourists spent a total of US$11.6 million to visit Tambopata, of which US$5.9 million were lodge revenues, US$5.2 million were airline revenues for airfares between Cusco and Puerto Maldonado, and US$0.5 million were TNR entrance fees and airport taxes. US$3.8 million (32.5%) of these revenues were in turn spent locally, in that the first-order transaction took place in Tambopata. This local spending could be further divided into low (12.2%, e.g., local staff, produce) and high leakage (20.3%, e.g., gasoline, boat motors) to the national economy. The TNR entrance fees exceeded the local park management budget, allowing US$172 530 to be transferred to the national budget. In 2005, the lodges earned a combined, after-tax profit of US$844 472. Additional (but only partly quantified) profits were distributed to some of the lodge owners in the form of above-market wages, possibly as much as doubling the above profit

figure. High levels of profitability, and the expectation of future profits, have created the incentive and the means for lodge owners to protect their businesses by protecting forest cover. Many lodges have taken advantage of government legislation, passed in 2002, that lets private businesses lease public forest outside of protected areas as concessions, for renewable 40-year terms. By 2005, lodges had taken control of 32 477 ha, with 90% of this area acquired by the four most profitable operators, together managing 8 lodges. Another 16 159 ha have been provisionally awarded as of early 2008, totalling 48 636 ha (Figure 11.A.1). Eight other lodges own and manage <100 ha each. In 2005, the pre-tax profit value of lodge-controlled lands corresponding to 12 fully operational lodges (for which economic and land-use data were available) was US$38.9 ha^{-1} [US$1 238 002/31 807 ha]. This figure exceeds the 2005 pre-tax profit value of titled lands, covering 10 2511 ha, that were managed by 200 generalist rural households for agriculture, fruit production, animal husbandry (cattle ranching, chickens, and pigs), and timber extraction, which was calculated at US$27.1 ha^{-1} [US$277 472/10 2511 ha]. However, when it comes to those households that specialize in a given land use (> 50% of household revenues), unsustainable cattle ranchers (with stocking rates >3.5 animals ha^{-1}, $N = 4$), who owned 262 ha, extracted a pre-tax profit value of US$39.0 ha^{-1}, whilst sustainable cattle ranchers (stocking <3.5 animals ha^{-1}, $N = 15$), who owned 1 154 ha, extracted US$35.8 ha^{-1}. We note, though, that cattle ranchers are generally located within 2 km of a road, for ease of transport, which limits the land area where this activity would compete with tourism. For households specialized in timber extraction or annual agriculture (i.e., rice, maize, manioc, and bananas), the pre-tax profit value of land is US$35.7 ha^{-1} and US$21.3 ha^{-1}, respectively. We expect that specialist producers, especially unsustainably stocked ranchers, will suffer reduced profits in the future as productivity drops, which should lower the present value of their profit stream below that expected for tourism. Our comparisons are also conservative in that we do not include the above-market-rate wages paid to lodge owners. One stated motive for lodges acquiring forest concessions is to exclude competitors from primary

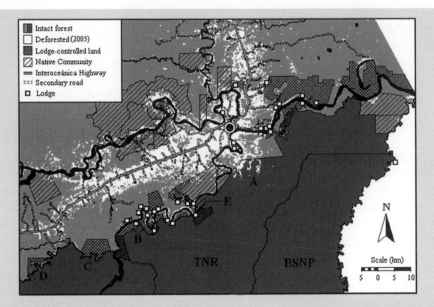

Figure 11.A.1   The location of lodge-controlled lands, made up of a mixture of designated ecotourism, conservation and Brazil-nut concessions, and concessions awaiting final approval, as of 1Q 2008, in relation to protected areas (TNR, Tambopata National Reserve; BSNP, Bahuaja-Sonene National Park), and areas of deforestation associated with Puerto Maldonado and the Interoceanica Highway. "A" depicts deforestation within the TNR, which is associated with the colonist communities of Jorge Chavez and Loero, corresponding to a 20-km wide gap between the two clusters of lodges. "B" and "C" are proposed ecotourism concessions and "D" is an ecotourism concession granted to a mestizo community that has historically mined alluvial gold deposits. "E" is a portion of forested land, located within the Native Community of Infierno, which though not controlled by a lodge has been set aside by the community for their ecotourism joint venture with a Lima-based tour company.

forest with valuable touristic features such as trail networks, oxbow lakes, and clay-licks. Another motive is that titles and concessions provide lodges with state-legitimized claims to forest parcels that they can defend via the legal system or direct action. Lodges have successfully sued and evicted loggers and miners from their concessions and have entered into benefit-sharing agreements with neighboring communities to cease extraction and hunting. Two lodges have entered into a joint venture with a community-based ecotourism project in return for monitored agreements to maintain forest cover and limit hunting in and around the proposed ecotourism concession (Figure 11.A.1, "B"). In one notable episode in 2007, the ecotourism industry added its weight to lobbying by Peru's conservation community and successfully reversed a government proposal to de-gazette a portion of the BSNP for oil exploration. The most serious threat to Tambopata's biodiversity is yet to come, however. In 2005, the government of Peru secured US$892 million to pave the Interoceanica Sur Highway, a westerly extension of the Trans-Amazon Highway that will connect Brazil to the Pacific Ocean. The highway will be completed in 2009 and will encourage deforestation along its length, thus directly threatening the lodges, which are on average only 18 km (8–62 km) from the highway, within the 50-km deforestation zone associated with paved roads in the Brazilian Amazon. Based on current behavior, Tambopata's ecotourism industry has the incentive and the means to continue protecting and even expand their concession holdings, but computer simulations indicate that even if lodges successfully maintain their concessions, deforestation will proceed through the unprotected gap between the two ecotourism clusters (Figure 11.A.1, "A"), ultimately threatening many of the lodges by degrading connectivity to the TNR. The gap area is less suitable for tourism, as it lacks oxbow lakes, so conservation in this section will require public investment, perhaps bolstered by collective action among the lodges.

production function (Eq. (11.1)). Loomis (2005) provides estimates of consumer surplus for US nature-based recreation and tourism, and Shrestha and Loomis (2001) provide estimates in other parts of the world. A roster of tourism consumer surplus databases is provided at http://recvaluation. forestry.oregonstate.edu/brief_history.html. Defenders of Wildlife and Colorado State University have constructed a "Toolkit" for benefit transfer that includes databases, meta-analyses and visitor use models: http://dare.colostate.edu/tools/benefittransfer.aspx. Finally, a database of values from around the world, called the Environmental Valuation Reference Inventory, can be found at http://www.EVRI.ca. The reader should be aware, however, and as we emphasize in Chapter 3, the evidence on the extent to which environmental benefit measures are transferable is mixed at best (Navrud and Ready 2007).

11.4.1 Example of tier 2 using data from the Willamette Basin

In this example we illustrate the estimation of a tier 2 visitation model and the value of a change in environmental attributes in state parks in the Willamette Basin. This model is very similar to the approach that Hill and Courtney (2006) use to estimate the effect of changes in landscape variables on visits to public and private forests in Great Britain.

First, we use Eq. (11.1) and data from 41 Willamette Basin state parks to assess the role of environmental attributes, infrastructure, and park location, on visitation rates. The model used in this example is a simplified form of Eq. (11.1) for two reasons: (1) visitation data are not stratified by activity combination a or tourist type, and (2) we did not model a substitution effect (S). We explain the number of visits to state park q in 2004, T_q, with seven factors. For the **EA** vector, we include presence/absence for four environmental attributes: (1) fishing possibilities, (2) wildlife viewing possibilities, (3) canyon features, and (4) boating facilities. We assume that a park with canyon features has dramatic views and good hiking opportunities. For the **X** vector, we include two infrastructure variables: the area of the park (assuming that larger sites have more access

points and may support more visits before becoming congested), and the presence/absence of historic sites. For G, we approximate the overall cost of accessing each site with a variable that measures the population living within 20 miles of the park (i.e., C_{ia} in Eq. (11.2) is replaced by C_q, the population living within 20 miles of q, $\sigma = 1$, $e^{-d_{iq}} = 1$ for all i, q, and $\beta_a = 1$ for all q).

We estimate the visitation model with ordinary least squares regression (Table 11.1). Fishing and wildlife opportunities are highly related (i.e., parks tend to have both or neither), so we estimate the model twice, once with each of these variables included. We found that canyon features increase park visitation. Somewhat surprisingly, fishing and wildlife viewing opportunities do not explain visitation in a statistically significant manner. Parks that are less costly for more people to access (as indicated by population around the park) are used more. Based on these results one might assume that important environmental attributes do not contribute to tourism use. What is more likely is that there is relatively little variation in these variables over the sites, particularly when they are measured using coarse indicators like presence/absence. This lack of variation in ecosystem-based variables is a common challenge in econometric studies of this type (e.g., Adamowicz et al. 1994). Improved information on levels of environmental attributes will help alleviate these problems.

Though this illustrative model is fairly simple, it reveals a great deal about the marginal value of environmental attributes at state parks in the Basin. For example, if a site that currently supports boating experienced a change in water delivery and/or quality to the degree that boating could no longer be supported, then, all else equal, the average site is predicted to lose between 65 000 (model 1) and 97 000 (model 2) visitors a year (the average $\Delta \hat{T}$ across all sites given the loss of boating). Given that the average visitation rate at the parks is about 200 000 visitors per year, this is a substantial loss. If we could obtain a measure of the consumer surplus in the Basin before and after a potential loss of boating in a state park, given by V_q and $\hat{V}_q$, respectively, we could derive coarse estimates of lost value using Eq. (11.6).

Table 11.1   Ordinary least squares regression model of annual visits to Oregon State Parks

Variable	Coefficient (*t*-stat)	
	Model 1	Model 2
Constant	36 870.47	59 669.76
	(1.34)	1.93***
Area of park (in km²)	14 328.77	16 504.51
	(3.72)*	(4.31)*
Population (in thousands) within 20 miles of park	109.68	109.19
	(2.13)**	(2.15)**
The presence of a historical site at the park (HISTST, 0/1 indicator variable)	155 545.83	157 442.03
	(2.66)*	(2.73)*
The presence of canyons as a natural feature of the park (CANYONS, 0/1 indicator variable)	222 907.70	203 305.19
	(3.24)*	(3.05)*
The opportunity for boating (BOATING, 0/1 indicator variable)	64 352.87	96 905.34
	(1.58)***	(1.96)**
The opportunities for wildlife viewing including birdwatching and other wildlife (WILDLFW, 0/1 indicator variable)	29 581.59 (0.78)	
The opportunities for fishing (FISHING, 0/1 indicator variable)		−43 391.51
		(−0.98)
R^2	0.78	0.78
R^2-adjusted	0.74	0.75
N	41	41

* Significant at a 0.01 level.
** Significant at a 0.05 level.
*** Significant at a 0.15 level.

11.5   State-of-the-art tourism model

The limitations associated with the simple overlay approach in tier 1 or regression model approach in tier 2 point to the utility of the development of state-of-the-art or tier 3 models. Tier 3 begins with the individual tourist as the core unit of analysis, and models human behavior from the standpoint of the decision on where and when to visit, rather than as a statistical analysis of aggregate visitation rates at parks. Note that the notation used in this section deviates somewhat from that above in order to be consistent with the published literature.

Consider an individual living in the Willamette Basin. This individual has a set of available choices of state parks, as well as other tourism and recreation options within and outside the Basin. The individual also has other non-tourism options competing for her time. The individual is assumed to make choices of how many tourism and recreation trips to make (in a season or a year) as well as the trip destination.

The basis for these choices is a preference function that is defined on characteristics of the individual (income, family size, etc.) as well as the attributes of the available options (distance to the parks, attributes at the parks, etc.). Two types of models have been examined in the literature: the Kuhn–Tucker model (Phaneuf 1999; Phaneuf *et al.* 2000) and the random utility model, or RUM (see Phaneuf and Smith 2005).

The Kuhn–Tucker model assumes that the individual *n* obtains utility from a set of trips to recreation sites where $q = 1, \ldots, Q$ indexes the set of sites available and $(x_1, \ldots, x_Q)$ indicates the number of visits to each site. Associated with these sites is a set of attributes $(b_1, \ldots, b_Q)$ where each element *b* contains a number of characteristics of the sites, including those related to environmental attributes. The individual's preference function includes a set of "other" activities that generate utility (all other activities are accumulated into a single element *z*). This results in a preference function $U(x_1, \ldots, x_Q, b_1, \ldots, b_Q, z)$. The individual *n* is assumed to maximize utility (U_n) subject to income (m) over the period and the prices of accessing the recreation alternatives ($p_1, \ldots, p_Q$; usually travel cost plus the opportunity cost of time plus site access costs) as well as the price of the "good" z (which is normalized to unity). The individual maximizes utility by choosing the number of trips to each site q, given by $x_1, \ldots x_Q$ where some or all x can be 0:

$$\max_{x_1, \ldots, x_Q, z} U(x_1, \ldots, x_Q, b_1, \ldots, b_Q, z) \qquad (11.7)$$

Subject to $p'x + z = m$.

The model provides estimates of the relationship between travel costs, attributes, income (and other demographic characteristics) and the frequency of visits to each of the defined alternatives. Typically, x_q^*, the number of visits to q that solves problem (11.7), will decrease as the environmental attributes at q decreases, all else equal.

The random utility model (Phaneuf and Smith 2005) examines the individual's choice on a particular occasion (a single trip). Rather than assessing the number of trips, the utility or satisfaction associated with visiting an alternative is described as a function of the travel costs and site environmental and infrastructure attributes (and potentially individual specific characteristics such as income, demographics, etc.). The utility (or preference) function for the choice to visit to site q (from individual n's viewpoint) takes the form:

$$V_{nq} = V(m - p_q, b_q). \qquad (11.8)$$

A common approach is to model the probability of individual n visiting a particular alternative q (or $0 \le \pi_{nq} \le 1$) as

$$\pi_{nq} = \frac{e^{V_{nq}}}{\sum_{k=1, \ldots Q} e^{V_{nk}}}, \qquad (11.9)$$

where the V_{nq} function would have to be estimated using observed data in a regression analysis (Train 2003).

Equation (11.9) expresses the probability that an individual visits a particular alternative (on a particular choice occasion—e.g., a day, a week) as a function of the attributes and prices. If the supply of

a desirable environmental attribute at site q declines then the probability of choosing site q for tourism declines. According to Eq. (11.9), as quality declines at one site, visitation will increase at other sites. This addresses one of the major limitations of the tier 2 model: the difficulty with satisfactorily modeling substitutability when explaining tourism visits (the index S_{qa}). Further, because this model incorporates the preferences of the individuals in the conditional indirect utility function (V) it can also be used to calculate the economic value of the change directly. In this case the value of the decline at site q is given by the amount of money it would take to make person n as well off as they were before the change at q; this can be calculated using the utility expression in (11.8) over the entire set of sites available. The relatively simple model can be combined with ecological models that describe the linkages between attributes to create an integrated ecological—economic model. An example of such a model is Naidoo and Adamowicz (2005) in which a behavioral model of tourism site choice is integrated with an ecological model of landscape change and bird diversity.

There have been many significant advances in the modeling framework outlined above. Some of the most significant include: (1) accounting for unobserved attribute effects in the model (Murdock 2006); (2) incorporation of congestion into the model as an example of interdependence between tourists (Timmins and Murdock 2007); (3) the incorporation of time (habits, variety seeking) into the framework (e.g., Swait *et al.* 2004); (4) inclusion of preference heterogeneity among tourists (Scarpa *et al.*, forthcoming; Boxall and Adamowicz 2002; Train 1998, 2003); and (5) the development of models that account for different alternatives in the choice set or set of sites that a tourist considers (Haab and Hicks 1997; von Haefen 2008) as well as a host of other advances in the modeling of choice data.

11.6   Limitations and next steps

We have outlined three approaches (tiers 1, 2, and 3) for assessing the value of nature-based tourism and recreation on the landscape and the changes in value that could be expected with a change in environmental attributes on the landscape.

Tier 1 provides an assessment of observed visitation rates to sites. If information on the value of a unit of visitation is available from other studies (benefits transfer) then these visitation rates can be used to approximate measures of economic value. The tier 1 methodology does not specify a relationship between changes in environmental attributes and changes in tourism value nor does it disentangle effects on visitation rates from environmental attributes, infrastructure, distance to population centers, and availability of substitute sites.

Tier 2 approximates visitation behavior by developing a statistical relationship between visitation and attributes at the sites. This approach can provide additional insight into the changes that may be expected in visitation rates and values if environmental attributes change on the landscape. Tier 2 approaches, however, suffer from several limitations.

First, the required data are usually highly correlated, and attributes often do not vary strongly among sites. Sites with good fishing quality also tend to have boating, picnicking, and other facilities. And sites often share many features, precluding the opportunity to identify the impact that these features have on visitation rates. Identifying the impact of the change in environmental attributes in such cases will be difficult. Increasing the number of parks or expanding the spatial extent of the range of parks may help, but these actions will also increase the complexity of the research task.

Second, there are few linkages to information about the tourists. Only the potential tourism population is included in the model. Factors such as specific travel and time costs of visiting (instead of an index), incomes, experience levels of the tourists, the substitution between tourism and other uses of time, and other demographic features are not included.

Third, tier 2 models will generally rely on benefits transfer procedures to provide estimates of economic value rather than estimating the values from the population of interest. Evidence on the applicability of values from one site to others is decidedly mixed, but may be the only option (Navrud and Ready 2007).

Finally, the models described here are static and do not reflect trends in preferences, demographics or other factors that might influence visitation rates. For example, Pergams and Zaradic (2006, 2008) argue that there have been widespread declines in nature-based recreation visits. Meanwhile Balmford *et al.* (2009) find that visits to protected areas in most parts of the world are in fact increasing. These are clearly areas for further research.

Tier 3 generates value estimates based on the behavior of the individuals. It is a fully integrated model of tourism and the environment. However, it is also very demanding in terms of data requirements and familiarity with sophisticated modeling techniques. As the model is individual based it requires information on the individual's residential location (for the determination of travel and time costs), the set of sites the individual considers when planning trips, as well as information on the attributes of the sites. There are several "scale" issues including assessment of the relevant geographical scale (how large is the area that is relevant for the demand for recreation tourism at a particular site) and the relevant time scale (is an annual time scale appropriate for decision making or are the trips more seasonal or perhaps a one-every-5-years type of trip; is there a broad trend of declining participation in recreation and preferences for nature?). Only a few regions will have the data available for such analysis.

Ideally the data for tier 3 models would be collected from general population surveys. Data collection of this type could provide excellent sampling properties and would provide information on the total number of visits taken by a population as well as the sites selected. However, such data are rarely collected. An alternative is to sample at the recreation sites and collect information on the participants. This is known as choice-based sampling (Ben-Akiva and Lerman 1985) and is commonplace in the transportation demand literature. This approach may provide the most practical solution for the development of tier 3 models.

An additional area that has not been investigated to any great extent is the feedback between visitation rates, congestion and environmental quality. There has been some assessment of the role of congestion in nature-based tourism, and examination of the impact of tourism on environmental quality, but little examination of the three elements jointly. These issues as well as the continuing evolution of niche markets for tourism and emerging trends form the basis for a rich research agenda.

Acknowledgments

Thanks to Isla Fishburn, University of Sheffield, for collecting much of the Willamette data used in this chapter.

References

Adamowicz, W. L., Louviere, J., and Williams, M. (1994). Combining stated and revealed preference methods for valuing environmental amenities. *Journal of Environmental Economics and Management*, **26**, 271–92.

Balmford, A., Beresford, J., Green, J., *et al.* (2009). A global perspective on trends in nature-based tourism. *PloS Biology*, **7**(6), e1000144. doi:10.1371/journal.pbio.1000144.

Ben-Akiva, M., and Lerman, S. R. (1985). *Discrete choice analysis: theory and application to predict travel demand.* MIT Press, Cambridge.

Bockstael, N. E., and McConnell, K. E. (2007). *Environmental and resource valuation with revealed preferences: a theoretical guide to empirical models.* Springer, Dordrecht, The Netherlands.

Boxall, P. C., and Adamowicz, W. L. (2002). Understanding heterogeneous preferences in random utility models: a latent class approach. *Environmental and Resource Economics*, **23**, 421–46.

Champ, P. A., Boyle, K. J., and Brown, T. C. (2003). *A primer on nonmarket valuation.* Kluwer, Dordrecht, The Netherlands.

Greene, W. H. (2003). *Econometric analysis,* 5th edn. Prentice-Hall, Upper Saddle River, NJ.

Haab, T. C., and Hicks, R. L. (1997). Accounting for choice set endogeneity in random utility models of recreation demand. *Journal of Environmental Economics and Management*, **34**, 127–47.

Hill, G. W., and Courtney, P. R. (2006). Demand analysis projections for recreational visits to countryside woodlands in Great Britain. *Forestry*, **79**, 185–200.

Hulse D., Gregory, S., and Baker, J., Eds. (2002). *Willamette River Basin planning atlas: trajectories of environmental and ecological change.* Oregon State University Press, Corvallis.

Loomis, J. (2005). Updated outdoor recreation use values on national forests and other public lands. General

Technical Report PNW-GTR-658. US Department of Agriculture, Forest Service, Pacific Northwest Research Station, Portland, OR.

Murdock, J. (2006). Handling unobserved site characteristics in random utility models of recreation demand. *Journal of Environmental Economics and Management*, **51**, 1–25.

Naidoo, R., and Adamowicz, W. L. (2005). Economic benefits of biodiversity exceed costs of conservation at an African rainforest reserve. *Proceedings of the National Academy of Sciences of the USA*, **102**, 16712–16.

Navrud, S., and Ready, R. (2007). *Environmental value transfer: issues and methods.* Springer, Dordrecht, The Netherlands.

Nelson, E., Mendoza, G., Regetz, J., *et al.* (2009). Modeling multiple ecosystem services, biodiversity conservation, commodity production, and tradeoffs at landscape scales. *Frontiers in Ecology and the Environment*, **7**, 4–11.

Oregon Fish and Wildlife Department (OFWD). (2005). 2005 Big Game Statistics.

Oregon Geospatial Enterprise Office (OGEO). (2008). Oregon geospatial data clearinghouse. Accessed at http://www.oregon.gov/DAS/EISPD/GEO/alphalist.shtml in 2008.

Pacific Northwest Ecological Research Consortium (PNW-ERC). (2008). The datasets. Accessed at http://www.fsl.orst.edu/pnwerc/wrb/access.html.

Pergams, O. R. W., and Zaradic, P. A. (2006). Is love of nature in the US becoming love of electronic media? 16-year downtrend in national park visits explained by watching movies, playing video games, internet use, and oil prices. *Journal of Environmental Management*, **80**, 387–93.

Pergams, O. R. W., and Zaradic, P. A. (2008). Evidence for a fundamental and pervasive shift away from nature-based recreation. *Proceedings of the National Academy of Sciences of the USA*, **105**, 2295–300.

Phaneuf, D. (1999). A dual approach to modeling corner solutions in recreation demand. *Journal of Environmental Economics and Management*, **37**, 85–105.

Phaneuf, D., Kling, C., and Herriges, J. (2000). Estimation and welfare calculations in a generalized corner solution model with an application to recreation demand. *Review of Economics and Statistics*, **82**, 83–92.

Phaneuf, D., and Smith, V. K. (2005). Recreation demand models. *Handbook of Environmental Economics*, **2**, 671–751.

Scarpa, R., Thiene, M., and Train, K. (forthcoming). Utility in willingness to pay space: a tool to address the confounding random scale effects in destination choice to the Alps. *American Journal of Agricultural Economics*, Appendices.

Shrestha, R. K., and Loomis, J. B. (2001). Testing a meta-analysis model for benefit transfer in international outdoor recreation. *Ecological Economics*, **39**, 67–83.

Swait, J., Adamowicz, W., and Van Bueren, M. (2004). Choice and temporal welfare impacts: Incorporating history into discrete choice models. *Journal of Environmental Economics and Management*, **47**, 94–116.

Timmins, C., and Murdock, J. (2007). A revealed preference approach to the measurement of congestion in travel cost models. *Journal of Environmental Economics and Management*, **53**, 230–49.

Towner, J., and Wall, G. (1991). History and tourism. *Annals of Tourism Research*, **18**, 71–84.

Train, K. (1998). Recreation demand models with taste variation. *Land Economics*, **74**, 230–9.

Train, K. (2003). *Discrete choice methods with simulation*. Cambridge University Press, Cambridge.

Von Haefen, R. H. (2008). Latent consideration sets and continuous demand systems. *Environmental and Resource Economics*, **41**, 363–79.

World Tourism Organization. (2008). *UNWTO World Tourism Barometer*, **6**(2, June). Accessed at http://www.unwto.org/facts/eng/barometer.htm.

CHAPTER 12

Cultural services and non-use values

Kai M. A. Chan, Joshua Goldstein, Terre Satterfield, Neil Hannahs,
Kekuewa Kikiloi, Robin Naidoo, Nathan Vadeboncoeur,
and Ulalia Woodside

12.1 Introduction

12.1.1 Defining cultural ecosystem services and non-use values

In the ongoing effort to better define ecosystem services (MA 2005; Boyd and Banzhaf 2007; Costanza 2008; Fisher and Turner 2008; Wallace 2008) and their valuation, few classes of value have been more difficult to identify and measure than those connected with the cultural and non-use dimensions of ecosystems. Whereas other ecosystem services make life possible in biophysical terms, cultural ecosystem services and non-use values inspire deep attachment in human communities. Accordingly, they need to be integrated into conservation and broader policy if societies are to sustain meaningful links between people and nature, and indeed many say if societies are to sustain themselves at all.

Cultural services and their connected values have come to represent nonmaterial benefits derived by those who use, might use, or share an attachment to a [natural] area. Costanza *et al.* (1997) defined cultural values-cum-service as "aesthetic, artistic, educational, spiritual and/or scientific values of ecosystems" (p. 254). The Millennium Ecosystem Assessment (2005, p. 894) expanded this definition to include the "nonmaterial benefits people obtain from ecosystems through spiritual enrichment, cognitive development, reflection, recreation, and aesthetic experience, including, e.g., knowledge systems, social relations, and aesthetic values." Others focus more fully on the educational benefits of cultural services. Chiesura and de Groot (2003), for example, delineate key socio-cultural dimensions of natural capital as scientific, cultural, historical, religious, and artistic educational benefits or flows.

More recently, the "cultural" class of value has come to encompass the concerns of indigenous peoples who might have a political and moral right to a natural area, such as a Treaty-based right to traditional territory. Hence, a number of studies in the values literature have begun to define such intangible things as place value (Basso 1996; Norton and Hannon 1997; Brown *et al.* 2002), spiritual value (Milton 2002), heritage value (Throsby 2001), and social-relational value (Lin 2001; Sable and Kling 2001).

Here we define cultural services inclusively as *ecosystems' contribution to the nonmaterial benefits (e.g., capabilities and experiences) that arise from human-ecosystem relationships.* For example, cultural services include the contribution of ecosystems to recreational experiences, to sense of place, and to the knowledge that a valued nonhuman species exists or will exist for future generations (knowledge that has existence value or bequest value, respectively). For clarity, and to align this chapter with other chapters in this book, we adopt the economics convention of differentiating values into use and non-use values. Use value refers to the direct (consumptive and non-consumptive) and indirect uses of ecosystem goods and services, while non-use encompasses all values separate from use (Goulder and Kennedy 1997). Our treatment of cultural ecosystem services includes both use and non-use values. Indeed, for all cultures, maintenance of cultural values requires the active use of biological resources and nonmaterial interaction with important sites, among other direct and indirect uses.

Non-use values can be categorized into component values, which often include existence value, bequest value, and option value. Krutilla (1967), an economist, first introduced the concept of "existence value" to capture the provision of "satisfaction from mere knowledge that part of wilderness North America [sic] remains" regardless of any intention the valuing agent has of ever visiting such destinations. More broadly, existence value can be designated geographically as well as in reference to specific biophysical units such as a particular species or ecosystem. While existence value is specific to an individual's satisfaction, bequest value relates to knowing that an environmental amenity will be available for future generations. Option value represents "the premium that people are willing to pay to preserve an environmental amenity, over and above the mean value (or expected value) of the use values anticipated from the amenity" (Goulder and Kennedy 1997; Chapter 2). While option value has a stronger connection to use values, non-use values is the broad category capturing non-users' preferences for the continued existence of ecosystem goods and services (Cicchetti and Wilde 1992).

The property of *intangibility* is central to cultural ecosystem services and non-use values and often renders them difficult to classify and measure. Classification and measurement of these values and services are nonetheless necessary as the risks associated with their loss are central to public thinking and discontent with land management decisions (Satterfield and Roberts 2008). An analogous "intangibility" case in point for many First Nations communities in Canada is enshrined in the concept "cultural keystone species" (Garibaldi and Turner 2004). The term refers to species, for example, salmon in the Pacific Northwest, which, if diminished, result in not just material loss of a provisioning service (the salmon), but a much larger suite of linked cultural losses or impacts including enduring practices. These may include food sharing and the social cohesion and alliances engendered by such exchanges; the transmission of traditional knowledge; and the maintenance of key cultural institu-

tions such as naming and gifting ceremonies and feasts wherein keystone foods are central (see also Turner *et al.* 2008).

The non-market nature of many cultural services and non-use values creates another important characteristic. Whereas many other services are the production of *things* or *conditions* that have value in relation to some market (supporting services do not produce marketed commodities, but they may be essential steps in the production of such commodities, such as the pollination of agricultural crops), cultural services and non-use values generally involve the production of *experiences* that are valued without entering markets. In such cases, the production and valuation are intimately linked, both occurring in part in the valuer's mind. These services are therefore co-produced by ecosystems and people in a deeper sense than other services. For example, if we consider the contribution to aesthetic experiences, there is no metric of production that does not involve the valuer. In other words, potential supply and realized supply cannot be differentiated as they can be for market services. There is no objective way to claim that a site provides a great aesthetic service except by appealing to people's behaviors and stated preferences regarding the object, which must generally be assessed for each site in turn.

While many assume that what matters gets measured (Meadows 2001), this is not so with cultural services. Their consideration is long overdue. In the following sections, we lay a conceptual background from which we identify components of cultural and non-use values and then link these values to ecosystem services.

12.1.2 From values to valuation: methodological conundrums

To the extent that the definition of cultural values has been controversial, the same may be said for methodological efforts to value them. While economics continues to provide important conceptual and methodological approaches for ecosystem service valuation (e.g., Champ 2003; National Research Council (US) 2005), there is another dialogue raising concerns that the common approach of expressing values in dollar metrics will result in the cultural and intangible dimensions of land management being improperly considered, or even left out altogether (Rees 1998; Gowdy 2001; Wunder and Vargas 2005; McCauley 2006). Most scholars agree that all dimensions of benefit should be recognized and valued, but they differ sharply on the accuracy and efficacy of using a single currency, the dollar, to appropriately measure multiple kinds of value (Brown 1984; Lockwood 1998; Martinez-Alier *et al.* 1998; Sagoff 1998).

The need to match valuation methods to relevant kinds of value is best illustrated through the controversy over the validity of "stated preference" approaches to valuation of ecosystem services (e.g., willingness to pay, WTP). For example, problems of validity have arisen with stated-preference approaches aimed at dollar valuation, because often survey respondents are expressing not a willingness to pay, but rather a willingness to contribute to a moral cause (Kahneman and Knetsch 2005). Opposing kinds of values are in operation in these judgments. Some kinds of value (virtues or principles) violate the assumptions of evaluation methods for preference value. By virtues and principles, we mean moral values associated with a person's intent, duties, and rights—notions of rightness or wrongness of people or actions *themselves* as opposed to their resulting consequences. Social scientists have provided evidence for this claim by demonstrating that expressions of value are rooted both in preferences (which address what a person values or will "pay for" because it produces outcomes that benefit him or her) and in virtues and principles (which address what a person believes to be right for nature and society) (Sagoff 1998). Two kinds of problems with WTP studies can be traced to the influence of principle-based values on people's responses to questions: "protest zeros" (which are often a rejection of dollars as the appropriate metric) and what are known as scaling or scoping problems (wherein respondents in a valuation survey are insensitive to quantity of the thing valued). In the former case, respondents resist the survey format by entering a 0, often stating that they find the question inappropriate. In the latter case, respondents do not distinguish between the value

of one lake versus five, in some cases because they feel that protecting lakes is the "kind of thing they should support." As such, these respondents offer a dollar amount that is in fact a proxy for a donation and not an expression of market value per se.

In order to match relevant values to appropriate methods and to differentiate values pertaining to cultural services from those not related to services, we distinguish eight dimensions of cultural and non-use values (Chan *et al.*, in prep.):

1. Preferences versus principles versus virtues (concern for ends versus means versus intent)
2. Market-mediated versus non-market-mediated (derived from contribution to a marketed commodity versus to something valued intrinsically)
3. Self-oriented versus other-oriented (for one's own versus for others' enjoyment)
4. Individual versus holistic/group (held by individuals versus groups or larger wholes; an example of the latter is the community value of cultural integrity)
5. Physical versus metaphysical (stemming from concrete physical experiences versus conceptual experiences; an example of the latter is existence value)
6. Supporting versus final (valued for its contribution to another value versus valued intrinsically; e.g., education might be valued for its contribution to other values, while spiritual value is intrinsically valuable; this distinction is also broadly relevant to biophysical ecosystem services—for example, pollination services support the production of agricultural products, a final service)
7. Transformative versus non-transformative (valuable because of its contribution to changes in values and perspectives versus valuable in relation to unchanging values/perspectives)
8. Anthropocentric versus bio/ecocentric (for human beings versus for all living organisms/biotic communities and ecological processes)

Not all kinds of values above pertain to cultural ecosystem services. From an ecosystem service perspective, values are a measure of the importance of a thing or experience; but some values cannot be understood as such because they are underlying ideals (virtues and principles, which overlap closely with held values—Brown 1984). Others are not considered to be products of ecosystem services because the services framework is focused on people (so bio- and ecocentric values are excluded). Whereas the eight dimensions of *kinds* of values are crucial for determining appropriate *kinds* of valuation, it is also worth distinguishing *categories* of values associated with cultural ecosystem services and non-use dimensions of ecosystems. These ten proposed general categories of values are a way of organizing the *nature of reason* that a thing or experience is valued (e.g., because it provides knowledge, or social capital). Some of the values are neither strictly non-use nor really "cultural" values (e.g., activity value); they are included here because they are crucial components of services in these categories (e.g., recreation, subsistence).

12.1.3   Distinguishing values from services

We distinguish values from services (the production of things of value) for two principal reasons (Table 12.2). First, values do not exist as entities for probing or characterizing as separate from objects and activities; rather, they are merely one way in which we organize our ideas about morality and preference. Accordingly, they generally require some concrete thing or activity to enable their elicitation (as in valuation): any attempt to ask people how much they value a landscape/seascape for its contribution to knowledge is likely to be far too abstract to get useful and meaningful answers. In our typology, we propose that services be the contribution to those concrete activities that ground values and provide a forum for their expression (or non-activities like contemplation, in the case of metaphysical services). So why not simply stop at services—why do we need to identify values at all? This brings us to our second reason.

Second, individual cultural ecosystem services simultaneously produce multiple intertwined values, so any attempt to valuate or characterize services must account for these interdependencies (see Section 12.2.1). With many cultural services, a one-to-one mapping from services to values is challenging, and sometimes inappropriate. Certain ecosystem services, such as the provision of water, can be distinguished in such a way that each service produces one value: e.g., one might distinguish the provision of water for irrigation from the provision of water for household use, calling each a separate ecosystem service with a separate use. With many cultural ecosystem services, a one-to-one mapping is awkward because component values of activities/objects generally cannot be properly separated in people's minds. For example, the physical-sustenance value of salmon to First Nations cultures of British Columbia cannot be effectively decoupled from the social and cultural values of the harvest and the ceremonies that depend upon this cultural keystone species (Garibaldi and Turner 2004). Perhaps more importantly, disentangling co-occurring values can be deeply offensive to people and antithetical to their cultures.

Another example of the difficulty of separating values can be drawn from the cultural world of Hawaiians. Kalo (taro), is referenced in the ancient cosmogonic history of the Hawaiian Islands not only as a prominent staple crop, but also as emerging from the body of Haloa-naka-lau-kapalili (Haloa-of-the-quivering-leaf), the first-born son of Wakea and Ho'ohokukalani. Stillborn, his body was returned to the earth and grew into the first kalo plant in Hawai'i. Their second-born son, named Haloa in honor of his older sibling, became the progenitor from whom all kanaka maoli (indigenous Hawaiians) descend. These are not just cosmological narratives, but are equally reflective of what, for lack of appropriate terms in English, reflect Hawaiians' genealogical or kin-like connections to both the human and nonhuman world. Such meanings—inscribed in and inextricably linked to physical places or goods of importance (in this case, kalo)—are crucial products of metaphysical ecosystem services. These may be as important to Hawaiian people today as they were to ancestors who planted the first fields of kalo over a hundred generations ago. They also represent an important sense of responsibility to all of one's ancestors: cosmological (Haloa), human, and nonhuman (e.g., kalo), and thus a

Table 12.1   Categories of cultural ecosystem services and associated benefits, and the site substitutability of each—with one possible mapping of ecosystem services to benefits

Service Contribution to... (experiences)	Benefits											Site Substitutability
	Place/heritage	Activity	Spiritual	Inspiration	Knowledge	Existence, Bequest	Option	Social capital and cohesion	Aesthetic	Employment	Identity	
Subsistence	x	x	x	x	x			x	x	x	x	varies
Recreation	x	x	x	x	x			x	x	x	x	depends
Education & research	x	x	x	x	x	x	x	x	x	x	x	depends
Artistic	x		x	x	x				x	x	x	varies
'Ceremonial'	x		x	x	x			x	x		x	varies
Site substitutability	*low*	*high*		*varies*		*depends*		*high*	*low*	*high*	*varies*	

Site substitutability is *low* if the service or value is linked directly to particular places, and *high* if not; it *depends* on whether there are clearly identifiable and understandable qualitative differences in instances within a category (e.g., existence value may be site-substitutable for valued species, but not for sacred sites); or it *varies* if the logic of the variation is more complex (e.g., the provision of subsistence opportunities is not site-specific for activity values, but it may be for place values). The contributions to employment and subsistence experiences are directly linked to provisioning services but the experiential portions are included here as a benefit that has not been (and likely would not be) captured effectively by strict market valuation. For example, the value of employment transcends its contribution to aggregate economic values: consider the intangible value of employment that allows a person to care for and sustain a relationship to a place or resources of cultural or spiritual significance.

need to care for this heritage plant as one would care for family. Reducing this complex, spiritual kinship to an economic currency is tantamount to pricing one's great-grandparent. Not only is the measure difficult to derive, the effort to derive it could be construed as deeply offensive.

Table 12.1 illustrates that an individual service can provide multiple benefits. While we agree that it would be simplest to map one service to one benefit, we do not see this being possible with cultural services. Spiritual, inspiration, and place values are not products of single kinds of experiences; rather these benefits are products of all manner of experiences associated with ecosystems (including metaphysical contemplation). Rather than expecting that we can partition separate "spiritual" and "inspiration" categories of ecosystem service (de Groot *et al.* 2005) and value each category separately, we must recognize that the production of these benefits is a deeply complex function of many activities and components. For any service, each kind of benefit denoted by an "x" may contribute to the values that people assign to the service's pertinent activities and components (e.g., a person may enjoy hiking and berry-picking partly—and inseparably—be-

cause of the spiritual experience and the inspiration of photography). Valuation exercises must account for these multiple benefits.

An example may help illustrate the point. In a Hawaiian forest, trees are regarded as manifestations or the embodiment of deities, such that they simultaneously provide use values and spiritual values. One source for this spiritual belief arises from a perceptive grasp of the role the forest plays in sustaining hydrologic ecosystem services. Hawaiians say, "*Hahai no ka ua i ka ulula`au.*" Rains always follow the forest. Knowing this and honoring the deities' embodiments, Hawaiians hewed only the trees that were needed and ensured that a sufficient stand was left in place to perform a key function in the hydrological cycle. Sacred sites in Tibet also seem to be revered and protected in part because of their contribution to ecosystem services such as flood mitigation (Box 12.1).

We distinguish between categories of benefits and services based on the potential for one site to substitute for another in production of the benefit/service (Table 12.1). There is a difference between the variation across sites and the possibility for substitution: for example, the provision of employment

Box 12.1 The sacred geography of Kawagebo

Jianzhong Ma and Christine Tam

Kawagebo Mountain, rising 6740 m along the eastern Tibetan Plateau, lies at the heart of a region dominated by sacred geography at the headwaters of three of the great rivers of Asia—the Yangtze, the Mekong, and the Salween. Sacred natural sites abound across this 1600 km² area, from springs to lakes, rocks to caves, trees to whole forests. Inhabited for over 2000 years, the region gained stature primarily in the thirteenth century when the living Buddhas Karma Pakshi and Karma Rinpo Dorje paid formal respect to Kawagebo, leaving behind offerings of scriptures that elevated the mountain to one of the most sacred in all the Tibetan world (Ma 2005). More recently, the surrounding mountains and valleys, dotted with small Tibetan villages, were included in the area declared a World Heritage site in 2002 as the Three Parallel Rivers. This strength of culture based on sacred geography, in fact, lays the foundation and framework for protection of the extraordinarily rich biodiversity, critical life-support functions, stunning landscapes, and cultural traditions that attracted UNESCO here in the first place.

Tibetan sacred geography is an embodiment of the integration of cultural sites and natural ecosystems, beliefs arising from a mixture of Tibetan Buddhism and the local Bon folk religion that attributed all natural things with spirit. In the Kawagebo region, sacred geography is expressed in four main forms: sacred mountains, "rigua," other sacred natural sites, and pilgrimage routes. Sacred mountains, or "rida" in Tibetan, are the main form of sacred geography, numbering over 70 in the region, and are venerated by local clans, specific communities, the broader region, and in some cases by all of Tibetan culture. The sacred mountain of Kawagebo is actually comprised of five major Gods that are the five highest peaks: Miancimu (Goddess, 6 054 m), Jiewarena (Crests of Five Buddhas, 5 470 m), Buqiongsongjiewuxie (Prince), Mabingzalawangdui (Fight God for Subduing the Devil, 6 400 m), and Kawagebo itself (Major God, 6 740 m) (Xiroa 2001). Reverence for these mountains includes restricting use of resources on their slopes.

"Rigua" are traditional zones established to seal off mountain areas especially for sustainable resource utilization, strictly controlling normal production activities. Use restrictions range from prohibitions on tree felling and hunting, to limitations on herding and non-timber product (NTP) collection. These zones often coincide with protection of ecosystem service provisioning for communities, such as prevention of mountain slope erosion or protection of water resources. Indeed the ties between the belief system and the provisioning of ecosystem services are strong. Ubiquitous local stories cite the importance of sacred sites in securing human well-being. Adong village is one such example. The floodwaters of Zhili-Rongqu of Adenggong inundated a large stretch of farmland and its households on the lower reaches of the mountain during the 1990s, while the main channel near Adong experienced three floods in the 1980s. The community organized all its resources for flood prevention, but was unsuccessful. It was only when the sacred mountain was established and sealed from direct human activity that the floods stopped. In another example, the people in and around Waha Village, who traditionally harvested the forest around their headwater springs, suffered from a decrease in water supply. In the end, it was after inviting the Living Buddha Da-De to establish a sealed sacred mountain area encompassing the springs that the water supply returned to normal (KCA 2004). Biologically, sacred sites have also been found to encompass greater biodiversity than similar habitats at similar elevations that are not sacred (Anderson *et al*. 2005), suggesting the special protection powers that fears of divine punishment can offer.

Other sacred natural sites may include temple forests, lakes, waterfalls, and forests or grasslands near sacred marks, and number over 300 in the Kawagebo region. Lastly, two pilgrimage routes circle Kawagebo, and host thousands of devotees each year from across the Tibetan region to pray for good fortune as they circumambulate the sacred mountain God.

Culturally, Tibetans draw three main functions from these types of sites. They solidify their religious core, structure moral thinking and behavior, and strengthen group identity (Ma 2005). Veneration of sacred mountains originated from the awe and mystery surrounding the ever-changing natural environment of Tibetans who "live at the roof of the world," spawning a psychology of fear and respect for nature. These early mountain gods were elevated under Buddhism to symbols of Buddhist doctrine. At the same time, sacred natural sites play a moral role by shaping fixed sets of rules to be obeyed related to the landscape and its resources, thereby stabilizing community action and thought. Hillside zones or mountain tops have designated rules of use, and those who violate the rules will be punished by divinities. Lastly, these sites strengthen

continues

Box 12.1 *continued*

community cultural identity through common religion, values, beliefs, and living habits. Community activities and rituals surround these sites, resulting ultimately even in institutionalization of commonly held boundaries, such as the delineation between national and community land in China, or the international borders between China-India or China-Nepal (Ma 2005).

Most recently, vineyard conversion, infrastructure construction, tourism expansion, and other associated development have been intensifying in this area. With the declaration of the UNESCO World Heritage site, plans have become even more ambitious to transform this remote mountainous landscape into a thriving

modern economy. The key to successful development, however, most likely lies in the sacred geography of the region, for it is this cultural and religious belief system that sets out a protected area network much stronger in many ways than the formal nature reserve networks established by government entities. The framework for sustainable utilization through zoned restrictions and strict protection directed by sacred geography also helps to ensure the continued provisioning of ecosystem services to the millions of people living downstream of this area, not only in China, but in the neighboring countries of Myanmar, Thailand, Vietnam, Laos, and Cambodia.

opportunities might be high (temporarily) for forestry in one woodlot due to the old-growth forests that occur there; but there is high potential for substitution in theory because—for the purposes of employment—the site could be replaced by other sites. In the case of place value, in contrast, particular places are of value, so site substitutability is low. In some cases, site substitutability *depends* on the kind of opportunity provided: e.g., it depends on whether recreation is place-based because of some site-specific (e.g., historical) significance (low substitutability), or mainly because of a site's aesthetic beauty (high substitutability). Variation in site substitutability is less easily pinned down in cases such as the provision of artistic or ceremonial opportunities, which are often complex functions of site-specific and other attributes.

While researchers might benefit from an understanding of the categories of ecosystem services at play, and the attendant categories and kinds of values, we do not recommend belaboring the application of any typology. Several points follow and explain this position. First, these typologies are intended for researchers and analysts, not for constituents and stakeholders. Those who are interviewed do not need to know how their values are being interpreted and organized in terms that have meaning principally in relation to the academic literature; what matters is that we as researchers can organize them in ways that will help us characterize the various ways that human values may be affected

by ecosystem change. Second, the principal purpose of identifying the categories and kinds of values at stake is to determine appropriate kinds of valuation and decision-making methods, and appropriate ways to apply and combine these (Figure 12.1). Some of these methods cover only a subset of the kinds of values (e.g., market valuation and economic non-market valuation), while others address many or all kinds of values (e.g., structured decision-making).

12.1.4   Environmental and social processes that affect service production

The production of cultural ecosystem services is affected by many of the environmental processes that affect other ecosystem services and biodiversity. Accordingly, principal environmental threats to service production are land-use change (including habitat loss and fragmentation), pollution, climate change, overharvesting, invasive species, and disease.

Because cultural ecosystem services are co-produced by ecosystems and people, they are deeply affected by social processes—at both the production and valuation stages. For example, if social processes have led people away from nature-based recreation (Pergams and Zaradic 2006; Louv 2008), then these processes also affect the ecosystem contribution to recreational experiences and the corresponding value.

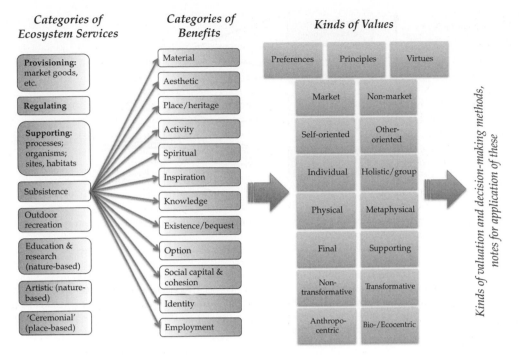

Figure 12.1 The suggested use of the typologies of ecosystem services and values.
First, identify the relevant categories of ecosystem services; map these services onto categories of benefits, and the benefit categories onto kinds of values; use the kinds of values at stake to inform choice and application of valuation and decision-making methods, to ensure appropriate representation of the full range of relevant values and to avoid double-counting. The arrows linking subsistence to categories of values are only one example of a mapping of one service onto values (other mappings are certainly possible).

The role of social processes in the production of cultural ecosystem services allows positive feedback cycles in which service production fosters habits and transfer of knowledge that in turn enhance service production. The negative flip side of this "positive" feedback is that external shocks that undermine the social side of service production can cause "vicious cycles" that fuel long-term loss of knowledge and practice (Turner and Turner 2008). For ecosystem services in which human use enhances—rather than degrades—biological production, service production is vulnerable to an even more pernicious socio-ecological spiral of lost use (Figure 12.2).

12.2   Methods: integrating cultural services and non-use values into decisions

In this section, we discuss methods for mapping and valuing cultural services and non-use values

spatially across a landscape. Our approach for tier 1 and tier 2 models is to provide an array of options for integrating these values into a comprehensive ecosystem services analysis with attention to linkages with models described in other chapters of this book, or their counterparts in the literature. Capturing the complex nature of cultural services (or any service) is challenging. We focus on generating summary information that decision-makers can use, with place-specific knowledge, to support improved decision-making about changes in ecosystems that affect cultural and non-use values. We hope that the approaches presented below will contribute to vibrant discussion about effective mapping and valuation tools, while recognizing that substantial work still lies ahead to cover the full array of values described in 12.1.2, as well as the needs of decision-makers in diverse socio-ecological contexts. Before discussing tier 1 and tier 2 methods, we first examine a set of issues that

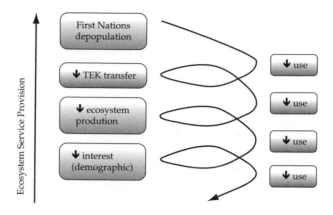

Figure 12.2 The downward spiraling of ecosystem service production for cultural ecosystem services co-produced by ecosystems and people, and for which use practices actually enhance future ecosystem production.

An example is the harvest of edible seaweed *Porphyra abbottiae* in coastal British Columbia, a practice initially reduced by the great depopulation of First Nations people by European colonization. While *Porphyra* provides a provisioning service, it is integrally linked to subsistence and ceremonial activities, among other cultural ecosystem services. *Porphyra* seems to be enhanced by harvest practices, which involve clearing competitors and herbivores, and spreading the reproductive gametes (although often not consciously) (White and Chan, in preparation). The harvest—as practice—depends on transmission of traditional ecological knowledge (TEK) from generation to generation, and TEK transfer in turn depends on use (Turner and Turner 2008). There are similar feedback cycles with use and both ecosystem production and loss of interest, and there are also feedback cycles between loss of knowledge, ecosystem production, and interest (White and Chan, in preparation). Other external influences can be positive (increased access by speedboats) or negative (pulp mill and domestic sewage pollution, new regularly scheduled commitments like school and wage jobs that interfere with the harvest cycle) (Turner and Turner 2008).

cut across all projects seeking to map and value cultural services and non-use values.

12.2.1 Cross-cutting issues: interdependency, double-counting, and trade-offs

The overlap of benefits across services raises the specter of interdependency and the double-counting that can result from interdependency. These issues have crucial implications for quantification and valuation methods. If we wish to assess the impact of a management decision on cultural values through its impacts on cultural ecosystem services, we have a choice. (A) If we attempt to evaluate impacts on each benefit separately (e.g., spiritual value, identity value), we are likely to encounter cognitive overload or dissonance (by which the sheer number and complexity of benefits affected by multiple services will overwhelm people's abilities to consider each benefit separately, or the artificial separation of apparently linked benefits will seem inappropriate or artificial) or other resistance (see earlier Section 12.1.3). (B) On the other hand, if we attempt to evaluate impacts by service, we encounter difficulties associated with the interdependency of services in producing value.

For example, both hiking (contribution to recreational experiences) and certain jobs (contribution to employment) may have value partly for their contribution to physical activity. In this context, these services are partly substitutable: an individual's need for physical activity from recreation would be lessened by a job that entails considerable physical activity. In either A or B, we must deal with the fact that some benefits are supporting and some are final, and the same is true for services. For example, if we value the contribution to subsistence activities, we must be aware that this service is important partly because it supports the contribution to "ceremonial" activities (e.g., by providing shellfish, salmon, and herbs to First Nations people in coastal British Columbia—Garibaldi and Turner 2004). It would be double-counting to add the value of ceremonial activities to an inclusive value of subsistence activities, unless the benefits are carefully parsed out, which may be impossible in practice.

These problems can be solved, though not easily. For intangible values, more inclusive valuation approaches should be favored over unconnected valuation of separate services. Separate valuation will only be possible when experiences provided by

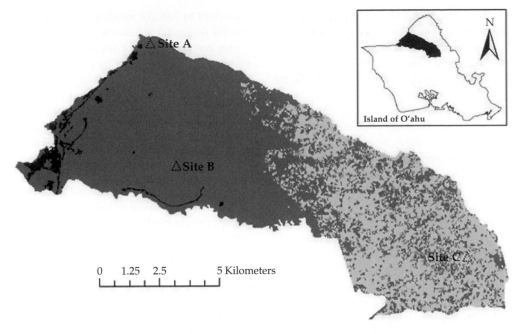

Figure 12.3   Land use/land cover (LULC) map from the Hawai'i Gap Analysis Program (2006) for the planning region on the north shore of the Island of O'ahu, Hawai'i.

See text for discussion of this planning region as an example of calculating the relative total landscape score of site quality. LULC categories are aggregated into three general classes: built areas (black), unbuilt areas dominated by non-native vegetation (dark gray), and unbuilt areas dominated by native vegetation (light gray). Dashed triangles denote location of the three hypothetical culturally important sites included in the calculation of the total landscape score.

a service stem from entirely distinct activities (e.g., in the provision of experiences for photography, exclusive photography trips are separate from hiking trips during which photos are taken). But we can expect interdependencies to be so pervasive that activities will rarely be truly separate from a value perspective. Many ecosystem services act simultaneously as both a supporting and a final service. At the supporting level, they are almost never independent of other services. For example, the provision of experiences for subsistence gathering is a supporting service in its contribution to ceremonial, educational, and identity values, but it is a final service in its provision of medicine and nutrition. These values are interdependent. Boyd and Banzhaf (2007) propose a framework for ecosystem services that would solve double-counting problems by recording only values of end products. Unfortunately, this useful framework does not apply easily to the valuation stage for values that cannot be easily monetized, for two reasons. First is the problem of inseparable end products. For instance, treating nutritional

and spiritual values as separate end products will likely generate considerable resistance from subsistence harvesters for whom the value of subsistence gathering arises through the concurrent co-production of multiple benefits. Second, if end products are considered at a high level (e.g., overall well-being) to avoid the inseparability problem, we are likely to encounter cognitive overload in people participating in value elicitation exercises. However, if we can use information from value elicitation to relate people's responses to high-level categories like overall well-being, then the Boyd and Banzhaf (2007) approach is promising.

In addition to concerns about interdependency and double-counting, some model users and stakeholders may be uncomfortable with the explicit manner in which our simple model deals with trade-offs in ecosystem service production. In particular, this concern may arise with communities for which all aspects of the cultural landscape are sacred. For example, stakeholders may be uncomfortable with or unwilling to evaluate alternative land-use change scenarios in which one cultural value increases at the expense of another, or in which carbon stock increases at the expense of cultural values (e.g., through reforestation with non-native species versus native species used for ceremonial purposes). At the same time, being aware of trade-offs and synergies is critical to understanding how our management of ecosystems affects all values of concern. Being sensitive to this issue is essential for ensuring that our model is used in a productive decision-support environment.

12.2.2 Tier 1 methods

The first step in any modeling effort is to prioritize the cultural ecosystem services and non-use values for a project's analysis. Any given modeling exercise is unlikely to have sufficient data or resources to consider all, or even a large fraction, of these services and the benefits at stake. It is critical to prioritize those cultural dimensions most important to people and likely to be impacted by ecosystem changes considered in modeled scenarios.

In this context, we present approaches for mapping and valuing from which users can choose based upon project goals and data availability. We note again that nature-based recreation and tourism activities are covered separately in Chapter 11.

12.2.2.1 *Mapping culturally important sites*
The ecosystem services framework presented in this book is inherently spatial, meaning that cultural services and non-use values must be derived from spatial data and presented spatially to be effectively integrated with other service model outputs. The most basic piece of information of possible use to decision-makers is a map of the location of sites or objects of cultural importance that are linked to ecosystem features. There is a wide range of features that could be mapped such as sites of subsistence activities (e.g., hunting and fishing grounds), recreational experiences (e.g., hiking or kayaking routes; these may already be covered by models in Chapter 11), educational and research experiences, ceremonial significance, and significance for identity value (e.g., historical trails such as the Grease Trail in Canada or uplands and ocean access trails in Hawaii) or existence value. In each case, it is

important to link the location being mapped to the specific service(s) of interest.

Our own experiences suggest that there will be wide variation in availability of even the most basic information to inform a spatial mapping exercise. This poses a practical challenge but also strong motivation to compile this information into a GIS. Doing so will require synthesizing information from diverse sources such as historical documents, interviews with community elders, and public agencies. In some cases, the location of sites may be sensitive information (e.g., sacred burial grounds), and care must be taken to respect the private nature of such information.

While mapping is a good first step, the ability to detect and represent change in sites across scenarios is limited to discrete addition or deletion of features. Extensions to mapping, as discussed below, and where feasible given data and resource constraints, could greatly improve the quality of information supporting tier 1 decision-making.

12.2.2.2 *Relative scoring of site quality across the landscape*
The value of culturally important sites is often affected by ecosystem conditions and other site characteristics, such as the surrounding land use/ land cover (LULC), proximity to population centers, and legal rights to access, among others. For a given cultural service, relevant factors could be integrated into a quantitative analysis that provides information on the relative improvement or deterioration of site quality in the context of land-use change scenarios. This approach would require that stakeholders identify which characteristics influence site quality and by how much. For example, in some cases, enhancing access to a site may be desirable (e.g., plant gathering area), while in other cases limiting access would be desirable (e.g., restricted ceremonial site).

To illustrate one approach for relative scoring of site quality across a landscape, we consider a case in which a landscape score of site quality (L) is (1) increasing in the number of sites (with flexibility, if appropriate, to assign weightings of relative importance across sites), and (2) affected by the LULC classes located in a buffer area surrounding the site. In certain cases, such as gathering areas, it may also be important to consider (3) the physical and legal accessibility of each site to users.

A general functional form to assess site conditions across the landscape could be expressed as follows:

$$L = f\{v_1, \ldots, v_N\}, \qquad (12.1)$$

where L is a landscape score of site quality and v_i is a relative score of site i calculated as follows:

$$v_i = w_i \cdot S_i \cdot B_i\{g_{i1}, \ldots, g_{in}\}, \qquad (12.2)$$

where w_i is the weighted importance of each site i; S_i is an indicator variable (0/1) representing the presence or absence of site i (in the current landscape, all existing sites would be given a value of "1." In scenarios, sites that are deleted would get a "0" and sites that are added would be given a "1"); B_i is the overall appropriateness score of surrounding land uses in the buffer around site i, most simply calculated as

$$B_i = \sum_j g_{ij} \,, \qquad (12.3)$$

where g_{ij} is the contribution of LULC class j to B_i, calculated as

$$g_{ij} = a_{ij}(r) \cdot \alpha_j, \qquad (12.4)$$

where a_{ij} is the fractional area of each LULC class j in the buffer area around site i; r is the radius of the buffer area (assuming a circular buffer; other shapes are possible); and α_j is the appropriateness scores for each LULC class j to reflect the cultural appropriateness/value of the LULC class being located in the site's buffer area.

All told, with the simplest versions of f and B (summations of v_i and g_{ij} respectively), we have the following:

$$L = \sum_{i=1}^{N} w_i \cdot S_i \cdot \left(\sum_{j=1}^{n} a_{ij}(r) \cdot \alpha_j \right) \qquad (12.5)$$

As a descriptive illustration of this approach, let's consider an example of computing a relative total landscape score (L) for culturally important sites in a region on the north shore of the island of O'ahu, Hawai'i (Figure 12.3). Kamehameha Schools, an educational trust for Native Hawaiians, is the primary landowner in this region, as well as the State of Hawai'i's largest private landowner (see Box in Chapter 14 for background information on Kamehameha Schools). Kamehameha Schools manages its diverse portfolio of lands to derive an overall balance of economic, educational, cultural, environmental, and community returns for current and future generations.

The region shown in Figure 12.3 covers approximately 10,500 ha from mountain top to the sea, including approximately 800 ha of developed rural community lands along the coast, 3,600 hectares of agricultural lands further inland, and 6,100 ha of rugged forested lands in the upper part of the watershed. Within this planning region, we have identified three culturally important sites for incorporation into the overall landscape score (L): Site A is near the coast, Site B is in the agricultural fields along a riparian corridor, and Site C is in the upper forested conservation lands (Figure 12.3). These sites are hypothetical but representative locations for the illustrative purpose of this example, though the approach would be identical with real sites. We assume equal weights (w_i) in this example suggesting that all sites are of equal importance from a decision-making perspective.

Evaluating the impacts of land-use change through scenarios is an important feature of our modeling approach (Chapter 3). The analysis described above would need to be rerun for each scenario, holding input values constant. If a site is removed in a scenario, then this should be reflected through a change in the indicator variable, S_i. Calculating the difference in the landscape score (L) for each scenario relative to the current landscape provides model users with a quantitative indication of the impacts of land-use change on sites of cultural importance. In this example, a scenario that expands residential development within the buffer region of site A would decrease the relative score for this site (B_i), as well as the overall landscape score (L). Conversely, investments in native plant restoration around sites B and C would increase each site's score and the landscape score.

This approach requires stakeholders to make judgments regarding the appropriate size and shape of the buffer region, the relative appropriateness of different LULC classes contained in the buffer, and potentially other factors. To provide more robust information to decision-makers, model users should perform analyses to determine the sensitivity of the landscape score to different specifications of model inputs. Sensitivity analysis could be straightforward, such as testing different buffer radius distances. Where useful, it could also consider different ways of quantifying the impacts of land-use change in the buffer or redefining the nature of the buffer. For example, an alternative buffer could contain two parts: a core area near the site and a doughnut-shaped area immediately surrounding this core.

12.2.2.3 Linking cultural services with models for other ecosystem services and biodiversity

Cultural services are produced, in part, by landscape features and processes quantified in other models described in this book for biophysical services and biodiversity. Agricultural models (Chapter 9) can provide information on the location, yield, and market value of culturally important agricultural crops (e.g., taro in Hawai'i). Pollination services (Chapter 10) support cultivation of insect-pollinated crops, and there may be overlap between habitat supporting wild pollinators and areas of cultural significance (e.g., plant gathering areas). Identifying this overlap is important for effective decision-making, as is an evaluation of how the relative benefits from pollination and cultural services would change given alternative land-use change scenarios.

Timber and non-timber forest products (Chapter 8) are often connected with the contribution to artistic and employment experiences, and with other cultural services. Output from these models could inform decisions about resource management and harvest levels for culturally important species. For example, decision-makers could use the non-timber forest products model to obtain information on the location, stock, and growth rate of species harvested for cultural purposes (e.g., plants used for traditional *hula* activities in Hawai'i) under different landscape scenarios and management regimes. This output could inform an assessment of the sustainability of varying levels of harvest in different gathering areas, which may in turn affect the

number of people who can sustainably participate in harvesting.

Hydrologic services including providing water for consumptive uses, regulation of water pollution, and mitigation of storm peak flow (Chapters 5–7) also impact cultural services and non-use values. Water flows and their corresponding quality affect the condition of aquatic cultural resources, such as wild fish or fish cultivated using traditional practices in fishponds. As such, changes in water flows could be used to quantify effects on cultural resources. Related to fish consumption, such an analysis could evaluate the fraction of a target population that could be served by the local fish harvest (wild or cultivated) under different management scenarios. In the context of mitigation of storm peak flow, model users could define regions of interest based upon culturally significant landscape features. By narrowing their analysis to these regions, they could specifically understand changes in flood risk to cultural assets across alternative landscape scenarios. These examples illustrate some of the many linkages with cultural services that can improve information supporting decision-making.

12.2.2.4 Integrating demographic and socio-economic data

Information related to demographic and socioeconomic conditions is another source of data for input in tier 1 models. It is obvious that different ethnic groups, even when not part of an indigenous culture, bring very different social and cultural backgrounds to their relationships with nature (Box 12.2). Census data are likely to be broadly available, and because this information is linked to geographic location, it is readily applicable to spatial ecosystem service analyses. The units of information (e.g., household, census block, county, state), however, will limit analysis to specific scales. As an illustration of integrating demographic and socioeconomic data, we consider the value of subsistence activities to communities in Alaska.

At the tier 1 level, it will often be difficult or impossible to distinguish between the various values discussed in this chapter that result from cultural activities such as subsistence harvest. At this level, it is therefore often desirable to calculate lower bound estimates of the importance of services to people.

Subsistence hunting and gathering is of great importance to Alaskans, but it is of much greater importance to some than to others. Those who benefit most from subsistence activities are often those who earn the least in monetary terms, which implies that averaging the surveyed value of subsistence activities across people will often under-represent the contribution to well-being. An example from five communities in Southeast Alaska, demonstrates this.

Vadeboncoeur and Chan (in preparation) calculated a lower bound estimate of the economic value of subsistence goods using publicly available harvest data for 48 species or groups of species in each community. The economic values for harvested species were derived from market prices for the species in question or the cheapest available local substitute meat. In each case, the lower bound valuation is supported by the logic that there is considerable trade and barter of such species, so if a harvest is not traded, this is likely because the harvester values it more than the possible exchange value. Income and population data were obtained from the 2000 US Census (US Census Bureau 2003).

Even with a tier 1 model drawing upon existing information sources, spatially resolved data paired with income data can provide insight into the degree to which people will be affected by changes in subsistence harvest. For example, based on per-capita economic values as a percentage of income, it is clear that the predominantly Alaskan Native community Hydaburg will be more severely affected by loss of subsistence harvests than would other communities. Similarly, within each community, subsistence permit holders will be affected much more than would others in the communities. Furthermore, assuming that the well-being impact of a foregone dollar of income is inversely proportional to a person's income, the relative value of the subsistence harvest to Hydaburg is 2.7 times greater (34% of local inclusive income) than it would have seemed if we had used nation- or state-wide income statistics.

This analysis clearly underestimates the value of the subsistence harvest to Alaskans, but it does illustrate (i) how readily available data can be used to calculate lower bound estimates of these values and (ii) how presenting these values in relation to a group's income can better track distributional concerns and provide a starkly different perspective on the potential well-being impacts of losses of subsistence harvest. This analysis also serves as an example of the interrelatedness of the models presented throughout this book; the service of subsistence harvest may also be relevant in analyses of the relationship of ecosystem services to poverty (see Chapter 15).

Table 12.2   The economic value of subsistence harvest in five sites in Southeast Alaska relative to income

Jurisdiction	Population	Adults in labor force (%)	Seasonal occupations (%)	Per-capita income (thousands)	Per-captia value of subs. harvest	Economic value of subs. harvest/ income (%)	Economic values of harvest/income—for subs. permit holders (%)
Sitka	8835	73.6	15.9	$23.9	$246	1.0	10
Petersburg	3224	70.8	24.6	$26.2	$368	1.4	7
Haines	2392	61.6	18.9	$22.3	$176	0.8	9
Yakutat	808	77.8	33.2	$23.0	$392	1.7	33
Hydaburg	382	49.1	21.1	$12.0	$602	5.0	34
United states	301 139 947			$30.5			
Alaska	670 153			$32.0			

Here, adults are considered to be all people over 16 years of age. Seasonal occupations include farming, fishing, forestry, resource extraction, construction, and maintenance. Economic values were calculated based on lower bound estimates of the values that could be obtained through reciprocal non-market exchange. Percentages of total income (including these subsistence values) are provided for average members of each community and—using more detailed harvest data—for those holding special subsistence permits. The low fraction of the population in the labor force demonstrates that in these communities time is not limiting; we assumed that time spent on subsistence harvesting is generally not detracting from time spent in gainful employment.

12.2.2.5 *General landscape composition and configuration*

Certain cultural services and intangible values may be linked to general landscape characteristics, such as the composition or configuration of LULC types within a planning region. Aesthetic values in rural areas, for example, may be linked with the amount or configuration of open space in agricultural or forested LULC types relative to built areas. This type of information can be quantified in a GIS using spatial analytical tools applied to the LULC layer and other relevant datasets. With this example, if aggregate values are sufficient for decision-makers, then several general values could be computed such as the fractional area in different LULC types or the degree of fragmentation of open space. A more detailed and information-rich approach would be to map viewsheds using three-dimensional projections in a GIS for locations important to stakeholders (Bishop and Karadaglis 1997; Lim and Honjo 2003; Grêt-Regamey *et al.* 2007), assessing how the composition of viewsheds changes under different scenarios of development.

This approach of quantifying general landscape characteristics must be partnered with consideration of cultural services that are highly site-specific. While the site substitutability of forest cover in a region may be high with regards to aesthetic values, it may be low with regards to its contribution to the heritage, spiritual, identity, and other values associated with a sacred site. As such, loss of forest cover around the site would substantially reduce these site values, but this conversion, all else equal, would have a lesser impact on the landscape's aesthetic value.

12.2.3 Tier 2 methods

Model users may have resources available to support new, on-site data collection. This is important for modeling cultural services and non-use values because existing data are likely to be scarce for cultural services in many locations. Indeed, one important advance at the tier 2 level will be to generate more refined, site-specific data inputs for the general methods described above. In addition, using tier 2 models for biophysical services and biodiversity will also provide more sophisticated output

that could be integrated into tier 2 analyses for cultural services and non-use values. In most cases, consideration of non-use values will need to occur at a tier 2 level, since collection of new, site-specific data will be necessary. Collecting data for each service and value will require specific methods. We describe below the process of a tier 2 analysis with emphasis on one value as an example.

12.2.3.1 *Mapping and quantifying existence value*

We discuss a possible tier 2 approach to mapping non-use values using existence value as an example. The few studies that have attempted to measure existence value have done so using stated preference survey techniques such as contingent valuation or choice experiments (Mitchell and Carson 1989; Louviere *et al.* 2000) to estimate willingness-to-pay (WTP) for conservation programs (Kramer and Mercer 1997; Rolfe *et al.* 2000; Horton *et al.* 2003).

Contingent valuation exercises ask survey respondents to express their WTP for a change in an environmental good or service that is described in a hypothetical scenario; resultant WTPs are "contingent" on the accurate description of the good being valued. Choice experiments present scenarios defined by quantitative or qualitative attributes (generally including a price variable), and respondents are asked to choose the scenario they prefer. Both contingent valuation (Kramer and Mercer 1997; Horton *et al.* 2003) and choice experiments (Rolfe *et al.* 2000) have been used to estimate WTP for the existence of tropical rainforests.

Evaluating existence value at fine spatial scales involves eliciting values for the existence of species or ecosystems at particular geographic locations. This is problematic for several reasons: (1) At within-country scales, individuals' WTP for a number of environmental non-use values declines with household distance from site of provision (Hanley *et al.* 2003; Bateman *et al.* 2006) (Figure 12.4d); (2) Individuals more familiar with a site will be better able to disaggregate their existence value across space, resulting in a more heterogeneous value surface than would come from individuals who are less familiar (Figure 12.4b); and (3) when values are elicited from an international group of respondents, individuals in richer countries are less income-

Box 12.2 People of color and love of nature

Hazel Wong

For the first 10 years of my life, the outdoors was my playground. Growing up in the Seychelles, without television, my sisters and I spent our vacation days and Saturdays immersed in creative play outdoors. Sundays were reserved for either a beach outing or a family lunch feast followed by seven people piling into a car to tour the island, making pit stops along the way for my parents to say hello to their friends and our extended family. I have memories of my paternal grandmother watching fourteen grand children at once. The house, a modest four-bedroom home, was by no means big enough to accommodate that many rambunctious children. Looking back, grand-mere had help, the outdoors. The only reasons to set foot inside the house were to eat lunch and use the bathroom.

On September 14, 1980, I woke up in the desert of Las Vegas, Nevada after my family left the Seychelles following a coup d'état and ensuing unrest. At ten years old, I remember feeling a tremendous loss for the beautiful island I left behind. Slowly, the desert became my new playground. Back then, there was plenty of desert in Las Vegas, the no-end-in-sight development boom that started in the late eighties had not taken place. Our introduction to television, even in a limited and supervised capacity, replaced playing outdoors, especially during the winter months, a new and not so welcome season for those born on an island with a year round average temperature of 80°F.

Life in the States was very different from the Seychelles; my parents had to adjust to a new culture, a lack of family support and raising five girls in an unfamiliar setting. Their number one priority at all times was our education, and they worked and saved to ensure we had a head start in life. This left little time and resources to explore the natural world in America. Yet, my father had a voracious appetite for nature books and television programs—fueling my knowledge, curiosity, and sense of amazement.

Years later, I am now a self identified conservationist and one of the 10% of people of color working in the conservation field. I also served on the board of the Nevada Conservation League for five years, serving as chair for over a year. My childhood exposure to the natural world, embedded in my psyche a love and passion for the environment, instilling in me a value system that firmly believes in the rights of nature.

Like many individuals, I am multifaceted and my appreciation of nature competes with other activities and passions. For example, I have driven to mountains, seeking nature's solitude and the comfort of the vast open space to gain clarity and strength for major decisions in my life. Yet, I have been camping only once in all my years, and I am not an avid hiker—I would rather do a two-hour martini lunch with friends than hike on most days. (I hope this confession is not grounds for revoking my conservationist card.) However, that does not negate my devotion and passion to protect the natural world.

Working in the conservation field for six years, I quickly learned that the Anglo Saxon culture defines the meaning of the word conservationist and narrowly interprets what constitutes the correct way of valuing and recreating in the natural world. I have concluded that the problem is not that people of color do not value and care about conservation issues. They are under represented in the conservation movement because of conservation's history of class, privilege, and a homogenous culture that has rendered a myopic view, perhaps a benign neglect and at worst actual resistance to conservation becoming a more inclusive movement.

Conservation's past is a history of white male privilege, conserving nature for the pleasure and enjoyment of the well-to-do class. In fact, conservation history books are filled with stories of how John Muir worked to save the redwoods and later founded the Sierra Club, or how Theodore Roosevelt established our national park system. While there is much to be proud of in the early conservation movement, there is also some disturbing history which rarely gets documented. When our national parks were founded, Native Americans were removed from their homes and displaced as contaminators of pristine nature. Cities were held with disdain as the dark polluted homes of industry, as well as society's undesirable poor minorities. John Muir, long heralded as the father of the conservation movement, was also known for his disparaging remarks about Native Americans and African Americans. Native Americans were "dirty," "savage," and needed the wilderness to cure "the grossness of their lives." African Americans "made a great deal of noise and did little work." While today's conservationists do not hold these views of Native Americans and African Americans, the legacy is that

continues

Box 12.2 *continued*

conservation organizations' appeal beyond the Anglo Saxon culture has been limited.

If you look at the surface statistics, reasonable people may draw the conclusion that people of color do not value conservation. Indeed, 90% of people working in conservation are white. To many, this could be interpreted to mean that people of color have no interest in this field. Furthermore, over 90% of members and donors to conservation organizations are white, again, leading many to conclude that people of color do not support conservation because they do not financially contribute. Lastly, people of color do not value nature because they do not recreate in nature as much, or in the same manner as the dominant culture.

The paradox, and what most conservationists are unaware of, is that people of color consistently vote overwhelmingly in favor of conservation measures placed on the ballot and often times they are taxing themselves to support public funding for conservation. Furthermore, qualitative and quantitative analysis through focus groups and public opinion research surveys show that communities of color value the natural world just as much (and in some cases more) than whites and that they support policies to address global climate change. Protecting land, air and water is a core value of communities of color. The challenge is to get these communities actively engaged in financing and advocating for conservation.

Today's conservation organizations' marketing and outreach efforts continue to target the dominant group in American society, a group of donors getting older and narrower in terms of overall population numbers. In the next three decades, communities of color will comprise 50% of the US population and the lack of cultivation of diverse groups means potential new donors and members are being neglected while nonprofits compete with each other for limited private resources. Furthermore, a lack of diversity in the work force continues to send a message of disconnect with communities of color, and ultimately leads to conservation losses as the talents of diverse people are not being harnessed. Unfortunately, the insular mode of hiring makes it that "who you know" is still the prevalent hiring practice. When you start with a 90% homogenous culture—that "who you know" way of hiring contributes to a lack of meaningful outreach for diverse candidates.

The rapid and profound demographic changes means we, the conservation community, are already behind in our outreach efforts to ensure a more diverse constituency base is working to protect our natural resources. Therefore, reaching out to communities to color should be a priority

strategy just as ecosystem services is a strategy to connect and drive home a key message with a key audience that nature has an economic value. Those of us who work in the conservation arena value nature for its intrinsic worth, and that is not the case for the vast majority of Americans including many of our decision-makers. The existence of ecosystem services as a field substantiates that fact. Ecosystem services is a strategy to get people to connect with nature who do not value nature for nature's sake but may value it for its economic benefits. The purist in me abhors the idea that the natural world must have a dollar value in order for us to protect it. However, given the lay of land, where economics reigns supreme, ecosystem services is pragmatic, strategic and will resonate with key influencers in the political arena. More importantly it will help us achieve our conservation goals.

Along that same line, conservationists must take a pro-active and strategic approach to ensure protecting our natural world is an investment that *everyone* is on board with politically, culturally and economically. It starts with researching and developing strategic messages and outreach plans to target communities of color and bring them into the membership and donor base. Although all humans depend on the benefits of a healthy ecosystem, delivering that message with an emphasis on biodiversity and wilderness protection to communities of color simply will not work.

Van Jones, an advocate for environmental and social justice, points out that the standard gloom and doom messages about species extinction, polar bears and melting glaciers do not resonate with a wide swath of the population. As an alternative message, he is touting ecosystem services as it relates to green jobs and the new green economy. Yet, he is not calling it ecosystem services—a term that only has resonance and meaning to a small number of insiders.

In conclusion, ecosystem services can play a pivotal role in the outreach to communities of color. For conservation to be relevant in the twenty-first century, we must frame our message beyond biodiversity protection to include nature's benefits to people such as green jobs, children's health, economic solutions, spirituality, quality of life issues such as clean air and clean water, and protecting places for families to gather. We need conservation messages to be more relevant to all demographics, and more directly connected to people's everyday lives. We also need to ensure that all people "see" themselves and those like them working in conservation. When we do that, we will find the common ground to build a movement that truly includes everyone. And the good news is, we do not have to change our mission.

constrained than those in developing countries; therefore their monetized values for the existence of biodiversity will be higher, all else equal (Figure 12.4c). Designing surveys to elicit comparable WTP estimates across large distance gradients that span multiple countries, languages, and cultures is challenging. Yet without such surveys, the empirical data necessary to produce spatially-explicit existence value estimates will remain elusive.

Choice experiments and contingent valuation have well known limitations, especially for ecosystem services that are unfamiliar to respondents and that vary spatially. Recent research on "learning design" contingent valuation may provide an appropriate methodology for situations in which preferences are formed through a process of repetition and experience with unfamiliar goods (Bateman *et al.* 2008), such as ecosystem services. Recent work has also demonstrated that the results of stated preference surveys can vary depending on whether individual valuation or deliberative, citizen-jury valuation techniques are used (Alvarez-Farizo and Hanley 2006), which we would expect given that the latter emphasizes different kinds of value (e.g., principle-based, group).

This latter distinction may be particularly relevant when trying to estimate existence values across geographic scales. For example, conservation of threatened Amazonian forests may be important to both local and international non-users. While international non-users are likely to hold existence

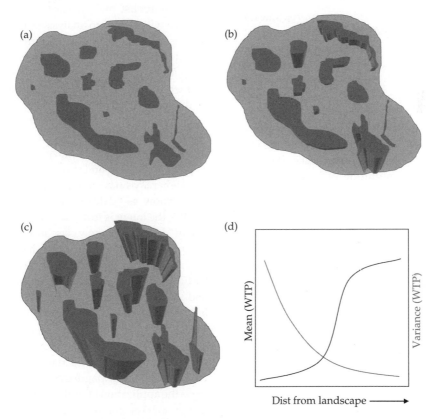

Figure 12.4 Variation in existence value across landscapes and stakeholder groups.
(a) Schematic of a conservation landscape in a developing country; dark gray units indicate tropical forest patches, light gray indicates anthropogenic land use. (b) Mean (proportional to height) WTP for conservation of patches from respondents living near the landscape. Variance in WTP is high as respondents are knowledgeable regarding the landscape and can discriminate among patches. (c) Mean WTP for conservation of patches from respondents living far from the landscape, in a developed country. Mean WTP is higher due to lower income constraints, and variance is lower due to less detailed knowledge of the landscape, as compared to (b).
(d) Relationship between WTP for patches in the landscape (mean and variance) and distance of respondents to the landscape.

values independent of other members of their community, locals may see forests as a community asset. As such, a deliberative group valuation approach may be the most appropriate to use when multiple perspectives like these exist, rather than an individual survey (Spash 2007).

Bearing in mind all of these issues and constraints, a program to evaluate the existence value of a particular biophysical feature across a landscape (e.g., tropical forests, endangered species) might follow the following steps:

1. Identify the group or groups of people whose existence values one wishes to elicit.
2. Consider the service being valued and use a variety of methods (focus groups, literature reviews, expert knowledge, etc.) to identify the features of the good that people care about.
3. Using similar methods, identify appropriate units of the service on the landscape. These units are liable to be larger/coarser for individuals who have little knowledge of the area.
4. Characterize each of the landscape units with regard to the important features identified in step 2.
5. Using maps and detailed descriptions, construct a survey that will elicit people's WTP for the existence of the feature across the landscape. We do not dwell here on the multitude of issues associated with constructing a stated preference survey; good references are Mitchell and Carson (1989) and Louviere et al. (2000).
6. Summarize WTP values for each particular unit of the service being valued across the landscape; inference can then be made about WTP across the population that the sampled group represents.
7. Modeling approaches (Mitchell and Carson 1989; Louviere et al. 2000) can be used to explain variation in WTP across both individuals and sites; this may be useful for predictive purposes in other contexts.

12.2.3.2 *Mapping and quantifying other values*

Many of the problems addressed in assigning measures to existence value apply, equally, to the larger class of cultural ecosystem services and values articulated here. While WTP methods continue to be employed, so too are expressed preference surveys drawn from the social sciences. Especially important, are techniques drawn from sociology (Dunlap et al. 2000), human ecology (Dietz et al. 2005), psychology (Gregory et al. 1993), anthropology (Kempton et al. 1995; Satterfield 2001), multi-attribute utility theory (Russell et al. 2001), and legal methods designed for the valuation of loss (Rutherford et al. 1998).

Ultimately newer practices must address abovementioned problems of human judgment (e.g., the instability in WTP judgments based on framing effects), and yet also achieve a balance between the benefits of group deliberation (collective and democratic conversations about what matters); the recognition and accommodation of political factors (justice, equity, land tenure, Treaty rights, etc.); and the kinds of quantitative and qualitative analyses necessary to arrive at robust and defensible valuations. Good practice with regard to analysis ensures that the valuations assigned reflect the full range of values pertinent to the case and are thereafter expressed as ordinal or cardinal scales amenable to trade-offs. It also requires that the constituents and/or the rights of different stakeholders are well defined thereby signaling important political concerns.

For these reasons, many valuation methods are moving away from the aggregation of willingness-to-pay judgments and toward deliberative processes whereby such values are arrived at collectively. Often know as "deliberative monetary valuation" (DMV), the intent is to arrive at a "social willingness to pay." In so doing, new classifications of value and concomitant valuation practices have arisen (Spash 2007, 2008). These include values and valuation based on exchange value, charitable contributions, and prices negotiated in reference to equity or linked social values. Others have moved toward the valuation of loss (for services since extinguished or degraded) by combining conjoint analyses (simple paired comparisons) and damage schedules (see, for example, Chuenpagdee et al. 2001). Also promising is the application of multi-attribute utility theory to non-market goods including the use of swing weighting (McDaniels and Trousdale 2005). Finally, practices derived from structured decision-making are particularly likely to be useful to the more intangible dimensions of cultural services (Gregory et al. 1993; Gregory 1999). This is due to the practice of constructive scaling, which assumes that a locally defined valuation metric (achieved through deliberation) is optimal when natural or proxy scales do not exist (Keeney and Gregory 2005) and when there are many different sub-components of a singularly important value. Compelling examples of this method can be found in work with First Nation/Native American communities in Canada (Gregory *et al.* 2008) and in environmental decision-making more broadly (O'Neill and Spash 2000).

12.3   Limitations and next steps

We have provided a framework for understanding, categorizing, mapping, and valuing cultural ecosystem services and non-use values. This contribution supports the notion that it is possible to quantify many of these services and their corresponding values through spatial landscape analysis that facilitates integration with output from other biophysical and economic service models. Such integration is essential for ensuring that cultural services and non-use values are considered on equal footing with other services and biodiversity in land management and policy decisions. At the same time, although we have presented a typology and categorization of values and services, the methods discussed above do not cover all of these values and services (Table 12.2). Expanding the set of mapping and valuation tools represents an important frontier for research and conservation practice; but as we have discussed, there are several obstacles to comprehensive valuation including interdependency between values and services and associated double-counting.

Modeling efforts explore ways to assign value to the cultural and non-use dimensions of ecosystems. For many stakeholders, an equally important consideration will be people's preferences regarding the processes by which decisions are made. This may be particularly true in the context of cultural services given the personal and community ties that they evoke. People's preferences or principles stem from their underlying or held values (Brown 1984), such as fairness, responsibility, autonomy, and sovereignty. For our proposed models to be effective tools guiding land-use decisions, the biophysical, economic, and non-economic values

produced must be integrated into broader decision-making processes considered legitimate by its participants and stakeholders.

Citizen juries and multi-criteria decision-making (MCDM) processes implicitly or explicitly include such process-oriented values, so they are possible alternatives (Chee 2004) or addenda to valuation of ecosystem services. Both approaches involve gathering groups of citizens intended to represent the spectrum of stakeholders affected by an issue, to carefully consider a body of evidence and collectively make decisions. They differ in that citizen juries involve competing teams of experts presenting alternative viewpoints (Shrader-Frechette 1985; Coote and Lenaghan 1997), and MCDM involves explicit structuring of the multiple dimensions relevant to the decision (Saaty 1996; Munda 2004). Both approaches could be undertaken based on (1) solely biophysical attributes, (2) biophysical attributes and a subset of valuation metrics (e.g., from market values), or (3) biophysical attributes and a full set of valuation results.

There are pros and cons to these various possibilities. On one hand, defensible valuation of the full set of ecosystem services at stake is likely to be time-consuming and difficult (e.g., to avoid double-counting and account for interdependencies). Not only is value elicitation time-consuming in its own right, but in order to have accurate results, these valuation exercises should be performed using realistic scenarios of outcomes from decision options—ideally, therefore, valuation will follow biophysical analysis and not run in parallel. Furthermore, explicit measurement of some intangible or sacred values will always trigger discomfort, whereas this discomfort may be lessened considerably if the decision is made without prices or similar metrics (as in citizen juries and MCDM).

On the other hand, citizen juries and MCDM are also time-consuming and difficult, and they, too, must follow biophysical analysis. Furthermore, they require that participants integrate in their minds the multitude of diverse values that will be affected by management scenarios, often indirectly—a considerable cognitive challenge. Finally, they involve giving a certain authority to the small subset of stakeholders invited to the decision-making table, authority likely to be questioned by parties unhappy with the resulting decision, perhaps even if considerable time is invested in choosing and vetting stakeholders involved.

The most thorough option (3), which can capture the full set of values and consider them in appropriate context through democratic, deliberative processes, is likely the most time-consuming and expensive. Fair, inclusive, and enlightened environmental decision-making is the Holy Grail. We hope this chapter has provided insights and approaches that move us partway toward this goal.

References

Alvarez-Farizo, B., and Hanley, N. (2006). Improving the process of valuing non-market benefits: Combining citizens' juries with choice modelling. *Land Economics*, **82**, 465–78.

Anderson, D., Salick, J., Moseley, R., *et al.* (2005). Conserving the Sacred Medicine Mountains: a vegetation analysis of Tibetan sacred sites in Northwest Yunnan. *Biodiversity and Conservation*, **14**, 3065–91.

Basso, K. H. (1996). *Wisdom sits in places: landscape and language among the Western Apache.* University of New Mexico Press, Albuquerque.

Bateman, I. J., Burgess, D., Hutchinson, W. G., *et al.* (2008). Learning design contingent valuation (LDCV): NOAA guidelines, preference learning and coherent arbitrariness. *Journal of Environmental Economics and Management*, **55**, 127–41.

Bateman, I. J., Day, B. H., Georgiou, S., *et al.* (2006). The aggregation of environmental benefit values: Welfare measures, distance decay and total WTP. *Ecological Economics*, **60**, 450–60.

Bishop, I. D., and Karadaglis, C. (1997). Linking modelling and visualisation for natural resources management. *Environment and Planning B: Planning and Design*, **24**, 345–58.

Boyd, J., and Banzhaf, S. (2007). What are ecosystem services? The need for standardized environmental accounting units. *Ecological Economics*, **63**, 616–26.

Brown, G. G., Reed, P., and Harris, C. C. (2002). Testing a place-based theory for environmental evaluation: an Alaska case study. *Applied Geography*, **22**, 49–76.

Brown, T. C. (1984). The concept of value in resource allocation. *Land Economics*, **60**, 231–46.

Champ, P. A., Boyle, K. J., and Brown, T. C., Ed. (2003). *A primer on nonmarket valuation: the economics of nonmarket goods and resources.* Kluwer Academic, Dordrecht, The Netherlands.

Chee, Y. E. (2004). An ecological perspective on the valuation of ecosystem services. *Biological Conservation*, **120**, 549–65.

Chiesura, A., and de Groot, R. (2003). Critical natural capital: a socio-culural perspective. *Ecological Economics*, **44**, 219–31.

Chuenpagdee, R., Knetsch, J. L., and Brown, T. C. (2001). Coastal management using public judgments, importance scales, and predetermined schedule. *Coastal Management*, **29**, 253–70.

Cicchetti, C. J., and Wilde, L. L. (1992). Uniqueness, irreversibility, and the theory of nonuse values. *American Journal of Agricultural Economics*, **74**, 1121–5.

Coote, A., and Lenaghan, J. (1997). *Citizens' juries: theory into practice.* Institute for Public Policy Research, London.

Costanza, R. (2008). Ecosystem services: Multiple classification systems are needed. *Biological Conservation*, **141**, 350–2.

Costanza, R., d'Arge, R., de Groot, R., *et al.* (1997). The value of the world's ecosystem services and natural capital. *Nature*, **387**, 253–60.

de Groot, R., Ramakrishnan, P. S., v. d. Berg, A., *et al.* (2005). Cultural and amenity services. in Millennium Ecosystem Assessment, Ed., *Ecosystems and human well-being: current status and trends.* Island Press, Washington, DC, pp. 455–76.

Dietz, T., Fitzgerald, A., and Shwom, R. (2005). Environmental values. *Annual Review of Environment and Resources*, **30**, 335–72.

Dunlap, R. E., Van Liere, K. D., Mertig, A. G., *et al.* (2000). Measuring endorsement of the new ecological paradigm: A revised NEP scale. *Journal of Social Issues* **56**, 425–42.

Fisher, B., and Turner, R. K. (2008). Ecosystem services: classification for valuation. *Biological Conservation*, **141**, 1167–9.

Garibaldi, A., and Turner, N. (2004). Cultural keystone species: Implications for ecological conservation and restoration. *Ecology and Society*, **9**(3), article 1.

Goulder, L. H., and Kennedy, D. (1997). Valuing ecosystem services: philosophical bases and empirical methods. In G. C. Daily, Ed., *Nature's services: societal dependence on natural ecosystems.* Island Press, Washington, DC, pp. 23–47.

Gowdy, J. M. (2001). The monetary valuation of biodiversity: Promises, pitfalls, and rays of hope. In V. C. Hollowell, Ed., *Managing human-dominated ecosystems: proceedings of the symposium at the Missouri Botanical Garden*, St. Louis, Missouri, 26–29 March 1998, pp. 141–9. Missouri Botanical Garden Press, St. Louis.

Gregory, R. (1999). Identifying environmental values. In V. H. Dale and M. R. English, Eds., *Tools to aid environ-*

mental decision making. Springer-Verlag, New York, pp. 32–58.

Gregory, R., Lichtenstein, S. and Slovic, P. (1993). Valuing environmental resources: A constructive approach. *Journal of Risk and Uncertainty,* **7**, 177–97.

Gregory, R., Failing, L., and Harstone, M. (2008). Meaningful resource consultations with first peoples: notes from British Columbia. *Environment,* **50**, 34–45.

Grêt-Regamey, A., Bishop, I. D., and Bebi, P. (2007). Predicting the scenic beauty value of mapped landscape changes in a mountainous region through the use of GIS. *Environment and Planning B: Planning and Design,* **34**, 50–67.

Hanley, N., Schlapfer, F., and Spurgeon, J. (2003). Aggregating the benefits of environmental improvements: distance-decay functions for use and non-use values. *Journal of Environmental Management,* **68**, 297–304.

Horton, B., Colarullo, G., Bateman, I. J., *et al.* (2003). Evaluating non-user willingness to pay for a large-scale conservation programme in Amazonia: a UK/Italian contingent valuation study. *Environmental Conservation,* **30**, 139–46.

Kahneman, D. and Knetsch, J. L. (2005). Valuing public goods: The purchase of moral satisfaction. In L. Kalof and T. Satterfield, Eds., *The Earthscan Reader in environmental values.* Earthscan, Sterling, pp. 229–43.

Kawagebo Culture Association (KCA). (2004). *Report on Sacred Sites in the Kawagebo Area.* Submitted to The Nature Conservancy China Program.

Keeney, R. L., and Gregory, R. S. (2005). Selecting attributes to measure the achievement of objectives. *Operations Research,* **53**, 1–11.

Kempton, W., Boster, J. S., and Hartley, J. A. (1995). *Environmental values in American culture.* MIT Press, Cambridge, MA.

Kramer, R. A., and Mercer, D. E. (1997). Valuing a global environmental good: US residents' willingness to pay to protect tropical rain forests. *Land Economics,* **73**, 196–210.

Krutilla, J. V. (1967). Conservation reconsidered. *American Economic Review,* **57**, 777–86.

Lim, E.-M., and Honjo, T. (2003). Three-dimensional visualization forest of landscapes by VRML. *Landscape and Urban Planning,* **63**, 175–86.

Lin, N. (2001). *Social capital: a theory of social structure and action.* Cambridge University Press, Cambridge.

Lockwood, M. (1998). Integrated value assessment using paired comparisons. *Ecological Economics,* **25**, 73–87.

Louv, R. (2008). *Last child in the woods: saving our children from nature-deficit disorder. Updated and expanded edition.* Algonquin Books, Chapel Hill.

Louviere, J. J., Hensher, D. A., and Swait, J. D. (2000). *Stated choice methods: analysis and applications.* Cambridge University Press, Cambridge.

Ma, J. (2005). Sacred natural sites and conservation in the Meili area. In J. Ma and J. Chen, Eds., *Tibetan Culture and Biodiversity Conservation.* Yunnan Nationality Press, China, pp. 33–40.

McCauley, D. J. (2006). Selling out on nature. *Nature,* **443**, 27–8.

McDaniels, T. L., and Trousdale, W. (2005). Resource compensation and negotiation support in an aboriginal context: Using community-based multi-attribute analysis to evaluate non-market losses. *Ecological Economics,* **55**, 173–86.

Martinez-Alier, J., Munda, G., and O'Neill, J. (1998). Weak comparability of values as a foundation for ecological economics. *Ecological Economics,* **26**, 277–86.

Meadows, D. H. (2001). Dancing with systems. *Whole Earth* (Winter).

Millennium Ecosystem Assessment. (2005). *Ecosystems and human well-being: synthesis.* Island Press, Washington, DC.

Milton, K. (2002). *Loving nature: towards an ecology of emotion.* Routledge, London.

Mitchell, R. C., and Carson, R. T. (1989). *Using surveys to value public goods: the contingent valuation method,* 3rd edn. Resources for the Future, Washington, DC.

Munda, G. (2004). Social multi-criteria evaluation: Methodological foundations and operational consequences. *European Journal of Operational Research,* **158**, 662–77.

National Research Council (US). Committee on Assessing and Valuing the Services of Aquatic and Related Terrestrial Ecosystems. (2005). *Valuing ecosystem services: toward better environmental decision-making.* National Research Council, Washington, DC.

Norton, B. G., and Hannon, B. (1997). Environmental values: A place-based theory. *Environmental Ethics,* **19**, 227–45.

O'Neill, J., and Spash, C. L. (2000). Conceptions of value in environmental decision-making. *Environmental Values,* **9**, 521–35.

Pergams, O. R. W., and Zaradic, P. A. (2006). Is love of nature in the US becoming love of electronic media? 16-year downtrend in national park visits explained by watching movies, playing video games, internet use, and oil prices. *Journal of Environmental Management,* **80**, 387–93.

Rees, W. E. (1998). How should a parasite value its host? *Ecological Economics,* **25**, 49–52.

Rolfe, J., Bennett, J. and Louviere, J. (2000). Choice modelling and its potential application to tropical rainforest preservation. *Ecological Economics,* **35**, 289–302.

Russell, C., Dale, V., Lee, J. S., *et al.* (2001). Experimenting with multi-attribute utility survey methods in a multidimensional valuation problem. *Ecological Economics,* **36**. 87–108.

Rutherford, M. B., Knetsch, J. L., and Brown, T. C. (1998). Assessing environmental losses: judgments of importance and damage schedules. *Harvard Environmental Law Review,* **22**, 51–101.

Saaty, T. L. (1996). *Multicriteria decision making: the analytic hierarchy process: planning, priority setting, resource allocation,* 2nd edn. RWS Publications, Pittsburgh.

Sable, K., and Kling, R. (2001). The double public good: A conceptual framework for "shared experience" values associated with heritage conservation. *Journal of Cultural Economics,* **25**, 77–89.

Sagoff, M. (1998). Aggregation and deliberation in valuing environmental public goods: A look beyond contingent pricing. *Ecological Economics,* **24**, 213–30.

Satterfield, T. (2001). In search of value literacy: suggestions for the elicitation of environmental values. *Environmental Values,* **10**, 331–59.

Satterfield, T., and Roberts, M. (2008). Incommensurate risks and the regulator's dilemma: considering culture in the governance of genetically modified organisms. *New Genetics and Society,* **27**, 201–16.

Shrader-Frechette, K. S. (1985). *Science policy, ethics, and economic methodology of social science: some problems of technology assessment and environmental impact analysis.* Springer, Dordrecht, The Netherlands.

Spash, C. L. (2007). Deliberative monetary valuation (DMV): Issues in combining economic and political processes to value environmental change. *Ecological Economics,* **63**, 690–9.

Spash, C. L. (2008). Deliberative monetary valuation and the evidence for a new value theory. *Land Economics,* **84**, 469–88.

Throsby, D. (2001). *Economics and culture.* Cambridge University Press, Cambridge.

Turner, N. J., Gregory, R., Brooks, C., *et al.* (2008). From invisibility to transparency: identifying the implications. *Ecology and Society,* **13**(2), 7.

Turner, N. J., and Turner, K. L. (2008). "Where our women used to get the food": cumulative effects and loss of ethnobotanical knowledge and practice; case study from coastal British Columbia. *Botany-Botanique,* **86**, 103–15.

US Census Bureau. (2003). *U.S. Census 2000.*

Wallace, K. J. (2008). Ecosystem services: Multiple classifications or confusion? *Biological Conservation,* **141**, 353–4.

Wunder, S., and Vargas, M. T. (2005). *Beyond "markets": Why terminology matters.* Ecosystem Marketplace. The Katoomba Group.

CHAPTER 13

Terrestrial biodiversity

Erik Nelson, D. Richard Cameron, James Regetz, Stephen Polasky, and Gretchen C. Daily

13.1 Introduction

Biodiversity is the variety of life at all levels of organization, from genetically distinct populations to species, habitats, ecosystems, and biomes (Leopold 1949; Wilson 1992). While biodiversity influences the provision of all ecosystem services, and is the basis for many (Sekercioglu *et al.* 2004; Díaz *et al.* 2005), it also inspires conservation for its own sake (e.g., Ehrlich and Ehrlich 1982, Chapter 2). To protect what remains of declining biodiversity (Hughes *et al.* 1997; Dirzo and Raven 2003; Worm *et al.* 2006) conservationists seek to identify habitat conservation networks that maximize habitat or species persistence. The design of these networks typically uses species distribution maps (e.g., Ceballos *et al.* 2005) and an understanding of the factors that affect species presence and persistence on the landscape (e.g., Sekercioglu *et al.* 2007). More recently, the design of conservation networks have considered the financial and opportunity cost of network conservation and implementation (Ando *et al.* 1998; Wilson *et al.* 2007; Polasky *et al.* 2008). Systematic conservation planning (SCP) (Margules and Pressey 2000) marries many of these network design principles and conservation organizations implement SCP principles in their work (Groves *et al.* 2002).

We present three relatively simple biodiversity models that spring from this conservation planning literature. The models are straightforward to implement and are designed for analysis at landscape scales. Our simplest model, tier 1, combines basic information about land cover and threats to biodiversity to produce habitat-quality and habitat rarity maps. In tier 1 we assume that protection of a variety of high-quality habitats will confer protection to their component species and populations (Groves *et al.* 2002).

Because the tier 1 biodiversity model uses data that are available virtually everywhere in the world and empirical data on the status of rare, endemic, and other species of conservation concern are unavailable for many places, a habitat analysis, instead of a species-based approach is commonly implemented as the first phase of a conservation assessment.

In tier 2 we assume data on potential distributions or ranges of species and on habitat suitability (as gauged by breeding and foraging activity) are available to us. We present two tier 2 models that rely on this species-specific data. The first model combines these data to calculate the relative contribution of a parcel's habitat to the overall quantity of suitable habitat across a landscape or region. This approach is similar to deductive species distribution modeling (Stoms *et al.* 1992), the rarity-weighted richness methodology (Williams *et al.* 1996), and the Biological Intactness Index (BII) system (Scholes and Biggs 2005). The second tier 2 model aggregates information on species distributions and habitat suitability into a single landscape score. This score is derived using species–area relationships to translate habitat area into a measure of landscape-wide biodiversity status, based on work by Sala *et al.* (2005) and Pereira and Daily (2006). In addition, we can incorporate tier 1 output into tier 2 modeling in order to weight suitable habitat area by quality. Such quality weighting of habitat is common in algorithms that are used to select networks of areas for conservation (e.g., Schill and Raber 2008).

13.2 Tier 1: habitat-quality and rarity model

Habitat-quality depends on its proximity to human land uses and the intensity of these land uses.

Generally, habitat-quality is degraded as the intensity of nearby human land use increases (Nelleman *et al.* 2001; Forman *et al* 2003). For example, a forest near a city in a developing country may be stripped of much of its timber and other non-timber forest products while forests isolated from people will tend to be less disturbed (see Chapter 8). Or a wetland near agricultural lands may have greater water quality issues then a wetland surrounded by other wetlands. These are both examples of the "edges" that human land use creates on the boundaries of near-by habitat. In general, edges facilitate entry of various degraders into habitat including predators, competitors, invasive species, toxic chemicals, and humans. In addition, a high density of human land use in an area means that any near-by habitat will tend to be isolated, further reducing the habitat's ability to contribute to species persistence. Our definition of habitat-quality is similar to the notion of habitat integrity as used by many conservation organizations. Habitat with high integrity, like high-quality habitat, is relatively intact and has structure and function within the range of historic variability.

This tier 1 model assumes that habitat areas with higher quality scores are better able to maintain their full complement of biodiversity over time than those areas with lower scores. This does not mean, however, that areas with lower quality scores are bereft of rare species or are not important sources of biodiversity (see Box 13.1). For example, in the U.S. some of the last remaining populations of the most threatened species are on or are immediately adjacent to

---

Box 13.1 Integrating biodiversity and agriculture: a success story in South Asia

Jai Ranganathan and Gretchen C. Daily

What is the long-term prospect for harmonizing food production and biodiversity conservation? Recent work in the Neotropics shows that native species across a wide range of taxa can persist in farming countryside, decades after land clearing, if critical landscape features are maintained (e.g., Medellín *et al.* 2000; Daily *et al.* 2003; Horner-Devine *et al.* 2003; Ricketts 2004; Mayfield and Daily 2005; Şekercioğlu *et al.* 2007).

But will these agricultural landscapes continue to support native species over centuries to millennia? We surveyed bird diversity in an ancient agricultural landscape, cultivated continuously for over 2000 years and inhabited by people for at least 20 000 years (Ranganathan *et al.* 2008). On the fringes of the Western Ghats in southwest India, the area retains many habitat features known to be important for biodiversity (landscape heterogeneity, vegetative structural complexity, and native vegetation). Most of the land covers—rice, peanut, cashew, arecanut palm, extremely degraded shrublands, and native forest—have been present for well over 200 years. The native forests are designated either as Reserve Forest (relatively intact, no extraction officially allowed) and Minor Forest (extraction of non-timber products permitted).

We found a rich bird fauna, of which only 4% of species were restricted to Reserve Forest. Arecanut palm plantations and Minor Forest together harbored a distinct bird community, including 90% of the forest-affiliated species of most conservation concern, such as the Great Hornbill and the Malabar Grey Hornbill.

Arecanut is consumed by over 10% of people, concentrated in south and south-east Asia. In traditional cultivation practices of the area, arecanut is intercropped (with pepper, vanilla, coffee, banana, cacao, etc.), increasing the economic return to farmers and the structural complexity so critical for forest birds. Further, as arecanut palm plantations have high water demands, they displace rice production, in effect trading a low conservation value land cover with a much higher one. There is a strong economic incentive to maintain the Minor Forests of the region as forest, since they provide a critical component for traditional arecanut cultivation: leaf litter, used as mulch in plantations (Figure 13.A.1).

Arecanut may be key to conservation in south and southeast Asia, a region with critical conservation challenges on the planet. This example shows that agricultural landscapes can sustain high levels of biodiversity over centuries to millennia, and offers hope that other such production systems can be found. More generally, a likely key to sustainable protection of biodiversity is harmonizing its protection with the delivery of as many other ecosystem services as possible, so that people reap rewards far beyond the iconic species and endemic species that have been the more traditional focus of conservationists.

Figure 13.A.1 Ox-cart loads of leaf litter, collected from Minor Forest, bound for arecanut palm plantations.

Photo credit: Jai Ranganathan.

heavily modified landscapes (this proximity to human activity may be why they are rare in the first place; Scott *et al.* 2006).

13.2.1 Calculating a parcel's habitat-quality score

The tier 1 model builds a habitat-quality score for each parcel x on the landscape ($x = 1, 2, \ldots, X$ where a parcel can be any user-defined land unit, including a grid cell, a hexagon, a polygon, etc.) by mapping the location and intensity of all human land uses in the neighborhood of the parcel and then estimating the impact of this human land use on the parcel's habitat. We index land-use types that can have a major impact on habitat by $r = 1, 2, \ldots, R$. These land-use types can be coarsely defined, such as roads, built areas, croplands, and so forth, and can be supplemented as warranted by much more refined land-use categories,

such as high vegetation density forests due to fire suppression, different classes of roads (e.g., primary versus secondary), or different densities of development. The maps of land uses R do not have to be comprised of the same parcel units as the parcel map but all maps need to overlap in space. We use generic spatial units $y = 1, 2, \ldots, Y$ to allocate land uses that can impact habitat-quality across the landscape. Let D_{yr} indicate the amount or density of land-use type r in spatial unit y. For example, D_{yr} can measure kilometers of road ha^{-1} in grid cell y, people ha^{-1} in parcel y, or the hectares of cropland in hexagon y.

The impact of D_{yr} on habitat-quality in parcel x depends on several factors. First, some human land-use types have more impact per unit than others. Let w_r be the relative impact weighting for r. For example, if built areas have been estimated to have twice the impact of roads on habitat-quality, then $w_{built}/w_{roads} = 2$. An equal weight can be used across all land uses R if information on the relative impact of each source on habitat-quality is not known.

Second, D_{yr}'s influence on habitat in parcel x is affected by x's institutional and structural features. For example, a fence along the edge of a protected parcel might reduce the impact of nearby human land uses on habitat in the area. Or extreme slopes may prevent the entry of predator and competitor species (the protection accorded by institutional and/or structural features in x will vary by land-use r). Let the resistance to r in parcel x due to the parcel's institutional and structural features be given by the parameter $\beta_{xr} \in [0,1]$ with $\beta_{xr} = 1$ indicating maximum resistance to influence.

Third, D_{yr}'s influence on habitat in parcel x is affected by the distance between x and y; generally, D_{yr}'s affect on habitat in x declines with distance. Let d_{xyr} represent the distance between parcel x and spatial unit y on the map of land-use type r as measured by Euclidean distance, road network, or any other relevant distance measure and let $\alpha_r \in [0,1]$ be the parameter that determines how quickly r's influence on habitat-quality decays with distance, all else equal.

To determine the potential impact of all land-use types R on habitat in parcel x, given by D_x, we consider all three factors that affect D_{yr}'s relationship with habitat in parcel x,

$$D_x = \sum_{r=1}^{R}\sum_{y=1}^{Y} w_r \times f_r\left(\beta_{xr}, \alpha_r, d_{xyr}\right) \times D_{yr} , \qquad (13.1)$$

where $w_r \times f_r(\beta_{xr}, \alpha_r, d_{xyr})$ translates the value of D_{yr} into an impact on parcel x. The higher D_x is, the greater the potential impact of human land uses on the quality of habitat in x. A standard way to model $f_r(\beta_{xr}, \alpha_r, d_{xyr})$, but by no means the only way, is with an exponential decay function,

$$f_r\left(\beta_{xr}, \alpha_r, d_{xyr}\right) = e^{-\alpha_r \beta_{xr} d_{xyr}}. \qquad (13.2)$$

In Eq. (13.2), f_r declines, and therefore, the impact of D_{yr} on habitat in parcel x declines, as α_r, β_{xr}, and/or d_{xyr} increases. A parcel with a D_x score near 0 is relatively unaffected by human land use within the context of the landscape while a D_x near $\max_{x=1,\dots,X}\{D_x\}$ indicates that a parcel is greatly affected by human land use within the context of the landscape.

When translating D_x into a habitat-quality score in parcel x we assume that not all habitat types are equally susceptible to sources of disturbance. Let $j = 1,\dots, J$ index habitat types on the landscape where habitat can include land covers and uses highly modified from their natural state, e.g., urban areas, high-intensity croplands, roads. In general, not every land use/land cover type found on the map has to be included in the set of J; the modeler is free to choose the subset of land use/land cover types in set J based on their definition of habitat. We scale habitat type j's resistance to human land uses by $L_j \in [0,1]$, where higher values of L_j means j is more resistant and L_j is close to or equal to 0 for j that are highly modified from their natural state. The values of L_j for all j need to be gauged empirically, ideally based on biodiversity response. The habitat-quality score for each habitat type j in a parcel, Q_{xj}, and the parcel's aggregate habitat-quality score, Q_x, is given by

$$Q_{xj} = q\left(L_j, D_x\right) \qquad (13.3)$$

and

$$Q_x = q\left(\sum_{j=1}^{J}\left(\frac{A_{xj}L_j}{A_x}\right), D_x\right), \qquad (13.4)$$

where q is any function that is increasing in L_j and decreasing in D_x, $Q_{xj} = 0$ when $L_j = 0$, A_{xj} is the area of parcel x in habitat type j, and A_x is the area of parcel x (land use/land cover types that are not part of the habitat set j are not given habitat-quality scores). A parcel with a Q_x score near 0 includes little habitat area and/or contains habitat that is severely degraded within the context of the landscape while a Q_x score near $\max_{x=1,\dots,X}\{Q_x\}$ indicates that a parcel is replete with habitat and is approximately of the highest quality within the context of the landscape.

13.2.2   Calculating a parcel's rarity score

While mapping habitat-quality can help identify the areas where biodiversity is likely to be more or less threatened on the landscape, it is also important to prioritize habitat types based on their relative rarity. We define a habitat type's rarity as the amount of the habitat type currently found on the landscape relative to the amount of that habitat type that existed on the landscape at some reference time. The ideal reference landscape would be from a period prior to substantial anthropogenic conversion of land (Scholes and Biggs

2005). For example, a description of the distribution of major habitat types in the Willamette Basin of Oregon, USA, from the year 1850 (Christy *et al.* 2000) would meet this criterion (even though Native Americans shaped their landscapes in many ways as well, see Mann 2005). If a habitat type that was relatively abundant on the reference landscape is now rare on the modern landscape, then species dependent on that habitat type will likely have declined.

Assuming we have a reference landscape, we calculate the relative rarity of habitat type j on a modern landscape as

$$Y_j = I_j \left(1 - \frac{\sum_{x=1}^{X} A_{xj}}{A_{jb}} \right), \tag{13.5}$$

where $Y_j = 0$ if $A_{jb} = 0$, I_j is equal to 1 if j is a natural land cover (as opposed to a j that is significantly managed) and is equal to 0 otherwise, and A_{jb} is the area of habitat type j on the reference landscape. The closer Y_j is to 1, the rarer the habitat type j is on the modern landscape vis-à-vis the reference landscape.

If appropriate reference maps are not available (and as a useful check even when they are), another option is to weight the relative scarcity of habitat types on the modern landscape according to a rarity metric from NatureServe. NatureServe measures the status of habitat types with a metric that ranges from 1 through 5 where a 1 indicates critical imperilment across a region or the globe, and a 5 indicates that the habitat type is abundant and secure (NatureServe 2008). In this approach, the relative rarity of habitat type j on the modern landscape is given by

$$Y_j = \left(\frac{6 - NS_j}{5} \right) \left(1 - \frac{\sum_{x=1}^{X} A_{xj}}{A} \right), \tag{13.6}$$

where NS_j is NatureServe's conservation status of habitat type j and A is the area of the landscape. Set $NS_j = 6$ for all j that are significantly managed. If possible, we use NatureServe's regional scores for NS_j instead of their global scores because a habitat type that is relatively secure from a global perspective may be relatively scarce in the particular landscape.

Once we have calculated Y_j for each habitat type, we quantify the overall rarity of habitat types in parcel x on the modern landscape by taking the area-weighted average of x's Y_j scores,

$$Y_x = \sum_{j=1}^{J} Y_j \left(\frac{A_{xj}}{A_x} \right), \tag{13.7}$$

where the more rare the habitat area in x the closer Y_x is to 1.

We can combine data on habitat-quality and rarity to provide a measure that can be used to prioritize conservation efforts. Rare habitats with high quality could represent one conservation priority. Such areas are identified by parcels with high $V_x \in [0,1]$ scores where

$$V_x = \frac{Q_x}{\max\limits_{m=1,\ldots,M} \{Q_m\}} \times \frac{Y_x}{\max\limits_{m=1,\ldots,M} \{Y_m\}}, \tag{13.8}$$

where $m = 1, 2, \ldots, M$ also index parcels. Another simple composite habitat-quality and rarity score is given by $VW_x \in [0,1]$,

$$VW_x = \gamma \frac{Q_x}{\max\limits_{m=1,\ldots,M} \{Q_m\}} + (1 - \gamma) \frac{Y_x}{\max\limits_{m=1,\ldots,M} \{Y_m\}}, \tag{13.9}$$

where $\gamma \in [0,1]$ determines the weight given to quality versus rarity when tracking the conservation value of each parcel.

13.3 Tier 2 models of terrestrial biodiversity

The major drawback with the tier 1 approach is that it does not necessarily indicate how well the landscape is meeting the specific needs of species of concern. For example, if the species of concern are generally located in areas of low-quality habitat then conserving the remaining patches of high-quality habitat may not generate as great a conservation return as restoring valuable degraded habitat. In tier 2 we complement tier 1 results by assessing how species react to and use the landscape and their spatial relationship with habitat-quality patterns.

To this end we consider two alternative formulations of tier 2 models that measure biodiversity patterns and status using species-specific data. The first tier 2 model measures the marginal contribution of each parcel to biodiversity on the entire landscape. The model can also track the change in a parcel's marginal contribution to biodiversity as the land use/land cover (LULC) pattern on the landscape changes over time. The second tier 2 model summarizes an entire landscape's ability to support a suite of species.

13.3.1 Parcel-level contribution to biodiversity conservation on the landscape

For each parcel x on the landscape we calculate a marginal biodiversity value (MBV_x) that measures the proportion of the landscape's total modeled biodiversity supplied by that parcel. The MBV of a parcel is a function of: (1) the number of modeled species whose potential ranges overlap with that parcel, and (2) the fraction of each species' suitable habitat area that the parcel contains. We deem a habitat type suitable for a species if the species has been observed intermittently or consistently using the habitat for breeding, foraging, migration, or other life-sustaining purposes. To understand MBV, consider a simple example for a single species. Five equally sized parcels on a landscape of 100 parcels comprise the species' geographic range. Each of the five parcels in the species' range is completely covered by equally suitable habitat. Then, in our MBV model, each of these five parcels has an MBV of 0.2, and all other parcels have an MBV of 0. Computing MBV on a large landscape with multiple species is a straightforward generalization of this simple example that allows for different parcel areas, species-specific weights, multiple habitat types within a parcel, and incorporation of tier 1 habitat-quality model scores.

The calculation of MBV requires a potential range map for each modeled species and a compatibility score for each species/LULC combination that indicates the degree of suitability of LULC j for species s (s = 1, 2,...,S) (in tier 1 j indexed the more narrowly defined habitat types). We set $H_{xs} = 1$ if parcel x is in the potential range of species s, and equals 0 otherwise. Ideally, each species' potential range map is based on its estimated geographic range in the pre-modern era. However, such data are only available for a limited subset of species and may not be particularly reliable. Therefore, we most often define H_{xs} with recently observed patterns of species distribution (however, see Rondinini *et al.* (2006) for a discussion of the biases introduced in biodiversity modeling and mapping when using maps of recently observed ranges).

We incorporate the degree to which LULC j supports species s by setting $C_{sj} = 0$ for unsuitable habitat and $C_{sj} > 0$ for suitable habitat where $C_{sj} = 1$ indicates the most preferred habitat. The per-unit-area marginal value of each LULC type j for species s on the landscape is,

$$\hat{C}_{sj} = \frac{C_{sj}}{\sum_{x=1}^{X} \sum_{k=1}^{K} C_{sk} A_{xk} H_{xs}} \qquad (13.10)$$

where k indexes LULC types as well, A_{kx} is the area of LULC k in parcel x, and the denominator gives species s' total suitable habitat area on the landscape.

Next, we use $\hat{C}_{sj}$ to calculate the MBV score on parcel x,

$$MBV_x = \sum_{s=1}^{S} w_s H_{xs} \left(\sum_{j=1}^{J} A_{xj} \hat{C}_{sj} \right), \qquad (13.11)$$

where MBV_x is an estimate of each parcel's contribution to the landscape's total supply of biodiversity, biodiversity consists of the S modeled species, and w_s is the weight assigned to species s. If we disregard w_s temporarily, a parcel will score highly on the MBV metric if it contains suitable habitat for the species that have little suitable habitat elsewhere on the landscape or if it contains reasonable shares of suitable habitat for many species. Not all species need be weighted equally. For example, threatened and endangered species may be given greater weight in conservation planning and implementation. If $\sum_{s=1}^{S} w_s = 1$ then $\sum_{x=1}^{X} MBV_x = 1$.

Alternative formulations of the per-unit-area marginal value that incorporates habitat-quality scores from tier 1 are

$$\hat{C}_{sj} = \frac{C_{sj}}{\sum_{x=1}^{X} \sum_{j=1}^{J} Q_{xj} C_{sj} A_{xj} H_{xs}} \qquad (13.12)$$

if we have habitat-quality scores for each LULC (habitat) type j in each parcel x or

$$\hat{C}_{sj} = \frac{C_{sj}}{\sum_{x=1}^{X} Q_x \sum_{j=1}^{J} C_{sj} A_{xj} H_{xs}} \qquad (13.13)$$

if we only have or prefer to use parcel-level habitat-quality scores. We assume quality affects suitable habitat in a linear manner; alternative rates of suitable habitat modification can be used if we have the data to support such relationships. If we use a version of $\hat{C}_{sj}$ given by Eq. (13.12) or (13.13) we calculate habitat-quality-adjusted MBV_x with a modified version of Eq. (13.11),

$$MBV_x = \sum_{s=1}^{S} w_s H_{xs} \left(\sum_{j=1}^{J} Q_{xj} A_{xj} \hat{C}_{sj} \right) \qquad (13.14)$$

or

$$MBV_x = Q_x \sum_{s=1}^{S} w_s H_{xs} \left(\sum_{j=1}^{J} A_{xj} \hat{C}_{sj} \right), \qquad (13.15)$$

where Eq. (13.14) uses the $\hat{C}_{sj}$ from Eq. (13.12) and Eq. (13.15) uses the $\hat{C}_{sj}$ from Eq. (13.13).

13.3.1.1 *Tracking changes in parcel-level marginal biodiversity value*

The *MBV* score compares the biodiversity value of parcels on a landscape for a given point in time, but it can be misleading to compare *MBV* scores of a particular parcel across time. Recall the simple example above for one species with five equally sized parcels in its range, each entirely covered with habitat preferred by the species, the *MBV* score for each parcel is 0.2. Consider a scenario in which all suitable habitat for the species in four of these parcels is lost, while half of the fifth parcel's area is converted to unsuitable habitat. In this new landscape, the *MBV* score of the fifth parcel would increase to 1 despite the loss of half of its suitable habitat. This change in score may provide useful information in terms of the parcel's contribution to remaining habitat, but it obscures determination of whether the amount of biodiversity supported by a parcel has increased or decreased over time.

We therefore use an alternative biodiversity statistic, the relative marginal biodiversity value (*RMBV*), to track the contribution of a parcel to the landscape's level of biodiversity through time. Calculation of *RMBV* scores and associated *RMBV* ratios use the same input data as *MBV*. The calculation of *RMBV* uses the per-unit-area value of each LULC type *j* as evaluated in the base landscape. This quantity is then applied to a future specification of the landscape.

Using a base landscape map (i.e., the first map in a chronological series of maps for the landcape), we calculate $\hat{C}_{sj}$ as in Eq. (13.10), (13.12), or (13.13). This statistic is denoted as $\hat{C}_{sjb}$ where the *b* subscript indicates the year associated with the baseline landscape. To obtain *RMBV* scores for parcel *x* at time *t* (where *t* > *b*), we plug $\hat{C}_{sjb}$ and *x*'s LULC mix from year *t* into,

$$RMBV_{xt} = \sum_{s=1}^{S} w_s H_{xst} \left(\sum_{j=1}^{J} A_{xjt} \hat{C}_{sjb} \right) \qquad (13.16)$$

$$RMBV_{xt} = \sum_{s=1}^{S} w_s H_{xst} \left(\sum_{j=1}^{J} Q_{xjt} A_{xjt} \hat{C}_{sjb} \right) \qquad (13.17)$$

or

$$RMBV_{xt} = Q_{xt} \sum_{s=1}^{S} w_s H_{xst} \left(\sum_{j=1}^{J} A_{xjt} \hat{C}_{sjb} \right) \qquad (13.18)$$

where we use Eq. (13.16) if $\hat{C}_{sjb}$ was calculated with Eq. (13.10), Eq. (13.17) if $\hat{C}_{sjb}$ was calculated with Eq. (13.12), Eq. (13.18) if $\hat{C}_{sjb}$ was calculated with Eq. (13.13), indexing H_{xs} by t H_{xst} acknowledges that species potential range can change over time due to climate change or other landscape-level disturbances, and A_{xjt} is the area of LULC *j* in a parcel *x* at time *t*.

Finally, for each time *t*, we calculate the ratio $RMBV_{xt}/MBV_x$ for each parcel *x*. This so-called $RMBV_{xt}$ ratio is greater than 1 if parcel *x*'s LULC and habitat-quality composition (if included in *RMBV*) has changed vis-à-vis the base landscape in a manner that produces a net increase in the parcel's suitable habitat across all species *S*.

13.3.2 Landscape level biodiversity model

The *MBV* and *RMBV* models described above are intended primarily for evaluating and comparing the biodiversity supplied by individual parcels, either within a landscape (the *MBV* model) or across time on the landscape (the *RMBV* model). The above methodologies do not yield a single, satisfactory landscape score, however, that can be used for assessing the trade-offs between landscape-level measures of biodiversity and ecosystem services under different landscape scenarios. To remedy this shortcoming of the *MBV* model, we use species–area relationships (SAR) to develop a landscape-level biodiversity score that requires many of the same data inputs used in the *MBV* or *RMBV* models.

The SAR of biogeography (MacArthur and Wilson 1967) specifies the following relationship between total habitat area (*A*) and species richness (*S*):

$$S = cA^z , \qquad (13.19)$$

where *c* is a constant and *z* indicates the rate of species accumulation as *A* increases and is typically between 0.1 and 0.7. Ideally, the values of the parameters c and z are calibrated to observed patterns on the studied landscape. This relationship between richness and habitat area has proven to be one of the most empirically robust patterns in all of ecology, though with important nuances (Rosenzweig 1995).

While the SAR is typically used to predict species richness, we use it to determine how well a landscape supports species. Each species receives a SAR score based on the proportion of area in that species' potential geographic range that is suitable habitat. The SAR score for species s at time t is

$$SAR_{st} = \gamma_{st} \frac{\left(\sum_{x=1}^{X} \sum_{j=1}^{J} A_{xjt} H_{xst} C_{sjt}\right)^{z_{st}}}{\left(\sum_{x=1}^{X} A_x H_{xst}\right)^{z_{st}}}, \quad (13.20)$$

where γ_{st} is a species-specific constant (the c in Eq. (13.19)), z_{st} is the species–area function power parameter for species s, and all other variables are as before. We index γ, H, C, and z with t to allow for changes in these parameters and variables over time. In Eq. (13.20) we normalize the observed species-area relationship value (the numerator) with the species–area relationship value that would hold if the species' entire potential range were in perfectly suitable habitat (the denominator). Therefore, SAR_{st} is a measure of the fraction of total potential support that the landscape provides for species s at time t where complete support on the landscape at time t is given by $SAR_{st} = 1$ (assuming $\gamma_{st} = 1$).

Species that have lost a greater proportion of their suitable habitat across their reference-era range are at a greater risk of extinction than species that have lost a smaller proportion, regardless of the absolute size of habitat loss (Channell and Lomolino 2000; Abbitt and Scott 2001; Scholes and Biggs 2005). Therefore, just like our MBV and RMBV models, our SAR model may be more accurate if each species' potential range map (i.e., H) is based on its estimated spatial distribution in the pre-modern era in lieu of recently observed ranges (though reliable estimates of pre-modern ranges are not available for most species and places).

Regardless of the potential range map used, research has also shown that, all else equal, range-restricted species tend to have higher extinction risks than large-range species (Newmark 1995; Purvis *et al.* 2000; Pimm

and Brooks 2000; Parks and Harcourt 2002; Cardillo *et al.* 2006). SAR_{st} reflects this tendency as a one-unit decrease in suitable habitat for a species (the numerator of Eq. (13.20)) with small range (a small denominator value) reduces SAR_{st} more than a similar one-unit decrease for a geographically widespread species.

We can modify Eq. (13.20) to include the habitat-quality as calculated in tier 1,

$$SAR_{st} = \gamma_{st} \frac{\left(\sum_{x=1}^{X} \sum_{j=1}^{J} Q_{xjt} A_{xjt} H_{xst} C_{sjt}\right)^{z_{st}}}{\left(\sum_{x=1}^{X} A_x H_{xst}\right)^{z_{st}}}, \quad (13.21)$$

$$SAR_{st} = \gamma_{st} \frac{\left(\sum_{x=1}^{X} Q_{xt} \sum_{j=1}^{J} A_{xjt} H_{xst} C_{sjt}\right)^{z_{st}}}{\left(\sum_{x=1}^{X} A_x H_{xst}\right)^{z_{st}}}, \quad (13.22)$$

where again habitat-quality modifies suitable habitat in a linear manner and the reference species range (the denominators of Eqs. (13.21) and (13.22)) assumes habitat of the highest quality is uniformly present across s' range ($Q_{xjt} = 1$ for all x, j, and t combinations and $Q_{xt} = 1$ for all x and t combinations). These equations are such that, all else equal, a smaller area of high-quality suitable habitat can generate a higher SAR_{st} score than a larger area of low-quality suitable habitat.

We can generate a single SAR_t score for the collection of modeled biodiversity on the landscape by taking a weighted sum of all SAR_{st} scores,

$$SAR_t = \sum_{s=1}^{S} w_s SAR_{st}, \quad (13.23)$$

where w_s is the weight attached to species s.

13.4 Tier 1 and 2 examples with sensitivity analysis

We illustrate the tier 1 and 2 models with projected LULC change in the region covered by the Sierra Nevada Conservancy (a California state agency) (see Figure 13.1). The $101\,000$ km^2 region encompasses the Sierra Nevada ecoregion and includes portions of six other ecoregions, with elevation ranging from 100 to 4421 m, including the highest peak in the contiguous United States. The region has approxiamtely 3500 species of vascular plants, including more than 400 endemics (Shevock 1996); 293 birds, 135 mammals, 46 reptiles, 37 amphibians, and 61 fish (CDFG-CIWTG 2007).

We used Sierra Nevada housing-density projections for 2030 from the Spatially Explicit Regional Growth Model (Theobald 2005) to create two 2030 landscape scenarios. The Conservation scenario assumes that new development will have a minimum housing density of one unit per 0.6 ha. The developed land footprint increases 140% (from 78 389 ha in 2000 to 187 769 ha in 2030) under this scenario. The Growth scenario includes lower density housing development options (a minimum of one unit per 4 ha). The developed land footprint increases 927% (from 78 389 ha in 2000 to 805 362 ha in 2030) under this scenario. The region's LULC map in 2000 is the base landscape. (See Davis *et al.* (2006) for a much more complex and thorough mapping of conservation priority areas in the Sierra given the spatial distribution of habitats, species, land tenure, and the predicted spatial pattern of development in the region, i.e., a tier 3 approach.)

For the purposes of the examples below, the areas that develop housing in these scenarios are always

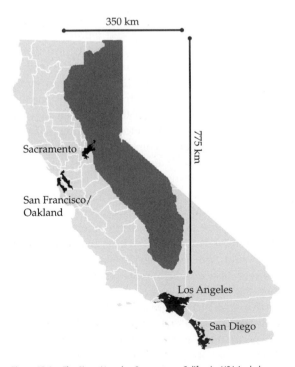

350 km

775 km

Sacramento

San Francisco/ Oakland

Los Angeles

San Diego

Figure 13.1   The Sierra Nevadas Conservancy, California, USA in dark gray.

referred to as "urban," though this will overstate the impact of low density development as exemplified by the Growth scenario on species that can tolerate lower density residential development. Further, we assume for simplicity's sake that no other land-use changes occur in either scenario, including no expansion of the region's transportation network. Obviously, no matter the actual future, new roads will be built in the region as its population and urban footprint expands.

13.4.1   Tier 1

We conducted twelve separate tier 1 habitat-quality mapping analyses; in each case we used a different combination of w_r and L_j values to map parcel-level habitat-quality scores, Q_x (the set of 1 habitat types remain constant across all analyses). In this illustrative example, human land uses that affect habitat-quality are roads, urban areas, and agricultural fields. In Figure 13.2 (Plate 6) we present tier 1 results for two such unique combinations of w_r and L_j values. In general the greatest difference between the two is the relative weight assigned to the human land uses of roads, urban areas, and agriculture. In the "Roads" parameter combination we assume roads are more deleterious to habitat-quality than urban areas and agriculture and in the "Urban" parameter combination we assume urban areas are more disruptive then the other two land uses. See the chapter's supplementary online material (SOM) for model details.

The "Roads" and "Urban" parameter combinations produce the two most extreme distributions of habitat-quality scores on the baseline and two future maps. Specifically, of the twelve parameter combinations, the Roads parameter combination produces a distribution of Q_x values that is most skewed to the left on the unit scale (low values) under both future scenarios whereas the Urban parameter combination produces a distribution of Q_x values that is the most skewed to the right under both future scenarios (high values).

Not surprisingly, regardless of the parameter combination used, the Growth scenario landscape consistently, because of its larger footprint change, produces lower Q scores when compared to the Conservation scenario landscape. Only a handful of parcels have lower Q scores under the Conservation scenario landscape than they do under the Growth

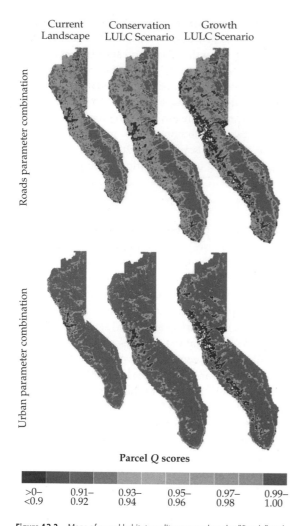

Parcel Q scores

>0–<0.9	0.91–0.92	0.93–0.94	0.95–0.96	0.97–0.98	0.99–1.00

Figure 13.2 Maps of parcel habitat-quality scores when the "Roads" and "Urban" parameter combinations are used in the Sierra Nevada illustrative example. We ran the tier 1 model on a grid map with a cellular resolution of 400 m × 400 m (16-ha grid cells). In these maps we present the mean habitat-quality score (Q) of all grid cells within 500 hectare hexagons. There are 23 042 500-ha hexagons in the Sierra Nevada Conservancy. In both future LULC scenarios the majority of residential development is centered on Sacramento, and generally along the western foothills. In the Growth scenario, montane hardwood is the land cover type that loses the most area to development (158 268 ha). In the Conservation scenario, annual grassland is the land cover type that loses the most area (17 798 ha). See the chapter's SOM for all tier 1 model details. (See Plate 6.)

landscape scenario (see Figure 13.S1 in the chapter's SOM) using either set of parameter combinations. Further, because the only LULC change on the landscape is development expansion, it is not surprising

that the habitat-quality trade-off between the two future scenarios seems starker when using the Urban parameter combination rather than the Road parameter combination (see Figure 13.S2 in the SOM).

Spatially, the Urban parameter combination leaves many more large patches of relatively high-quality habitat (the darkest green on the maps) than the Roads parameter combination (Figure 13.2; Plate 6). Given our uncertainty regarding the direction of development in the future and the relative impact of these sources of human land use on habitat-quality, it is appropriate to conclude that the areas that are the darkest green (high-habitat-quality) on all four future scenario-parameter combination variations in Figure 13.2 (Plate 6) are the areas most likely to contain high-quality habitat in the future.

13.4.2 Tier 2 analyses

We calculated MBV, $RMBV$, and $RMBV$ ratio scores (without considering habitat-quality) for various subgroups of herpetofauna—federally endangered herpetofauna (FE), federally threatened herpetofauna (FT), California herpetofauna of special concern (CSC), amphibians (A), and reptiles (R)—that have at least part of their range in the Sierra Nevada Conservancy. In Figure 13.3 histograms of the distribution of parcel $RMBV$ ratio values across the landscape (i.e., the relative change in parcels' MBV scores from 2000 to 2030) under each scenario for each subgroup of herpetofauna using mean C_{sj} values for each s and j combination are given. The FT ($N = 5$), FE ($N = 3$) and CSC ($N = 25$) subgroups experience severe reductions in effective habitat area in many parcels under the Growth scenario (recall that a $RMBV$ ratio value near 0 indicates a significant loss of habitat in the parcel over time for the modeled species). These histograms also show that the reduction in habitat under the Conservation scenario most acutely affects the same subgroups.

We map the FT subgroup species year 2000 MBV and RMBV ratio values under both scenarios using minimum and maximum C_{sj} values in Figure 13.4 (Plate 7). The MBV maps (Figure 13.4; Plate 7) indicate that the most important habitat for FT

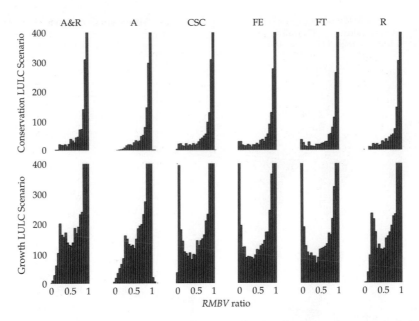

Figure 13.3   Frequency of *RMBV* ratio parcel scores using mean species-habitat suitability scores for all herpetofauna subgroups in the Sierra Nevada illustrative example. The frequency axis of each histogram has been truncated at 400. The California herpetofauna of special concern (CSC), federally endangered herpetofauna (FE), and federally threatened herpetofauna (FT) subgroups experience the greatest drop in effective habitat area relative to the 2000 landscape. The amphibian (A) herpetofauna subgroup has the least number of parcels with significant reductions in effective habitat area under both future LULC scenarios. The other herpetofauna subgroups include all (A&R) and reptiles (R). See the chapter's SOM for all tier 2 model details.

species in 2000 were in the western foothills. It is in this area that most urbanization is expected to occur by 2030 under both scenarios, albeit to a much greater extent in the Growth scenario. The maps appear to be fairly insensitive to the range in C_{sj}. Finally, these maps explain why the A ($N = 35$) subgroup has the most right-skewed histogram of parcel *RMBV* ratio scores: species in subgroup A are more likely than any other subgroup of species to be found in the higher elevation areas, the areas experiencing the least development.

Data in Figure 13.5 corroborates the decrease in FT and CSC subgroup suitable habitat as suggested by the *RMBV* ratio histograms. These two subgroups experience the greatest relative decrease in their *SAR* scores (without considering habitat-quality) under both scenarios no matter which C_{sj} and z values we use (low, mean, or high for the C_{sj} scores for each s and j combination and 0.11, 0.25, 0.64, and 1 for the z scores). Again, this is due to the fact that many of the taxa included in these subgroups are primarily found in the foothill areas and the Central Valley to the west, the

zone in which most of the projected development is expected to occur. At what point a decline in *SAR* indicates an immediate and imminent threat to the persistence of a subgroup is an empirical question. See the chapter's SOM for tier 2 model details.

13.4.3 Incorporating tier 1 results in a tier 2 analysis

In many cases we will not be able to perform a tier 2 analysis due to a lack of species-specific data. However, if we do have the wherewithal to perform a tier 2 analysis, as we do in this example, if would be judicious on our part to perform it twice, once with tier 1 output incorporated and another time without (e.g., the results above). Such an analysis will indicate some of the spatial relationships between habitat-quality, species habitat, and the pattern of habitat loss on the landscape over time.

By definition, MVB_{xt} and SAR_{st} scores calculated with habitat-quality scores will always be equal to

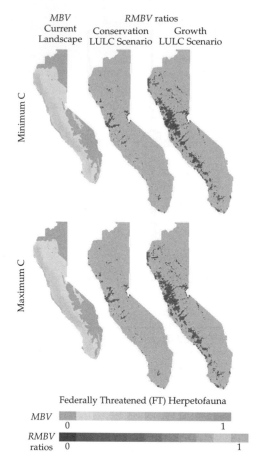

Federally Threatened (FT) Herpetofauna

Figure 13.4   The spatial distribution of MBV and RMBV ratio scores for federally threatened herpetofauna (FT) using minimum and maximum species-habitat suitability scores in the Sierra Nevada illustrative example. The Growth scenario creates a much greater loss in FT subgroup species effective habitat area in the foothills of the Sierra Nevada than the Conservation scenario does. See the chapter's SOM for all tier 2 model details. (See Plate 7.)

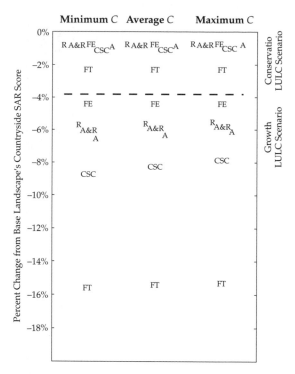

Figure 13.5   Percent change in a herpetofauna subgroup's SAR scores from 2000 to 2030 under both future LULC scenarios in the Sierra Nevadas illustrative example. The subgroups in the graph include all herpetofauna (A&R), amphibians (A), California herpetofauna of special concern (CSC), federally endangered herpetofauna (FE), federally threatened herpetofauna (FT), and reptiles (R). We conduct the analysis using minimum, average, and maximum C_{sj} values for all s,j combinations. See the chapter's SOM for all tier 2 model details.

or lower than those without. However, habitat-quality-modified $RMBV_{xt}$ ratio scores and the change in habitat-quality-modified SAR_{st} scores over time may be more or less than their unmodified counterparts. In general, habitat-quality-modified $RMBV$ ratio scores will be higher and decreases in habitat-quality-modified SAR scores will be less severe than their unmodified counterparts if the habitat that is lost over time tends to be of low quality on the base landscape.

In Table 13.1 we present the change in the CSC and FT herpetofauna subgroups' SAR statistics

when and when not incorporating Q_x scores from the Roads and Urban parameter combinations (using mean C_{sj} values for all s and j combinations). The inclusion of Q_x scores in the SAR calculations suggests: (1) that habitat conversion under the Growth scenario tends to occur more frequently on higher quality habitat than is does under the Conservation scenario, (2) if we use the Urban parameter combination to measure the relative impact of human land uses on habitat-quality then a greater proportion of higher quality habitat is developed by 2030 under either scenario than if we used the Roads parameter combination, and (3) a greater fraction of CSC subgroup's high-quality habitat is developed than that of the FT subgroup's.

While we do not perform the analysis for this illustrative example, we expect a comparison of the

Table 13.1 Percent change in *SAR* statistic from 2000 to 2030 where z = 1

2030 LULC scenario	Conservation			Growth		
	Q scores from tier 1 are not included (%)	Q scores from the tier 1 "Roads" parameter combination are included (%)	Q scores from the tier 1 "Urban" parameter combination are included (%)	Q scores from tier 1 are not included (%)	Q scores from the tier 1 "Roads" parameter combination are included (%)	Q scores from the tier 1 "Urban" parameter combination are included (%)
Herpetofauna subgroup						
California herpetofauna of special concern (CSC)	−1.11	−0.88	−1.11	−8.26	−9.95	−10.14
Federally threatened herpetofauna (FT)	−2.35	−1.76	−2.15	−15.58	−17.91	−18.16

RMBV ratio scores with and without habitat-quality scores to corroborate our initial finding that the Growth scenario converts a greater portion of high-quality habitat and that under either scenario a greater portion of high-quality CSC subgroup habitat than high-quality FT subgroup habitat is developed between 2000 and 2030.

13.4.4 Sensitivity analyses and analysis limitations

A parcel's habitat-quality scores are explained by variables w_r, α_r, β, and L_j, that are uncertain and functions, f_r and q, that are simplistic and in many cases, unverifiable representations of complicated ecological processes. In tier 2 analyses, the data used in the H and C matrices and to set the z exponents are often derived from a limited number of field-based studies.

In this illustrative example we attempt to quantify some of the ramifications of this uncertainty with a limited data sensitivity analysis. First, for the baseline and the two scenario maps, we used 12 different combinations of w_r and L_j values to find the combinations that produced the most left-skewed and right-skewed distributions of Q_x across all 3 maps, respectively (Figure 13.2; Plate 6). We did not vary the values of α_r or β_{xr} nor did we experiment with the structures of f_r and q; otherwise, we would have calculated an even greater range in habitat-quality results.

In our tier 2 illustration we found *MBV*, *RMBV*, and *RMBV* ratio scores for all parcels and *SAR* values on all three landscapes using a range of C_{sj} values for each s,j combination. For example, in Figures 13.S3 and 13.S4 in this chapter's SOM we illustrate which subgroup's distribution of *RMBV* ratios is most sensitive to uncertainty in C_{sj} scores under both the Growth and Conservation scenarios. In Figures 13.4 and 13.5 the ramifications of some of this uncertainty is mapped and graphed, respectively.

Placing reasonable bounds on model outputs is the simplest uncertainty analysis we can perform. A more thorough method for determining how sensitive a model's output is to variable and parameter value and functional form uncertainty is to run a Monte Carlo simulation analysis. Under the Monte Carlo method, one or more variable and parameter values are randomly drawn from distributions that describe the variables and parameters range of possible values. In addition, functional forms can also be varied in a random manner. The process of randomly drawing variable and parameter values or functional forms and then running the model is repeated many times. Thus, this method calculates a distribution of potential solutions, including, if the process is iterated enough times, approximations of the worst and best case scenarios. The more variables, parameters, and functions that are simultaneously varied, the more robust the uncertainty analysis. See the SOM for all data used in this illustrative example.

13.5 Limitations and next steps

13.5.1 Limitations

Tier 1 and 2 models should be viewed as complements and not as alternatives. Because of data limitations, we presume that tier 1 modeling will be much more widely implemented. If species-level data on habitat compatibility and potential ranges are available, however, we believe it is important to determine if both tiers suggest the same trends and spatial patterns of biodiversity. If tier 1 and 2 analyses produce spatially correlated results (e.g., areas with high Q scores also have high *MBV* scores; the parcels that exhibit the greatest decline in habitat-quality over time have the lowest *RMBV* ratio scores; the relative change in SAR scores over time are the same with and without habitat-quality included; etc.), then the quality of habitat can be used as a proxy for the status of species of concern on the landscape.

The sources of anthropogenic threat that we have presented here as candidates for use in the tier 1 habitat-quality model are all land use related, but many factors affect biodiversity that are difficult to map, such as the presence of exotic species or an altered disturbance regime. Further, our tier 1 model could be extended to allow habitat resistance L_j to vary across the R sources of human land uses (i.e., we could use L_{jr} instead of L_{jr}). For example, some forested areas may be resistant to the chemicals applied to nearby farm fields but particularly affected by a nearby road's provision of easy access to gatherers of timber and non-timber forest products. In addition, the source of human land use and its ecological impact may be thousands of miles apart. This is most strikingly demonstrated by the pronounced effect of fertilizer use on farm fields of the upper Midwest US on water quality in the Gulf of Mexico. Such broad-scale impacts will not be characterized by our tier 1 model.

Integrating landscape structure and connectivity analyses in the tier 2 model would allow for more explicit population viability modeling (for examples of such modeling see Hanski and Ovaskainen 2000; Vos *et al.* 2001, Schumaker *et al.* 2004). Other than the spatial relationship between potential range space and habitat type, the tier 2 biodiversity models presented here do not consider how the spatial configuration of suitable habitat on a landscape may affect species. Further, the models do not consider the size of habitat patches or the ability of animals to move from patch to patch (see Polasky *et al.* 2005; Winfree *et al.* 2005; Nelson *et al.* 2008; Polasky *et al.* 2008 for examples of species conservation models that do consider species movement among patches). In addition, issues of minimum viable population size, population stochasticity, competition and other species interactions, are not addressed in our models.

By treating a landscape as an island surrounded by *terra incognito* we may under- or overestimate species status in the broader region. For example, a species with a small amount of suitable habitat in its potential range on the studied landscape will have a low SAR_{st} score. However, if the species has effective habitat in a significant portion of its potential range outside of the landscape in question then a low *SAR* score may not be indicative of the species' overall status. On the other hand, a species can have a high SAR score in the studied landscape yet only have a small portion of its regional or global potential range space in suitable habitat.

13.5.2   Next steps

All of the models presented in this chapter can be extended in fruitful ways. For example, the set of habitat types J and L_j values in a tier 1 analysis can be defined according to the habitat needs and suit-

abilities of some specific objective of biodiversity conservation; for example, a species guild that shares habitat or ecological roles. We would then be calculating guild-specific Q_{xj} and Q_x scores. For example, if mapping the habitat-quality of interior forest-dependent species, we would construct the set of habitat types J and L_j values according to their needs, suitabilities, and reaction to different sources of disturbance. Then a species *MBV*, *RMBV*, and *RMBV* ratio scores for these interior forest-dependent species, as well as the countryside *SAR* score for the entire landscape, could be modified by Q_{xj} or Q_x scores that are more descriptive of the habitat limitations facing this particular guild.

Both tiers of the biodiversity model can be easily integrated with systematic conservation planning processes, such as ecoregional assessments or biodiversity visions. The habitat-quality model in tier 1 could be used to ensure that the portfolio of selected sites includes those features with the highest quality. In addition, ecosystem service planning processes can use these models to calculate the biodiversity "costs" or trade-offs, if any exist at all, associated with LULC changes designed to enhance or conserve ecosystem service provision where landscape-level biodiversity costs are given by changes in SAR values and spatial biodiversity costs by RMBV ratio maps (e.g., Chan *et al.* 2006; Nelson *et al.* 2008). In comparing spatial patterns of biodiversity and ecosystem services we may identify areas where conservation investments can benefit species *and* increase the supply and value of ecosystem services (Balvanera *et al.* 2001; Naidoo and Ricketts 2006; Turner *et al.* 2007).

Finally, the output from either tier could be used as input into a site-selection algorithm, such as Marxan, to design alternative conservation area networks (Ball and Possingham 2000). For example, a conservation objective could be to find LULC changes that increase as many RMBV ratio parcel scores, or alternatively minimize the number of RMBV ratio parcel scores below 1, subject to some conservation budget constraint (see Nelson *et al.* 2008 for details on a optimization model that maximized the gain in SAR across a suite of species for a given conservation budget). Then, once a network of conservation areas is selected to meet targets, the habitat-quality model could be used to conduct a threats assessment on proposed or existing protected areas.

References

Abbitt, R., and Scott, J. (2001). Examining differences between recovered and declining endangered species. *Conservation Biology*, **15**, 1274–84.

Ando, A., Camm, J., Polasky, S., *et al.* (1998). Species distributions, land values, and efficient conservation. *Science*, **279**, 2126–8.

Ball, I., and Possingham, H. (2000). *MARXAN (v1.8.2): Marine reserve design using spatially explicit annealing.*

Balvanera, P., Daily, G., Ehrlich, P. *et al.* (2001). Conserving biodiversity and ecosystem services. *Science*, **291**, 2047.

California Department of Fish and Game. California Interagency Wildlife Task Group (CDFG-CIWTG). (2007). *CWHR personal computer program.*

Cardillo, M., Mace, G., Gittleman, J. *et al.* (2006). Latent extinction risk and the future battlegrounds of mammal conservation. *Proceedings of the National Academy of Sciences*, **103**, 4157–61.

Ceballos, G., Ehrlich, P., Soberon, J., *et al.* (2005). Global mammal conservation: What must we manage? *Science*, **309**, 603–7.

Chan, K., Shaw, M., Cameron, D., *et al.* (2006). Conservation planning for ecosystem services. *Plos Biology*, **4**, 2138–52.

Channell, R. and Lomolino, M. (2000). Dynamic biogeography and conservation of endangered species. *Nature*, **403**, 84–6.

Christy, J., Alverson, E., Dougherty, M., *et al.* (2000). *Presettlement vegetation for the Willamette Valley, Oregon, version 4.0, compiled from records of the General Land Office Surveyors (c.1850).*

Daily, G., Ceballos, G., Pacheco, J., *et al.* (2003). Countryside biogeography of neotropical mammals. *Conservation Biology*, **17**, 1814–26.

Davis, F., Costello, C., and Stoms, D. (2006). Efficient conservation in a utility-maximization framework. *Ecology and Society*, **11**, 33.

Diaz, S., Tilman, D., and Fargione, J. (2005). Biodiversity regulation of ecosystem services. In: R. Hassan, R. Scholes, and N. Ash, Eds., *Ecosystems and human wellbeing: current state and trends.* Island Press, Washington, DC, pp. 297–329.

Dirzo, R., and Raven, P. (2003). Global state of biodiversity and loss. *Annual Review of Environment and Natural Resources*, **28**, 137–67.

Ehrlich, P. R., and Ehrlich, A. H. (1982). *Extinction.* Ballantine, New York.

Forman, R., Sperling, D., Bissonette, J., *et al.* (2003). *Road Ecology.* Island Press, Washington, DC.

Groves, C. R., Jensen, D. B., Valutis, L. L., *et al.* (2002). Planning for biodiversity conservation. *Bioscience*, **52**, 499–512.

Hanski, I., and Ovaskainen, O. (2000). The metapopulation capacity of a fragmented landscape. *Nature*, **404**, 755–8.

Horner-Devine, M. C., Daily, G. C., Ehrlich, P. R., *et al.* (2003). Countryside biogeography of tropical butterflies. *Conservation Biology*, **17**, 168–77.

Hughes, J. B., Daily, G. C., and Ehrlich, P. R. (1997). Population diversity: Its extent and extinction. *Science*, **278**, 689–92.

Leopold, A. (1949). *Sand county almanac, and sketches here and there.* Oxford University Press, New York.

MacArthur, R. H., and Wilson, E. O. (1967). *The theory of island biogeography.* Princeton University Press, Princeton, NJ.

Mann, C. C. (2005). *1491: new revelations of the Americas before Columbus.* Vintage Books, New York.

Margules C. R., and Pressey, R. L. (2000). Systematic conservation planning. *Nature*, **405**, 243–53.

Mayfield, M. M., and Daily, G. C. (2005). Countryside biogeography of neotropical herbaceous and shrubby plants. *Ecological Applications*, **15**, 423–39.

Medellin, R. A., Equihua, M. and Amin, M. A. (2000). Bat diversity and abundance as indicators of disturbance in neotropical rainforests. *Conservation Biology*, **14**, 1666–75.

Naidoo, R., and Ricketts, T. H. (2006). Mapping economic costs and benefits of conservation. *PLoS Biology*, **4**, e360.

NatureServe. (2008). NatureServe Explorer: Ecological Communities and Sytems. Accessed at http://www.natureserve.org

Nelleman, C., Kullered, L., Vistnes, I., *et al.* (2001). GLOBIO. Global methodology for mapping human impacts on the biosphere. United Nations Environment Program (UNEP); Report UNEP/DEWA/TR.01–3.

Nelson, E., Polasky, S., Lewis, D., *et al.* (2008). Efficiency of incentives to jointly increase carbon sequestration and species conservation on a landscape. *Proceedings of the National Academy of Sciences.* **105**, 9471–6.

Newmark, W. (1995). Extinction of mammal populations in western American national parks. *Conservation Biology*, **9**, 512–26.

Parks, S., and Harcourt, A. (2002). Reserve size, local human density, and mammalian extinctions in U.S. protected areas. *Conservation Biology*, **16**, 800–8.

Pereira, H., and Daily, G. (2006). Modeling biodiversity dynamics in countryside landscapes. *Ecology*, **87**, 1877–85.

Pimm, S., and Brooks, T. (2000). The sixth extinction: How large, where, and when? In: T. Raven and P. H. Williams, Eds., *Nature and human society*, pp. 46–62. National Academy Press, Washington, DC.

Polasky, S., Nelson, E., Lonsdorf, E., *et al.* (2005). Conserving species in a working landscape: land use with biological and economic objectives. *Ecological Applications*, **15**, 1387–1401.

Polasky, S., Nelson, E., Camm, J. *et al.* (2008). Where to put things? Spatial land management to sustain biodiversity and economic returns. *Biological Conservation*, **141**, 1505–24.

Purvis, A., Gittleman, J., Cowlishaw, G., *et al.* (2000). Predicting extinction risk in declining species. *Proceedings of the Royal Society of London, Series B: Biological Sciences*. **267**, 1947–52.

Ranganathan, J., Daniels, R., Chandran, S., *et al.* (2008). Sustaining biodiversity in ancient tropical countryside. *Proceedings of the National Academy of Sciences*, **105**, 17852–4.

Ricketts, T. (2004). Tropical forest fragments enhance pollinator activity in nearby coffee crops. *Conservation Biology*, **18**, 1262–71.

Rondinini, C., Wilson, K., Boitani, L., *et al.* (2006). Tradeoffs of different types of species occurrence data for use in systematic conservation planning. *Ecology Letters*, **9**, 1136–45.

Rosenzweig, M. (1995). *Species diversity in space and time.* Cambridge University Press, Cambridge.

Sala, O., Vuuren, D. v., Pereira, H., *et al.* (2005). Biodiversity across scenarios. In: S. Carpenter, P. Pingali, E. Bennett, and M. Zurek, Eds., *Ecosystems and human well-being*, vol. 2: *Scenarios*, pp. 375–408. Island Press, Washington, DC.

Schill, S. and Raber, G. (2008). Protected Area Tools (PAT) for ArcGIS 9.2 (version 2.0) software.

Scholes, R. and Biggs, R. (2005). A biodiversity intactness index. *Nature*, **434**, 45–9.

Schumaker, N., Ernst, T., White, D., *et al.* (2004). Projecting wildlife responses to alternative future landscapes in Oregon's Willamette Basin. *Ecological Applications*, **14**, 381–400.

Scott, J. M., Goble, D., and Davis, F. (2006). *The Endangered Species Act at thirty: conserving biodiversity in human-dominated landscapes*, vols 1 and 2. Island Press, Washington, DC.

Şekercioğlu, Ç., Daily, G., and Ehrlich, P. (2004). Ecosystem consequences of bird declines. *Proceedings of the National Academy of Sciences*, **101**, 18042–7.

Şekercioğlu, Ç., Loarie, S., Brenes, F., *et al.* (2007). Persistence of forest birds in the Costa Rican agricultural countryside. *Conservation Biology*, **21**, 482–94.

Shevock, J. (1996). Status of rare and endemic plants. In: D. C. Erman, Ed., *Sierra Nevada ecosystem project: final report to Congress*, vol. II, pp. 691–707. University of California—Davis, Centers for Water Centers for Water and Wildland Resources, Davis.

Stoms, D., Davis, F., and Cogan, C. (1992). Sensitivity of wildlife habitat models to uncertainties in GIS data. *Photogrammetric Engineering and Remote Sensing*, **58**, 843–50.

Theobald, D. (2005). Landscape patterns of exurban growth in the USA from 1980 to 2020. *Ecology and Society*, **10**, 32.

Turner, W., Brandon, K., Brooks, T., *et al.* (2007). Global conservation of biodiversity and ecosystem services. *BioScience*, **57**, 868–73.

Vos, C., Verboom, J., Opdam, P., *et al.* (2001). Toward ecologically scaled landscape indices. *American Naturalist*, **157**, 24–41.

Williams, P. H., Gibbons, D., Margules, C., *et al.* (1996). A comparison of richness hotspots, rarity hotspots, and complementary areas for conserving diversity of British birds. *Conservation Biology*, **10**, 155–74.

Wilson, E. O. (1992). *The diversity of life.* Harvard University Press, Cambridge, MA.

Wilson, K., Underwood, E., Morrison, S., *et al.* (2007). Conserving biodiversity efficiently: What to do, where, and when. *PLoS Biology*, **5**, e223.

Winfree, R., Dushoff, J., Crone, E., *et al.* (2005). Testing simple indices of habitat proximity. *American Naturalist*, **165**, 707–17.

Worm B., Barbier, E., Beaumont, N. *et al.* (2006). Impacts of biodiversity loss on ocean ecosystem services. *Science*, **314**, 787–90.

SECTION III

Extensions, applications, and the next generation of ecosystem service assessments

Putting ecosystem service models to work: conservation, management, and trade-offs

Stephen Polasky, Giorgio Caldarone, T. Ka'eo Duarte,
Joshua Goldstein, Neil Hannahs, Taylor Ricketts, and Heather Tallis

14.1 Introduction

Changing land use or land management affects the provision and value of a range of ecosystem services as well as biodiversity. The large number of potentially competing objectives can complicate decisions about landscape management. In rare cases the best choice among management alternatives will be obvious because one alternative delivers higher levels of all ecosystem services and biodiversity compared to other alternatives ("win–win" solutions). In most cases, however, comparing among management alternative requires evaluating trade-offs among various ecosystem services and biodiversity conservation.

The models described in this book generate predictions about the provision of multiple ecosystem services and biodiversity for any given pattern of land use and management across a landscape. In this chapter we illustrate how one might use these predictions collectively to analyze alternative conservation and management strategies. By comparing maps of ecosystem services and biodiversity, managers can locate areas that, if managed correctly, can provide high levels of both. By comparing outcomes across different management alternatives, conservationists and managers can gain insight into which alternatives may be most desirable. Further, the analyses can be used to suggest and investigate new strategies that may improve results for key ecosystem services or biodiversity conservation objectives.

Trade-offs and potential win–win solutions are illustrated in Figure 14.1, which shows a simple styl-ized example for two ecosystem services, carbon sequestration and water quality, under four management alternatives (A, B, C, and D). For simplicity of the illustration, suppose that the cost of implementing each management is the same and that managers care only about water quality (measured on the vertical axis) and carbon sequestration (measured on the horizontal axis). Alternatives B and C are preferred to alternative A because both carbon sequestration and water quality scores are higher under these alternatives. Choosing among B, C, or D (or between A and D), however, involves a trade-off with each alternative providing more of one service and less of the other. Which of these alternatives (B, C, or D) will be preferred depends on the relative value of the two ecosystem services. If carbon sequestration is highly valued relative to water quality, then alternative D will be preferred to the other alternatives. As water quality increases sufficiently in value relative to carbon sequestration, alternative B or C will be the most preferred. Alternative A, however, will never be the most preferred option regardless of the value judgment about how to weight the value of carbon sequestration relative to water quality because it is dominated by other alternatives (B and C).

Here we describe four different approaches to analyzing conservation and management alternatives that illustrate potential application of models of the type described in this book. We start with an example that builds from conservation planning in which the planner chooses sites to include in a reserve network. We expand upon the traditional

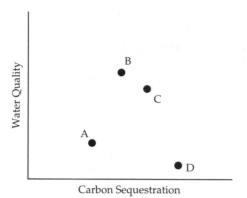

Figure 14.1 Stylized example of output of two ecosystem services, carbon sequestration and water quality, evaluated under four hypothetical management alternatives (A, B, C, and D). Moving from A to either B or C increases both carbon sequestration and water quality. Comparison among all other management alternatives involves trade-offs of an increase in one service and a decrease in the other.

conservation planning approach by including the effect of choosing conservation reserves on the provision and value of ecosystem services. Second, we evaluate the provision of multiple ecosystem services from alternative scenarios of land use and land management, illustrating synergies and trade-offs among ecosystem services and biodiversity conservation. Third, we combine the models with optimization to define an efficiency frontier that shows the maximum possible combinations of ecosystem services provision and biodiversity conservation that are feasible from a landscape. Fourth, we illustrate how one can include estimates of monetary value of ecosystem services to provide a benefit–cost analysis of management alternatives. At the end of the chapter we offer some brief concluding comments on the current state of the art and important next steps.

14.2 Applying ecosystem service and biodiversity models in management and conservation contexts

14.2.1 Site selection for conservation

Conservation managers typically face a situation in which they have a large number of worthwhile conservation projects but only have resources sufficient to fund a small fraction of these projects. The systematic conservation planning field developed to provide advice to conservation managers on how best to con-

serve biodiversity with limited resources (Margules and Pressey 2000; Sarkar *et al.* 2006). The conservation planning literature has developed a set of methods for choosing which sites to include in a conservation reserve network in a range of applications (e.g., Kirkpatrick 1983; Margules *et al.* 1988; Cocks and Baird 1989; Camm *et al.* 1996; Willis *et al.* 1996; Possingham *et al.* 2000) incorporating such factors as varying land cost (e.g., Ando et al. 1998; Naidoo *et al.* 2006), species persistence in reserves (e.g., Cabeza and Moilanen 2001; Nicholson *et al.* 2006) and sequential choice and threats of habitat loss (e.g., Costello and Polasky 2004; Meir *et al.* 2004; Wilson *et al.* 2006).

Even in the well-studied context of conservation site selection, spatially explicit models of ecosystem services and biodiversity can expand the type of information available to conservation managers and improve conservation decision-making. Such models can identify areas of high and low value for a variety of ecosystem services that can be compared spatially to areas of high and low value for biodiversity (Chan *et al.* 2006). In areas of high overlap, conservation organizations can partner with other groups interested in water quality, carbon sequestration or other services to affect outcomes, effectively increasing the resources available for conservation (Goldman *et al.* 2008). Conservation organizations can then concentrate their own resources on areas of high biodiversity value but that do not have high values for services.

An example of this type of analysis is shown in Naidoo and Ricketts (2006). They map the monetary values of five ecosystem services (bushmeat harvest, timber harvest, bioprospecting, existence value, and carbon storage) in the Mbaracyau Forest Biosphere Reserve in Paraguay (see Figure 14.2; Plate 8). (Some conservationists express concern about putting monetary values on nature; we discuss these issues in Section 14.2.4. Also, see Chapter 2). Naidoo and Ricketts (2006) develop maps that show areas where conservation benefits are high and would more than cover the costs of conservation and other areas where the converse is true. Naidoo and Ricketts (2006) also use these maps to evaluate three alternative locations for a proposed corridor linking two protected areas, and find that one corridor would provide much higher benefits relative to costs than the other two. This is an example of how such maps can help direct conservation efforts to high benefit areas.

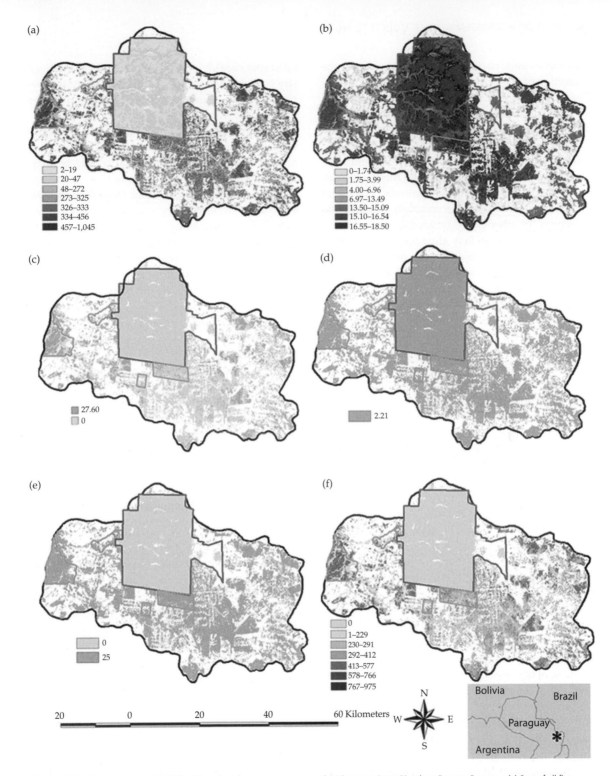

Figure 14.2 Net present values in US$ha⁻¹ for selected ecosystem services in the Mbaracyau Forest Biosphere Reserve, Paraguay. (a) Sum of all five services; (b) sustainable bushmeat harvest; (c) sustainable timber harvest; (d) bioprospecting; (e) existence value; and (f) carbon storage. (See Plate 8.)

Source: Naidoo and Ricketts (2006).

In some cases, particularly in cases involving the protection of municipal drinking water supply (e.g., Bogota, New York City, Quito), the value of ecosystem services is high enough to choose management decisions that also support conservation (Chichilnisky and Heal 1998; NRC 2000; Echevarria 2002). In such cases, payments for ecosystem services can be more than sufficient and there is little or no need for a conservation organization to spend their scarce resources to accomplish conservation objectives. In other cases, promoting the provision of ecosystem services may align with conservation, but the services themselves may not be valuable enough to tip the balance toward biodiversity-friendly management. In this case, conservation organizations can usefully partner with other groups interested in the provision of ecosystem services. Finally, there will be other cases where management for ecosystem services does not align with conservation objectives. In these cases, conservation organizations will be on their own, just as they would be with no consideration of ecosystem services.

Spatially explicit information on ecosystem services can also be integrated with conservation planning exercises in other ways. For example, if ecosystem services are valued in monetary terms, the cost of including a particular site could be reduced by the increase in value of ecosystem services provided if the site is chosen as a reserve. Doing so would shift conservation priorities toward sites that generate valuable ecosystem services in a fashion similar to priority given to inexpensive sites. It is also possible to require that targets could be specified for certain ecosystem services and only reserve networks that met these targets would be considered as potential solutions in the conservation planning exercise.

Using spatially explicit models that incorporate both ecosystem services and biodiversity is powerful because conservation decisions are often inherently spatial: Where to protect? How much area is needed? Where to allow development? In this way, adding maps of ecosystem services broadens an existing approach to conservation planning that is used and understood by many conservation practitioners (Groves 2003). Using spatially explicit models of ecosystem services expands the set of

outcomes considered in planning beyond biodiversity conservation targets. Doing so can show areas on the landscape that are of high priority for conservation targets and various ecosystem services.

The main disadvantage of using spatially explicit models in this manner is that results do not necessarily indicate how the landscape should be managed. Management to promote a particular ecosystem service might differ from management to promote another service or biodiversity conservation. For example, carbon sequestration may be maximized by planting trees but this may decrease surface water runoff and reduce availability for downstream users (Jackson *et al.* 2005). In classic conservation site selection the management choice is simple—either protect a site or don't—and a protected site is assumed to benefit all species. With the inclusion of ecosystem services, however, the choice of management options is of greater interest and complexity. When different management options at the same spatial location are best for different objectives just highlighting high priority areas on the landscape is not enough. What is needed in this case is an analysis that shows outcomes for ecosystem services and biodiversity under different types of management.

14.2.2   Analysis of management alternatives

The spatially explicit models defined in earlier chapters are designed to evaluate multiple ecosystem services and biodiversity objectives under alternative conservation or management plans. In this section we highlight the use of these models to analyze the effect of alternative land-use plans on the provision of ecosystem services and biodiversity conservation. The first case study involves evaluating alternative future scenarios for land use in the Willamette Basin in Oregon. The second case study involves evaluating alternative land uses for a watershed owned by Kamehameha Schools on O`ahu, Hawai`i.

14.2.2.1   *Alternative future scenarios in the Willamette Basin, Oregon*
Nelson *et al.* (2009) applied several of the spatially explicit models described in previous chapters, or their precursors. These models were used to predict changes in ecosystem services and conservation of

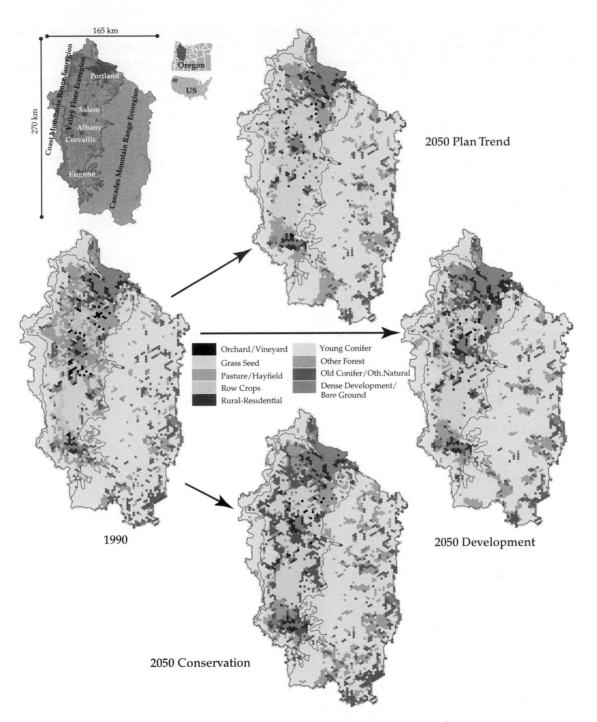

Figure 14.3 Maps of the Willamette Basin with the land-use pattern for 1990 and three land-use change scenarios for 2050.

Source: Nelson *et al.* (2009).

terrestrial vertebrate species for the Willamette Basin in Oregon, USA (Figure 14.3). Using stakeholder-defined land-use change scenarios for the period 1990 to 2050, they compared outcomes for the basin in terms of carbon storage, water quality (reduction of phosphorus discharge), soil conservation (reduction of erosion), storm peak mitigation, terrestrial vertebrate conservation and value of marketed commodities (agriculture, forestry and rural residential housing development). Basin-wide maps for the three land-use change scenarios and the 1990 land-use pattern for the Willamette Basin (Figure 14.3) were developed by the Pacific Northwest Ecosystem Research Consortium, an alliance of government agencies, non-government organizations, and universities (Hulse *et al.* 2002; USEPA 2002; Baker *et al.* 2004). The three land-use change scenarios were: (i) "plan trend" that extended current policies and trends into the future, (ii) "development" that relaxed current land-use policies and allowed greater freedom for market forces, and (iii) "conservation" that gave greater emphasis to ecosystem protection and restoration (USEPA 2002, pp. 2–3).

Of the three scenarios, the conservation scenario produces the best results for all ecosystem services and biodiversity conservation (Figure 14.4). The results for the conservation scenario were significantly better than for either the plan trend or development scenarios for carbon sequestration, water quality, and soil conservation. Only the market value of commodity production was higher in the plan trend and development scenarios than in the conservation scenario. Under the plan trend and development scenarios, more land was devoted to housing development and to timber production increasing the value of market returns (but lowering the scores for biodiversity conservation and many ecosystem services).

The trade-off between the value of marketed commodities and ecosystem services changes if we expand the set of marketed commodities to include the possibility of markets in carbon credits. Nelson *et al.* (2009) calculated the aggregate market value of carbon sequestration under the three scenarios using a price of $43 per metric ton of carbon, which is the mean of estimates of the social value of carbon reduction (either from emissions reductions or from seques-

tration) from peer-reviewed studies (Tol 2005). Because there was more carbon sequestered under the conservation scenario, adding the carbon sequestration value to the market value of commodities meant that the conservation scenario generated the highest monetary returns of the three scenarios (Figure 14.5). A carbon market that rewarded carbon sequestration could turn a trade-off curve with a negative slope (Figure 14.5, circles) into one with a positive slope (triangles), converting a trade-off into a win–win. Making payments for other ecosystem services would further increase the value of the conservation scenario relative to the other two scenarios.

14.2.2.2   *Kamehameha Schools, O`ahu, Hawai`i*

A subset of the spatially explicit models described in earlier chapters were also used to evaluate impacts on ecosystem services for local land-use planning in Hawai`i in collaboration with Kamehameha Schools, an educational trust and the largest private landowner in the state (see Box 14.1 for additional information on Kamehameha Schools). The analysis focused on Kamehameha Schools' land holdings on the north shore of the island of O`ahu (Figure 14.6; Plate 9). This region contains approximately 26 000 acres stretching from ocean to mountain tops, including ~2000 acres of coastal rural community lands, ~9000 acres of agricultural lands in the middle section (once a sugarcane plantation, now largely abandoned and invaded by exotic species), and ~15 000 forested acres in the upper part.

With extensive input from Kamehameha Schools, three spatially explicit scenarios were created to explore contrasting directions that could be taken with the agricultural lands:

(1) *Sugarcane ethanol*—returning the plantation lands to sugarcane cultivation to produce ethanol biofuel;
(2) *Diversified agriculture and forestry*—using the lower irrigated fields for diversified agriculture, establishing vegetation buffers to reduce field runoff, and undertaking native forestry plantings on the remaining higher elevation fields;
(3) *Residential subdivision*—selling coastal and plantation lands for a residential housing development. These scenarios were compared in terms of effects

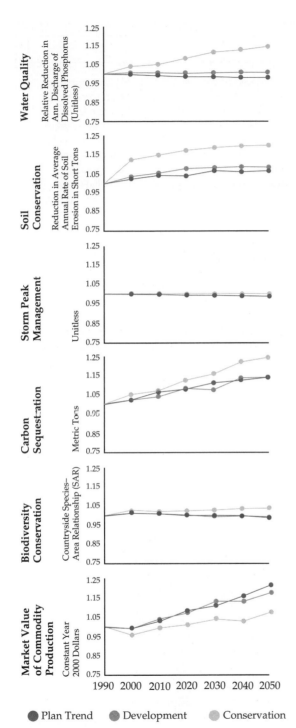

Figure 14.4 Trends in landscape-scale ecosystem services levels, biodiversity conservation status, and market values of commodity production for the three land-use change scenarios. All scores are normalized by their 1990 levels.

Source: Nelson *et al.* (2009).

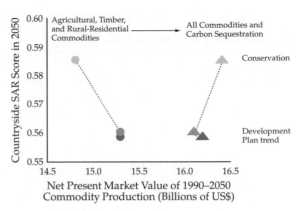

Figure 14.5 Trade-offs between market values of commodity production and biodiversity **conservation** on the landscape between 1990 and 2050 excluding the market value of carbon sequestration (circles) and including the market value of carbon sequestration (triangles). The *x*-axis measures the total discounted value of commodities and the *y*-axis measures the biodiversity conservation score.

Source: Nelson *et al.* (2009).

on water quality (for nitrogen discharge), carbon storage, and income generation.

All three scenarios are projected to generate positive income streams that exceed the current negative returns (Figure 14.7). The residential subdivision scenario, not surprisingly, has the greatest net present value of income. This income boost, however, is linked with reductions in carbon stock (6.8%) and water quality (21.1%) relative to the current landscape. Impacts on carbon stock and water quality are even more pronounced for the sugarcane ethanol scenario with reductions of 12.6 and 44.2%, respectively. In both cases, losses in carbon stock are driven by clearing invasive woody vegetation on abandoned fields. While both scenarios lead to reductions in carbon stock, the sugarcane ethanol scenario has the potential to "pay off" the lost carbon stock through use of ethanol to offset more carbon-intensive energy sources. Following the biofuel carbon debt methodology of Fargione *et al.* (2008), the estimated payback period is approximately 10 years to return to baseline conditions.

The remaining scenario, diversified agriculture and forestry, is projected to improve carbon stock (9.8%) and water quality (7.0%) relative to the current landscape, while also generating positive income. These improvements are driven by plantings to restore native forest cover and establishing

Box 14.1 Plight of a people

Neil Hannahs

Disease and change exacted a horrific toll on the native people of Hawaii throughout the nineteenth century. The thriving population of more than half a million Hawaiians at the beginning of the century had dwindled to a mere 40 000 by the 1880s. To address these desperate conditions and to assure the perpetuation of Hawaiian culture and welfare of her people, Princess Bernice Pauahi Bishop and her husband left over 400 000 acres of Hawaii land, as well as personal resources, in a perpetual charitable trust dedicated to improving the wellbeing of Hawaiian people through educational services offered by Kamehameha Schools (KS).

Since its inception in 1884, the endowment of KS' founders has been managed to produce financial resources to build and maintain campuses and educational programs. For much of the School's history, the trust was considered land rich, but cash poor. To fund construction, operation and growth, an asset management strategy was adopted to maximize economic productivity. This provided the means for KS to become one of the largest private educational institutions in the world and afforded the Schools the opportunity to greatly expand its educational reach.

However, the commercial, residential, and agricultural land developments that brightened KS' economic prospects were often conducted with insufficient regard for cultural resources, environmental impacts and community values. This tendency, coupled with rapid population growth fueled by in-migration and the introduction of invasive exotic species of flora and fauna, resulted in displacement of Hawaiian communities and degradation of indigenous resources and the cultural practices that thrived upon them. These circumstances produced a tragic conundrum: Hawaiians being helped by KS suffered the most from land-use changes implemented to provide resources for their educational programs.

Concern for Hawaii's ecosystems and traditional lifestyles mounted over the past four decades as Hawaiian culture experienced a renaissance and natural resources became increasingly stressed. Resource supply has declined in the face of rising demand and ravaging impacts of invasive plants and ungulates. Conflicts manifested as resistance to new development and Western concepts of property rights, as well as advocacy for constitutional and regulatory protections of the environment, at-risk species and Hawaiian cultural practices. Consequently, KS' efforts to apply economic maximization strategies to undeveloped

An Indigenous Worldview:
Focus on Living systems

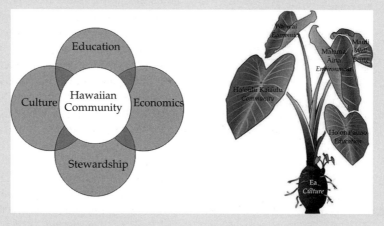

Figure 14.A.1   Kamehameha Schools' representation of the indigenous Hawai'i worldview.

lands faced increasing resistance in the latter twentieth century.

Paradigm shift

The "Kamehameha Schools' Strategic Plan 2000–2015" promised an organization that would align itself to the values of the founders, incorporate the views of stakeholders and set new directions. The Plan established the following goals for the management of the endowment. Kamehameha Schools will optimize the value and use of the current financial and non-financial resources and actively seek and develop new resources; and to practice ethical, prudent and culturally appropriate stewardship of lands and resources.

These goals provide an opportunity to re-think the value of land and each asset's role in fulfilling the mission as part of a dynamic portfolio. The emergent Integrated Management Strategy has attracted the interest of cultural stakeholders and other First Nations peoples, as well as the conservation and business communities, including the Natural Capital Project (NatCap).

NatCap's InVEST tool, the software framework for several of the models described in earlier chapters, has helped to inform courses of action and land management decisions that propel a shift from one dimensional returns to a balance of desirable outcomes. KS, owner of the Kawailoa lands to which InVEST is now being applied, has depicted its efforts to achieve an optimal balance of multi-value returns as an image of over-lapping spheres.

A risk inherent in this view, as well as in using a tool like InVEST, is that the challenge might be met by assembling indiscriminate and disconnected considerations in each value domain. An alternative approach is to maintain focus on holistic, living systems. This is depicted in the taro (kalo) image (Figure **14.A.1**).

Kamehameha Schools is now monitoring several key performance metrics of sustainability to determine whether this high standard is being achieved. These include carbon footprint; assessments of ecosystem services; financial values and returns; and various measures of well-being impact. InVEST is playing an integral role in helping KS and others in projecting the outcome of land-use decisions on many of these indicators of vitality.

---

field buffers to reduce nutrient runoff. As such, the diversified agriculture and forestry scenario has the greatest potential to provide balanced, positive returns across the modeled services.

Cultural values are also important to the north shore community and to Kamehameha Schools' approach to land management. While they were not assessed quantitatively in this analysis, the scenarios are likely to have differing impacts. Many residents prize the north shore's rural character, and maintaining active agricultural lands is one key part of this. These lands provide jobs and income to the local community, as well as contribute to a sense of place and connection with previous generations. These benefits would be best captured by the sugarcane ethanol and diversified agriculture and forestry scenarios, and not the residential subdivision scenario. The north shore also contains sacred burial grounds and other historic remains that must be considered in land-use planning. Integrating these and other cultural dimensions into the formal modeling effort is an important next step, and the

methods discussed in Chapter 12 provide a template for doing so.

14.2.3   Generating an efficiency frontier

In the previous section, we showed how to use multiple spatially explicit models to analyze specific scenarios (management alternatives) of interest to users. Such analyses can show which of the considered management alternatives generates better performance in terms of provision of ecosystem services or meeting biodiversity targets. Another use of these models is to show what is possible to achieve on the landscape by considering all potential land-use scenarios. In reality, of course, not all land-use scenarios will be politically or socially acceptable. But considering all possible alternatives can often identify solutions that are far superior to the narrow range of options currently being considered. Providing this evidence can broaden the perspective of users and begin a dialog about what options should be on the table.

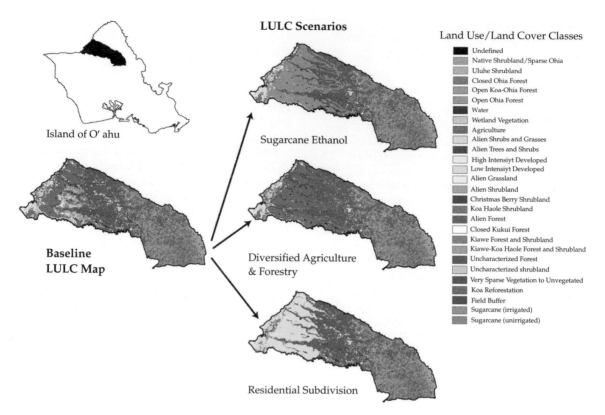

Figure 14.6   Land use/land cover maps on the north shore of O'ahu. The area shown here includes all of Kamehameha Schools' north shore land holdings, as well as small adjacent parcels that make for a continuous region. The baseline map is from the Hawai'i Gap Analysis Program's land cover layer for O'ahu (Hawai'i Gap Analysis Program 2006). (See Plate 9.)

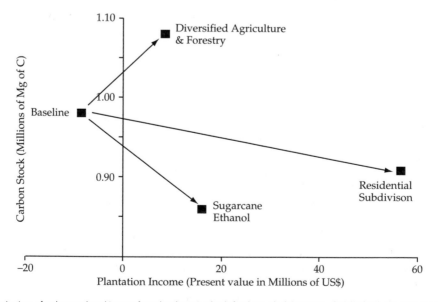

Figure 14.7   Projections of carbon stock and income from the plantation lands for the north shore region of O'ahu for the baseline land use/land cover map and the three planning scenarios (sugarcane ethanol, diversified agriculture and forestry, residential subdivision).

By combining ecosystem service models with optimization methods, one can determine the maximum feasible combinations of ecosystem services and biodiversity that can be achieved on a landscape. The results of this analysis can be presented with an efficiency frontier, which is defined as the outcomes for which it is not possible to improve on any particular objective (ecosystem service or biodiversity conservation) without decreasing performance on some other objective.

Polasky *et al.* (2008) estimated such an efficiency frontier for conservation of terrestrial vertebrates and the value of marketed commodities (timber, agricultural output and housing) for the Willamette Basin in Oregon. They developed models that used a land-use plan for the basin as input and reported output in terms of the expected number of terrestrial vertebrate species that would persist in the basin (biological score) and the value of marketed commodities (economic score). Using optimization methods from operations research, they searched for land-use plans that maximized the biological score for a given economic score. Then by repeating this analysis across the full range of economic scores ($0–27.6 billion) they traced out an efficiency frontier (Figure 14.8; Plate 10). The results show that it is possible to achieve both high biological and economic scores by thinking carefully about the spatial pattern of land use in the basin. For example, the land-use plan that for point D in Figure 14.8 (Plate 10) generates a biological score of 248.5 species and an economic score of $25.8 billion. This outcome is far better than the outcome generated by the current land use (point I in Figure 14.8; Plate 10), a biological score of 238.6 and an economic score of $17.1 billion (Polasky *et al.* 2008).

Analyses such as these can demonstrate what is possible for a given region and how much improvement can be made by careful planning. Because of political, social and economic complications, it may not be possible to reach efficiency frontiers. Still, knowing what it is possible can provide a spark to ignite efforts to improve upon current performance.

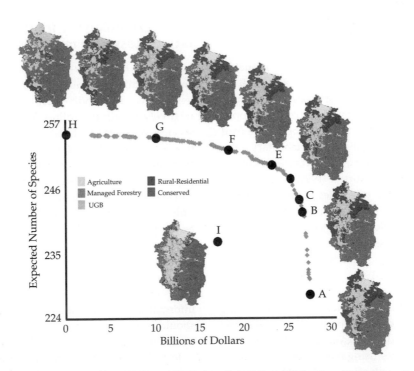

Figure 14.8 Efficiency frontier showing maximum feasible combinations of economic returns and biodiversity scores. Land-use patterns associated with specific points along the efficiency frontier (points A–H) and the current landscape (point I). (See Plate 10.)

Source: Polasky *et al.* (2008).

14.2.4 Benefit–cost analysis

Results from ecosystem service models can be reported in biophysical units or in monetary values. Much of the analysis on ecosystem services to date has been reported in biophysical units, including most of the case studies discussed above. In some settings, such as dealing with government or private sector managers used to thinking in monetary terms, it may be advantageous to report results of the analysis in terms of monetary values. Doing so may also make it easier to compare management options. Because results are reported in a single metric (i.e., dollars), managers can compare apples with apples rather than with oranges.

Economists have developed a variety of market and non-market valuation methods that can be applied to estimate the monetary value of ecosystem services (Freeman 2003). The estimates of monetary value can be incorporated into benefit–cost analysis to analyze the net benefits of alternative management alternatives. Naidoo and Ricketts (2006) in Section 14.2.1 and Nelson et al. (2009) in Section 14.2.3 are examples of how monetizing ecosystem service values can result in benefit–cost analyses that can inform managers and potentially improve management decisions.

Translating from biophysical units to monetary value units, however, is problematic for biodiversity targets and some types of ecosystem services. In some cases trying to convert oranges into apples will result in pulp rather than a recognizable fruit. For example, trying to estimate the monetary value of cultural and spiritual values is controversial (see Chapters 2 and 12; Norton 1991; Sagoff 1988). Valuing the existence of species is viewed as morally objectionable and inherently misguided by some (e.g., Ehrenfeld 1988; McCauley 2006), and even some economists who have tried to value biodiversity admit to the practical difficulties of doing so (e.g., Stevens et al. 1991). Other economists think that all values, including the value of biodiversity, can be measured using economic methods as long as the analysis is done properly (e.g., Loomis and White 1996). For some ecosystem services, estimating monetary values using market prices or applying non-market valuation techniques may be relatively uncomplicated and uncontroversial.

Examples include the value of provisioning services such as timber or fish (e.g., Naidoo and Ricketts 2006; Barbier 2007; Polasky et al. 2008; Nelson et al. 2008, 2009; Chapter 8), or crop pollination, which is an input to a priced commodity (e.g., Ricketts et al. 2004; Chapters 9, 10). Depending on the ecosystem services and the decision context at hand, the user of these models can decide whether it is better to use biophysical units or monetary values.

Chapters in this book typically aim to monetize the value of ecosystem services, but we do not attempt to translate biodiversity targets to a monetary measure of value (Chapter 13). That is because biodiversity is a fundamental attribute of natural systems, which may contribute to the provision of various ecosystem services but which also has intrinsic value (i.e., value in and of itself). Even without attempting to put monetary value on biodiversity, one can still show feasible combinations of biodiversity and services, along with potential trade-offs between them (as shown in Section 14.2.3). Then managers can decide for themselves what trade-offs are acceptable.

14.3 Extending the frontier: challenges facing ecosystem management

Integrated landscape-level analysis that tracks changes across a number of dimensions of ecosystem services and biodiversity conservation is still a relatively young discipline. Models of ecosystem services and geographically explicit data sets are developing rapidly, offering the prospect of further improvements in the near future. To date, applications in the USA, South Africa, Paraguay, and elsewhere have demonstrated the power and utility of an integrated spatially explicit landscape-level approach. Application of such models can generate information for decision-makers showing the consequences of choices for a range of important ecosystem services and biodiversity conservation objectives. In principle, putting this information in the hands of decision-makers should lead to improved landscape planning and management.

To fully realize the promise of spatially explicit integrated modeling approaches, further improvements will be necessary. As discussed in Chapter 15, more work on improving and validating the component models of particular services is needed. Our understanding of the links between management actions and provision of ecosystem services is limited for many services. Additional empirical research on provision of services in a wide variety of circumstances will improve understanding and accuracy of models. Additional understanding of ecosystem functions and conditions that link together provision of multiple services, such as connections between land cover, water availability, nutrient cycling and local climate, will also improve the overall modeling effort. Perhaps the greatest need on the biophysical modeling side, however, is improved understanding and inclusion of system dynamics and feedback effects. Coupled human and natural systems may exhibit threshold effects and non-linear responses in which provision of ecosystem services might change suddenly as conditions in the system evolve.

Even with knowledge of biophysical systems, understanding the provision of ecosystem services also requires detailed understanding of what is of value to people. For example, the provision of clean drinking water in areas without people will not provide an ecosystem service of value while the same provision in a watershed providing water to a major city will have great value. Understanding the value of ecosystem services requires integration of natural and social science. Such integrated work, partly in response to the focus on ecosystem services, has begun to expand rapidly in recent years but is still limited relative to what is needed to seamlessly integrate the supply of services (primarily the province of natural science) with the demand for services (primarily the province of social science). Integrated understanding of ecosystem services has progressed to the point where we can highlight important areas on a landscape for ecosystem services. In many cases, however, we cannot yet provide the level of certainty, either in terms of biophysical or economic modeling, to underpin payments for ecosystem services or other policy approaches that require numerical estimates of value (see Chapters 15 and 19 for further discussion).

An important aspect of integrated spatially explicit models is the ability to show not only the total value of ecosystem services to society but also the distribution of benefits to various groups in society. Such distributional analysis is important for understanding the effects of conservation and management decisions on the poor (see Chapter 16). Distributional analysis of the people who benefit and bear the costs of alternative conservation and management is also important for the design of policy approaches to ensure that those who make decisions affecting ecosystems have incentives to provide ecosystem services of value to society (see Chapter 19).

It is important to recognize that estimates of value, spatial priorities, trade-off analyses, and most other results reviewed in this chapter depend strongly on the choice of ecosystem services to include. Ecosystem management affects a large range of ecosystem services, not all of which may be feasible to model given limited time, resources, data or scientific understanding. In many cases, water quantity and quality, carbon sequestration, and the market value of commodities will be of great importance. Biodiversity conservation will be of primary importance in many conservation applications. However, other services may also be important in particular applications (e.g., non-timber forest products, pollination services, effects on poverty, number of jobs). Early and continuous engagement with people potentially impacted by ecosystem management is the best approach to ensuring that the most important ecosystem services and other policy dimensions (e.g., number of jobs) are included in the analysis.

Finally, the analysis of integrated spatially explicit models is but one step in a much larger and longer process needed to implement real change on the ground. As Knight *et al.* (2006) and Cowling *et al.* (2008) emphasize, there are plenty of analyses and reams of plans but far less action, and that "our understanding of these techniques currently far exceeds our ability to apply them effectively to pragmatic conservation problems" (Knight *et al.* 2006, p. 408). Spatially explicit integrated models can provide useful information but unless they are embedded in a larger policy process that involves those who use land and resources the information will not be utilized to improve ecosystem management or conservation outcomes.

References

Ando, A., Camm, J. D., Polasky, S., *et al.* (1998). Species distributions, land values and efficient conservation. *Science*, **279**, 2126–8.

Baker, J. P., Hulse, D. W., Gregory, S. V., *et al.* (2004). Alternative futures for the Willamette River Basin, Oregon. *Ecological Applications*, **14**, 313–24.

Barbier, E. B. (2007). Valuing ecosystem services as productive inputs. *Economic Policy*, **22**, 177–229.

Cabeza, M., and Moilanen, A. (2001). Design of reserve networks and the persistence of biodiversity. *Trends in Ecology and Evolution*, **16**, 242–8.

Camm, J., Polasky, S., Solow, A., *et al.* (1996). A note on optimization algorithms for reserve site selection. *Biological Conservation*, **78**, 353–5.

Chan K. M. A., Shaw, M. R., Cameron, D. R., *et al.* (2006). Conservation planning for ecosystem services. *PLoS Biology*, **4**, 2138–52.

Chichilnisky, G., and Heal, G. (1998). Economic returns from the biosphere. *Nature*, **391**, 629–30.

Cocks, K. D., and Baird, I. A. (1989). Using mathematical programming to address the multiple reserve selection problem: an example from the Eyre Peninsula, South Australia. *Biological Conservation*, **78**, 113–30.

Costello, C., and Polasky, S. (2004). Dynamic reserve site selection. *Resource and Energy Economics*, **26**, 157–74.

Cowling, R. M., Egoh, B., Knight, A. T., *et al.* (2008). An operational model for mainstreaming ecosystem services for implementation. *Proceedings of the National Academy of Sciences*, 105, 9483–8.

Echevarria, M. (2002). Financing watershed conservation: The FONAG water fund in Quito, Ecuador. In: S. Pagiola, J. Bishop, and N. Landell-Mills, Eds., *Selling forest environmental services: market-based mechanisms for conservation and development*, pp. 91–102. Earthscan, London.

Ehrenfeld, D. (1988). Why put a value on biodiversity? In: E. O. Wilson, Ed., *Biodiversity*, pp. 212–16. National Academy Press, Washington, DC.

Fargione, J., Hill, J., Tilman, D., *et al.* (2008). Land clearing and the biofuel carbon debt. *Science*, **319**, 1235–8.

Freeman, A. M. III. (2003). *The measurement of environmental and resource values*. Resources for the Future, Washington, DC.

Goldman, R. L., Tallis, H., Kareiva, P., *et al.* (2008). Field evidence that ecosystem service projects support biodiversity and diversify options. *Proceedings of the National Academy of Sciences of the USA* 105(27), 9445–8.

Groves, C. R. (2003). *Drafting a conservation blueprint: a practitioner's guide to planning for biodiversity*. Island Press, Washington, DC.

Hawai'i Gap Analysis Program. (2006). *Land cover*. US Geological Survey, Honolulu.

Hulse, D., Gregory, S., and Baker, J., Eds. (2002). *Willamette River Basin Planning Atlas: trajectories of environmental and ecological change*. Oregon State University Press, Corvallis.

Jackson, R. B., Jobbagy, E. G., Avissar, R., *et al.* (2005). Trading water for carbon with biological carbon sequestration. *Science*, **310**, 1944–7.

Kirkpatrick, J. B. (1983). An iterative method for establishing priorities for the selection of natural reserves: an example from Tasmania. *Biological Conservation*, **25**, 127–34.

Knight, A. T., Cowling, R. M., and Campbell, B. M. (2006). An operational model for implementing conservation action. *Conservation Biology*, **20**, 408–19.

Loomis, J. B., and White, D. S. (1996). Economic benefits of rare and endangered species: summary and meta-analysis. *Ecological Economics*, **18**, 197–206.

McCauley, D. (2006). Selling out on nature. *Nature*, **443**, 26–7.

Margules, C. R., Nicholls, A. O., and Pressey, R. L. (1988). Selecting networks of reserves to maximize biological diversity. *Biological Conservation*, **43**, 63–76.

Margules, C. R. and Pressey, R. L. (2000). Systematic conservation planning. *Nature*, **405**, 242–53.

Meir, E., Andelman, S., and Possingham, H. P. (2004). Does conservation planning matter in a dynamic and uncertain world? *Ecology Letters*, **7**, 615–22.

Naidoo, R., Balmford, A., Ferraro, P. J., *et al.* (2006). Integrating economic costs into conservation planning. *Trends in Ecology and Evolution*, **21**, 681–7.

Naidoo, R., and Ricketts, T.H. (2006). Mapping the economic costs and benefits of Conservation. *PLoS Biol*, **4**, 2153–64.

National Research Council (NRC). (2000). *Watershed management for potable water supply: assessing the New York City strategy*. National Academies Press, Washington, DC.

Nelson, E., Polasky, S., Lewis, D. J., *et al.* (2008). Efficiency of incentives to jointly increase carbon sequestration and species conservation on a landscape. *Proceedings of the National Academy of Sciences*, **105**, 9471–6.

Nelson, E., Mendoza, G., Regetz, J., *et al.* (2009). Modeling multiple ecosystem services, biodiversity conservation, commodity production, and tradeoffs at landscape scales. *Frontiers in Ecology and the Environment*, **7**, 4–11.

Nicholson, E., Westphal, M. I., Frank, K., *et al.* (2006). A new method for conservation planning for the persistence of multiple species. *Ecology Letters*, **9**, 1049–60.

Norton, B. G. (1991). *Toward unity among environmentalists*. Oxford University Press, New York.

Polasky, S., Nelson, E., Camm, J., *et al.* (2008). Where to put things? Spatial land management to sustain biodiversity and economic returns. *Biological Conservation*, **141**, 1505–24.

Possingham, H. P., Ball, I. R., and Andelman, S. (2000). Mathematical methods for identifying representative reserve networks. In: S. Ferson and M. Burgman, Eds., *Quantitative methods for conservation biology*, pp. 291–305. Springer-Verlag, New York.

Ricketts, T. H., Daily, G. C., Ehrlich, P. R., *et al.* (2004). Economic value of tropical forest to coffee production. *Proceedings of the National Academy of Sciences*, **101**, 12579–82.

Sagoff, M. (1988). *The economy of the earth*. Cambridge University Press, New York.

Sarkar, S., Pressey, R. L., Faith, D. P., *et al.* (2006). Biodiversity conservation planning tools: present status and challenges for the future. *Annual Review of Environment and Resources*, **31**,123–59.

Stevens, T. H., Echeverria, J., Glass, R. J., *et al.* (1991). Measuring the existence value of wildlife: what do CVM estimates really show? *Land Economics* **67**, 390–400.

Tol, R. S. J. (2005). The marginal damage costs of carbon dioxide emissions: an assessment of the uncertainties. *Energy Policy*, **33**, 2064–74.

US Environmental Protection Agency (USEPA). (2002) *Willamette Basin alternative futures analysis: environmental assessment approach that facilitates consensus building*. EPA 600/R-02/045(a). US Environmental Protection Agency, Office of Research and Development, Washington, DC.

Willis C. K., Lombard, A. T., Cowling, R. M., *et al.* (1996). Reserve systems for the limestone endemic flora of the Cape lowlands: iterative vs linear programming techniques. *Biological Conservation*, **77**, 53–62.

Wilson, K. A., McBride, M. R., Bode, M., *et al.* (2006). Prioritizing global conservation efforts. *Nature*, **440**, 337–40.

CHAPTER 15

How much information do managers need? The sensitivity of ecosystem service decisions to model complexity

Heather Tallis and Stephen Polasky

15.1 Introduction

Natural systems and human systems are inherently complex. When they are examined as coupled systems, which is necessary for the valuation of ecosystem services, the complexity can be daunting. Because it is usually impractical to apply experiments to human systems, models represent the primary tool for studying coupled human-natural systems. We use models in two ways: (1) to simplify the complexity and extract the key dynamics, and (2) to run simulations or experiments that can only be done on a computer, but that are necessary for anticipating the consequences of different policies and different economic activities.

Although models can never fully represent the intricacies of the real world, they can yield insights into how systems work. Or, as modeler George Box famously put it, "All models are wrong but some are useful." Modelers of ecosystem services face the same challenge all modelers face: deciding how much complexity and detail to include. The major question is how to create models that are sufficiently complex to represent system dynamics, yet simple enough to be understood and appropriately parameterized with often limited data (see Van Nes and Scheffer 2005 for a discussion of the issue). This balancing act is a long standing dilemma in fields as diverse as hydrology (e.g., Gan *et al.* 1997) and population biology (e.g., Pascual *et al.* 1997, Stephens *et al.* 2002).

There is no formulaic recipe for determining the degree of complexity and detail needed. In practice, researchers have found that different levels of com-plexity are needed for different types of questions. For example, in water resource management, simple models have proven suitable for predicting run-off in relatively dry watersheds (Gan *et al.* 1997). Similarly, in fisheries and wildlife management, setting optimal fishing effort levels in systems with relatively little variability (Ludwig and Walters 1985) and identifying area requirements for sustaining individual species (such as the marmot (*Marmota marmot*; Stephens *et al.* 2002)) are exercises well addressed with simple models. However, more complex models may be required for other management decisions, such as accurately predicting the consequences of harvesting when species interact or predicting the effect of climate change on species persistence (Stephens *et al.* 2002).

Understanding when simpler models can be used is critical to managers since they often function with short deadlines and limited information (Box 15.1). Simple models typically have reduced data requirements, are easier to set up and run and are more transparent and easier to explain than more complex models. If simple models give the same answers as more complicated models, they would obviously be the preferred option. Elsewhere (e.g., Chapter 4) we compare models to actual field data. Here we focus on comparing models to models, asking to what extent simpler models are able to do the same things and give similar answers to more data-hungry, complex models. To supplement our general discussion we focus in particular on the correspondence between simple and complex models in providing information about carbon sequestration and crop pollination.

Box 15.1   How much data do we need to support our models: a case study using biodiversity mapping and conservation planning

Craig Groves and Edward Game

Governments and non-governmental organizations worldwide commonly develop regional-scale conservation plans that identify areas important for biodiversity conservation. The amount of data—both biodiversity data and other important related information—is highly variable around the world. A few areas have been well surveyed for elements of biodiversity but most areas have limited data sets. As a result, planners are often confronted with the challenging issue of deciding how many data are needed and whether or not to spend resources gathering additional information.

There are at least two reasons why additional investments in biodiversity data may not deliver a commensurate increase in the quality of the planning effort. First, the selection of priority areas for conservation is usually driven by many factors in addition to biodiversity information. For example, active use of lands and waters for production that forecloses the opportunity for conservation, degraded habitats or ecosystem processes of available lands and waters, or prohibitively high costs of conservation due to escalating land prices can be overriding factors that restrict options without much need for complete biodiversity data.

The second and more obvious reason why additional data may not be worth the investment is related to the time and money needed to acquire the additional information as well as conservation opportunities that may be lost while this information is being gathered. Several studies, mostly from South Africa, have evaluated the influence of amounts of data on conservation planning results. Perhaps the best example is a recent paper by Grantham and colleagues (2008) who took advantage of an extensive and nearly complete data set on the distribution of *Protea* species in South Africa's Fynbos biome. They asked the question of whether additional investments in surveys, mapping, and modeling would lead to better planning decisions given the costs of undertaking these surveys and the fact that ongoing land clearing would continue to lead to some losses while the surveys were carried out. Their results clearly showed that while investing in minimal survey data can improve a plan substantially, especially if there is little biodiversity data available to begin with, there are rapidly diminishing returns in additional data (Figure 15.A.1).

Other studies using different data have reached a similar conclusion. For example, in a widely cited study, Gaston and Rodrigues (2003) used data on bird distribution and abundances from a South African bird atlas to evaluate different levels of data (abundance data, presence/absence data, low sampling effort, no data) on the design of reserve networks. Their results showed that even low sampling efforts can be reasonably effective at representing bird species in a network of reserves, suggesting that reserve designs based on methods of complementarity can be effective in regions of the world with limited biological data.

The one disadvantage of fewer data that stands out is a less efficient design of conservation areas, meaning that fewer options are identified that can achieve conservation and that it often takes a greater amount of area to achieve the same conservation result. For example, a study by Grand *et al.* (2007) using the same data set as Grantham and colleagues reached this very conclusion. As a result, there is a trade-off that planners must balance between a reduction in efficiency of the planning effort and the costs of getting better information.

Can results from these sorts of studies be generalized more broadly? That is difficult to know with confidence. The distribution and quantity of available data, assumptions about land-use change, the timing and amount of new conservation areas established, the costs of implementation, and a host of other factors all affect the outcomes of conservation plans (Grantham *et al.* 2008). At the very least, these efforts suggest that the amount of biodiversity data used in conservation plans may not be the limiting factor that we once thought it was.

A related but equally important question is whether relatively simple plans with limited data can have conservation impact. Evidence, again from South Africa, clearly suggests that simple, straightforward plans can indeed have conservation impact. Reyers *et al.* (2005) prepared a biodiversity profile and priority assessment of South Africa's grasslands, a large biome in eastern South Africa. Like most grasslands worldwide, South Africa's are under-represented in protected areas, highly threatened, and easily converted to farmland. Their assessment, which aimed to integrate knowledge about this large area and use it to direct investment in mainstreaming conservation efforts in the grassland, was completed in just two months without collecting additional data (B. Reyers, personal

continues

Box 15.1 *continued*

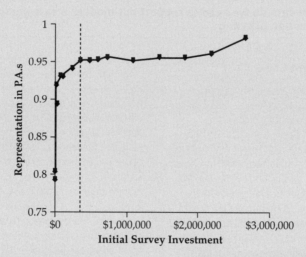

Figure 15.A.1 Representation of *Protea* species in protected areas in the Fynbos biome, South Africa, as a function of investments in biological surveys. Adapted from Grantham *et al.* (2008).

communication), but capitalizing on data that had been gathered as part of a National Spatial Biodiversity Assessment (Driver *et al.* 2005). It was based on analyses of three different types of priority areas: terrestrial biodiversity areas, river biodiversity areas, and ecosystem service priority areas.

National-level data on ecosystem status, threatened species status, and critical ecological processes were used in a summary fashion to define terrestrial biodiversity areas. Simple measures of river integrity such as the percentage of a catchment containing natural land cover were used to identify river biodiversity areas. Ecosystem services were mapped for water production, soil protection, carbon sequestration, grazing and services supporting harvestable products to aid in pinpointing ecosystem service priority areas. As a final step, priority "clusters" of catchments for grassland conservation were identified based largely on a summing and rescaling of priority areas for terrestrial biodiversity, freshwater biodiversity, and ecosystem services. Profiles of these clusters were developed in order to identify the opportunities and constraints for mainstreaming conservation into the sectors (e.g., agriculture, mining or forestry) within each cluster.

This grassland assessment, based largely on existing data and developed over a short-time period, has resulted in coordinated and strategic investment in conservation in the grassland biome (A. Stephens, personal communication). This includes directing the selection of and investment in demonstration project sites. Together with other

assessments conducted on agriculture, economics, urban development and the mining sector, it helped to provide a foundation for the implementation phase, galvanizing stakeholder buy-in into what needs to be done and where, ensuring commitment by the national government to the program, and securing additional funding from national agencies. To what factors can we attribute this success? First, this was the first time that knowledge, data and expertise from across the biome were brought together into a single assessment through workshops, data collation and interviews. Second, this assessment benefitted from the lessons learned in the National Spatial Biodiversity Assessment and other regional plans about the value of using simple, visual metrics of terrestrial biodiversity, freshwater biodiversity, and ecosystem services which are easy to understand and interpret, user friendly, and compelling to their target audience (Pierce *et al.* 2005; Reyers *et al.* 2007). Finally, the assessment learned from the lessons of the conservation planning community in South Africa that data collation and assessment, although of critical importance, are very small components of the long and complicated process of implementation, and should therefore not dominate the resources available (Cowling *et al.* 2004). A tentative conclusion we may take from these South African planning efforts is that not only are more and better data not always necessary and helpful, but how data are translated and conveyed to key stakeholders and decision-makers is every bit as important as the amount of data.

15.1.1   Predicting ecosystem service provision levels

There are two major types of ecosystem service model outputs that are useful to decision-makers: (1) predictions regarding the aggregate or overall quantity or value of ecosystem services provided under different scenarios and (2) the spatial distribution of ecosystem services across the landscape. The type of outputs needed depends on the type of decisions being made and the institutional framework for the decisions (see Table 15.1). In general, quantities or values of ecosystem services are useful for answering the question of "how much?" Government regulators responsible for allocating permits or requiring offsets need to know *how much* of each particular service is being generated or impacted by actions of regulated entities (Table 15.1). For example, in one of the first major cap-and-trade programs, the US Environmental Protection Agency required continuous emissions monitoring that recorded the level of SO_2 emissions for all large point sources. This information was essential for knowing if firms had the required number of permits to cover their emissions. The Colombian Ministry of the Environment,

Housing and Territorial Development is establishing a system through which they will assess all major sectoral projects (mining, agriculture, transportation, oil and gas) for the next 5 years to identify how much ecosystem service and biodiversity loss is expected and consequently what level of mitigation should be required.

Governments may also face choices over the kind of management action to take to reach a regulatory goal. Knowing *how much* of a service each action will yield can help managers choose among different actions. For example, the city of Santiago (Chile) has some of the worst particulate matter air pollution in Latin America. They used a production function model to determine how much air filtration would be provided by urban forests. This information was used in a cost-effectiveness analysis that revealed that urban forest management was a more cost-effective pollution control approach than other methods such as using alternative fuels (Escobedo *et al.* 2008). Their ability to identify *how much* of a service different management actions would yield helped them make the most efficient decision.

Conservation priorities are also shaped by knowing *how much* ecosystem service or biodiversity return

Table 15.1   Types of ecosystem service information used to inform management by diverse decision-makers

Decision-maker	Ecosystem service trait	
	Spatial distribution	**Absolute level of production**
Regulator of permits, mitigation and offsets	• Determine where offsets would be most efficient • Inform calculation of offset ratio	• Determine how many permits or offsets are required • Determine how much area is required to meet offsets
Regulator of subsidies or fees	• Determine where to target programs for most efficient outcomes	• Determine levels of payment or fee required
Conservation planner	• Determine where investments would be most efficient • Identify win–win areas where services overlap with biodiversity	• Design action plans to meet quantitative goals
Corporation	• Determine where investments would be most efficient	• Determine level of impacts that require offsets or fees or level of services created that generate permits or payments
Financial institution	• Determine where investments would be most profitable and least risk prone	• Determine likely level of return on investment
Designer of payment for ecosystem service program	• Determine where payments would be most efficient	• Determine level of payment

they can be expect from different geographic patterns of protection or restoration (Box 15.1). The Nature Conservancy is looking at this question in the Willamette Basin in Oregon, by asking how much ecosystem service return they are likely to get if they are successful in protecting their entire portfolio of conservation priority sites. Corporations trying to meet either internally set goals or government regulations need to know *how much* offsetting their proposed activities will cost for budgeting and strategic planning. BC Hydro, the third largest electric utility in Canada, has set an internal goal to become "environment neutral" in the next 20 years, meaning that they intend to reduce environmental impacts as much as possible and then offset unavoidable damages. To achieve this goal, they need to know how much impact their activities will have on biodiversity and ecosystem services so that they can reduce these effects or mitigate them elsewhere.

In general, when the amount of an ecosystem service is of primary interest, then questions of appropriate model complexity should focus on comparing levels of ecosystem services.

15.1.2 Predicting ecosystem service patterns in space

Ecosystem services are not delivered or valued uniformly across landscapes. Certain locations harbor more biodiversity, or provide more carbon storage or other ecosystem services than other locations. Delivery of some services, such as flood control or provision of clean drinking water, is more valuable if it takes place near large cities. In general, the effectiveness of any management action aimed at delivering an ecosystem service will be determined by the location of management activity. This means that mapped predictions of the relative value of different locations in terms of ecosystem service provision can be very useful when implementing investments, incentives or regulations. For example, in the United States, the spatial orientation of damaged areas and corresponding mitigation areas is used to determine the required mitigation ratio (specified as how many acres of restored or protected habitat are required for each acre damaged). For example, under the Clean Water Act, selecting a mitigation site outside the impacted watershed

doubles the required mitigation ratio (e.g., from 2 acres restored/protected: 1 acre damaged to 4:1) (ACOE 2004). An obvious goal for conservation is efficiency which translates into identifying where conservation activities should be distributed across a landscape to get the greatest possible biodiversity and ecosystem service returns (Box 15.1). Similarly, regulators developing new subsidy or fee programs can achieve efficiency by considering *where* in space to target payments or levy charges (e.g., Nelson *et al.* 2008).

15.2 Testing agreement between simple and complex ecosystem service models

In statistical theory it is relatively straightforward to compare the performance of models when the models are nested in terms of complexity, the easiest form of which is a multiple linear regression with a succession of possible additional predictors. However, deciding between simple versus complex models is more difficult when the alternatives are not nested and use different data sets (Pascual *et al.* 1997), as is the case with most of our tier 1 (simple) and tier 2 (complex) models. Here we focus on assessing how well the simple tier 1 models agree with more complex tier 2 models, and if models disagree, how using predictions from one particular model are likely to influence management decisions.

15.2.1 Carbon stock and carbon sequestration

The tier 1 and tier 2 models for estimating carbon stocks both predict the amount of carbon stored in a land use and land cover type on the landscape in Mg carbon ha^{-1} (see Chapter 7 for details). Both tier 1 and tier 2 models can also predict the amount or value of carbon sequestered over time (Mg carbon ha^{-1} yr^{-1} or \$US (or other currency) ha^{-1} yr^{-1}). The two models differ in their treatment of carbon storage as vegetation grows and decays, and in their treatment of carbon changes with land use and cover transitions. The simpler tier 1 model assumes that all parcels within a specified land use and cover type hold the same amount of carbon, regardless of the age of vegetation. The tier 1 model also assumes that carbon sequestration rates in a given land cover class are not related to previous land-use practices or cover types, or to the time since transition. Tier 1 is thus most appropriate for analysis of average or equilibrium "steady-state" values of carbon storage. The tier 2 model uses a coefficient to adjust the amount of carbon stored in a land cover type for its age, relative to the maximum possible storage of carbon in that land cover class. The tier 2 model also accounts for the previous land use and cover type and the amount of time that has passed since the transition to the current type occurred.

15.2.1.1 Predicting carbon storage and sequestration levels

Predicting the total amount of carbon stored or sequestered by a parcel can be important in the design of payment for ecosystem service programs or in the process landowners or corporations go through to determine whether they should participate in voluntary carbon markets. Carbon storage or sequestration levels may also be useful when companies or organizations are trying to assess the value of their current land assets. For instance, The Nature Conservancy or Exxon-Mobil may want to know how much carbon sequestration benefit is being provided by lands currently under their management.

We used the application of both tier 1 and tier 2 models in the Willamette Basin, Oregon (USA) to ask whether these differences in model complexity lead to important differences in the prediction of absolute levels of carbon storage and sequestration on the landscape for conditions in 1990 and three future land-use scenarios. All maps used to represent land use and land cover future scenarios were developed by a multi-stakeholder alliance (Hulse *et al.* 2002; US EPA 2002; Baker *et al.* 2004) that scripted three different management pathways: (1) conservation, where greater emphasis is placed on ecosystem protection and restoration, (2) development, where market forces are given greater free rein across all components of the landscape, and (3) planned trend, where current policies are implemented as written and recent trends continue.

In all cases, there was close agreement between the simple and complex models (Figures 15.1a, b, c). As predictions progressed farther into the future, the tier 2 model predicted higher carbon storage values than the tier 1 model, which was expected since the tier 2 model accounts for changes in carbon storage with age while the tier 1 model does not. As forests mature or carbon builds up in soils over decades, estimates of their carbon stocks stay static in the tier 1 model, while they move closer to the maximum possible storage capacity of each land cover type in the tier 2 model. Managers interested in using our models to track carbon sequestration where land use and cover types have changed in the recent past, or may change in the future, should use the tier 2 model if possible. In the Willamette application, conservation planners or landowners using the tier 1 model to report the contribution of their holdings can be confident that they are likely presenting conservative estimates.

15.2.1.2 Predicting carbon sequestration patterns in space

Information describing the spatial configuration of carbon storage or sequestration is useful when investors, governments, corporations or conservation groups are trying to identify the best places for action, or the best course of action. Policy makers designing new subsidy or fee programs, or conservation groups choosing new sites for action, may want to know where carbon will be sequestered or lost as a result of a proposed management program. We asked how well the tier 1 and tier 2 models agree on these kinds of spatial questions by taking the difference between the carbon storage maps produced by the two models for the Willamette Basin (USA) (Figure 15.2). We found differences in the prediction of carbon storage for initial conditions in the 1990 landscape. These differences, however, show no clear pattern and seem to be fairly randomly distributed across the landscape. By 2050, the age class and landscape transition considerations included in the tier 2 model lead to much higher estimates of carbon storage in forested areas than those given by the tier 1 model (especially for the conservation scenario in 2050; Figure 15.2). Tier 2 shows lower estimates for carbon storage in converted areas.

The type and age of land use and land cover classes included in the scenario being examined affected how much the models disagreed. This can be seen most clearly by comparing the estimates of carbon sequestration (or change in carbon storage) over time

(a)

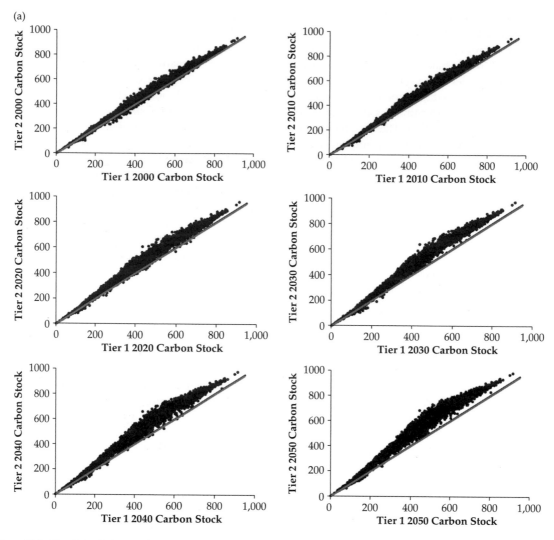

Figure 15.1 The relationship between the tier 1 and tier 2 estimates of carbon stock (MgC ha⁻¹) in the Willamette Basin (USA) for the conservation scenario (a), development scenario (b). and plan trend scenario (c) across decades from 1990 to 2050. The solid line shows a 1:1 relationship between tier 1 and tier 2 results. Data points are values for individual parcels in the Willamette Basin. The models agree relatively well over short time projections, but the tier 2 model consistently predicts higher levels of carbon storage than the tier 1 model at mid- to high carbon storage values as time progresses.

between 1990 and 2050 under each scenario for the two tiers (Figure 15.3). The models show the greatest disagreement in prediction of change under the development scenario where some forested areas are allowed to grow to maturity, yielding higher tier 2 estimates, and where other forested areas and some agricultural lands are converted to commercial timber harvest and housing development, yielding lower tier 2 estimates of change.

Managers interpreting spatial maps of carbon storage estimates should be aware that the tier 2 model will depict a more heterogeneous landscape, with bigger differences in carbon stocks among land cover classes and thus, bigger apparent differences in the opportunities for investment. With the tier 2 model, old forests will store much more carbon than young agricultural lands, whereas the difference will not be so stark in the tier 1 model. In

(b)

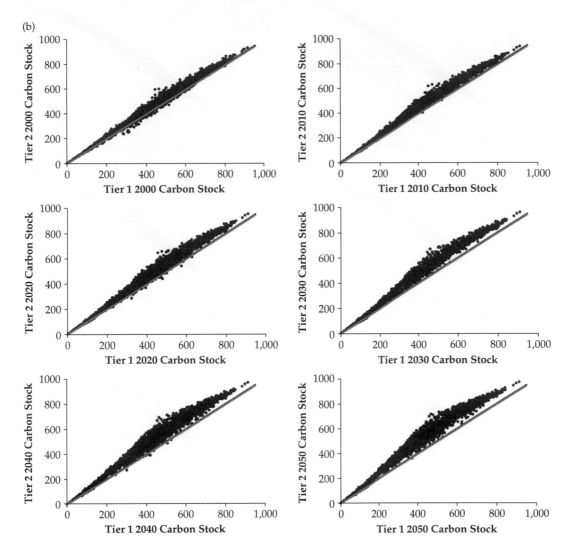

Figure 15.1 *continued*

other words, managers looking for the best places to invest, or the places to avoid, will have an easier (or clearer) decisions when the tier 2 model is used. As with the absolute predictions, this discrepancy in model predictions gets more severe the farther one looks into the future (although the gap will stop growing at some point in the future when all land cover classes reach their maximum storage capacity).

Rather than looking at the full spectrum of carbon values found on the landscape, many man-

agement questions are focused specifically on finding the places with the highest levels of carbon storage or sequestration. These "priority areas" are the places where the greatest return on investment can be expected either in terms of storage or sequestration, if acquisition prices and accessibility are consistent across the landscape. We asked if the simple and complex models identify the same "priority areas" in the Willamette Basin. For carbon storage, we defined priority areas as the best 25% of the landscape for storing

(c)

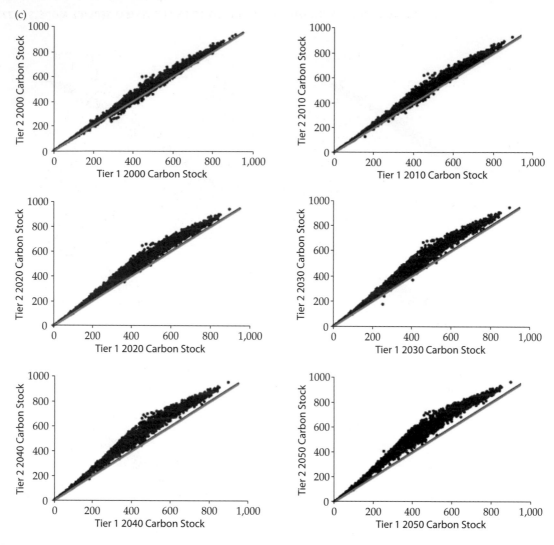

Figure 15.1 *continued*

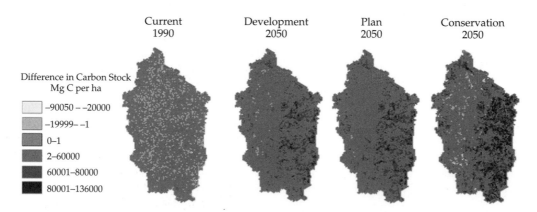

Figure 15.2 Difference between tier 1 and tier 2 estimates of carbon stock distributions across space under current and future conditions in the Willamette Basin (USA). Tier 2 estimates are higher than tier 1 estimates in dark areas and lower than tier 1 estimates in light areas.

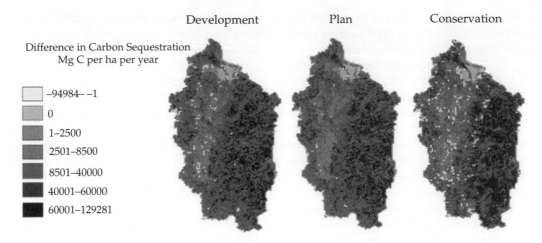

Figure 15.3   The difference between tier 1 and tier 2 estimates of carbon sequestration rates across space between 1990 and 2050 under three possible scenarios for the Willamette Basin (USA). Tier 2 estimates were higher than tier 1 estimates in dark areas and lower than tier 1 estimates in light areas.

carbon. We then used the Kappa statistic to calculate the degree of overlap between the priority areas identified by tier 1 and tier 2 models. The Kappa statistic ranges from -1 to 1, with -1 indicating no overlap of high service areas in space, 0 indicating overlap purely due to chance, and 1 indicating perfect overlap. We found that the two carbon models identified nearly identical locations as high priority carbon storage areas under both current (1990) and future conditions with Kappa values of 0.9 or higher (Table 15.2). Given these results, decision-makers can be confident that the selection of priority areas for carbon storage will not be significantly affected by the level of model complexity used.

Table 15.2   Spatial overlap of high carbon storage areas identified by tier 1 and tier 2 models

Carbon Stock 75th percentile (tons C parcel⁻¹)			
Scenario	Tier 1	Tier 2	Kappa
1990 Base Scenario	232,138	244,450	0.94
2050 Conservation Trend	242,935	309,201	0.90
2050 Development Trend	219,713	275,167	0.90
2050 Plan Trend	217,968	278,128	0.93

High storage areas were identified as those having per parcel carbon stocks greater than or equal to the 75th percentile value. The Kappa statistic identifies the degree of spatial overlap between tiers with a value of 1 indicating perfect overlap.

15.2.2   Crop pollination

The tier 1 and tier 2 pollination models both produce an index of pollinator abundance in native habitat areas and on farm fields. This index is a relative metric that gives a sense of how many pollinators are likely to be present on the landscape, but it is not an estimate of actual pollination rates. The main difference between the simple and complex pollination models is how accurately they represent the composition of the native pollinator community and the characteristics of each species present on the landscape. The tier 1 model assumes there is only one generic species of pollinator on the landscape while the tier 2 model considers each species separately and incorporates species specific characteristics for foraging distance, nest site choice, and floral resource specificity (see Chapter 10 for more details).

15.2.2.1   Predicting crop pollination levels
In agricultural areas, many conservation groups recommend the retention of islands of native habitat within the production landscape to serve as reservoirs for native pollinators. Farmers considering this approach might want to weigh the economic value of the additional yields they receive from native pollination against the income they lose from taking an area out of crop production. Being able to predict the level

of pollination (and thus additional crop yield) that can be expected from restoring or protecting a parcel is essential to this and other types of decisions.

Using California's major agricultural region, the Central Valley, as a case example, we asked whether the differences between simple and complex pollination models are important for predicting pollination levels across a landscape. We used watermelon as the focal agricultural crop for this study. We assessed the agreement of simple (tier 1) and complex (tier 2) pollination models by generating predictions for 40 000 randomly selected parcels (entire landscape consists of 600 000 parcels) (Figure 15.4). The predictions of relative abundance of pollinators on watermelon crops in each parcel were very highly correlated ($R^2 =$ 0.99). The tier 1 pollination model predicts lower levels of pollination than the tier 2 model at the low end of the pollination spectrum and higher relative pollination levels at the high end of the spectrum. The difference in predictions of pollination provision can be explained by the difference in pollinator community composition represented by the two models. The tier 2 model depicts several different species of pollinators with potentially different preferences for each habitat type, while the tier 1 model depicts one generic pollinator representing average characteristics of the pollinator community. A land use and cover class that is very compatible for the generic species in the tier 1 model will generate a high score in tier 1. At the other

end of the spectrum, the tier 1 model will predict a lower level of pollination from a land use and cover class that the generic species finds incompatible.

The tier 1 model is likely most appropriate where native pollinator communities are not diverse, or where pollinator preferences for habitat types are similar among all pollinator species. When these conditions do not hold, it will be more appropriate to use the tier 2 model. The models differ the most when the landscape of interest contains land use and cover classes that vary dramatically in their suitability for different pollinators.

15.2.2.2 Predicting crop pollination patterns in space

Pollination is a service driven by relatively small scale processes (on the order of meters to kilometers), so the question of where services are provided is especially important for management decisions with implications for pollination services. For investments in pollinator habitat to be fruitful, habitat patches need to be within pollinator foraging range distance of crops that require insect pollination. Of course, different crops sell at different prices, so providing viable pollinator habitat near the most lucrative crops will also be preferred from an investment perspective.

We analyzed how well model predictions of the distribution of pollination services agreed by subtracting tier 1 estimates of the relative pollinator

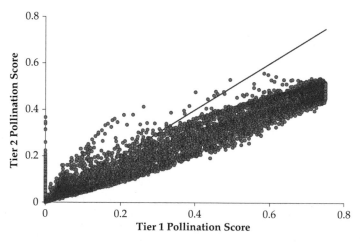

Figure 15.4 Relationship between tier 1 and tier 2 predictions of a pollinator abundance index in California's Central Valley for watermelon crops. The black line shows a 1:1 relationship, or perfect agreement between the models.

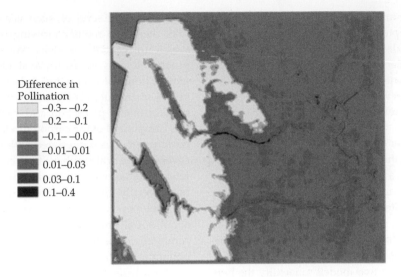

Figure 15.5 Difference in spatial patterns of the pollinator abundance index predicted by tier 1 and tier 2 models in California's Central Valley for watermelon crops. The tier 2 model predicted higher pollinator levels than the tier 1 model in dark areas and lower levels in light areas.

abundance index from tier 2 estimates across space (Figure 15.5). The tier 2 model predicts lower pollination provision in a large area in the western part of the Central Valley. This area is dominated by high quality pollinator habitat. The generic bee species represented in the tier 1 model sees this high quality habi-

tat as very good for nesting and feeding, whereas the tier 2 model includes some species that do not view this habitat as perfectly compatible, yielding an overall lower pollination score. Similarly, the green areas, located along thin riparian buffers, are seen as incompatible habitat by the single insect species in the tier 1

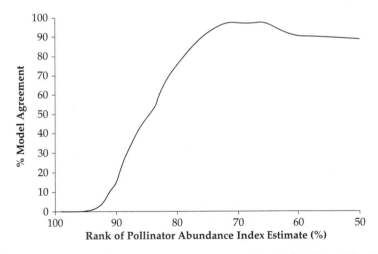

Figure 15.6 Summary of spatial agreement between tier 1 and tier 2 pollinator abundance index models applied in California's Central Valley for watermelon crops. The *x* axis shows the percentile cutoff used to define "high priority areas" and the *y* axis shows how often the models chose the same parcels within that percentile. If one was interested in finding the parcels on the landscape that have the highest 10% of abundance estimates (the 90th percentile), the models do not agree well (only 15% of those parcels overlap in space). If one was interested in locating the highest 25% of abundance estimates on the landscape (75th percentile), the models agree very well (92% overlap of parcels in space).

model, but they are seen as viable habitat for some other species represented in the tier 2 model. The same recommendations given in the previous section apply here; the tier 1 model will be most appropriate for landscapes with few pollinator species, pollination communities with similar habitat preferences, or landscapes with relatively homogenous, high quality pollinator habitat.

As with other services, targeting investments towards parts of the landscape with the most pollination potential is a common approach. Here, rather than using the Kappa statistic to look at agreement among priority sites defined by one cutoff value, we look at how well predictions agree across space at every percentile from the 50th to the 100th (Figure 15.6). If we choose the same threshold as we did for carbon, and ask the two models to identify the best quarter of the landscape for pollination service provision (75th percentile), we see again that the models agree very well (92% overlap in space) (Figure 15.6). However, if we had a more limited budget and needed to focus on the best 10% of the landscape (90th percentile), the models pick quite different parts of the landscape, agreeing only 15% of the time. So while the selection of priority areas focusing on the best 25% of the landscape for pollinator habitat or below can be done equally well with either pollination model, priority setting that chooses to set more selective standards will be affected by model complexity.

15.3 Future directions and open questions

It is encouraging that the simple and complex models of carbon storage and crop pollination agree well in terms of rank order predictions of ecosystem service level and in patterns of ecosystem service distribution across space. The tests used in this chapter could be applied to tier 1 and tier 2 model pairs for all of our ecosystem service models (see Chapters 4–6, 8, 9, 11, 13). Model results should also be compared in landscapes with different biophysical and socio-economic characteristics to ensure that patterns of similarity seen in our sample landscapes are robust.

Although understanding how simple and complex models relate to each other is important, confronting models with empirical data is essential for validating model results (Van Nes and Scheffer

2005). For some services, such as carbon sequestration, there is a rapidly growing body of data on which to validate models. Measures of above ground biomass are readily available for many systems. Empirical results for carbon storage in soil, however, are subject to considerable heterogeneity making model prediction and validity testing for this component of the carbon model difficult. How rapidly carbon is sequestered with land use or land cover change is also subject to uncertainty. We have found validation of ecosystem service models to be especially challenging because model outputs are often new metrics that incorporate information about social and biological systems that are not commonly measured today. For instance, our water pollution regulation model combines information on hydrological processes that determine ecosystem service supply with information from water treatment plants that determine service demand and value. We can validate individual parts of the model based on readily available information (see Chapter 4), but it is difficult to validate the final model outputs. Improvements in this area will only come as we amass data sets on metrics that relate to integrated socio-ecological systems. For other services, the simplicity of the tier 1 models leads to estimation of a proxy for the service of interest. For example, the crop pollination model returns an index of pollinator abundance based on habitat availability. We can relate this proxy to empirical, field-based observations of pollinator abundance or visitation rates to ask if the model does well in predicting the rank order of pollination levels or spatial distribution, but we cannot verify the absolute level of pollination provided.

Even when data become available, identifying which model is most appropriate for representing system dynamics can be complex (Pascual *et al.* 1997). Pascual *et al.* (1997) suggest that instead of trying to find a single model that does well in dealing with all components of a system, a more fruitful approach may be to focus on finding management choices that are robust to the model chosen for use. We have attempted to take that approach here for a subset of decisions, suggesting that identifying the best 25% of a landscape for service provision is generally robust to model choice for both carbon sequestration and pollination. Decisions regarding the placement of investments by conservation groups, financial institutions or others in the private sector, along with decisions about where to direct mitigation and offsetting or payments for services could all be made with tier 1 or tier 2 models with equal confidence. This approach can easily be applied to the many other models discussed in this book.

The biggest challenge in terms of model selection and testing entails questions of "how much" as opposed to "where are the best places." There need to be national ecosystem monitoring programs that report spatially explicit data on water quality, carbon sequestration and other vital ecosystem outputs if we are to understand which models to use and how much confidence to have in model results.

References

Army Corps of Engineers (ACOE). (2004). *Chicago District 2004 mitigation requirements*. Chicago District of the U.S. Army Corps of Engineers, Chicago.

Baker, J. P., Hulse, D. W., Gregory, S. V., *et al.* (2004). Alternative futures for the Willamette River Basin, Oregon. *Ecological Applications*, **14**, 313–24.

Cowling, R. M., Knight, A. T., Faith, D. P., *et al.* (2004). Nature conservation requires more than a passion for species. *Conservation Biology*, **18**, 1674–7.

Driver, A., Maxe, K., Rouget, M., *et al.* (2005). *National spatial biodiversity assessment 2004: Priorities for biodiversity conservation in South Africa*. South African National Biodiversity Institute, Pretoria.

Escobedo, F. J., Wagner, J. E., Nowak, D. J., *et al.* (2008). Analyzing the cost effectiveness of Santiago, Chile's policy of using urban forests to improve air quality. *Journal of Environmental Management*, **86**, 148–57.

Gan, T. Y., Dlamini, E. M., and Biftu, G. F. (1997). Effects of model complexity and structure, data quality, and objective functionson hydrologic modeling. *Journal of Hydrology*, **192**, 81–103.

Gaston, K. J., and Rodrigues, A. S. I. (2003). Reserve selection in regions with poor biological data. *Conservation Biology*, **17**, 188–95.

Grand, J., Cummings, M. P., Rebelo, T. G., *et al.* (2007). Biased data reduce efficiency and effectiveness of conservation reserve networks. *Ecology Letters*, **10**, 364–74.

Grantham, H. S., Moilanen, A., Wilson, K. A., *et al.* (2008). Diminishing return on investment for biodiverstiy data in conservation planning. *Conservation Letters*, **1**, 190–8.

Hulse, D., Gregory, S., and Baker, J. (2002). *Willamette River Basin planning atlas: trajectories of environmental and ecological change*. Oregon State University Press, Corvallis.

Ludwig, D., and Walters, C. J. (1985). Are age-structured models appropriate for catch-effort data? *Canadian Journal of Fisheries and Aquatic Sciences*, **42**, 1066.

Nelson, E., Polasky, S., Lewis, D. J. *et al.* (2008). Efficiency of incentives to jointly increase carbon sequestration and species conservation on a landscape. *Proceedings of the National Academy of Sciences of the USA*, **105**, 9471–6.

Pascual, M. A., Kareiva, P., and Hilborn, R. (1997). The Influence of Model Structure on Conclusions about the Viability and Harvesting of Serengeti Wildebeest. *Conservation Biology*, **11**, 966–76.

Pierce, S. M., Cowling, R. M., Knight, A. T., *et al.* (2005). Systematic conservation assessment products for landuse planning: interpretation for implementation. *Biological Conservation*, **125**, 441–8.

Reyers, B., Nel, J., Egoh, B., *et al.* (2005). *A conservation assessment of South Africa's grassland biome: Integrating terrestrial, river, ecosystem services and fine scale priorities*. SCIR, Cape Town.

Reyers, B., Rouget, M., Jonas, Z., *et al.* (2007). Developing products for conservation decision-making: Lessons from a spatial biodiversity assessment for South Africa. *Diversity and Distributions*, **13**, 608–19.

Stephens, P. A., Frey-Roos, F., Arnold, W., *et al.* (2002). Model complexity and population predictions, The alpine marmot as a case study. *Journal of Animal Ecology*, **71**, 343–61.

United States Environmental Protection Agency (US EPA) (2002). *Willamette Basin alternative futures analysis. Environmental assessment approach that facilitates consensus building*. US EPA Office of Research and Development, Washington, DC.

Van Nes, E. H., and Scheffer, M. (2005). A strategy to improve the contribution of complex simulation models to ecological theory. *Ecological Modelling*, **185**, 153–64.

CHAPTER 16

Poverty and the distribution of ecosystem services

Heather Tallis, Stefano Pagiola, Wei Zhang, Sabina Shaikh, Erik Nelson, Charlotte Stanton, and Priya Shyamsundar

16.1 Introduction

The planet's stocks of natural assets continue to diminish while widespread poverty persists throughout the developing world (Chen and Ravallion 2008). Recognizing the link between humans and nature has never been more salient for policy-makers and researchers concerned with conservation and poverty alleviation (Sanderson and Redford 2003; Millennium Ecosystem Assessment 2005). Information detailing the specific links between the poor and the environment is building (e.g. Albla-Betrand 1993; Scherr *et al.* 2003; Delang 2006) and efforts are underway to determine how international agreements concerned with poverty alleviation can, and should, incorporate conservation as a means to their ends. For example, Roe and Elliott (2004) note that United Nations (UN) Millennium Development Goals (MDGs), adopted by the General Assembly of the UN, have direct ties to environmental condition and stability. Of the eight goals, only one is explicitly environmental, but the success of five others will rely on healthy ecosystems (Roe and Elliott 2004).

Incorporation of these advances in local and regional decision-making has proven challenging. Creating policies that account for interactions and trade-offs among environmental, economic and social values is difficult today because many of the connections between humans and the environment are not formally recognized by political and economic systems. Instead, decisions made today based on costs and benefits to society leave out many of the public goods and services provided to the poor by the environment, such as the provision

of clean water, food from bushmeat and native plants, medicinal plants, and protection from "natural disasters," such as storms and floods (Millennium Ecosystem Assessment 2005). For example, conventional household economic surveys fail to include directly the contributions of natural assets to rural household welfare (Cavendish 2000), making it difficult to assess how they will be affected by public or private programs that change the status of these resources. Similarly, ecosystem services that make up a country's aggregated natural capital are largely absent in national accounting. Progress in economic valuation of ecosystem services is needed to support the development of a standardized methodology for the inclusion of ecosystem values into Standard National Accounts (Mäler *et al.* 2008) and other policy decisions. The fundamental problem with incorporating ecosystem services into the balance sheets used in decision-making is the lack of tools that easily track the status of or changes in these services, and their distributional effects on human well-being.

In this chapter, we examine how the emerging field of modeling and mapping ecosystem services can address this gap. We begin by discussing the need for a detailed understanding of the linkages between poverty and ecosystem services, and provide a brief review of the literature on these links (Section 16.2). We then examine the opportunities for and difficulties with an integrated approach to mapping poverty with ecosystem services (Section 16.3). Finally, we demonstrate how several of the mapping and valuation models described in earlier chapters (Chapter 8) and developed elsewhere can be applied in this context using case studies of the Amazon Basin and Guatemala's highlands (Section 16.4).

16.2 Ecosystem services and the poor

16.2.1 Dependence of the poor on ecosystem services

Earth's ecosystems provide myriad goods and services that are essential to the well-being of all people, but natural capital contributes disproportionately to the welfare of the poor (see Section 16.3.2 for a discussion of "poor"), in some cases significantly, because often the poor have limited capacity to purchase goods and services from elsewhere (Table 16.1) (World Bank 2008). For instance, Vedeld *et al.* (2004) found that about 22% of rural household income can be attributed to the harvest of goods from forests, contributing almost twice as much to the incomes of the poor as to the non-poor (Table 16.1). This finding cannot be generalized to all rural households, but it is relevant to those living on the fringes of forests, those that are largely dependent on natural resources for subsistence, or those who are engaged in agricultural activities that rely heavily on natural capital and ecosystem processes (Box 16.1).

In addition to providing income, natural resources can serve as safety nets for the poor during times of stress (Justice *et al.* 2001; World Bank 2008). The poor have been shown to respond to known agricultural risks and sudden agricultural shocks by increasing their dependence on natural resources

(Pattanayak and Sills 2001), and the insurance provided by natural resources can make it feasible for households to recoup after natural disasters (McSweeney 2005). Large changes in access to or availability of these services are likely to have significant effects on the poor (Ferraro 2002; World Bank 2008), making it essential that policy-makers recognize the possible costs that choices related to resource sectors impose on the poor through changes in services (World Bank 2008).

16.2.2 The poor as agents or victims of environmental degradation

Depending on the context, the poor may be agents of degradation, its victims, or both. When the poor are agents of environmental degradation, they can cause declines in ecosystem service provision for themselves and for others. For example, the use of forests for fuel wood and other non-timber forest products by the poor has been the main cause of forest degradation in India (Baland *et al.* 2006) and Tanzania (Luoga *et al.* 2000; Ndangalasi *et al.* 2007). Similarly, degradation in Peru's Pacaya-Samiria National Reserve was largely the result of the subsistence use of resources by households living around the reserve (Takasaki *et al.* 2004). Degradation can affect the functions of local ecosystems, increase ecological fragility, and increase the vulnerability of the poor to natural shocks (Shyamsundar 2001).

Despite its apparent irrationality, one of the reasons that households degrade ecosystem services

Table 16.1 Environmental income as percentage of total income in resource-poor and resource-rich areas

Study	Resource-rich areas		Resource-poor/ low access areas		Average	
	Poor	Rich	Poor	Rich	Poor	Rich
Jodha (1986)					9–26	1–4
Cavendish (2000)			44	30		
Vedeld et al. (2004)[a]					32	17
Narain et al. (2005)	41	23	18	18		
Chettri-Khattri (forthcoming)[b]	20	14	2	1		

[a] Data reported are from multiple earlier studies.
[b] Nontimber forest product (NTFP) income only.
In most cases, "poor" refers to poorest 20% and "rich" to the richest 20% of households.
Adapted with permission from World Bank (2008).

Box 16.1   Can the natural capital of agroecosystems provide a pathway out of poverty?

C. Peter Timmer

Historically, the major pathway out of poverty has been the structural transformation of economies, where the share of agriculture in employment and value added to the national economy declines as the share of urban industry and modern services rises. Labor productivity is higher in urban activities, and migration from rural to urban jobs raises wages at both ends. In a broad sense, this transformation has been a transition from dependence on biological processes of production—especially in agriculture—to physical processes of production—primarily in manufacturing processes for metals, chemicals, and automobiles. Ultimately, the main source of economic growth—and poverty reduction—has been in knowledge-intensive processes such as finance, information technology, and communications. In short, reducing poverty has meant reducing reliance on natural capital.

This evolution away from apparent dependence on natural capital as the source of economic growth and livelihoods for the poor obviously missed a key point: we all have to eat. It was easy to miss the point: in rich countries with highly productive agriculture, there are more lawyers than farmers. The structural transformation has as its endpoint "a world without agriculture," or at least a world where the farm sector behaves economically like all other sectors in the economy.

But the importance of our dependence on agriculture as the most efficient way for human society to capture solar energy in a form that we can consume has returned with a vengeance, in the form of high food prices. There is vigorous debate over the causes of the price resurgence, but most analysts feel that the link between energy prices and food prices that has been established by bio-fuel programs in the Unites States (ethanol from corn) and Europe (bio-diesel from vegetable oils) now means that high fuel prices mean high food prices. "Renewable energy" largely means capturing it from the sun, and photosynthesis remains the most efficient way of doing that over large expanses of land.

Natural capital means far more than agriculture, of course, and most researchers in the field spend most of their energy understanding the value to local economic productivity of biodiversity from natural ecological systems. But this focus is too narrow. Unless we understand the broader context in which natural capital has value, especially in the link between capturing solar energy as an agricultural activity and the subsequent role of agriculture

in economic development and poverty reduction, we will not be able to understand the role of natural capital in poverty reduction directly.

The major dilemma, for economists and policy-makers, is in placing a monetary value on the output from agriculture. For many decades rich countries have sought mechanisms to place a higher value on their agricultural sectors than market prices would indicate, and thus, implicitly, value the underlying natural resources committed to agricultural production more highly. At least three rationales for supporting agriculture in rich countries at taxpayer and consumer expense are increasingly accepted by mainstream policy analysts as reflecting appropriate public action in the face of market failures. These are: support for the multiple functions that agriculture performs, beyond the commodity production that is offered for sale ("multi-functionality"); support for "local" food systems that might offer reduced carbon footprints for most food consumers and possibly even fresher and healthier food; and support for bio-fuel production as a mechanism to break dependence on imported fossil fuels and slow emissions of greenhouse gases.

Multi-functionality and the non-market contributions of agriculture

Bucolic landscapes, green buffers to urban density, preservation and development of rural societies, domestic food security, and flood alleviation through proper land management all have economic value even if there is no market price for their "production." These non-commodity outputs, although essential to economic, environmental and social well-being, are unpaid by-products of commodity production. If farmers are paid only the market price for their commodities, the by-products will not be produced in optimal amounts, and may be lost altogether if farmers are forced out of business because of international competitive pressures.

Efforts to value in economic terms the flow of multiple services from natural ecosystems, including agriculture, need far more analytical research and empirical testing. With better valuation will come better designed initiatives to conserve natural resources and better mechanisms to pay the provider of these services, including farmers. From an economic perspective, simply paying farmers to do more of what they do anyway cannot be an efficient use of fiscal or natural resources. Agriculture performs multiple functions, but finding ways for the market to value, and pay for, these functions will be essential to sustainable production.

Local food systems

Buying food that is produced "locally" is the current agenda for two related causes: the anti-globalization movement and the sustainability movement. The anti-globalization movement has its roots in a clear sense of lost control over something as deeply felt as where the food on our tables comes from. Modern supply chains seem impervious to consumer desires to control what they eat. The sustainability movement has its roots in the broader environmental movement that now links to climate change as the key challenge to quality of life in rich and poor countries alike. Can transporting food thousands of miles, often on jet freighters, possibly be a sustainable way of eating? Will buying and consuming foods produced locally make any difference to either of these agendas?

Economic efficiency has a hard time entering these debates. Both the anti-globalization and sustainability movements specifically reject market prices as the basis for evaluating decisions about what consumers should consume, because these prices have too many subsidies and distortions to reflect real opportunity costs in terms of natural resources used. There is some merit to these arguments. The question is, should the "local food movement" receive more policy support?

Consumers, especially wealthy consumers, like to know where their food comes from and buy from producers who are neighbors. The rapid growth of farmers' markets, of organic food, and of "local food" sections in supermarkets is testimony to this basic desire. The trend bears watching, because it is the ultimate form of agricultural protection. Expanded trade has been the basis of much economic growth, and restricting it could have serious and unforeseen consequences.

Bio-fuels and the potential to reverse the structural transformation

Bio-fuels are not exactly new. Although coal, the first fossil fuel, was known in China in pre-historic times, and was traded in England as early as the 13th century, it was not used widely for industrial purposes until the 17th century. Until then, bio-fuels were virtually the only source of energy for human economic activities, and for many poor people they remain so today. But the widespread use of fossil fuels since the Industrial Revolution has provided a huge subsidy to these economic activities—because coal and later petroleum were so cheap—a subsidy which seems to be nearing an end. Are bio-fuels the answer to growing scarcity of fossil fuels?

Not surprisingly, the answer depends on the role of agriculture in individual countries, the pattern of commodity production and the distribution of rural assets, especially land. It is certainly possible to see circumstances where small farmers respond to higher grain prices by increasing output and reaping higher incomes. These incomes might be spent in the local, rural non-farm economy, stimulating investments and raising wages for non-farm workers. In such environments, higher grain prices could stimulate an upward spiral of prosperity.

An alternative scenario seems more likely however, partly because the role of small farmers has been under so much pressure in the past several decades. If only large farmers are able to reap the benefits of higher grain prices, and their profits do not stimulate a dynamic rural economy, a downward spiral can start for the poor. High food prices cut their food intake, children are sent to work instead of school and an intergenerational poverty trap develops. If the poor are numerous enough, the entire economy is threatened, and the structural transformation comes to a halt. The share of agriculture in both employment and GDP starts to rise, and this reversal condemns future generations to lower living standards.

that they rely on is that the impact of slow and incremental reductions in resource availability on welfare is small due to substitution effects. Households adapt to destruction of one resource over time by obtaining their resources from alternate areas or switching to alternate resources. As long as the opportunity cost of spending time to alter resource-use patterns is low, the welfare impact of degradation is likely to be small. In cases where the poor are agents of degradation, programs and policies designed to incentivize management activities that enhance ecosystem service provision would help advance goals of both poverty alleviation and conservation.

The poor often are victims of changes in ecosystem services caused by the activities of other sectors or people in other locations (Box 16.2). In these cases, the poor may benefit indirectly from regulations (such as protection, see Andam *et al.* 2010) or incentive

---

Box 16.2   Poverty and ecosystem service mapping at work in Kenya

Norbert Henninger and Florence Landsberg

A new atlas of Kenya, designed to improve understanding of the relationships between poverty and the environment, was released in 2007 (World Resources Institute 2007). The atlas and its 96 different maps include significant policy and economic development analyses that will be useful to policy-makers worldwide. This collection of maps is a step forward from the landmark findings of the 2005 Millennium Ecosystem Assessment—that 15 of the world's 24 ecosystem services are degraded. It will help enable other countries to develop their own similar maps.

Professor Wangari Maathai, 2004 Nobel Peace Laureate, said of the Atlas, "As a result of this type of work, we will never be able to claim that we did not know. Planting trees has been a way to break the cycle of diminishing resources for the women of the Green Belt Movement. I see the ideas and maps in this Atlas to be much like a small seedling. If nurtured, if further developed and grown, and if used by both government and civil society, this seedling carries the promise of breaking the cycle of unenlightened decision-making that is not accountable to the people most affected by economic or environmental changes; that does not consider the impact on our children and grandchildren."

As an example, one map from *Nature's Benefits in Kenya* outlines the upper watersheds of the Tana River

and combines that with poverty rates in 222 administrative areas (Figure 16.B.1; Plate 12). Most of the poorer communities are located in the drier plains downstream of the foothills of the Aberdare Range and Mount Kenya. The quantity and quality of the surface water supply for these poorer communities is highly dependent upon the use of land and water resources by the upstream communities. If upstream users withdraw large quantities of water, little is left for families downstream. If upstream users contaminate the water supply, families downstream bear the consequences. Communities and decision-makers need to be aware of these relationships to make better management and policy decisions. For example, upstream investments in improved watershed management to reduce water pollution and water shortages could yield two benefits: improved ecosystem health and benefits to poor downstream communities. However, any mechanism to pay for these changes in the supply of ecosystem services cannot rely on the downstream communities because of their lack of resources.

Similarly, other maps in the Atlas show how and where people derive benefits from the land and how that relates to the spatial pattern of human well-being. The Atlas is designed to inspire improved analysis of poverty-environment relationships and informed decision-making.

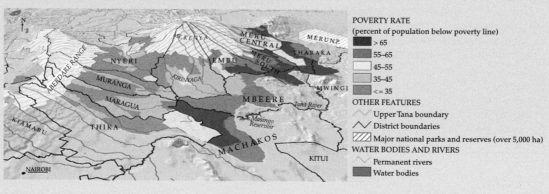

Figure 16.B.1   Map of the Tana River headwaters in Kenya, and the distribution of poor communities. (See Plate 12.)

---

programs targeted at government or wealthier actors to control ecosystem service degradation in poorer areas. For example, through the Clean Water Act, the US Government has unintentionally redistributed wetland services (e.g., fish for food, flood mitigation) away from urban poor to areas with significantly lower population densities (Ruhl and Salzman 2006; BenDor *et al.* 2008). The equity impacts of this regulation are rarely addressed in the Clean Water Act decision-making process (BenDor *et al.* 2008). An alternative regulatory approach could intentionally direct wetland mitigation activities towards areas with poor populations, thereby improving the poor's access to wetlands' many services.

16.3 Mapping poverty and ecosystem services

The relationship between the poor and natural resources is mediated by factors at various scales, such as labor and credit markets, property rights and other institutions, and information about best practices (Bluffstone 1995; Duraiappah 1998; Wunder 2001; Adhikari 2005). In some cases, a lack of markets contributes to degradation of natural systems; in other cases growth in markets can lead to ecosystem declines. Weak governance institutions, ill-defined property rights or lax enforcement, high discount rates, and population growth will all likely continue to contribute to degradation of local natural capital.

Given the complexities of connections between the poor and the environment, it is not easy to map poverty and ecosystem services together in a way that is robust and practically useful. Ideally, poverty and ecosystem service mapping would be done such that (1) the resolution of both poverty and ecosystem service information is sufficient to represent patterns in each accurately, (2) the poverty indicators chosen are directly tied to the component of human well-being of interest and well matched with the ecosystem service(s) of concern, and (3) the institutions that control the provision of services are represented. Here, we discuss what can be done today as first steps toward this ideal and what challenges remain.

16.3.1 Data resolution

Robust poverty analyses require uniform and high-quality data that are often unavailable, especially in developing countries. In most cases, indicators such as poverty rates are only available for relatively large administrative units. These data are often of little use for detailed analysis because administrative boundaries seldom match those of the ecosystems of interest, and because neither ecosystem services nor the poor are likely to be distributed uniformly within these boundaries.

More detailed poverty maps are typically created by using comprehensive data from small-sample household budget surveys to obtain a predictive relationship for poverty rates that is then applied to data from a national census at the highest available level of disaggregation (Poggi et al. 1998; Elbers et al. 2002). This approach has been used to generate poverty maps for several countries (Hentschel et al. 2000; Minot 2000; Müller et al. 2006; Bedi et al. 2007; Nelson and Chomitz 2007). However, even this finer scale analysis still produces maps based on administrative divisions (census tracts) and not ecosystem boundaries. Nelson and Chomitz (2007) dealt with this limitation by converting a poverty map of Guatemala to a watershed map. Even with poverty maps aligned to ecosystem boundaries, the assumption of uniform distribution within an ecosystem can seriously distort analysis. Within a watershed, for example, it may matter whether the poor are concentrated in the steeper upper slopes or the flatter riparian zones (Box 16.2).

16.3.2 Poverty indicators

In addition to challenges with mapping poverty at an appropriate resolution for analysis, we also face the challenge of defining poverty in different settings. Poverty historically has been defined in strictly economic terms, with income as the common indicator. Some analysts now argue that consumption is a better measure of poverty, as it is more closely related to well-being and reflects capacity to meet basic needs through income and access to credit. It also avoids the problem of income flows being erratic at certain times of the year, especially in poor agrarian economies where fluctuations can cause reporting errors.

All money-based indicators have the limitation that they cannot reflect individuals' feelings of well-being and access to basic services. In recent years, a broader understanding has developed in which poverty encompasses not only deprivation of materially-based well-being, but also a broader deprivation of opportunities (World Bank 2001; UNEP 2004). The Millennium Ecosystem Assessment (MA) recognized five linked components of poverty: the necessary material for a good life, health, good social relations, security, and freedom of choice (Millennium

Ecosystem Assessment 2005). Consider just one of these, security. A household's ability to address risks and threats can change dramatically even as income or consumption remain stable. Factoring in the effect of vulnerability, analysts estimate that monetary-based indicators can understate poverty and inequality by around 25% (World Bank 2001). In response, efforts have been made to develop non-monetary poverty indicators related to health, nutrition, or education, as well as composite indices of wealth (Wodon and Gacitúa-Marió 2001).

Poverty measures used in mapping are typically defined relative to a poverty line which is a cut-off separating the poor from the non-poor. For instance, the headcount index is a measure of poverty incidence that computes share of the population below the poverty line. An important distinction must be made between poverty rate, which is the proportion of people in an area that are poor, and poverty density, which is the number of people in an area that are poor. Many previous efforts to map poverty have found that areas with high poverty rates are often areas with low population density, and thus a small absolute number of poor people. Poverty rates may be most relevant if an analysis aims to locate segments of a population that are worst off, but poverty density may be most relevant if the analysis aims to find regions with the greatest number of poor.

When choosing the appropriate poverty indicator(s) to map, we should also consider the ecosystem service(s) of interest and how the poor relate to those services. Pairings between ecosystem services and poverty indicators can be either direct or indirect. Pairings are "direct" if a change in the ecosystem service directly influences the poverty indicator of choice (Table 16.2). If there is not a causal link between the service and the indicator, the pairing is "indirect" (Table 16.2).

In cases where the poor are agents of ecosystem service change, indirect pairings can be useful in mapping and modeling exercises used to design new programs. Consider a carbon offset program where a private sector buyer from a developed country wants to make payments to landholders in the tropics to plant trees in order to offset the buyer's carbon emissions. Mapping exercises could combine information on carbon sequestration potential and any indirect indicator of poverty to identify areas where sequestration projects could meet economic and conservation goals. An indirect indicator is appropriate here because poverty in this location is not directly related to the ecosystem service in question. The only requirement for the desired welfare transfer to the poor is that people who are poor have ownership rights or control over the deforested or degraded lands where carbon sequestration potential is high.

Table 16.2 Pairing ecosystem services with poverty indicators

Ecosystem service	Poverty indicator						
	Water poverty index	Child hunger	Infant mortality	HDI	Income	UBN	Literacy
Water purification	D	I	D[b]	I	I	D[c]	I
Provision of food	I	D	D	I	I	I	I
Provision of medicinal plants	I	D	D	I	I	I	I
Timber production	I	I	I	I	I	I	I
Carbon sequestration	I	I	I	I	I	I	I
Crop pollination	I	D[a]	I	I	I	I	I
Erosion control	I	I	I	I	I	I	I

[a] If local food crops need insect pollination.
[b] If diarrhea from waterborne disease is significant cause of infant mortality.
[c] If one of the unsatisfied basic needs is clean water.
Pairs where the poverty indicator could be directly influenced by a change in the ecosystem service are "direct" (D). Pairs where there is not a causal linkage between the service and the indicator are "indirect" (I). HDI = Human Development Index. UBN = Unsatisfied Basic Needs.

When the poor are victims of environmental degradation, or beneficiaries of services, improvements in well-being occur through improvements in service delivery, not through payments or incentives to the poor. In these cases, directly pairing ecosystem service provision with poverty indicators tied to the service(s) of interest is appropriate. For example, consider a change in the wetland mitigation example given above where the government requires developers to target offsets to benefit the poor. The most appropriate poverty indicators to use are those associated with wetland benefits: hunger where wetlands provide fish for consumption, access to clean drinking water where wetlands provide water purification, or flood vulnerability where wetlands provide storm surge protection. Using an income-based indicator, such as the percent of the population below the poverty line, to recommend the allocation of improved wetland services would be inappropriate if household income is not sensitive to changes in wetland services.

16.4   Case studies

The following two case studies represent some of the analyses we can do today. They highlight the types of decisions that can be informed by currently available data and methods, while identifying remaining challenges.

16.4.1   Deforestation in the Amazon Basin

The Amazon Basin is one of the world's most threatened and most diverse ecosystems, in terms of both biological and cultural diversity (~380 ethnic groups; Porro *et al.* 2008). Many of the people who reside here, poor and non-poor alike, rely directly on ecosystem services for their subsistence and livelihoods (e.g., Clement 1993; Barham *et al.* 1999; Pattanayak and Sills 2001). Alarming rates of forest loss in the basin (Malhi *et al.* 2008) cause great concern for biodiversity loss, but we still have little sense of what this loss means for society beyond species extinction and climate change (e.g., Laurance 1998; Ferraz *et al.* 2003). In this case study, we address current and future provision of non-timber forest products (NTFP) and implications for the poor.

First, we used a mapping and valuation model (see Nelson *et al.* 2009) to predict current levels of non-timber forest product harvest in the Amazon (Porro *et al.* 2008; see final model in Chapter 8; results presented here are output from an earlier version of the NTFP model). The model estimates the relative level of current NTFP harvest as a function of the association between harvested species and habitat types, current NTFP stock (assuming that more pristine forests were harvested less in the past and have a higher stock of products for harvest today), travel time from population centers (≥ 1000 people) along roads and waterways, and ease of product harvest and current legal protection from harvest (assuming enforcement) (Porro *et al.* 2008; Peralvo *et al.* 2008). Forest regions with greater stocks of an NTFP, that were easier to reach, and were currently open to all households were assumed to be harvested the most; the region with the highest likely harvest has an index score of 1.0 for that NTFP. We focused on the provision of wood, fruits and nuts, and medicinal plants sold in the market and wood, fiber, hunted bushmeat, fruits and nuts, and medicinal plants used for subsistence (Figure 16.1).

Next, we used both direct and indirect pairings of these NTFP harvest projections with poverty indicators to determine if current NTFP harvest is aligned spatially with areas of poverty. We also investigated the impact of projected road expansion and deforestation over the next 20 years on the provision of NTFPs (deforestation scenario from the Instituto de Pesquisa Ambiental da Amazonia, road development scenario from the Initiative for Integration of Regional Infrastructure in the South (Porro *et al.* 2008)). These analyses are the first step in understanding whether the poor are at a greater risk of losing their livelihood and well-being as a result of new roads and deforestation.

We represented current poverty with percentage of underweight (UW) children (Figure 16.2a; Plate 11a) and the percentage of the population with unsatisfied basic needs (UBN) (Figure 16.2c; Plate 11c). The resolution of UW children data varied by country and was generally very coarse (census blocks range in area from 13 202 km^2 to 3 778 690 km^2). Basic needs refer to any human need where lack of satisfaction is considered to be an indicator of deprivation or poor living conditions (Abaleron 1995). Basic needs, and the appropriate

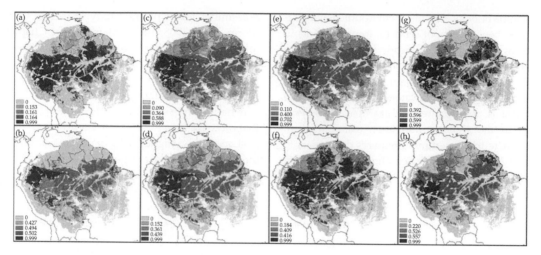

Figure 16.1 Distribution of forest product harvest index in the Amazon Basin in 2000.

The relative harvest index for bushmeat for subsistence (a), fiber for subsistence (b), fruits and nuts for subsistence (c) or market (d), medicinal plants for subsistence (e) or market (f), and wood for subsistence (g) or market (h) varied across the Basin. All units are a relative harvest index in which the parcel with the highest likely harvest received a score of 1.0.

measures used to define their satisfaction, vary dramatically across geographies. Therefore, a standardized method for measuring UBN exists and involves defining the basic needs of the population of interest, the appropriate measures for those needs, and the thresholds below which each need is considered unmet and people in such conditions can be considered poor (Abaleron 1995).

In this case, the basic needs considered were: access to housing (type of material used for house flooring, walls and roof and the number of people per room), access to sanitation (type of water supply source in the house and type and accessibility of bathrooms), access to education (presence or absence of at least one school-aged child not enrolled in school) and economic capacity (a calculation based on the age and number of household members, their highest level of education and their condition) (Schuschny and Gallopin 2004). Calculations were made using census data (at the scale of municipalities) from 1993 for Peru, Ecuador, Brazil and Colombia and from 2001 for Bolivia. The size of municipalities varies dramatically across the region and many important patterns in UBN are likely missed by the often large regions included in a single census block. Threshold levels for each need were set according to Feres and Mancero (2001).

We mapped two direct pairings of ecosystem services with a poverty indicator: the percentage of underweight (UW) children with the provision of food (fruits and nuts) for subsistence (Figure 16.2b; Plate 11b) and underweight children with the provision of bushmeat for subsistence. All other pairings were indirect, linking UBN to harvest of marketed or subsistence NTFPs. In these cases we assumed that households near areas of greater NTFP harvest would be in a better position to satisfy their basic needs by directly consuming or selling harvested NTFPs than households not located near these bases of consumption and income supplementation.

In 2000, we estimated that forest product harvest for both subsistence and market sale was relatively higher in places where people had high UBN or UW children (Figure 16.3). Superficially, this finding may suggest that forest products do not improve well-being. However, even though people in these regions are poor, forest products do make up a critical part of local incomes, supporting the health and nutritional well-being of many Amazonian forest-related households (Shanley *et al.* 2002). Deforestation leads to the loss of many of the most prominent and most profitable fruit and medicinal species that are not found in secondary forests (Shanley *et al.* 2002), thus

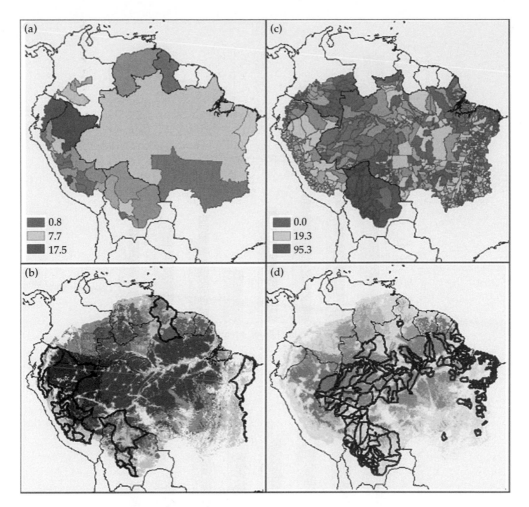

Figure 16.2   Poverty indicators and representative forest product harvest distributions in the Amazon Basin.
The incidence of underweight children is highest in northern Peru and eastern Ecuador (a) while unsatisfied basic needs are highest in Bolivia (c). High poverty
areas defined as those above the 75th percentile for underweight children (outlined in dark black) are shown in a direct pairing, overlaid with the harvest index of
fruits and nuts for subsistence (b). High poverty areas defined as those above the 75th percentile for unsatisfied basic needs are shown in an indirect pairing,
overlaid with the harvest index of wood for market sale (d). Units for underweight children are percentage of the population under the age of 5 that is
underweight. Units for unsatisfied basic needs are the percentage of the population with unsatisfied basic needs. The legends and units for (b) and (d) are the same
as in Figure 16.1 (c) and (h). (See Plate 11.)

continued forest loss is likely to contribute to declining household health and nutrition.

Finally, we found that over 20 years, the poor and
non-poor alike will lose access to nearly all products analyzed (Figure 16.3a, c). However, the greatest losses in harvest are likely to occur in areas
inhabited by the non-poor, or people with *lower*
rates of underweight children and UBN
(Figure 16.3b, d). That is to say, this preliminary

analysis suggests that proposed expansion of roads
and development in the Basin will not likely result
in greater losses to the poor through decreases in
forest products. This is probably because the region
is so large and the future time window so relatively
short that the overall change in the Basin is relatively small. Most change also happens near already
developed areas that tend to be far from indigenous
regions where rural poverty is generally higher

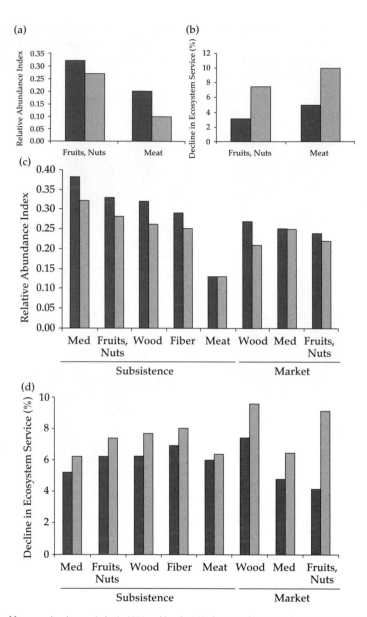

Figure 16.3 Distribution of forest product harvest index in 2000 and loss by 2020 between the poor and non-poor in the Amazon Basin.
The harvest index for fruits, nuts, and bushmeat collected for subsistence use was higher in areas with high rates of underweight children (a, black bars) than in areas with low rates (a, gray bars). The harvest index of products for subsistence or market sale were also relatively more abundant in areas with high rates of unsatisfied basic needs (c, black bars) than in areas with low rates (b, gray bars). Both the poor and the non-poor are projected to lose non-timber forest product provision in the next 20 yrs, but the non-poor will lose more, on average (b, d, gray bars) than the poor (b, d, black bars). High rates are defined as the upper 25th percentile. Meat = bushmeat, Med = medicinal plants.

(Porro *et al.* 2008). However, those currently enjoying low poverty rates are likely to experience the greatest losses and may be pushed across an important threshold as they lose access to forest products, resulting in a growing population of poor as defined by these indicators. Further, small losses to the poor may do greater harm than larger losses to the non-poor. Information on the importance of specific forest products to different groups, and an assessment of additional ecosystem services and revenue streams is needed to explore the impact predicted losses will have.

If we look not at the entire basin but at each country separately, we see a somewhat different story. In Brazil and Peru, regions with high UBN are pre-dicted to see substantially greater declines in the provision of forest products than areas that are bet-ter off (Figure 16.4). Losses in the provision of medicinal plants harvested for market sale were also higher in underprivileged areas of Colombia

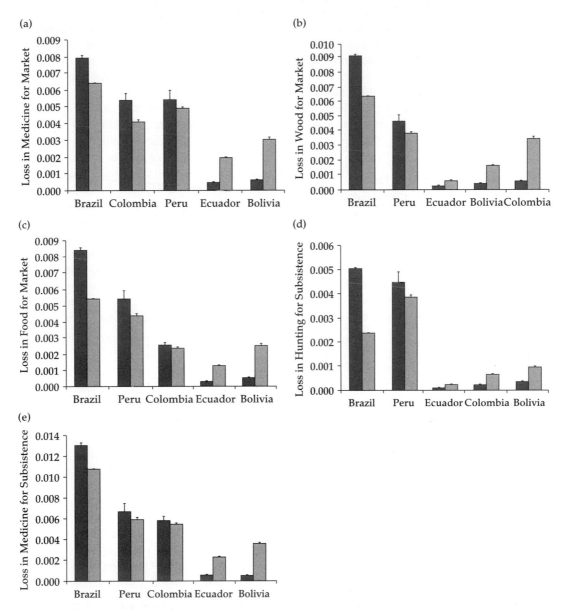

Figure 16.4   Predicted losses of forest product harvest after 20 years of deforestation and road expansion in individual countries in the Amazon Basin. Patterns were different at the country scale. Areas with high rates of unsatisfied basic needs (black bars) showed greater declines in the harvest index for medicinal plants for market sale in Brazil, Peru and Colombia (a). This same pattern of disproportional loss in areas of high unsatisfied basic needs (black bars) held for wood (b) and food for market sale (c) as well as hunted bushmeat (d) and medicinal plants for subsistence use (e) in Brail and Peru. High rates of underweight children or unsatisfied basic needs are those above the 75th percentile value for the country. Units in all graphs are the decline in relative abundance index. Error bars show standard error.

(Figure 16.4a). We need additional information about resource access, demand and changes in other ecosystem services and other factors associated with well-being to make sound conclusions about how these changes are likely to affect the poor, but these trends suggest that the rural poor in certain Amazonian regions may be most vulnerable to declines in natural capital in these countries over the next 20 years. In addition, these analyses only pertain to the poor living within the Amazon basin, and do not include poor in urban settings in each country. Our analyses highlight the utility of simple mapping in identifying areas where the coupling of institutional information with ecosystem service information should be pursued.

16.4.2   Potential for payments for environmental services in highland Guatemala

Recent years have seen considerable interest in the development of programs of Payments for Environmental Services (PES). The PES approach aims to address the classic problem of environmental externalities by establishing a mechanism through which service users can compensate land users that provide the desired service, or that adopt land uses that

are thought to improve provision (Wunder 2001; Pagiola and Platais 2007). Although the approach was conceptualized as a mechanism to improve the efficiency of natural resource management, most land users in upper watersheds are thought to be poor (Heath and Binswanger 1996; CGIAR 1997), and because most ecosystem services are thought to come from such areas (Nelson and Chomitz 2007), many have assumed that most potential PES recipients are poor. Others indicate that the linkages between potential PES recipients and poverty are more complex and show mixed results (Grieg-Gran et al. 2005; Pagiola et al. 2005; Ravnborg et al. 2007; Nelson and Chomitz 2007; Pagiola et al. 2008). Further, while PES can benefit poor landowners, other poor populations may be negatively affected by the changes in land use through higher prices or lost employment (Zilberman et al. 2008).

To examine whether PES approaches benefit the poor in practice, Pagiola et al. (2007) analyzed the spatial distribution of poverty in areas that are important to the provision of water services in highland Guatemala (Figure 16.5). With about 56% of its population under the poverty line, Guatemala has one of the highest poverty rates in Central America (World Bank 2004). Guatemalan poverty

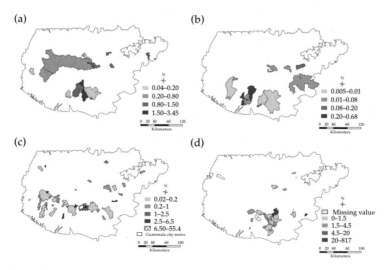

Figure 16.5   Water supply areas for principal surface water users in highland Guatemala.
Major uses highlighted are hydroelectric power generation in generating capacity per hectare of upstream area (kW ha⁻¹) (a), large-scale irrigation as irrigated area per hectare of upstream area (ha ha⁻¹) (b), domestic water supply as households served per hectare upstream area (households ha⁻¹) (c) and coffee mills as production quantity per hectare of upstream area (quintal ha⁻¹) (d).

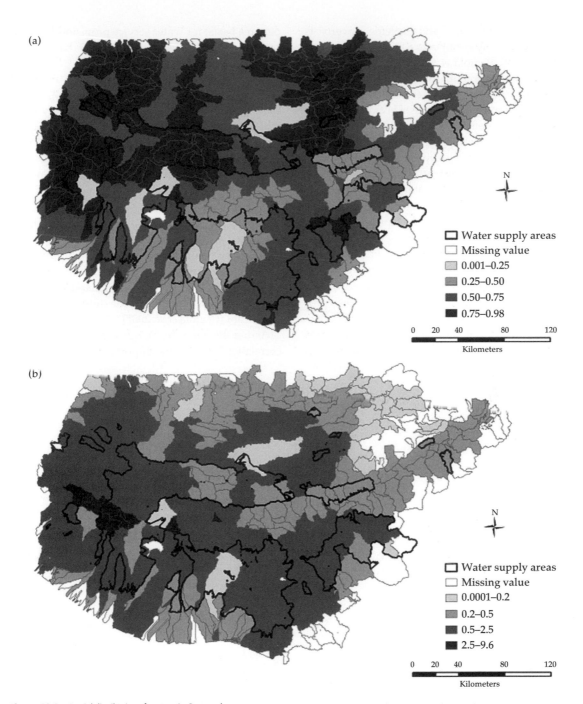

Figure 16.6 Spatial distribution of poverty in Guatemala.

Patterns are shown for both the poverty rate (number of poor) (a) and poverty density (number of poor ha^{-1}) (b). The water supply areas for major water use points are outlined in black. The poor are defined as those under Guatemala's official poverty line, which is estimated using data from the 1994 census and consumption data from a household survey 1998–9. The household-unit imputations are aggregated to small statistical areas to estimate the percentage of households living below the poverty line (Nelson and Chomitz 2007).

is predominantly rural and extreme poverty is almost exclusively rural: over 81% of the poor live in rural areas. Pagiola *et al.* (2007) asked whether these poor could potentially benefit from payments for the provision of water services by comparing water provision areas to the distribution of poverty in highland Guatemala (Nelson and Chomitz 2007) (Figure 16.6). They mapped the areas that provide water services ("water supply areas") by identifying the specific location of the intakes used by major users to obtain their water and then delineating the portions of the watershed that contribute water to those intakes. For example, Figure 16.5 shows a sample map, highlighting the water supply areas serving hydroelectric power plants. The relative importance of each water supply area was estimated by constructing an index based on the measures of the magnitude of the service they provide (in this example, installed generating capacity) and the size of the water supply area. This analysis showed that the most valuable areas (on a per area basis) for water service provision are those serving mid-size users (who often have much smaller water supply areas).

The poverty rates in the water supply areas varied substantially (Figure 16.6, Figure 16.7). While some water supply areas had high poverty rates, others had low poverty rates. The water supply areas for hydroelectric power production had relatively high poverty rates of 67% on average, but with a very high variance (Figure 16.7a). Poverty rates were lowest in the water supply areas that provide Guatemala City's domestic water supplies (Figure 16.6). The poverty density in water supply areas also varied substantially (Figure 16.6b). The average poverty density within water supply areas was 103 poor km^{-2}. This is slightly more than the average poverty density in the highland areas of the country of 83 poor km^{-2}, but the difference was not significant (Pagiola *et al.* 2007).

Across the entire highlands region, there was no correlation between the importance of a water sup-

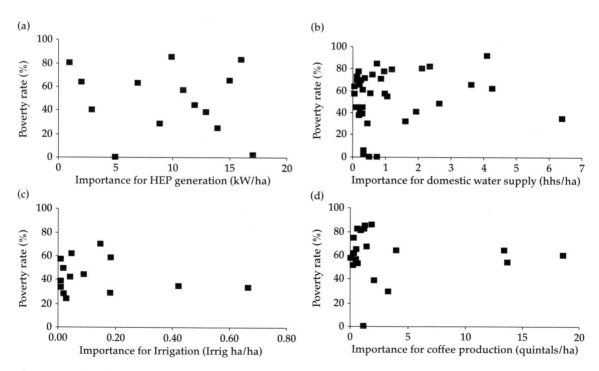

Figure 16.7 Relationship between poverty rate and importance of water supply areas.
Patterns are shown for water used to generate hydropower (a), for domestic consumption (b), for general irrigation (c), and for coffee production (d). Importance is defined by allocating the amount of supply delivered to a use point equally across the supplying watershed.

ply area and the poverty rate of people living within it (Figure 16.7). However, there were at least some areas with high water service provision and a high incidence of poverty. This suggests that payments targeted to these areas would have the potential to meet the joint goals of ecosystem service provision and poverty alleviation. However, these analyses do not include information about the nature of institutions in the region governing control of the water supply or the relationship between the poor and water-regulating land-use activities. The poor in important water supply watersheds may have no control over land-use practices that alter the provision of water-related services, and as such, payments made to the poor may not result in ecosystem service returns. If the poor control water resources and PES mechanisms were developed in all the water supply areas, 1.76 million people, or 73% of the poor in highland Guatemala could potentially be reached. This figure engenders enough promise that taking the next step to identify water resource institutions in highland Guatemala would be well worth the investment.

16.5   Including institutions: the way forward

As the case studies demonstrate, overlaying maps of poverty indicators and ecosystem services can be an informative first step in understanding the complex relationships between the poor and the environment, but it is not sufficient. Current poverty mapping seldom makes explicit connections between the poor and the resources they rely on. For example, poverty maps rarely distinguish between land owners and the landless, simply because the necessary information is not available. Land users that are renters have less ability to dictate how resources are managed. Similarly, even if land users have title to their land, their tenure may be insecure or not supported by local institutions. Again, this will inhibit their ability to change practices and alter the level of ecosystem services provided by the landscape. Information about institutions such as property rights, ownership, or management responsibility, and their stability must be incorporated into poverty mapping exercises to maximize their utility for understanding who

currently provides and benefits from ecosystem services, and who will gain or lose from future management changes.

References

Abaleron, C. A. (1995). Marginal urban space and unsatisfied basic needs: the case of San Carlos de Bariloche, Argentina. *Environment and Urbanization*, **7**, 97–116.

Adhikari, B. (2005). Poverty, property rights and collective action: understanding the distributive aspects of common property resource management. *Environment and Development Economics*, **10**, 7–31.

Albla-Betrand, J. M. (1993). *The political economy of large natural disasters*. Clarendon Press, Oxford.

Andam, K. S., Ferraro, P. J., Sims, K. R. E., *et al.* (2010). Protected areas reduced poverty in Costa Rica and Thailand. *Proceedings of the National Academy of Sciences of the USA*, **107**, 9996–10001.

Baland, J., Bardhan, P., Das, S., *et al.* (2006). *Managing the environmental consequences of growth: Forest degradation in the Indian mid-Himalayas*. National Council of Applied Economic Research, India, New Delhi.

Barham, B. L., Coomes, O. T. and Takasaki, Y. (1999). Rainforest livelihoods: income generation, household wealth and forest use. *Unasylva*, **50**, 34–42.

Bedi, T., Coudouel, A. and Simler, K. (2007). *More than a pretty picture: using poverty maps to design better policies and interventions*. World Bank, Washington, DC.

BenDor, T., Brozovic, N., and Pallathucheril, V. G. (2008). The social impacts of wetland mitigation policies in the United States. *Journal of Planning Literature*, **22**, 341–57.

Bluffstone, R. (1995). The effects of labor markets on deforestation in developing countries under open access: an example from rural Nepal. *Journal of Environmental Economics and Management*, **29**, 42–63.

Cavendish, W. (2000). Empirical regularities in the poverty-environment relationship of rural households: Evidence from Zimbabwe. *World Development*, **28**, 1979–2003.

CGIAR. (1997). *Report of the study on CBIAR research priorities for marginal lands.* Consultative Group on International Agricultural Research, Technical Advisory Committee Secretariat, Food and Agriculture Organization of the United Nations, Rome.

Chen, S., and Ravallion, M. (2008). *The developing world is poorer than we thought, but no less successful in the fight against poverty*. World Bank, Washington, DC.

Clement, C. (1993). *Native Amazonian fruits and nuts: Composition, production and potential use for sustainable development*. UNESCO, New York.

Delang, C. O. (2006). The role of wild food plants in pover4ty alleviation and biodiversity conservation in tropical countries. *Progress in Development Studies*, **6**, 275–86.

Duraiappah, A. K. (1998). Poverty environment degradation: A review and analyses of the nexus. *World Development*, **26**, 2169–79.

Elbers, C., Lanjouw, J. O., and Lanjouw, P. F. (2002). *Microlevel estimation of welfare*. World Bank, Washington, DC.

Feres, J. C. and Mancero, X. (2001). *El metodo de las necesidades basicas insatisfechas (NBI) y sus aplicaciones en America Latina*. Division de Estadistica y Proyecciones Economicas, CEPAL-ECLAC, Bogota.

Ferraro, P. J. (2002). The local costs of establishing protected areas in low-income nations: Ranomafana National Park, Madagascar. *Ecological Economics*, **43**, 261–75.

Ferraz, G., Russel, G. J., Stouffer, P. C., *et al.* (2003). Rates of species loss from Amazonian forest fragments. *Proceedings of the National Academy of Sciences of the USA*, **100**, 14069–73.

Grieg-Gran, M., Porras, I., and Wunder, S. (2005). How can market mechanisms for forest environmental services help the poor? Preliminary lessons from Latin America. *World Development*, **33**, 1511–27.

Heath, J., and Binswanger, H. (1996). Natural resource degradation effects of poverty and population growth are largely policy-induced: the case of Colombia. *Environment and Development Economics*, **1**, 65–84.

Hentschel, J., Lanjouw, J. O., Lanjouw, P. F., *et al.* (2000). Combining census and survey data to trace the spatial dimensions of poverty: A case study of Ecuador. *World Bank Economic Review*, **14**, 147–65.

Jodha, N. S. (1986). Common property resources and the rural poor in dry regions of India. *Economic and Political Weekly*, **21**, 1169-81.

Justice, C., Wilkie, D., Zhang, Q., *et al.* (2001). Central African forests, carbon and climate change. *Climate Research*, **17**, 229–46.

Laurance, W. F. (1998). A crisis in the making: responses of Amazonian forests to land use and climate change. *Trends in Ecology and Evolution*, **13**, 411–15.

Luoga, E. M., Witkowski, E. T. F., and Balkwill, K. (2000). Economics of charcoal production in miombo woodlands of eastern Tanzania: some hidden costs associated with commercialization of the resources. *Ecological Economics*, **35**, 243–57.

McSweeney, K. (2005). Natural insurance, forest access and compounded misfortune: forest resources in smallholder coping strategies before and after Hurricane Mitch, north eastern Honduras. *World Development*, **33**, 1453–71.

Mäler, K., Aniyar, S., and Jansson, A. (2008). Accounting for ecosystem services as a way to understand the requirements for sustainable development. *Proceedings of the National Academy of Sciences of the USA*, **105**, 9501–6.

Malhi, Y., Roberts, J. T., Betts, R. A., *et al.* (2008). Climate change, deforestation, and the fate of the Amazon. *Science*, **319**, 169–72.

Millennium Ecosystem Assessment. (2005). *Ecosystems and human well-being: synthesis*. Island Press, Washington, DC.

Minot, N. (2000). Generating disaggregated poverty maps: an application in Vietnam. *World Development*, **28**, 319–31.

Müller, D., Epprecht, M., and Sunderlin, W. D. (2006). *Where are the poor and where are the trees? Targeting of poverty reduction and forest conservation in Vietnam*. CIFOR, Bogor.

Narain, U., Gupta, S., and Van 't Veld, K. (2005). Poverty and the environment: exploring the relationship between household incomes, private assets, and natural assets. Poverty Reduction and Environmental Management (PREM) Working Paper 05/09. Institute for Environmental Studies, Vrije University, Amsterdam.

Ndangalasi, H. J., Bitariho, R., and Dovie, D. B. K. (2007). Harvesting of non-timber forest products and implications for conservation in two montate forests of East Africa. *Biological Conservation*, **134**, 242–50.

Nelson, A., and Chomitz, K. (2007). The forest-hydrology-poverty nexus in Central America: An heuristic analysis. *Environment, Development and Sustainability*, **9**, 369–85.

Nelson, E. N., Mendoza, G. M., Regetz, J. *et al.* (2009). Modeling multiple ecosystem services, biodiversity conservation, commodity production and tradeoffs at landscape scales. *Frontiers in Ecology and the Environment*, **7**, 4–11.

Pagiola, S., Arcenas, A. and Platais, G. (2005). Can payments for environmental services help reduce poverty? An exploration of the issues and the evidence to date from Latin America. *World Development*, **33**, 237–53.

Pagiola, S., and Platais, G. (2007). *Payments for environmental services: from theory to Practice*. World Bank, Washington, DC.

Pagiola, S., Zhang, W. and Colom, A. (2007). *Assessing the potential for payments for watershed services to reduce poverty in highland Guatemala*. World Bank, Washington, DC.

Pagiola, S., Rios, A. R., and Arcenas, A. (2008). Can the poor participate in payments for environmental services? Lessons from the Silvopastoral Project in Nicaragua. *Environment and Development Economics*, **13**, 299–325.

Pattanayak, S. K., and Sills, E. O. (2001). Do tropical forests provide natural insurance? The microeconomics of non-timber forest product collection in the Brazilian Amazon. *Land Economics*, **77**, 595–612.

Peralvo, M., Benitez, S., Nelson, E., *et al.* (2008). *Mapping spatial patterns of supply and demand of ecosystem services in the Amazon Basin*. The Nature Conservancy, Quito.

Poggi, J., Lanjouw, J. O., Hentschel, J., *et al.* (1998). *Combining census and survey data to study spatial dimensions of poverty*. World Bank, Washington, DC.

Porro, R., Borner, J., Jarvis, A., *et al.* (2008). *Challenges to managing ecosystems sustainably for poverty alleviation: securing well-being in the Andes/Amazon*. Amazon Initiative Consortium, Belem.

Ravnborg, H. M., Damsgaard, M. G., and Raben, K. (2007). *Payment for ecosystem services-issues and pro-poor opportunities for development assistance*. Danish Institute for International Studies, Copenhangen.

Roe, D., and Elliott, J. (2004). Poverty reduction and biodiversity conservation: rebuilding the bridges. *Oryx*, **38**, 137–9.

Ruhl, J. B., and Salzman, J. (2006). The effects of wetland mitigation banking on people. *National Wetlands Newsletter*, **28**, 9–14.

Sanderson, S. E., and Redford, K. H. (2003). Contested relationships between biodiversity conservation and poverty alleviation. *Oryx*, **37**, 389–90.

Scherr, S. J., White, A., and Kaimowitz, D. (2003). *A new agenda for forest conservation and poverty reduction*. Forest Trends, Washington, DC.

Schuschny, A. R., and Gallopin, G. C. (2004). *La distribucion espacial de la pobreza en relacion a los sistemas ambientales en America Latina*. Division de Desarrollo Sostenible y Asentamientos Humanos, United Nations, Santiago.

Shanley, P., Luz, L., and Swingland, I. R. (2002). The faint promise of a distant market: a survey of Belem's trade in non-timber forest products. *Biodiversity and Conservation*, **11**, 615–36.

Shyamsundar, P. (2001). *Poverty-environment indicators*. World Bank, Washington, DC.

Takasaki, Y., Barham, B. L., and Coomes, O. T. (2004). Risk coping strategies in tropical forests: Flood, illnesses and resource extraction. *Environment and Development Economics*, **9**, 203–24.

UNEP. (2004). *Human well-being, poverty and ecosystem services: exploring the links*. Premier Printing, Winnipeg.

Vedeld, P., Angelsen, A., Bojö, J., *et al.* (2004). Forest environmental incomes and the rural poor. *Forest Policy and Economics*, **9**, 869–79.

Wodon, Q., and Gacitúa-Marió, E. (2001). *Measurement and meaning: combiing quantative and qualitative methods for the analysis of poverty and social exclusion in Latin America*. World Bank, Washington, DC.

World Bank. (2001). *World Development Report 2000/2001: attacking poverty*. Oxford University Press, Oxford.

World Bank. (2004). *Poverty in Guatemala*. World Bank, Washington, DC.

World Bank. (2008). *Poverty and the environment: understanding linkages at the household level*. World Bank, Washington, DC.

World Resources Institute. (2007). *Nature's benefits in Kenya: an atlas of ecosystems and human well-being*. World Resources Institute, Nairobi.

Wunder, S. (2001). Poverty alleviation and tropical forests: What scope for synergies? *World Development*, **29**, 1817–33.

Zilberman, D., Lipper, L., and McCarthy, N. (2008). When could payments for environmental services benefit the poor? *Environment and Development Economics*, **13**, 255–78.

CHAPTER 17

Ecosystem service assessments for marine conservation

Anne D. Guerry, Mark L. Plummer, Mary H. Ruckelshaus, and Chris J. Harvey

17.1 Introduction

Humans always have benefited from marine ecosystems—either obviously in the form of food resources, or more subtly in the form of cultural and recreational opportunities. For example, 80–85 million tons of fish were landed in marine capture fisheries worldwide in 2006, and fish account for approximately 15% of the annual animal protein consumption by humans (FAO Fisheries Department 2009). A growing recognition of the degradation of global marine ecosystems has led to numerous calls for a shift toward more holistic, ecosystem-based management of marine systems (Pew Oceans Commission 2003; US Commission on Ocean Policy 2004; Council on Environmental Quality 2009). Ecosystem-based management is a coordinated effort to manage the diverse human impacts that affect an ecosystem to ensure the sustainability of the ecosystem services it provides (Rosenberg and Mcleod 2005). Two key aspects of ecosystem-based management are relevant here. First, ecosystem-based management fundamentally recognizes the inseparability of human and ecological systems. Human well-being is derived from ecosystems through ecosystem services and, in turn, human behavior affects natural systems. Second, ecosystem-based management is inherently multifaceted, encompassing suites of services, rather than the traditional approach of sector-by-sector management. Importantly, the framework of ecosystem services can provide performance metrics for different management strategies that attempt to balance multiple objectives by allowing for the explicit examination of trade-offs in ecosystem services provided under alternative management scenarios (National Research Council 2004).

Using an ecosystem services framework also has the potential to draw a larger and more diverse population of people to marine and other conservation efforts, beyond those who value the environment purely for its direct uses. For example, many residents are drawn to the Puget Sound region, USA because of the sound's physical beauty and concomitant aesthetic benefits to their well-being. Indeed, existence values have been found to be among the "most important" benefits provided by the Puget Sound system (Iceland *et al.* 2008). If such cultural values can be included in tallies of the consequences of ecosystem protection, conservation efforts are likely to engage a greater fraction of the population. Helping people to see the many ways their well-being is affected by marine and coastal environments is key to the success of conservation.

In principle, marine ecosystem services are not fundamentally different from their terrestrial counterparts. In practice, however, the valuation and mapping of ecosystem services in marine environments is not as well developed as it is for terrestrial ecosystems. As described in Chapters 4–13, there have been some early successes applying ecosystem service mapping and modeling tools in diverse terrestrial and freshwater settings. These approaches and models all start with basic land cover and land-use data layers. The same approach can work for marine environments—marine systems have patchy habitats that provide flows of ecosystem services, and management actions can alter those habitats and flows. Several challenges must be addressed, however. Maps of habitat type and habitat use are much harder to come by in marine systems than they are on land. We cannot readily "see" many parts of the marine ecosystem and its habitat types using satellite imagery or other remote sensing technology. Moreover, marine habitats and the processes that maintain them are more transient and three-dimensional than their terrestrial counterparts, and associations between particular species and habitats are harder to document.

Another challenge stems from the ways in which humans interact with marine environments. While fishery harvest, one of the most important marine ecosystem services, is straightforward to measure, its ecological effects and potential impacts on other ecosystem services are harder to discern. In addition, many of our actions that affect the marine environment take place on land. For example, coastal development; land-use practices that produce nutrient, sediment, and pathogen inputs to freshwater; and increases in impervious surfaces can severely degrade nearshore marine systems (Carpenter *et al.* 1998; Mallin *et al.* 2000; Diaz and Rosenberg 2008). Incorporating an ecosystem service perspective into marine management, then, facilitates integration of ter-restrial and marine policies, which have been historically disconnected.

Fortunately, there are advanced aspects of marine science that will provide a good foundation for ecosystem service analyses. In particular, although basic mapping data are less refined in marine environments, marine science has a rich ecosystem-based modeling tradition to draw on for quantifying ecosystem services. For example, modeling for fisheries management (e.g., Christensen and Walters 2004; Pauly *et al.* 2000; Fulton *et al.* 2004a, b) and water use impacts in the Everglades and Florida Bay (e.g., US Geological Survey 1997) provide sophisticated system and food web models that can be extended to evaluate a more comprehensive suite of human activities and ecosystem services.

17.2   Ecosystem services provided by marine environments

Global oceans provide a wealth of ecosystem benefits that span all four major categories of services identified by the Millennium Ecosystem Assessment (Millennium Ecosystem Assessment 2005): provisioning, regulating, cultural, and supporting services (Table 17.1). Marine ecosystems provide goods and

Table 17.1   Ecosystem services provided by oceans and coasts[††]

	Subcategory	Examples
Provisioning services		
Food	Capture Fisheries	Tuna, mahi-mahi, crab, scallops
	Aquaculture	Salmon, oysters, shrimp, seaweed
	Wild foods	Mussels, clams, seaweed
Fiber		Mangrove wood (construction, boat-building), seagrass fiber
Biomass fuel		Mangrove wood (charcoal), biofuel from algae
Water		Shipping, tidal turbines
Genetic resources		Individual salmon stocks, marine diversity for bioprospecting
Biochemicals, natural medicines, and pharmaceuticals	Medicines	Anti-viral and anti-cancer drugs from sponges
	Food additives	Seaweed harvest for carrageenans
Regulating services		
Air quality regulation		Sea salt and spray help cleanse the atmosphere of air pollution*
Climate regulation		Major role in global CO_2 cycle
Water regulation		Natural stormwater management by coastal wetlands and floodplains
Erosion regulation		Nearshore vegetation stabilizes shorelines

(continues)

Table 17.1 continued

	Subcategory	Examples
Water purification and waste treatment		Uptake of nutrients from sewage wastewater, detoxification of PAH's by marine microbes, sequestration of heavy metals
Disease regulation		Natural processes may keep harmful algal blooms and waterborne pathogens in check
Pest regulation		Grazing fish help keep algae from overgrowing coral reefs
Pollination/assistance of external fertilization		Innumerable marine species require seawater to deliver sperm to egg
Natural hazard regulation		Coastal and estuarine wetlands and coral reefs protect coastlines from storms
Cultural services		
Ethical values	Novn-use	Spiritual fulfillment derived from estuaries, coastlines, and marine waters
Existence values	Non-use	Belief that all species are worth protecting, no matter their direct value to humans
Recreation and ecotourism	Non-consumptive use	SCUBA diving, beachcombing, whale watching, boating, snorkeling
	Consumptive use	Fishing, clamming
Supporting services		
Nutrient cycling		Major role in carbon, nitrogen, oxygen, phosphorus, and sulfur cycles
Soil formation		Many salt marsh surfaces vertically accrete; eelgrass slows water and traps sediment
Primary production		~40% global NPP**
Water cycling		96.5% of earth's water is in oceans***

* Rosenfeld *et al*. (2002).
** Schlesinger (1991).
** Melillo *et al*. (1993).
*** Gleick (1996).
††The taxonomy of services is adapted from the Millennium Ecosystem Assessment (2005).

services from both biotic (e.g., depend on food webs) and abiotic (e.g., depend on the presence of seawater) aspects of natural capital. Assessment reports within (Agardy *et al.* 2005) and based on the Millennium Ecosystem Assessment (UNEP 2006) and other synthesis documents (Peterson and Lubchenco 1997; Costanza 2000; Patterson and Glavovic 2008; Wilson and Liu 2008) provide useful overviews of these services, as do descriptions of the particular services provided by fish populations (Holmlund and Hammer 1999), coral reef ecosystems (Moberg and Folke 1999), and mangroves (Ronnback 1999).

Provisioning services include the most high-profile marine ecosystem services such as food from capture fisheries, aquaculture, and wild foods. On average, each person alive in 2006 ate 16.7 kg of fish that year (18% of that total came from marine aquaculture; the proportion from capture fisheries is difficult to calculate given non-food uses of wild fish) (FAO Fisheries Department 2009). Some of the less

obvious provisioning services include timber and fiber from mangroves and seagrass beds, and biochemicals for cosmetics and food additives. The potential also exists for developing novel natural products from marine species with medical applications (Carté 1996). In addition, the ocean may become an important energy source: biofuels from algae and power generation from wave and tidal energy have potential for more widespread use. And finally, the world's oceans provide the highways for the global shipping trade.

Marine systems also are responsible for a wide range of regulating services. Most prominent of these is natural hazard regulation. As was vividly highlighted by the human losses wrought by the 2004 Asian tsunami and 2005 hurricanes on the Gulf Coast of the USA, coastal and estuarine wetlands have value for their ability to reduce storm surge elevations and wave heights (Danielsen *et al.* 2005; Travis 2005; Box 17.1). Other regulating services

Box 17.1 Nonlinear wave attenuation and the economic value of mangrove land-use choices

Edward B. Barbier

Although most ecologists have concluded that ecosystem size and functional relationships are non-linear, the lack of data or mapping of these relationships has often precluded estimating how the value of an ecosystem service varies across an ecological landscape. However, recent collaborations between ecologists, hydrologists and economists have demonstrated this effect for the wave attenuation function of mangroves, which in turn impacts on the land-use choices for conserving or developing mangrove forests.

Barbier (2007) conducted a comparison of land-use values between various mangrove ecosystem benefits and conversion of the mangrove to shrimp ponds in Thailand. He found that all three ecosystem services - coastal protection, wood product collection and habitat support for off-shore fisheries—have a combined value ranging from $10158 to $12392 ha^{-1} in net present value terms over the 1996 to 2004 period of analysis, and that the highest value of the mangrove by far is its storm protection service, which yields an annual benefit of $1879 ha^{-1} annually, or a net present value of $8966 to $10821.

But what if these per hectare values for mangroves were used to inform a land-use decision weighing conversion of

an entire mangrove ecosystem to shrimp aquaculture? For example, deciding how much of a mangrove forest extending 1000 m seaward along a 10-km coastline to convert to shrimp aquaculture may depend critically on whether all the mangroves in the 10 km^2 ecosystem are equally beneficial in terms of coastal storm protection (Barbier *et al.* 2008).

Suppose that it is assumed initially that the annual per ha values for the various ecosystem benefits are "uniform," and thus vary linearly, across the entire 10 km^2 mangrove landscape. Following this assumption, a mangrove area of 10 km^2 would have an annual storm protection value of 1000 times the $1879 ha^{-1} "point estimate," which yields an annual total benefit estimate of nearly $1.9 million. Barbier *et al.* (2008) show how this assumption translates into a comparison of the net present value (10% discount rate and 20-year horizon) of shrimp farming to the three mangrove services - coastal protection, wood product collection and habitat support for off-shore fisheries—as a function of mangrove area (km^2) for the example of a 10 km^2 coastal landscape. Figure 17.A.1 shows the comparison of benefits.

The figure also aggregates all four values to test whether an "integrated" land-use option involving some conversion and some preservation yields the highest total value. When all values are linear, as shown in the figure, the outcome is

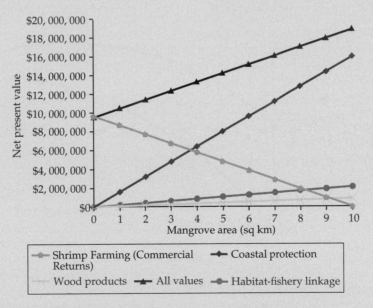

Figure 17.A.1 Linear ecosystem service returns from mangroves.

continues

Box 17.1 *continued*

a typical "all or none" scenario; either the aggregate values will favor complete conversion or they will favor preserving the entire habitat. Because the ecosystem service values are large and increase linearly with mangrove area the preservation option is preferred. The aggregate value of the mangrove system is at its highest ($18.98 million) when it is completely preserved, and any conversion to shrimp farming would lead to less aggregate value compared to full preservation, thus any land-use strategy that considers all the values of the ecosystem would favor mangrove preservation and no shrimp farm conversion

However, not all mangroves along a coastline are equally effective in storm protection. It follows that the storm protection value is unlikely to be uniform across all mangroves. The reason is that the storm protection "service" provided by mangroves depends on their critical ecological function in terms of "attenuation" of storm waves. That is, the ecological damages arising from tropical storms come mostly from the large wave surges associated with these storms. Ecological and hydrological field studies suggest that mangroves are unlikely to stop storm waves that are greater than 6 m (Forbes and Broadhead 2007; Wolanski 2007; Alongi 2008; Cochard *et al.* 2008). On the other hand, where mangroves are effective as "natural barriers," against storms that generate waves less than 6 m in height, the wave height of a storm decreases quadratically for each 100 m that a mangrove forest extends out to sea (Mazda *et al.*

1997; Barbier *et al.* 2008). In other words, wave attenuation is greatest for the first 100 m of mangroves but declines as more mangroves are added to the seaward edge.

Barbier et al (2008) employ the non-linear wave attenuation function for mangroves based on the field study by Mazda *et al.* (1997) to revise the estimate of storm protection service value for the Thailand case study. The result is depicted in Figure 17.A.2.

The storm protection service of mangroves still dominates all values, but small losses in mangroves will not cause the economic benefits of storm buffering by mangroves to fall precipitously. The consequence is that the aggregate value across all uses of the mangroves, shrimp farming and ecosystem values, is at its highest ($17.5 million) when up to 2 km^2 of mangroves are allowed to be converted to shrimp aquaculture and the remainder of the ecosystem is preserved.

Taking into account the "nonlinear" relationship between an ecological function and the value of the ecosystem service it provides can therefore have a significant impact on a land-use decision at the landscape scale. Other ecosystem services, including those for mangroves, are likely to have similar effects. For example, a study of the nursery habitat function of mangroves in the Gulf of California, Mexico reveals that the function's influence on the productivity of off-shore fisheries does not scale-up in direct proportion to the area of the mangrove forests in the nearby lagoons (Aburto-Oropeza *et al.* 2008).

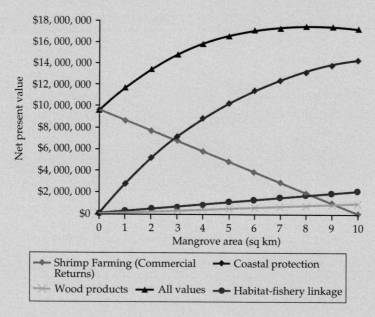

Figure 17.A.2 Nonlinear ecosystem service returns from mangroves.

provided by marine systems include the transformation, detoxification, and sequestration of wastes (Peterson and Lubchenco 1997).

Rich cultural services are provided by marine systems. Human coastal communities—both native and non-native—often define their identities in relation to the sea. In the U.S., people love to live near the ocean; one study predicts average increases of 3,600 people a day moving to coastal counties through 2015 (Culliton 1998). Globally, coastal tourism is a key component of many economies (Box 17.2). It is one of the fastest growing sectors

of tourism, and is one of the world's most profitable industries (United Nations Environment Programme 2006).

Finally, the oceans provide essential supporting services that underpin many of the world's ecological functions. The oceans are the center of the global water cycle; they hold 96.5% of the earth's water (Gleick 1996) and are a primary driver of the atmosphere's temperature, moisture content, and stability (Colling 2001). Oceans are also key players in the global cycles of carbon, nitrogen, oxygen, phosphorus, sulfur, and other key elements (Peterson

---

Box 17.2 Valuation of coral reefs in the Caribbean

Emily Cooper and Lauretta Burke

In the Caribbean, nearly 70% of coral reefs are threatened by human activities—including over-fishing, dredging, sewage discharge, increased runoff from agricultural activities, and coastal development (Burke and Maidens 2004). Degradation of reefs not only results in a tremendous loss of biodiversity but also leads to a decline in the services they provide to coastal communities, resulting in lost revenue from declining tourism and fishing, increased poverty and malnutrition, and increased coastal erosion.

Many of these damaging activities occur because an individual or group seizes an immediate benefit, without knowing or caring about the long-term consequences. Quantifying the value of ecosystem services provided by reefs can help to facilitate more sensible, far-sighted decision-making by drawing attention to the economic benefits associated with reefs, and by demonstrating the true costs of poor coastal management. In 2005 the World Resources Institute (WRI) launched a project to assess the economic contribution of three reef-related ecosystem services to countries in the Caribbean: reef-related fisheries, tourism, and shoreline protection. These three services were chosen because they are (a) relatively easy to measure using published information, (b) easily understood by politicians and decision-makers, and (c) especially important to local economies. National-level studies have been completed for St. Lucia, Tobago, and Belize. In Tobago, one of two pilot sites, the project estimated the value of these three services at US$62–78 million per year (Burke *et al.* 2008).

Reef-related tourism and fisheries

Tourism is Tobago's largest economic sector, contributing 46% of GDP and employing 60% of the workforce (WTTC 2005). WRI conducted a financial analysis of reef-related tourism, including net revenues from all reef-related activities, accommodation, and other spending on reef-related days. In addition, the study drew on a local-use survey to estimate recreational use of the reefs and coralline beaches by local residents each year. In total, coral reef-associated tourism and recreation contributes an estimated US$43.5 million to the national economy per year.

Revenues from reef-associated fisheries tend to be dwarfed by tourism in the Caribbean, but fishing is an important cultural tradition, safety net, and livelihood for many people. Coastal fishing communities are often among the most vulnerable groups to degradation of the ecosystem, as they may have fewer income alternatives. A financial analysis of reef-related fisheries in Tobago found that annual economic benefits are between US$0.8–1.3 million (0.7–1.1 million).

The role of coral reefs in protecting the shoreline

As part of this valuation effort, WRI developed an innovative method for evaluating the shoreline protection services provided by coral reefs. By integrating data on coastal characteristics, storm events, and coral reef location and type into a Geographic Information System, we are able to evaluate the role of coral reefs in maintaining the

continues

Box 17.2 *continued*

stability of a country's shoreline. In Tobago (Fig. 17.B.1), the relative reef contribution is zero in areas not protected by a coral reef, and ranges from 27% where the shoreline has relatively good protection due to other factors, to 42% where the shoreline would be most vulnerable without the reef. The relative share of protection provided by coral reefs is particularly high behind the Buccoo Reef in southwest Tobago, as well as along several portions of the windward coast. After assessing the relative protection provided by coral reefs, we integrate property values for vulnerable areas to arrive at an estimate of US$18–33 million in "potentially avoided damages" per year.

Policy relevance

This type of valuation produces a picture of the current estimated value of these three services. The method has the advantage of being simple, replicable, and transparent, and it is a useful exercise for drawing attention to reef-related

benefits that are often undervalued or unnoticed. Even ballpark values help to support an economic case for including these types of ecosystem services in decision-making processes. Going forward, policy-makers in many Caribbean countries may find it worthwhile to invest in economic valuation to support decision-making, including conducting cost-benefit analyses and assessing the effects of coral reef degradation on the value of these services over time.

Working with local partners, WRI has tied the economic findings to some clear opportunities for improved coastal management in Tobago. For instance, the Buccoo Reef Marine Park (BRMP) in the southwest of the country is a cornerstone of the tourism industry—60% of international visitors take trips into the park—and provides significant coastal protection to a heavily developed and low-lying section of the island. Applying the same methods as at the national level but looking over a 25-year period, we estimate that tourism associated with BRMP contributes

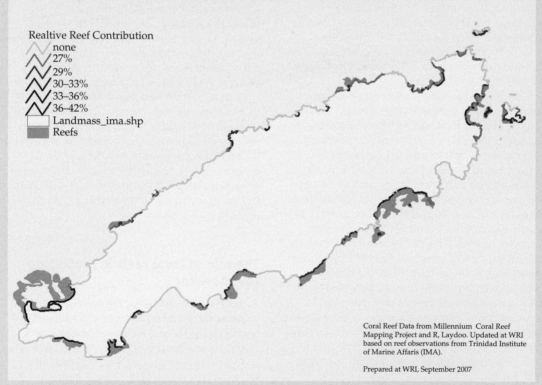

Shoreline Protection by Coral Reefs—Relative Reef Contribution

Realtive Reef Contribution
- none
- 27%
- 29%
- 30–33%
- 33–36%
- 36–42%
- Landmass_ima.shp
- Reefs

Coral Reef Data from Millennium Coral Reef Mapping Project and R, Laydoo. Updated at WRI based on reef observations from Trinidad Institute of Marine Affaris (IMA).

Prepared at WRI, September 2007

Figure 17.B.1 Shoreline protection by coral reefs.

between US$128 and 156 million in net present value over 25 years. The reef offers protection valued at between US$140 and 250 million over the same period (using a 3% discount rate). The park is meant to be a no-take zone, so fishing is not assessed.

There is little enforcement of park regulations, and the reefs suffer from over-fishing as well as sediment and nutrient runoff. Steps to preserve the reefs could include (from lowest cost to highest): increasing

enforcement, re-routing a sewer drain out of the lagoon, improving watershed management and installing sediment traps, or building a sewage treatment plant for the area. A cost-effectiveness assessment of these interventions was outside the scope of this study, but would be fairly straightforward. Local conservation groups hope the valuation findings will draw attention to a threatened and valuable resource, and point to the need for change.

and Lubchenco 1997) and are responsible for approximately 40% of global net primary productivity (Schlesinger 1991; Melillo *et al.* 1993). The oceans are home to vast reservoirs of genetic and ecological diversity, arguably the most fundamental of supporting services as it is directly linked to the rate of evolution and therefore the ability to adapt to a changing climate (Pergams and Kareiva, in press).

The valuation of marine ecosystem services lags behind efforts aimed at terrestrial systems, although coastal wetlands (Batie and Wilson 1978; Lynne *et al.* 1981; Farber 1988; Bell 1989), coral reefs (Spurgeon 1992; Moberg and Folke 1999; Cesar 2000; Brander *et al.* 2007), and mangroves (Bennett and Reynolds 1993; Gilbert and Janssen 1998; Ronnback 1999; Ruitenbeek 1994; Barbier 2000; Sathirathai and Barbier 2001; Barbier 2003, 2007; Barbier *et al.* 2008) are notable exceptions. Marine and coastal ecosystem services have been included in a few comprehensive valuation exercises. Costanza *et al.* (1997) used a (mostly) benefits-transfer approach, applying estimates of ecosystem service values for specific terrestrial and marine habitats to extrapolate the global value of ecosystem services. Without careful matching of sites to ensure that the benefits can and should be transferred, however, this approach can be misleading (Plummer 2009). One of the most interesting discussions of ecosystem service valuation in the marine environment entails four case studies that demonstrate how valuing a suite of ecosystem services has the potential to inform decision-making in the Swedish coastal zone (Soderqvist *et al.* 2005). One of these case studies explores the costs (increased water treatment and reduced fertilizer

use) and benefits (recreational and other cultural benefits) of improved water quality in the Stockholm Archipelago. Recent examinations of shoreline stabilization and trade-offs with aquaculture are illustrative of a general growing interest in services from coastal environments (Box 17.1, Barbier *et al.* 2008).

17.3 Mapping and modeling the flow of marine ecosystem services: a case study of Puget Sound

Ecosystem services are a useful currency for cost-benefit analyses or assessments of the trade-offs among alternative strategies for achieving multiple objectives in marine systems. This is especially true when those ecosystem objectives explicitly include human well-being in addition to traditional conservation goals, which is the situation in Washington's Puget Sound region. In this section, we present a small step forward in applying the concept of quantifying dynamic flows of ecosystem services to the management of Puget Sound. The Puget Sound ecosystem in Washington State is home to 3.8 million people encompassed in a 42 000-km^2 basin, including temperate-latitude lands and rivers from the crests of the Cascade and Olympic mountains through a deep, fjord-type estuary to the Pacific Ocean. The region's marine environment produces basic provisioning services such as commercial and tribal subsistence fisheries for salmon (*Oncorhynchus* spp.) and other species, as well as clam, oyster, crab, and other shellfish harvests. It provides regulating services as global as the carbon cycle, and as local as waste treatment through the breakdown of PAHs (polycyclic aromatic hydrocarbons) and PCBs (polychlorinated biphenyls) by eelgrass (Huesemann *et al.* 2009). It offers numerous cultural services through bird and whale watching, recreational fishing, water recreation, educational opportunities, and simply the human value placed on the existence of the region's biodiversity. Puget Sound also provides a rich cultural heritage for native Indian tribes. And underlying all of these are basic supporting services such as primary production and the provision of habitat for the Pacific Northwest icons salmon and orcas (*Orcinus orca*).

Using Puget Sound as a case study is motivated by the region's move toward an ecosystem-level management approach. In 2007, the Washington State Legislature mandated formation of a new State agency guided by a public-private council— the Puget Sound Partnership (Partnership)— whose charge is to recover the ecosystem by 2020. The Partnership's governance structure and mandate for ecosystem recovery presents an opportunity to apply principles from ecological theory and the science of ecosystem services to help prioritize management actions for Puget Sound. The Partnership recognizes that ecosystem recovery will require changes in the way local, state, Federal and tribal governments act and—just as importantly—changes in choices human residents make about how they commute to work, where they buy their food, homes, and so forth (Puget Sound Partnership 2006). To meet these challenges, the Partnership has adopted a systemwide approach to restoring the ecosystem, and they have explicitly defined their multiple objectives in terms of what ecosystem services people in the region care the most about (Puget Sound Partnership 2008).

Identifying these public values is an essential step toward making an ecosystem services framework of practical use. In Puget Sound, a diverse group of stakeholders including those from fisheries and aquaculture, tourism, ports and shipping, cities, counties, tribal governments, environmental interests, agriculture, forestry, homebuilding, and business sectors were interviewed to identify those services they believe to be "most important." The interviewers first educated the participants about the concept of ecosystem services and offered them

a list of 24 services translated from the Millennium Ecosystem Assessment into locally relevant terminology. Across 12 different sectors, there was broad agreement that water quantity and water regulation, recreation and ecotourism, and ethical and existence values were of the utmost importance; capture fisheries, aquaculture, water purification and waste treatment also ranked highly (Iceland *et al.* 2008). Trade-offs are likely to occur among services even in this short list of valued ecosystem benefits. Representing outcomes of management choices in terms of multiple benefits, in currencies related to human well being, has promise for engaging a broader spectrum of the public in charting a path forward.

An early focus of the Partnership's effort is nearshore habitats. This builds on the work of a number of previous planning efforts in Puget Sound, which identified specific actions aimed at either protecting existing nearshore habitats or restoring degraded areas to provide improved function for species, habitat maintenance, or human access (Shared Strategy 2007; Puget Sound Nearshore Ecosystem Restoration Project 2008; Alliance for Puget Sound Shorelines 2008). These nearshore recovery schemes in Puget Sound have broadly similar objectives in their common desire to increase the amount of functioning nearshore habitat.

In the remainder of this chapter, we focus on the suite of ecosystem services that nearshore habitats produce and support in Puget Sound, and how those services could change in response to a set of possible management actions. To illustrate this approach, we quantify the outcomes of nearshore protection or restoration for three distinct kinds of services that flow from an important foundation species—eelgrass (*Zostera marina*): (1) carbon sequestration for climate regulation (a regulating service with global reach), (2) marine commercial harvest (a provisioning service), and (3) non-consumptive values (recreation and existence values) associated with species that belong to the Puget Sound food web. We examine how changes in ecosystem services are created by changes in eelgrass itself (carbon sequestration) and how changes in eelgrass produce changes in services provided through higher levels of the food web (harvest, recreation, and existence values).

17.3.1   Eelgrass

Eelgrass (*Zostera marina*) is a widely distributed, clonal seagrass that forms large, often monospecific stands in shallow temperate estuaries worldwide. Much of the vegetative biomass of eelgrass is below the surface of sediments and the above-ground biomass tends to be highly seasonal. Ecosystem services attributed to seagrass beds include the sequestration of carbon, the provision of habitat for fish and invertebrates, and the control of erosion through sediment stabilization (Williams and Heck 2001). Threats to eelgrass in Puget Sound are similar to those facing this habitat type elsewhere, including mechanical damage (such as through dredging and anchoring), eutrophication, some aquaculture practices, siltation, coastal construction, invasions by non-native species, alterations to coastal food webs, and climate change (Williams and Heck 2001; Duarte 2002; Bando 2006). Eelgrass beds currently occur along approximately 37% of the coast of Puget Sound, where they provide habitat for mobile organisms such as crabs and small fishes and feeding habitat for larger consumers such as seabirds, salmon, and marine mammals (National Marine Fisheries Service 2007). They also provide spawning and rearing habitat for Pacific herring (*Clupea pallasi*), a key species in the regional food web (Penttila 2007; National Marine Fisheries Service 2007).

To evaluate policy scenarios for the nearshore, we ask how management actions are likely to affect eelgrass, and then how multiple ecosystem services provided by eelgrass are likely to change as a result. We estimate ecosystem services using simple approaches for the sake of illustration, and do not include ecological nuances such as spatial or temporal variation in their production and delivery. Also, it is important to note that since we are examining changes in flows of services that are likely to result from changes in a foundation species, we focus on bottom-up effects; future scenario work will include the examination of top-down effects such as changes in harvest of key fish species.

We know from previous work that the total area of eelgrass in Puget Sound has declined from historical levels (Thom and Hallum 1990). Although the reasons for this decline are not well understood (Thom and Albright 1990; Thom and Hallum 1990), we assume for the sake of illustration that the Partnership can identify and implement policies capable of protecting eelgrass and halting this decline. In addition, we examine other policies aimed at restoring eelgrass in areas where it used to occur. To assess the potential for restoration, we built a spatially explicit habitat suitability model for eelgrass in Puget Sound, with the aim of identifying locations in which eelgrass has the potential to grow but where its current status is unknown (Figure 17.1; Plate 13, Davies *et al.*, in preparation). Our model suggests that an additional 36 877 ha of Puget Sound's benthic area has the potential to be occupied by eelgrass. If even half of this were to be occupied in the future due to restoration projects, it would represent nearly a doubling of the current area of eelgrass and would yield area similar to that estimated to be present historically (41 239 ha compared to 47 328 ha summarized from Thom and Hallum 1990).

17.3.2   How do changes in eelgrass habitats affect carbon storage and sequestration?

Estimates of carbon storage and sequestration in marine systems are rare. For example, a marine analog does not exist that is similar to the look-up tables for carbon storage and sequestration values for various land use/land cover categories available in terrestrial systems (Intergovernmental Panel on Climate Change 2006). Our approach to estimating the ecosystem service value of eelgrass carbon sequestration is similar to that of Chapter 7 (this book) and is summarized in Table 17.2a. We first estimate the amount of carbon stored in eelgrass biomass and soils. We then estimate the rate of carbon sequestration for eelgrass in Puget Sound, noting that carbon also flows through eelgrass to be consumed by herbivores, decomposed within the system, and exported from the system. These two estimates yield ecosystem service values for the scenarios examined. In the end, our approach illustrates how changes in eelgrass habitats result in changes in the amounts and values of carbon storage and sequestration.

17.3.2.1   *Carbon storage*
Our estimates of carbon storage in Puget Sound eelgrass beds and sediments range from 1–6.3 TgC (Table 17.2a). The amount of C in the sediment pool

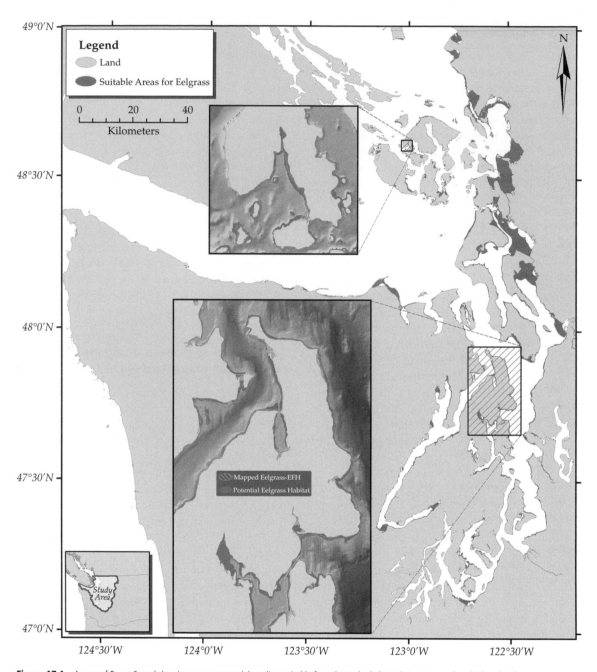

Figure 17.1 A map of Puget Sound showing areas our model predicts suitable for eelgrass beds (green). Inset maps show higher detail; orange represents currently mapped eelgrass from the NOAA Essential Fish Habitat data (TerraLogic GIS Inc. 2004). (See Plate 13.)

Table 17.2a Summary of methods used for estimating how changes in ecosystem structure and function give rise to changes in services provided for carbon storage and sequestration, commercial fisheries, and food web support

Ecosystem service	Estimation approach	Parameter values	Key assumptions
Carbon storage by eelgrass for climate regulation	$C = (A)*(C_B + C_S)$ A is area (ha) of eelgrass C_B is carbon stored in biomass per ha C_S is soil organic carbon content per ha	$A = 21\ 140$ ha[1] $C_B = 0.03–0.4$ MgC/ha[2] $C_S = 50–300$ MgC/ha[3]	40% of eelgrass biomass is C; C in soil is attributed to eelgrass stabilizing and trapping sediments and preventing decomposition.
Carbon sequestration by eelgrass for climate regulation	$\Delta C = NPP*S*A$ NPP is net primary productivity of eelgrass S is proportion of NPP stored in sediments A is area (ha) of eelgrass	$NPP = 300–600$ gC/m²/yr[4] $S = 5–15\%$[5] $A = 21\ 140$ ha[1]	Carbon exported to other systems (including the deep sea) has the potential to be sequestered, but because its ultimate fate is unknown it is not considered further here.
Commercial fisheries for food: current	Observed landed biomass by species for commercial fisheries in different marine sub-basins	Pounds per species/year landed in Puget Sound[6]	
Food web support mediated through eelgrass-herring interaction	EwE food web model[7] Mediation function defines changes in herring egg vulnerability due to changes in eelgrass biomass	Sigmoid mediation function[7]	Eelgrass affects herring biomass through egg survival; food web responses to changes in eelgrass are mediated through herring
Wildlife viewing and existence mediated through eelgrass-herring interaction	EwE food web model[7] Mediation function defines changes in herring egg vulnerability due to changes in eelgrass biomass EwE food web model	Sigmoid mediation function[7]	Eelgrass affects herring biomass through egg survival; food web responses to changes in eelgrass are mediated through herring

1. Gaeckle et al. (2007).

2. In one subtidal Puget Sound meadow Nelson and Waaland (1997) estimated annual above- and below-ground eelgrass biomass to average 256.3 gdw/m² (seasonal range: 72.2 gdw/m² in January to 445.0 gdw/m² in July). These are similar to Webber et al. (1987) from another Puget Sound location. Yang et al. (unpublished data) surveyed 17 sites around the sound in the spring and found above- and below-ground biomass to range 17–217 gdw/m². Because we are interested in estimating the C in relatively steady-state pools, we used winter biomass and chose a range 8–100 gdw/m².

3. Eelgrass sediments in Rhode Island have been characterized as having up to 300 MgC/ha (Payne 2007). Jesperson and Osher (2007) found soils to a depth of one meter in an estuary in Maine to average 136 MgC/ha with a range for different (generally unvegetated) habitats of 67–177 MgC/ha. We used a range of 50–300 MgC/ha for the sediment C estimates. To put this in context, the global average for wetland soils is 720 MgC/ha (US Department of Energy 1999) and Pacific Northwest old-growth forest soils are estimated to hold 30–400 MgC/ha (Homann et al. 2004). For comparison, Pacific Northwest forests have been estimated to store 180 MgC/ha above-ground (Lippke et al. 2003).

4. Globally, Mann (1982) estimated the NPP of coastal systems to range 300–1000 gC/m²/yr. Duarte and Cebrian (1996) estimated seagrass NPP to be 548gC/m²/yr; Mateo et al. (2006) estimated it to be 817 gC/m²/yr. Estimates for Z. marina in Europe and Asia range 620–2600 gC/m²/yr (summarized by Stevenson 1988). Estimates of NPP of Z. marina in Alaska and Oregon are 1000–1500 and 316–450 gC/m²/yr, respectively (McRoy 1974; Kentula and Mcintire 1986). An estimate of NPP for above-ground Z. marina and epiphytes at one location in Puget Sound is 344 gC/m²/yr (Thom 1990). Thom (unpublished data) used an estimate of 600 gC/m²/yr in Puget Sound for above- and below-ground biomass.

5. A carbon budget for generalized seagrass systems estimated that 15.9% of NPP is stored in sediments (Duarte and Cebrian 1996); because most studies have been conducted on a tropical genus that forms large mats of organic material, we assumed that 15 was an upper bound for Z. marina.

6. PacFIN (Pacific Fisheries Information Network), unpublished data.

7. See the text for description of the EwE model. We varied the shape of the mediation function from nearly linear to steeply sigmoid; as model outputs were qualitatively similar across all steepness terms, we discuss the results for an intermediate function.

Table 17.2b Summary of methods for estimating how changes in ecosystem services result in changes in their value for carbon storage and sequestration, commercial fisheries, and food web support

Ecosystem service	Estimation approach	Parameter values	Key assumptions
Carbon storage and sequestration by eelgrass for climate regulation	Eelgrass protection: Reduction in expected damage from climate change through carbon storage and sequestration. Estimate difference in C stocks with and without eelgrass protection.[1,2,3] T = Years of carbon sequestration C_B = Initial carbon stock for biomass (MgC/ha) C_S = Initial carbon stock for sediments (MgC/ha) ΔC_B = Annual carbon sequestration for biomass (MgC/ha/yr) ΔC_S = Annual carbon sequestration for sediments (MgC/ha/yr) C_{min} = Minimum carbon stock for sediments if eelgrass is lost(MgC/ha) t_1 = Time when eelgrass is lost and sediment carbon begins release t_2 = Time when sediment carbon ends release p = Social value ($/MgC) r = Social discount rate c = Carbon discount rate	$T = 50$ $C_B = 0.21$ $C_S = 175$ $\Delta C_B = 0$ $\Delta C_S = 0.525$ $C_{min} = 43.75$ $t_1 = 10$ $t_2 = 20$ $p = \$25$ $r = 3\%$ $c = 3\%$	Eelgrass is mature; sediment C losses when eelgrass biomass is lost span 3 periods: An initial period before eelgrass is lost in which sequestration continues; a second period in which eelgrass sediment carbon is lost (at a constant rate); and a third period in which sediment carbon remains stable at its minimum level.
Carbon storage and sequestration by eelgrass for climate regulation	Eelgrass restoration: Reduction in expected damage from climate change through carbon storage and sequestration. Estimate difference in C stocks with and without eelgrass restoration.[4,5] T = Years of carbon sequestration C_B = Carbon stock for biomass (MgC/ha) C_S = Carbon stock for sediments (MgC/ha) ΔC_B = Annual carbon sequestration for biomass (MgC/ha/yr) ΔC_S = Annual carbon sequestration for sediments (MgC/ha/yr) T_B = Years for restored eelgrass to reach "maturity" π = Probability of successful restoration p = Social value ($/MgC)	$T = 50$ $C_B = 0$ $C_S = 0$ $\Delta C_B = 0.0432$ $\Delta C_S = 0.525$ $T_B = 5$ $\pi = 0.5$ $p = \$25$ $r = 3\%$ $c = 3\%$	C stored and sequestered in area to be restored is 0.

Ecosystem service	Estimation approach	Parameter values	Key assumptions
	r = Social discount rate		
	c = Carbon discount rate		
Commercial fisheries for food: current	Net revenues by species for commercial fisheries in different marine sub-basins	Pounds and dollars per species/year landed in each marine sub-basin[6]	Harvest rates for all species do not change; non-trophic relationships between eelgrass and other species (e.g., Chinook salmon and Dungeness crab) are not examined.
Commercial fisheries for food: mediated through eelgrass-herring interaction	Use food web model to examine changes in commercial harvest due to changes in eelgrass biomass[7]		Non-commercial value is related to population size (lbs are used as a proxy for value); non-trophic relationships between eelgrass and other species (e.g., Chinook salmon) are not examined.
Wildlife viewing and existence mediated through eelgrass-herring interaction	Use food web model to examine changes in biomass of species groups due to changes in eelgrass biomass[7]		

[1.] With protection, the amount of carbon in year t is

$$C_{with}(t) = C_B + C_S + t\Delta C_S.$$

[2.] Without protection, the amount of carbon in year t follows a step function:

$$C_{w/o}(t) = C_B + C_S + t\Delta C_S, \quad t < t_1$$
$$= [C_S + t_1\Delta C_S] - [C_S + t_1\Delta C_S - C_{min}]\left[\frac{(t - t_1)}{(t_2 - t_1)}\right], \quad t_1 \leq t < t_2,$$
$$= C_{min}, \quad t_2 \leq t \leq T$$

[3.] The economic value of eelgrass protection is derived by first considering the difference between the carbon stock in each year with and without protection:

$$C_{with}(t) - C_{w/o}(t) = (C_B + C_S + t\Delta C_S) - (C_B + C_S + t\Delta C_S), \quad t < t_1$$
$$= (C_B + C_S + t\Delta C_S) - \left([C_S + t_1\Delta C_S] - [C_S + t_1\Delta C_S - C_{min}]\left[\frac{(t - t_1)}{(t_2 - t_1)}\right]\right), \quad t_1 \leq t < t_2$$
$$= (C_B + C_S + t\Delta C_S) - C_{min}, \quad t_2 \leq t \leq T$$

Note that for the first period, before the eelgrass biomass has been lost, there is no difference between the two stock levels, so that the value of protection (in those years) is zero. We then attach an economic value to the difference in each year:

$$V(Protection) = \sum_{t=1}^{T} \frac{p(C_{with}(t) - C_{w/o}(t))}{(1+r)^t(1+c)^t}.$$

[4.] With restoration, the amount of carbon in year t is

$$C(t) = \Delta C_B + \Delta C_S, \quad t < T_B$$
$$= \Delta C_S, \quad T_B \leq t \leq T$$

[5.] Without successful restoration, the amount of carbon in any year is zero, and so the expected value of eelgrass restoration is

$$V(Restoration) = \pi\left(\sum_{t=1}^{T_B} \frac{p\Delta C_B}{(1+r)^t(1+c)^t} + \sum_{t=1}^{T} \frac{p\Delta C_S}{(1+r)^t(1+c)^t}\right)$$

[6.] PacFIN (Pacific Fisheries Information Network), unpublished data.
[7.] See the text for a description of the EwE model.

dwarfs that in the biomass pool (such that the biomass pool is truly negligible). In comparison, total US forest carbon stocks are estimated to be in the range of 40 000–50 000 TgC. Pacific Northwest (western OR and WA) forests are estimated to contain approximately 351 TgC (US Environmental Protection Agency 2007).

17.3.2.2   Carbon sequestration

Seagrasses stabilize sediments, slow water motion, and cause the deposition of organic matter from the water column (Gacia and Duarte 1999; Gacia *et al.* 1999). Below the sediment surface, anoxia and light-limitation inhibit microbial processing and photo-degradation (Jesperson and Osher 2007), allowing for the build-up of C. Soil carbon generally has long residence times—particularly when submerged—and is therefore considered "sequestered carbon" (Wang and Hseih 2002). Despite ideal conditions for production and preservation of organic matter, the C-sequestration capacity of the soils of coastal ecosystems has been under-studied (Chmura *et al.* 2003; Thom *et al.* 2003; Jesperson and Osher 2007).

Our initial estimate of a sequestration rate in Puget Sound is 3 171–19 026 MgC/yr, or 11 627–69 762 Mg CO_2/yr (Table 17.2a). This represents 0.02–0.1% of the emissions of Washington State, 0.06–0.36% of the emissions of King County (Seattle's home), and up to 72% of the annual emissions of all transit busses in King County (King County 2007). For comparison, the carbon contained in all US forests offset approximately 10% of total US CO_2 emissions in 2005 (Woodbury *et al.* 2007).

If a restoration policy is being pursued, particular habitat types would be changing *to* eelgrass from a previous habitat type, and the original state would have had its own C-storage/sequestration values. Similarly, if eelgrass habitat is being lost, it is being replaced with another habitat type, and a similar comparison could be made. This makes marginal changes impossible to calculate without going through the same exercise for all habitat types. Among possible habitat types, however, we expect eelgrass to have the greatest capacity for carbon storage and sequestration, compared to other non-vegetated intertidal and shallow subtidal habitats such as rocky reefs, cobble, mud-, or sand-flats.

17.3.2.3   *Valuing ecosystem service value changes for carbon*

The ecosystem service value of carbon storage and sequestration is based on the reduction in the expected damage from climate change. Increasing levels of carbon dioxide and other greenhouse gases are linked to harmful changes in temperature and other aspects of climate. Controlling carbon dioxide by sequestering carbon therefore mitigates those harmful effects, which counts as an economic benefit. This value is enjoyed by society at large, and so it is referred to as the *social* value of carbon storage and sequestration.

For eelgrass, this ecosystem service value can be generated either by protecting current eelgrass or investing in eelgrass restoration, as described above. Protection provides value if existing eelgrass areas are threatened and the projected amount without protection decreases over time; restoration provides value when eelgrass would otherwise decline or remain stable in the future. In either case, the economic value is based on the difference in carbon stocks over time for two scenarios, one with the appropriate action (protection or restoration) and one without that action. Thus, it is important to understand what form of carbon sequestration and storage (if any) would either replace eelgrass (for the case of protection) or be replaced by eelgrass (in the case of restoration). Calculating this economic value is relatively straightforward (Chapter 7), but settling on the values for some of the parameters is fraught with controversy (Nordhaus 2007; Weitzman 2007). For the purposes of this chapter, we pick values merely to illustrate how the calculation and resulting value depends on the scenarios described above (Table 17.2b).

Based on the methods outlined in Table 17.2b, the social value of eelgrass protection is much higher than that of restoration: $1 496 to 4 585 versus $104 ha^{-1}. The range in values for protection reflect different assumptions about the loss of carbon from sediments, with the high value representing an assumed total loss of stored carbon when eelgrass is left unprotected and the low value resulting from a ten-year lag before carbon is released, a ten-year carbon release period, and a 25% minimum carbon stock that remains in the sediments without eelgrass. The large disparity is due in large part to the

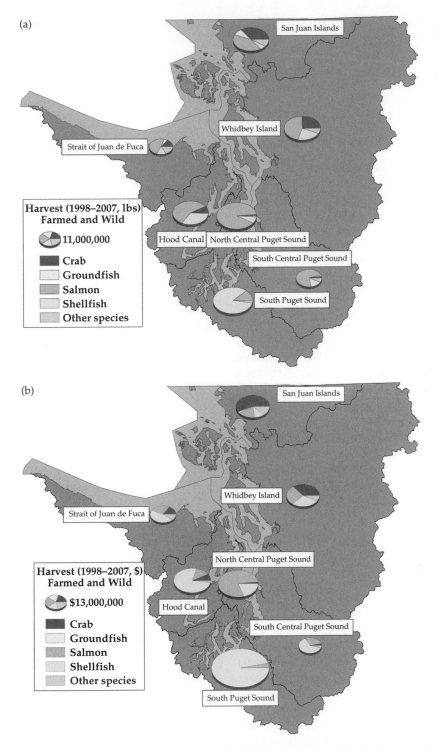

Figure 17.2   A map of the Puget Sound Partnership's action areas showing the distribution of (a) landings (in UK£) and (b) revenue (in US$) of farmed and wild seafood from 1998 to 2007. (See Plate 14.)

assumed loss of stored carbon when eelgrass is left unprotected. If leaving eelgrass unprotected does not produce a significant loss of stored carbon in the sediments, the value of protection and its advantage over restoration is diminished accordingly.

17.3.3 Marine harvest and non-consumptive values

Puget Sound's living marine resources, though depleted relative to historic times, remain a bountiful source of provisioning and other ecosystem services. Commercial fisheries harvest over 35 species of finfish and shellfish, and generate more than $50 million in annual revenue (Pacific States Marine Fisheries Commission, Pacific Coast Fisheries Information Network (PacFIN), unpublished data). Recreational harvest concentrates on Pacific salmon and steelhead (*Oncorhynchus mykiss*) but also includes shellfish such as Dungeness crab (*Cancer magister*) and butter clams (*Saxidomus giganteus*) (Washington State Department of Fish and Wildlife (WDFW), unpublished data). Puget Sound Indian tribes enjoy a rich tradition of ceremonial harvest. Aquaculture uses the ecological functioning of Puget Sound to produce more than $30 million in annual revenue for shellfish and almost $20 million for Atlantic salmon (*Salmo salar*) (WDFW, unpublished data). Non-consumptive activities, such as recreational whale and bird watching, also provide an important flow of services, while just the existence of some species produces value for Puget Sound residents and visitors.

In order to provide a perspective with which to gauge the effects of changes in the ecosystem, we tallied the commercial harvest coming from Puget Sound in biological (lbs.) and monetary ($) units (PacFIN; Figures 17.2a, b; Plate 14). This snapshot of the seafood provisioning service allows us to examine where particular types of harvest are highest. For example, shellfish, which are particularly lucrative, are predominately produced in southern Puget Sound.

The mobility of fish and their use of multiple habitat types necessitates a food web-based modeling approach rather than a habitat-based one. To understand how changes in nearshore environments and

eelgrass are likely to affect changes in the food web-based ecosystem service flows, we focus on the link between eelgrass, a foundation species, and one other species with wide-ranging food web interactions, Pacific herring.

17.3.3.1 How might changes in eelgrass habitats affect marine harvest and non-consumptive values?

In order to begin to understand how changes in nearshore environments are likely to affect changes in the flows of harvest and other services, we examined the habitat associations of the top 25 species harvested in the sound (including: geoducks (*Panopea abrupta*), salmon, Dungeness crab, and oysters), and categorized their dependence on nearshore habitats. Only 5 of these species (spiny dogfish, (*Squalus acanthias*), and 4 salmon—steelhead; sockeye (*Oncorhynchus nerka*), coho (*Oncorhynchus kisutch*), and pink (*Oncorhynchus gorbuscha*)) did not rely on nearshore habitats for at least one part of their life cycle. Thus, harvest levels of most species in the top 25 are likely to be sensitive to changes in nearshore habitats, but further modeling is necessary to understand how.

Pacific herring are a key food web species that interacts with eelgrass—they aggregate in the nearshore prior to reproducing then spawn in shallow water, usually on submerged vegetation (eelgrass or algae). Submerged vegetation provides spawning substrate, food resources, cover, and nursery habitat (Thayer and Phillips 1977; Dean *et al.* 2000; Penttila 2007). Survivorship of eggs is higher with lower spawn density (Galkina 1971; Taylor 1971). To survive, planktonic larvae must have sufficient supplies of microplankton; blooms of which are believed to be earlier, more dense and more consistent in sheltered bays (Penttila 2007). The survival of larval herring (determined particularly by food availability and predation) is thought to have a significant impact on the future abundance of the year-class (Alderdice and Hourston 1985). Juveniles spend several months inshore before moving into deeper waters (Penttila 2007). Herring are important prey to seabirds, crabs, salmon, marine mammals, and numerous other groups (Haegele 1993a, b; Penttila 2007). Given the strong connection between herring and nearshore habitat, we focus here on the consequences of how changes in nearshore habitat give rise to changes in herring, and how such effects can propagate through the food web.

17.3.3.2   Puget Sound food web model

Biomass dynamics of eelgrass and herring take place in the context of a broader community of interacting species, and resulting feedbacks within the food web are difficult to anticipate without the benefit of models. We used the Ecopath with Ecosim (EwE; Christensen and Walters 2004) software to construct a food web model for the central basin of Puget Sound (Figure 17.3a; Plate 15; Harvey et al. 2010). EwE models trophically and reproductively link biomass pools using a mass-balance modeling approach that satisfies two master equations describing production (as a function of catch, predation, migration, and biomass) and consumption (as a function of production, respiration, and unassimilated food). An initial mass-balanced snapshot of the ecosystem can then be used to explore dynamic simulations by expressing biomass flux rates among pools through time. The model of Puget Sound's Central Basin (Harvey et al. 2010) has functional groups ranging from primary producers to marine mammals and seabirds, as well as several fisheries. The model results we present below are preliminary outcomes that illustrate the complex and often unforeseen nature of community responses to perturbations (Christensen and Walters 2004).

Manipulating eelgrass production in EwE has negligible effects on the food web through consumption—a result that reflects our current understanding of eelgrass as a relatively unimportant direct food source (Mumford 2007). In contrast, non-trophic effects of eelgrass—such as habitat provisioning—are known to be very important (Thayer and Phillips 1977; Orth et al. 1984; Hosack et al. 2006; Mumford 2007). Such effects have important positive effects on other species and can be reflected in EwE through density-dependent mediation functions (Ainsworth et al. 2008). Mediation functions quantitatively link the vulnerability of a group (in this case, herring eggs) to the biomass of a mediating group (in this case, eelgrass): in other words, the less eelgrass is present, the more vulnerable herring eggs are to their predators. This is but one of numerous mechanisms by which changes in eelgrass could lead to changes in herring populations. Here we present only indirect effects that act through herring (Figure 17.3b; Plate 15) and focus on species and/or groups whose relationship with eelgrass is either trophic (i.e., they consume it) or is mediated through direct or indirect trophic interactions with herring.

We simulated a 50% decrease, a 50% increase, and a doubling of eelgrass biomass, and linked this to herring egg vulnerability. Depending on eelgrass biomass, herring eggs became either more or less vulnerable to predation by several groups (ducks and brants, gulls, ratfish, Dungeness crabs, small nearshore fishes, and small crustacean omnivores). Increases in eelgrass biomass yielded increases in herring and in turn increases in harbor seals (*Phoca vitulina*, whose primary prey is herring), ducks and brants (consumers of eelgrass), and greenlings (consumers of small crustaceans who feed on herring eggs). Increases in eelgrass yielded decreases in gulls and terns, skates, gadoids, lingcod (*Ophiodon elongatus*), and numerous flatfish. Most of these decreases result from competition with increased herring populations. Skates, however, likely decline with increases in eelgrass and herring because their primary predators are harbor seals, a species that increases with eelgrass and herring. Results for decreases in eelgrass biomass generally mirrored those of increases.

17.3.3.3   Valuing commercial harvest and non-consumptive services from food web changes

The species in the Puget Sound food web model provide consumptive services that include commercial harvest, and non-consumptive services such as whale and bird watching and existence value (because of data limitations on recreational fishing values, we do not consider recreational harvest in this section). Of the functional groups in the Puget Sound food web model, 15 are harvested commercially (Table 17.3). By weight, the most important fisheries are salmon (sockeye; chum, *Oncorhynchus keta*; coho; and Chinook, *Oncorhynchus tshawytscha*, both wild and hatchery stocks), geoduck, Pacific herring, and Dungeness crab, in that order. By value, the same fisheries dominate but the geoduck's high price per pound makes it the most economically

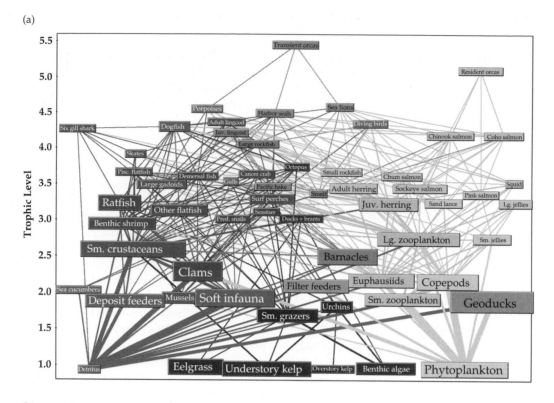

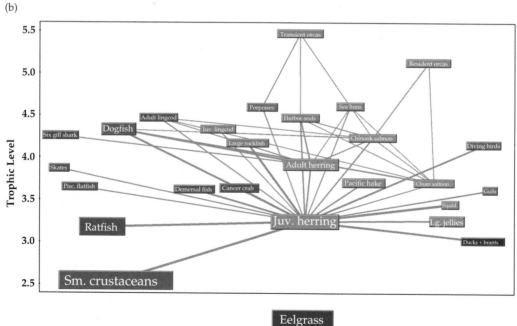

Figure 17.3 (a) The structure of the EwE food web model of the Central Basin of Puget Sound (without fisheries) and (b) a subset of the EwE food web model focusing on eelgrass and herring.

(a) Box size is proportional to standing stock biomass; line thickness is proportional to the flow of energy/material from the prey to the predator. Red colors represent detritus and the portion of the food web it supports, blues are benthic primary producers and those they support, and greens are phytoplankton and phytoplankton-supported groups. Consumers' colors are a mix proportional to the amount of production that ultimately stems from those sources. In (b) dashed arrows indicate groups whose predation on herring eggs is mediated by the biomass of eelgrass. Colors are as those in (a). (See Plate 15.)

valuable commercial species ($5.4 million, average annual revenue, 2005–7; PacFIN data). This value is split about evenly between the northern and southern parts of the Central Basin. Salmon and herring harvest, valued annually at $3.8 million and $159.3 thousand respectively, however, occur predominantly in the southern part of the Central Basin. Fifteen functional groups in the food web, including orcas, seals, ducks, sea stars, and the three groups of wild salmon, arguably have non-consumptive economic values based on outdoor recreation or simply for their existence (Table 17.3).

How are these values affected by the changes in nearshore conditions we have modeled? In biological terms, the food web model results show that the abundance of some species increases while it decreases for others. An ecosystem service value framework provides us with a way of evaluating these trade-offs. Ideally, expressing values in a common metric (dollars) enables one to make a grand aggregation of all the changes, producing a bottom

Table 17.3 Groups with significant commercial and non-consumptive values in the Puget Sound food web

Commercial harvest value	Non-consumptive value (group)
Geoducks	Transient orcas (cetacean)*
Sockeye and chum salmon (wild & hatchery)	Resident orcas (cetacean)*
Chinook and coho salmon (wild & hatchery)	Porpoises (cetacean)*
Dungeness crab	Gray whales (cetacean)*
Clams (various spp.)	Harbor seals (pinniped)*
Pacific herring*	Sea lions (pinniped)*
Shrimp (various spp.)	Gulls (bird)*
Sea cucumbers	Piscivorous diving birds (bird)*
Dogfish*	Murrelets (bird)*
Burrowing shrimp	Ducks and brants (bird)*
Wild pink salmon	Seastars (invertebrate)
Surf smelt*	Wild Chinook and coho salmon
Sea urchins	Wild pink salmon salmon
	Wild sockeye and chum salmon
	Pacific hake (species of concern)*

* Species whose primary interaction with eelgrass is mediated through their interactions with herring (or who directly consume eelgrass).
The commercial list includes all species groups with more than $1000 annual harvest in 2003–7. The non-consumptive group is subjectively chosen to represent species humans care about. The "group" for species of non-consumptive values indicates assignments to taxonomic groups for analysis of responses to eelgrass and herring perturbations in the EwE model.

line in terms of how nearshore conditions determine Puget Sound ecosystem service values. Two primary roadblocks prevent such an aggregation at this point: (1) we lack complete data on these values—commercial species are relatively easy to value, non-consumptive value species are not; and (2) we have not modeled the non-trophic relationships between eelgrass and a number of important species (e.g., Chinook salmon, chum salmon, and Dungeness crab).

To address these two issues, we assume that the non-consumptive value of a species is related to population numbers, using pounds as a proxy metric for these values; and we limit our discussion below to species whose primary interaction with eelgrass is through direct consumption of eelgrass or is mediated through their interactions with herring (Table 17.3). Among these species is Pacific hake (or whiting), *Merluccius productus*, which is currently considered a "species of concern" by NOAA Fisheries. This status indicates some concern about the viability of the species but insufficient information is available to make a formal ruling (National Marine Fisheries Service 2004). We consider this species separately, then, because further decreases in its status might trigger additional legal protections. Other species that currently have legal protection under the ESA (i.e., Chinook and summer chum salmon, steelhead, orca) are not examined here because they neither consume eelgrass nor primarily interact with eelgrass indirectly through their interactions with herring.

For the limited set of commercial species identified in this way, the herring fishery is the most important (in pounds harvested and revenue). Assuming the harvest rates for all species do not change, total herring harvest responds positively to changes in eelgrass, approximately doubling as eelgrass ranges from 50 to 200% of its baseline level. In contrast, spiny dogfish harvest increases by about 24% over that range, and surf smelt harvest decreases by about 23%; however, harvest yields for both of these species are considerably less than that of herring. Over the modeled range of eelgrass changes and using the average prices for each fishery (2005–7; PacFIN), the total harvest revenue for this limited set of commercial species would increase by 82% or $942 000 as eelgrass increases to 200% of its baseline.

For non-consumptive value species, the aggregate weight of this group is negatively related to increases in eelgrass biomass, decreasing by 15% across the range of eelgrass levels. Expressing their total value as a simple summation of pounds, however, implicitly assumes that these species have an equal per-lb economic value. Although data are not available to provide any guidance to differentiate these values, dividing the group into subgroups defined by taxonomy and legal status reveals potentially important differences (Figure 17.4). Pinnipeds have a strong positive relation with eelgrass biomass, birds and cetaceans have a very weak positive relation, and Pacific hake exhibits a negative relation.

These results are heavily qualified, of course, by the absence in our modeling to date of ecological relations between eelgrass and species other than herring. It is not yet possible to assess the overall direction of the change in values captured in the Puget Sound food web in response to changes in

nearshore habitat conditions. Our point here, however, is not so much to "accurately" depict the ecology of central Puget Sound, as to illustrate some important issues for using ecosystem service values. Commercial fisheries harvests are one of the most straightforward and easily measured ecosystem service values, and so producing a credible "bottom line" for this ecosystem service is possible as long as there are credible food web models. The modeling also allows us to understand that trade-offs among individual fisheries are still possible, and so improvements in ecological conditions may not be universally supported. The same can be seen in the trade-offs among non-consumptive value species.

17.3.4 Suites of ecosystem services in space

Overlaying carbon storage and sequestration services, and marine harvest and non-consumptive values is another way to consider spatial variation in ecosystem services. Herring spawn in twenty to twenty-one locations around Puget Sound and

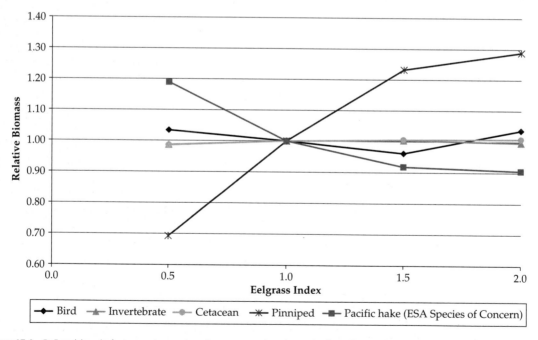

Figure 17.4 EwE model results for taxonomic groupings of non-consumptive value species (see Table 17.3 for group membership). The eelgrass index is eelgrass biomass/initial eelgrass biomass so values to the right of 1 represent increases in eelgrass biomass from the original baseline and values to the left represent decreases.

observations during years of relatively high abundance suggest that they may expand their spawning activities adjacent to currently used meadows, rather than colonizing new beds (Penttila 2007). Therefore, eelgrass restoration, if undertaken, will likely produce more value in areas adjacent to documented herring spawning sites where the benefits of increased carbon storage and sequestration are most likely to be complemented by the benefits of increased herring spawn, increased herring populations, and associated benefits derived from the food web (e.g., Figure 17.5).

This example illustrates the need to consider ecosystem services en suite, rather than one-by-one. It might be, for example, that the most productive areas for eelgrass restoration in terms of carbon services are not adjacent to existing herring spawning locations. The question is then whether the additional services derived from herring popula-

tions are worth the lesser carbon services. An ecosystem service framework can answer this question easily if a common metric (e.g., dollars) is used to measure the value of both sets of services. Even absent that information, the framework can illustrate where trade-offs may exist among services. In this way, policy-makers gain an understanding of the nature and extent of such trade-offs and can better set priorities in accordance with public values.

17.4   Future directions

The analyses presented here represent an initial step in developing an ecosystem services framework to support ecosystem-based management in Puget Sound. In this final section, we sketch out additional steps that can move such a framework closer to fruition.

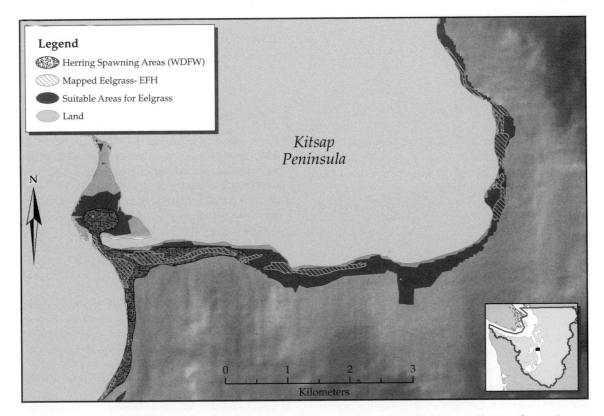

Figure 17.5   A portion of the Kitsap peninsula in Central Puget Sound, showing current eelgrass beds (hatched), areas used by herring for spawning (stippling), and areas predicted to be suitable for eelgrass restoration (dark gray).

A more complete evaluation of protection or restoration strategies will incorporate spatial variation in carbon storage and sequestration or food web functions and assess specific locations in terms of their current or potential production of these benefits. This information can provide useful guidance for recovery of the Puget Sound nearshore by providing a map with sites categorized according to their likely ecosystem service benefits under protection (for currently intact sites) or restoration (for sites that are currently degraded but with high intrinsic potential) strategies.

Such an evaluation also needs to expand the modeling of marine ecosystem services beyond the current set covered. Incorporating more links between nearshore habitat conditions and the marine food web will allow us to investigate other potential trade-offs among the provisioning services of commercial fisheries for salmon and other finfish, clam, oyster, crab, and other shellfish harvests, as well as the numerous cultural services that include bird and whale watching and recreational fishing. Waste treatment through the breakdown of PAHs (polycyclic aromatic hydrocarbons) and PCBs (polychlorinated biphenyls) by eelgrass should be added to the list of marine ecosystem services included in the analysis.

It also is possible in Puget Sound to extend the ecosystem services approach to include upland activities and their associated ecosystem services—basin-wide maps exist of current provisioning of water yields, water retention for floods, water purification potential, carbon storage, and commercial values of working landscapes in watersheds (Aukema *et al.* 2009; Rogers and Cooke 2009). The Partnership is interested in understanding how those watershed-based ecosystem benefits affect nearshore services provided to inform how and where to encourage different land-use practices around the region.

Clearly, developing a framework for assessing marine ecosystem services useful to policy-makers is an ambitious undertaking. Marine systems lack the commonly available spatial data that inform assessments of terrestrial services. As a result, building an assessment toolkit for marine environments may always be reliant on a richer set of local data and models developed for a particular location. We are using lessons learned from working in the Puget Sound region to develop models for multiple marine ecosystem services. The Marine Initiative of the Natural Capital Project is developing a marine InVEST tool that, like its terrestrial counterpart, will be an ecosystem services scenario assessment tool for application in ecosystem-based management processes with diverse stakeholders and across multiple scales. Building on the success of InVEST on land, we will connect existing models through the land-sea interface to new and existing marine models.

Ultimately, quantifying, mapping, and valuing marine ecosystem services has the potential to fundamentally change the ways in which decisions about marine and coastal environments are made. Making explicit the connections between human activities in one sector and their effects on a broad range of other sectors forces decision-makers and the human communities they represent to think about whole ecosystems and to manage them accordingly. By making clear the life-sustaining services oceans and coasts provide, appropriately valuing marine natural capital can help human communities make better choices about how we use these treasured environments.

References

Aburto-Oropeza, O., Ezcurra, E., Danemann, G., *et al.* (2008). Mangroves in the Gulf of California increase fishery yields. *Proceedings of the National Academy of Sciences*, **105**, 10456–9.

Agardy, T., Alder, J., Dayton, P., *et al.* (2005). *Coastal systems: assessment report*, Chapter 19. Millennium Ecosystem Assessment.

Ainsworth, C., Varkey, D., and Pitcher, T. (2008). Ecosystem simulations supporting ecosystem-based fisheries management in the Coral Triangle, Indonesia. *Ecological Modeling*, **214**, 361–74.

Alderdice, D. F., and Hourston, A. S. (1985). Factors influencing development and survival of Pacific Herring (*Clupea harengus pallasi*) eggs and larvae to beginning of exogenous feeding. *Canadian Journal of Fisheries and Aquatic Sciences*, **42**, 56–68.

Alliance for Puget Sound Shorelines. (2008). http://www.shorelinealliance.org/.

Alongi, D. M. (2008). Mangrove forests: Resilience, protection from tsunamis, and responses to global climate change. *Estuarine, Coastal and Shelf Science*, **76**, 1–13.

Aukema, J., Vigerstol, K., and Foster, J. (2009). *Application of InVEST models in Puget Sound*, March 20, 2009. Unpublished report available from authors upon request.

Bando, K. J. (2006). The roles of competition and disturbance in a marine invasion. *Biological Invasions*, **8**, 755–63.

Barbier, E. B. (2000). Valuing the environment as input: Applications to mangrove-fishery linkages. *Ecological Economics*, **35**, 47–61.

Barbier, E. B. (2003). Habitat-fishery linkages and mangrove loss in Thailand. *Contemporary Economic Policy*, **21**, 59–77.

Barbier, E. B. (2007). Valuing ecosystem services as productive inputs. *Economic Policy*, **22**, 177–229.

Barbier, E. B., Koch, E. W., Silliman, B. R., *et al.* (2008). Coastal Ecosystem-Based Management with Nonlinear Ecological Functions and Values. *Science*, **319**, 321–3.

Batie, S., and Wilson, J. (1978). Economic values attributable to Virginia's coastal wetlands as inputs in oyster production. *Southern Journal of Agricultural Economics*, **10**, 111–18.

Bell, F. (1989). *Application of wetland valuation theory to Florida fisheries*. Florida Sea Grant Program, Florida State University, Tallahassee.

Bennett, E. L., and Reynolds, C. J. (1993). The value of a mangrove area in Sarawak. *Biodiversity and Conservation*, **2**, 359–75.

Brander, L. M., Van Beukering, P., and Cesar, H. S. J. (2007). The recreational value of coral reefs: a meta-analysis. *Ecological Economics*, **63**, 209–18.

Burke, L., and Maidens, J. (2004). *Reefs at risk in the Carribean*. World Resources Institute, Washington, DC. Available online at: http://pdf.wri.org/reefs_caribbean_front.pdf

Burke, L., Greenhalgh, S., Prager, D., *et al.* (2008). *Coastal capital—economic valuation of coral reefs in Tobago and St. Lucia*. World Resources Institute, Washington, DC. Online at: http://www.wri.org/project/coral-reefs.

Carpenter, S., Caraco, N., Correll, D., *et al.* (1998). Nonpoint pollution of surface waters with phosphorus and nitrogen. *Ecological Applications*, **8**, 559–68.

Carte, B. K. (1996). Biomedical potential of marine natural products. *Bioscience*, **46**, 271–86.

Cesar, H. S. J., Ed. (2000). *Collected essays on the economics of coral reefs*. CORDIO, Kalmar Univeristy, Sweden.

Chmura, G., Anisfeld, S., Cahoon, D., *et al.* (2003). Global carbon sequestration in tidal, saline wetlands soils. *Global Biogeochemical Cycles*, **17**, 1111.

Christensen, V., and Walters, C. (2004). Ecopath with Ecosim: methods, capabilities and limitations. *Ecological Modeling*, **172**, 109–39.

Cochard, S. L., Ranamukhaarachchi, G. P., Shivakotib, O. V., *et al.* (2008). The 2004 tsunami in Aceh and Southern Thailand: A review on coastal ecosystems, wave hazards and vulnerability. *Perspectives in Plant Ecology, Evolution and Systematics*, **10**, 3–40.

Colling, A. (2001). *Ocean circulation*. Butterworth Heineman/Open University, Milton Keynes.

Costanza, R. (2000). The ecological, economic and social importance of the oceans. In: C. R. C. Sheppard, Ed., *Seas at the millenium: an environmental evaluation*, vol. 3: *global issues and processes*. Pergamon Press, New York.

Costanza, R., dArge, R., deGroot, R., *et al.* (1997). The value of the world's ecosystem services and natural capital. *Nature*, **387**, 253–60.

Council on Environmental Quality. (2009). *Interim Report of the Interagency Ocean Policy Task Force*. September 10, 2009. available at: http://www.whitehouse.gov/administration/eop/ceq/initiatives/oceans/interimrep.

Culliton, T. (1998). *Population: distribution, density, and growth*. National Oceanic and Atmospheric Administration, Silver Spring, MD.

Danielsen, F., Sorensen, M., Olwig, M., *et al.* (2005). The Asian tsunami: a protective role for coastal vegetation. *Science*, **370**, 643.

Davies, J., Guerry, A., Ruckelshaus, M. (In preparation). Mapping the potential distrubution of shallow-subtidal eelgrass in the greater Puget Sound region of Washington state.

Dean, T., Haldorson, L., Laur, D., *et al.* (2000). The distribution of nearshore fishes in kelp and eelgrass communities in Prince William Sound, Alaska: Associations with vegetation and physical habitat characteristics. *Environmental Biology of Fishes*, **57**, 271–87.

Diaz, R., and Rosenberg, R. (2008). Spreading dead zones and consequences for marine ecosystems. *Science*, **321**, 926–9.

Duarte, C. M. (2002). The future of seagrass meadows. *Environmental Conservation*, **29**, 192–206.

Duarte, C., and Cebrian, J. (1996). The fate of marine autotrophic production. *Limnology and Oceanography*, **41**, 1758–66.

FAO Fisheries Department. (2009). *The state of world fisheries and aquaculture—2008* (SOFIA). FAO, Rome.

Farber, S. (1988). The value of coastal wetlands for recreation—an application of travel cost and contingent valuation methodologies. *Journal of Environmental Management*, **26**, 299–312.

Forbes, K., and Broadhead, J. (2007). *The role of coastal forests in the mitigation of tsunami impacts*. RAP Publication 2007/1, Food and Agricultural Organization of the United Nations, Regional Office for Asia and the Pacific, Bangkok.

Fulton, E. A., Smith, A. D. M., and Johnson, C. R. (2004a). Biogeochemical marine ecosystem models I: IGBEM - a model of marine bay ecosystems. *Ecological Modeling*, **174**, 267–307.

Fulton, E. A., Parslow, J. S., Smith, A. D. M., *et al.* (2004b). Biogeochemical marine ecosystem models II: the effect of physiological detail on model performance. *Ecological Modeling*, **173**, 371–406.

Gacia, E., Granata, T. C., and Duarte, C. M. (1999). An approach to measurement of particle flux and sediment retention within seagrass (*Posidonia oceanica*) meadows. *Aquatic Botany*, **65**, 255–68.

Gacia, E., and Duarte, C. (1999). Sediment retention by a Mediterranean *Posidonia oceanica* meadow: the balance between deposition and resuspension. *Estuarine and Coastal Shelf Science*, **52**, 505–14.

Gaeckle, J., Dowty, P., Reeves, B., *et al.* (2007). *Puget Sound Submerged Vegetation Monitoring Project, 2005 Monitoring Report*. Puget Sound Assessment and Monitoring Program, Washington State Department of Natural Resources, Nearshore Habitat Program, Aquatic Resources Division, Olympia.

Galkina, L. A. (1971). Survival of spawn of the Pacific herring (*Clupea harengus pallasii val.*) related to the abundance of the spawning stock. *Rapports et Proces Verbaux des Reunions*, **160**, 30–3.

Gilbert, A. J., and Janssen, R. (1998). Use of environmental functions to communicate the values of a mangrove ecosystem under different management regimes. *Ecological Economics*, **25**, 323–46.

Gleick, P. (1996). Water resources. In: S. Schneider, Ed., Encyclopedia of climate and weather. Oxford University Press, New York.

Haegele, C. W. (1993a). Seabird predation of Pacific herring, *Clupea pallasi*, spawn in British Columbia. *Canadian Field Naturalist*, **107**, 73–82.

Haegele, C. W. (1993b). Epibenthic invertebrate predation of Pacific herring, *Clupea pallasi*, spawn in British Columbia. *Canadian Field Naturalist*, **107**, 83–91.

Harvey, C. J., Bartz, K.K., Davies, J., et al. (2010). A mass-balanced model for evaluating food web structure and community-scale indicators in the central basin of Pudget Sound. U.S. Dept. Commer., NOAA Tech. Memo. NMFS-NWFSC-106, 180.

Holmlund, C. M., and Hammer, M. (1999). Ecosystem services generated by fish populations. *Ecological Economics*, **29**, 253–68.

Homann, P. S., Remillard, S. M., Harmon, M. E., *et al.* (2004). Carbon storage in coarse and fine fractions of Pacific Northwest old-growth forest soils. *Soil Science Society of America Journal*, **68**, 2023–30.

Hosack, G. R., Dumbauld, B. R., Ruesink, J .L., *et al.* (2006). Habitat associations of estuarine species: Comparisons of intertidal mudflat, seagrass (*Zostera marina*), and oyster (*Crassostrea gigas*) habitats. *Estuaries and Coasts*, **29**, 1150–60.

Huesemann, M., Hausmann, T., Fortman, T., Thom, R. and Cullinan, V. (2009). In-situ phytoremediation of PAH and PCB contaminated marine sediments with eelgrass (*Zostera marina*). *Ecological Engineering*, **35**, 1395–1404.

Iceland, C., Hanson, C., and Lewis, C. (2008). *Identifying important ecosystem goods and services in Puget Sound*. World Resources Institute, Washington, DC.

Intergovernmental Panel on Climate Change—National Greenhouse Gas Inventories Programme. (2006). *IPCC Guidelines for National Greenhouse Gas Inventories*, vol. 4: *agriculture, forestry and other land use*. Available at: http://www.ipcc-nggip.iges.or.jp/public/2006gl/vol4.html

Jesperson, J., and Osher, L.. (2007). Carbon storage in the soils of a mesotidal Gulf of Maine estuary. *Soil Science Society of America Journal*, **71**, 372–9.

Kentula, M. E., and Mcintire, C. D. (1986). The autecology and production dynamics of eelgrass (*Zostera marina L*) in Netarts Bay, Oregon. *Estuaries*, **9**, 188–99.

King County. (2007). *King County climate plan*. Available at: http://www.metrokc.gov/exec/news/2007/pdf/climateplan.pdf

Lippke, B., Garcia, J. P., and Manriquez, C. (2003). *Executive summary: the impacts of forests and forest management on carbon storage*. Rural Technology Initiative, College of Forest Resources, University of Washington.

Lynne, G. D., Conroy, P., and Prochaska, F. J. (1981). Economic valuation of marsh areas for marine production processes. *Journal of Environmental Economics and Management*, **8**, 175–86.

McRoy, C. (1974). Seagrass productivity: carbon uptake experiments in eelgrass, Zostera marina. *Aquaculture*, **4**, 131–7.

Mallin, M. A., Williams, K. E., Esham, E. C., *et al.* (2000). Effect of human development on bacteriological water quality in coastal watersheds. *Ecological Applications*, **10**, 1047–56.

Mann, K. (1982). *Ecology of coastal waters*. University of California Press, Berkeley.

Mateo, M. A., Cebrian, J., Dunton, K., *et al.* (2006). Carbon flux in seagrass ecosystems. In: A. W. D. Larkum, R. J. Orth, and C. M. Duarte, Eds., *Seagrasses: biology, ecology, and conservation*. Springer, Dordrecht.

Mazda, Y., Magi, M., Kogo, M., *et al.* (1997). Mangroves as a coastal protection from waves in the Tong King Delta, Vietnam. *Mangroves and Salt Marshes* **1**, 127–35.

Melillo, J. M., Mcguire, A. D., Kicklighter, D. W., *et al.* (1993). Global climate-change and terrestrial net primary production. *Nature*, **363**, 234–40.

Millennium Ecosystem Assessment. (2005). *Ecosystems and human well-being: wetlands and water synthesis.* World Resources Institute, Washington, DC.

Moberg, F., and Folke, C. (1999). Ecological goods and services of coral reef ecosystems. *Ecological Economics*, **29**, 215–33.

Mumford, T. F. J. (2007). *Kelp and eelgrass in Puget Sound.* Seattle District, US Army Corps of Engineers, Seattle, WA.

National Marine Fisheries Service. (2007). *Sound science: synthesizing ecological and socio-economic information about the Puget Sound ecosystem.* US Department of Commerce, National Oceanic and Atmospheric Administration, Northwest Fisheries Science Center, Seattle, WA.

National Marine Fisheries Service. (2004). Endangered and threatened species; establishment of species of concern list, addition of species to species of concern list, description of factors for identifying species of concern, and revision of candidate species list under the Endangered Species Act. *Federal Register*, **73**, 19975–9.

National Research Council, Committee on Assessing and Valuing the Services of Aquatic and Related Terrestrial Ecosystems. (2004). *Valuing ecosystem services: toward better environmental decision-making.* National Academies Press, Washington, DC.

Nelson, T., and Waaland, J. (1997). Seasonality of eelgrass, epiphyte, and grazer biomass and productivity in subtidal eelgrass meadows subjected to moderate tidal amplitude. *Aquatic Botany*, **56**, 51–74.

Nordhaus, W. (2007). A review of the Stern Review on the economics of climate change. *Journal of Economic Literature*, **45**, 685–702.

Orth, R. J., Heck, K. L., and Vanmontfrans, J. (1984). Faunal communities in seagrass beds—a review of the influence of plant structure and prey characteristics on predator prey relationships. *Estuaries*, **7**, 339–50.

Patterson, M., and Glavovic, B. Eds. (2008). *Ecological economics of the oceans and coasts.* Edward Elgar, Cheltenham.

Pauly, D., Christensen, V., and Walters, C. (2000). Ecopath, Ecosim, and Ecospace as tools for evaluating ecosystem impact of fisheries. *ICES Journal of Marine Science*, **57**, 697–706.

Payne, M. K. (2007). *Landscape-level assessment of subaqueous soils and water quality in shallow embayments in Southern New England.* Masters thesis, Department of Natural Resource Science, University of Rhode Island, Kingston, RI.

Penttila, D. (2007). *Marine forage fishes in Puget Sound.* Seattle District, US Army Corps of Engineers, Seattle, WA.

Pergams, O., and Kareiva, P. (In press). Support services: A focus on genetic diversity. In: A. Kinzig, Ed., *Ecosystem services.* Princeton University Press, Princeton, NJ.

Peterson, C., and Lubchenco, J. (1997). Marine ecosystem services. In: G. Daily, Ed., *Nature's services.* Island Press, Washington, DC.

Pew Oceans Commission. (2003). *America's living oceans: charting a course for sea change. A report to the nation.* Pew Oceans Commission, Arlington, VA.

Plummer, M.L. (2009). Assessing benefit transfer for the valuation of ecosystem services. *Frontiers in Ecology and the Environment*, **7**, 38–45.

Puget Sound Nearshore Ecosystem Restoration Project. (2008). *Puget Sound Future Scenarios.* Draft report prepared by University of Washington Urban Ecology Research Lab. Available at: http://www.pugetsound-nearshore.org/index.htm

Puget Sound Partnership. (2006). *Interim Report to the Governor*, Puget Sound Partnership.

Puget Sound Partnership. (2008). *Action Agenda.* Puget Sound Partnership. December, 2008. Available at: http://www.psp.wa.gov/aa_action_agenda.php

Rogers, L. W., and Cooke, A. G. (2009). *The 2007 Washington State Forestland Database: Final Report.* Prepared for the USDA Forest Service. Available at: http://www.ruraltech.org/projects/wrl/fldb/

Ronnback, P. (1999). The ecological basis for economic value of seafood production supported by mangrove ecosystems. *Ecological Economics*, **29**, 235–52.

Rosenberg, A., and McLeod, K. (2005). Implementing ecosystem-based approaches to management for the conservation of ecosystem services. *Marine Ecology Progress Series*, **300**, 270–74.

Rosenfeld, D., Lahav, R., Khain, A., *et al.* (2002). The role of sea spray in cleansing air pollution over ocean via cloud processes. *Science*, **297**, 1667–70.

Ruitenbeek, H. J. (1994). Modeling economy ecology linkages in mangroves—economic evidence for promoting conservation in Bintuni Bay, Indonesia. *Ecological Economics*, **10**, 233–47.

Sathirathai, S., and Barbier, E. B. (2001). Valuing mangrove conservation in southern Thailand. *Contemporary Economic Policy*, **19**, 109–22.

Schlesinger, W. (1991). *Biogeochemistry: an analysis of global change.* Academic Press, San Diego.

Shared Strategy for Puget Sound. (2007). *Puget Sound salmon recovery plan.* Plan adopted by the National Marine Fisheries Service. January 19, 2007. Seattle, WA.

Soderqvist, T., Eggert, H., Olsson, B., *et al.* (2005). Economic valuation for sustainable development in the Swedish coastal zone. *Ambio*, **34**, 169–75.

Spurgeon, J. P. G. (1992). The economic valuation of coral reefs. *Marine Pollution Bulletin*, **24**, 529–36.

Stevenson, J. C. (1988). Comparative ecology of submersed grass beds in freshwater, estuarine, and marine environments. *Limnology and Oceanography*, **33**, 867–93.

Taylor, F. H. C. (1971). Variation in hatching success in pacific herring (*Clupea pallasii*) eggs with water depth, tempertature, salinity and egg mass thickness. *Rapports et Proces Verbaux des Reunions*, **160**, 34–41.

TerraLogic GIS Inc. (2004). *Public seagrass compilation for west coast Essential Fish Habitat (EFH) Environmental Impact Statement*. Pacific States Marine Fisheries Commission, Portland, OR.

Thayer, G., and Phillips, S. (1977). Importance of Eelgrass Beds in Puget Sound. *Marine Fisheries Review*, **39**, 18–22.

Thom, R. M., Borde, A. B., Williams, G. D., *et al.* (2003). Climate change and seagrasses: climate-linked dynamics, carbon limitation and carbon sequestration. *Gulf of Mexico Science*, **21**, 134.

Thom, R. M. (1990). Spatial and temporal patterns in plant standing stock and primary production in a temperate seagrass system. *Botanica Marina*, **33**, 497–510.

Thom, R. M., and Albright, R. G. (1990). Dynamics of benthic vegetation standing-stock, irradiance, and water properties in Central Puget Sound. *Marine Biology*, **104**, 129–41.

Thom, R. M., and Hallum, L. (1990). *Long-term changes in the areal extent of tidal marshes, eelgrass meadows and kelp forests of Puget Sound*. FRI-UW-9008.

Travis, J. (2005). Hurricane Katrina: scientists' fears come true as hurricane floods New Orleans. *Science*, **309**, 1656–9.

United Nations Environment Programme. (2006). *Marine and coastal ecosystems and human well-being: a synthesis report based on the findings of the Millennium Ecosystem Assessment*. UN Environment Programme, Nairobi.

US Department of Energy. (1999). *Carbon sequestration: the state of the science: carbon sequestration research and development*. US Department of Energy, Office of Fossil Energy, Washington, DC.

US Environmental Protection Agency. (2007). *Inventory of U.S. greenhouse gas emissions and sinks: 1990–2005, Annex 3.12, Methodology for estimating net carbon stock changes in forest lands remaining forest lands*. US Environmental Protection Agency, Washington, DC.

US Geological Survey. (1997). *ATLSS: across-trophic-level system simulation: an approach to analysis of South Florida ecosystems*. Biological Resrources Division USGS, Miami, FL.

US Commission on Ocean Policy. (2004). *An ocean blueprint for the 21st century*. US Commission on Ocean Policy, Washington, DC.

Wang, Y., and Hseih, Y. (2002). Uncertainties and novel prospects in the study of the soil carbon dynamics. *Chemosphere*, **49**, 703–24.

Webber, H. H., Mumford, T. J., and Eby, J. (1987). *Remote sensing inventory of the seagrass meadow of the Padilla Bay National Estuarine Research Reserve: areal extent and estimation of biomass*. NOAA technical report series OCRM/MEMD, Reprint series 6. Padilla Bay National Estuarine Research Reserve, Seattle.

Weitzman, M. (2007). A review of the Stern Review on the economics of climate change. *Journal of Economic Literature*, **45**, 703–24.

Williams, S., and Heck, K. L. (2001). Seagrass communities. In: M. Bertness, S. Gaines, and M. Hay, Eds., *Marine community ecology*. Sinauer, Sunderland.

Wilson, M. A., and Liu, S. (2008). Evaluating the non-market value of ecosystem goods and services provided by coastal and nearshore marine system. In: M. G. Patterson and B. C. Glavovic, Eds., *The ecological economics of the oceans and coasts*. Edward Elgar, Cheltenham.

Wolanski, E. (2007). *Estuarine ecohydrology*. Elsevier, Amsterdam.

Woodbury, P. B., Smith, J. E., and Heath, L. S. (2007). Carbon sequestration in the US forest sector from 1990 to 2010. *Forest Ecology and Management*, **241**, 14–27.

World Travel and Tourism Council (WTTC). (2005). *Trinidad and Tobago: the impact of travel and tourism on jobs and the economy*.

CHAPTER 18

Modeling the impacts of climate change on ecosystem services

Joshua J. Lawler, Erik Nelson, Marc Conte, Sarah L. Shafer, Driss Ennaanay, and Guillermo Mendoza

18.1 Introduction

The Earth's climate has already changed significantly in response to increased greenhouse-gas concentrations from human activities, and future climatic changes are projected to be even more dramatic, with global average temperatures expected to rise between 1.1 and 6.4°C by 2100, depending on future emissions from human activities (Alley *et al.* 2007). With the rise in temperature will also come changes in precipitation patterns, with some areas becoming drier, and others wetter. Climate change of this magnitude is certain to have major impacts on basic ecosystem processes such as primary production, hydrological cycles, and nutrient cycles (Cramer *et al.* 2001; Stewart *et al.* 2005; Betts *et al.* 2007). In this chapter, we discuss how models that link land cover and land use to ecosystem services can also be used to assess the impact of climate change on ecosystem services. Many studies have addressed how specific ecosystem services, particularly water availability and agricultural production or food security, are expected to respond to climate change. Here, we go one step beyond and demonstrate how climate impacts can be integrated by examining several ecosystem services within the same framework.

18.2 Previous analyses of climate-driven changes in ecosystem services

Much of the world's population relies on surface water for drinking and irrigation. Given the clear links between climate and hydrology, projected changes in climate will likely have significant effects on water availability. Climate change is projected to increase surface-water availability at far northern latitudes, across much of Asia, and parts of eastern Africa and southeastern South America (Milly *et al.* 2005). In contrast, water availability is projected to decrease in the Middle East, mid-latitude western North America, southern Africa, and southern Europe (Milly *et al.* 2005). However, not all studies agree on the geographic distribution of potential future hydrologic changes, particularly when human population growth is taken into account (Vörösmarty *et al.* 2000; Arnell 2004). Furthermore, the effects of changes in temperature and precipitation on water availability are not necessarily simple. For example, across Africa, decreases in precipitation are projected to have very different effects on runoff in areas with different annual rainfall regimes (de Wit and Stankiewicz 2006). A 10% decrease in precipitation was projected to result in a 20% decrease in runoff in Ouagadougou, Burkina Faso, whereas the same decrease in precipitation was projected to produce a 77% decrease in runoff in Okavango, Botswana (de Wit and Stankiewicz 2006).

Changes in water availability as well as changes in temperature and atmospheric CO_2 concentrations will affect water demand, agricultural production, and, in some cases, food security (Alcamo and Henrichs 2002; Arnell 2004; Battisti and Naylor 2009). For example, there are clear linkages between recent changes in temperature and crop production (Box 18.1)—increases in temperature have been linked to overall decreases in corn and soybean yields in the USA (Lobell and Asner 2003). As one would expect, however, the direction of this trend differs by geographic region—US corn and soybean yields in cooler areas have responded positively to warming whereas yields in warmer areas have responded negatively. Globally, crop yields have been projected to increase with a doubling of atmospheric CO_2 concentrations (Rosenzweig and Parry 1994). Again, however, some regions will likely experience gains in crop yields while others will experience significant losses.

Individual wild species will also respond to climate change, resulting in changes in the ecosystem services they provide. The most frequently documented effects of climate change on wild species include shifts in distributions and shifts in phenology (Walther *et al.* 2002; Parmesan and Yohe 2003; Root *et al.* 2003; Parmesan 2006). Most directly, the abundance or location of species that are themselves directly harvested for food may be altered by climate change. For example, increases in temperature will likely reduce the overall range of many cool- and coldwater fish, including salmonids. Furthermore, reductions in snowpack have the potential to alter flow regimes potentially affecting spawning habitat and egg survival of salmonids. Battin *et al.* (2007) explored the potential combined effects of climate change and stream restoration on Chinook salmon in a watershed in western Washington in the western United States. Projected changes in the modeled salmon population for 2050 ranged from a decrease of 40% to an increase of 19% depending on the climate-change scenario that was used as well as the extent of habitat restoration that was assumed to take place. It is likely that studies of other economically important wild species will also reveal changes in their productivity as a result of climate change.

18.3 Using ecosystem-service models to evaluate the impact of climate change on natural and human systems

Water supply is dependent on climate, geology, topography, land cover and land management (Chapter 4, this volume). By combining land-use and land-cover data with climate projections it is possible to examine likely changes in interconnected ecosystem services (see Chapters 5 through 13). The simplest models (tier 1) for water-related

services are based on average annual inputs of climatic conditions and provide annual average outputs of service provision. Therefore, in regions where climate change is expected to affect annual precipitation and temperature patterns, the tier 1 water-related service models provide informative but temporally-coarse projections. However, in many cases, climate change is expected to shift seasonal precipitation patterns, but leave annual total precipitation relatively unchanged. For these cases, tier 1 models will be inappropriate and the ramifications of climate change on water services can only be estimated with water-related service models that make projections based on monthly or daily inputs (tier 2 models).

In most cases, the agricultural production models described in Chapter 9 can also be directly applied to climate simulations. If the tier 1 agriculture production model employs yield functions that include average temperature or total water budget during the growing season as an input, then the effect of changes in these variables on agriculture production are easily simulated. The tier 2 agriculture production model discussed in Chapter 9 uses climate inputs at a daily or monthly resolution. In most cases, tier 2 models will more accurately capture changes in agricultural production because growth in most agricultural plants is a process that operates on the scale of hours to months, not seasons to years.

Climate will also indirectly alter ecosystem services by altering vegetation type, land cover, and even land management (Bachelet *et al.* 2001; Ramankutty *et al.* 2002). Climate is also known to influence the abundance and distribution of plants and animals and in turn biodiversity. Several different approaches can be taken to modeling climate-driven shifts in land cover, land management, and individual species. For example, dynamic global vegetation models can be used to project changes in the distribution of plant functional types which can be translated into maps of potential vegetation change (e.g., Sitch *et al.* 2003). Other approaches take both climate-driven changes in potential vegetation and climate-change influenced land-use decisions into account (e.g., Bouwman *et al.* 2006). There are likewise different approaches to modeling climate-driven changes in individual species

Box 18.1   An estimate of the effects of climate change on global agricultural ecosystem services

David Lobell

To estimate climate impacts on agricultural services, we must consider two questions. First, how much is food worth? A simple measure is the world market price for an incremental ton of food (this ignores ethical aspects such as the value of a human life). As of August 2008, the prices per ton of wheat, rice, and maize—the three most widely grown crops—were roughly $330, $730, and $200, respectively. Second, we must consider how climate changes are affecting production of these crops, a task made difficult because farmers are constantly adjusting to their environment. Numerous studies have shown that incremental warming is clearly beneficial to production in some regions (e.g. Siberia) and harmful in others (e.g. India), even after accounting for farmer adjustments. Thus, the answer to the question at hand clearly depends on the region and scale of interest.

Focusing on the global scale, one approach to estimating how climate changes are affecting agricultural production is to compare year-to-year changes in crop yields with year-to-year changes in average temperatures over the areas and seasons where the crop is grown (Figure 18.A.1). This approach ignores the many important aspects of climate other than average temperature, but trends for most of these other aspects are much less pronounced than for temperature. As an average, global production of several key crops exhibit a linear decline with warming of roughly 5–8% per degree, a number that agrees well with many site-level studies. According to the IPCC, global warming in the past 25 years has proceeded at approximately 0.18 °C per decade. Combining these two numbers we see that warming costs roughly 1% of global production per decade. Global cereal production is roughly 2 billion tons per year, so that 1% represents 20 million

tons. If we use $300 per ton as an average price for cereals, this amounts to $6 of annual losses for each decade of warming.

These impacts relate only to warming. For a more complete accounting, one would also want to evaluate changes in atmospheric carbon dioxide, ozone, and patterns of precipitation. For example, future increases in the frequency and intensity of droughts or heavy rains could hamper agriculture in many regions. In general, the largest net impacts are likely to be in systems with C4 crops (maize, sorghum, millet, sugarcane) that do not benefit greatly from higher CO_2. These crops are especially important to millions of poor in Africa, as well as global livestock production and the rapidly growing biofuels industry.

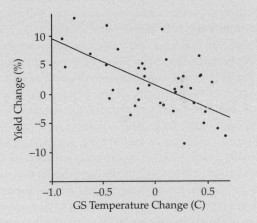

Figure 18.A.1   The relationship between year-to-year changes in globally averaged maize yields and average growing season (GS) temperatures over maize growing regions, 1961–2002 (Data from Lobell and Field (2007)).

distributions or abundances—both correlative and mechanistic models with a range of complexities have been used (e.g., Pearson *et al.* 2002; Carroll 2007). The outputs of these models can also be taken as inputs to ecosystem service models.

A primary concern of conservation is the spatial patterning of biodiversity and critical ecosystem services. Previous studies that assessed climate impacts on ecosystem services did not also include impacts on biodiversity (e.g., Alcamo *et al.* 2005).

Conversely studies of the impact of climate on biodiversity have neglected concordant impacts on ecosystem services. Here we examine the Willamette Basin in Oregon, USA, and consider climate impacts on irrigation water demand among select crop production systems, carbon storage in forests, and biodiversity. As described below, we used a limited set of climate simulations, vegetation model outputs, and projected species distributions as inputs to the ecosystem service models. These limited inputs do not take many of the important uncertainties in future climate-change projections or system responses into account. Furthermore, these input data are of relatively coarse spatial and/or temporal resolution, further limiting the conclusions that can be drawn from the results. Although the accuracy and generality of our model projections are limited, the examples we develop are illustrative of a general but useful approach to combining projections of climate impacts with spatially explicit maps of ecosystem services to determine how those services change under different climate scenarios.

18.4 Climate impacts on ecosystem services in the Willamette Basin of Oregon

We focus on climate impacts on three services and the biodiversity conservation provided by the Willamette Basin of Oregon in the western United States. The Willamette Basin is a 2.93 million hectare watershed that includes the urban areas of Portland, Salem, Albany, Corvallis, and Eugene-Springfield as well as significant areas of agricultural land on its central floor and forests in the surrounding mountains. The lowlands once supported expansive areas of oak savannah and grasslands but little of these habitats remain today. Most of the uplands are covered by conifer forest (Figure 18.1). We have chosen this valley as the site for our case studies because it has a diversity of land uses and it allows us to take advantage of the extensive data sets that have been created for this area (Schumaker *et al.* 2004; Polasky *et al.* 2005, 2008; Nelson *et al.* 2009). To draw our conclusions, we rely on a single, middle of the twenty-first century, climate simulation. Projected spatial patterns of future temperature and precipitation were derived from the UKMO-HadCM3 coupled atmosphere–ocean general circulation model (AOGCM; Gordon *et al.* 2000) driven by the Intergovernmental Panel on Climate Change (IPCC) Special Report on Emissions Scenarios (SRES) mid-high A2 emission scenario (Nakicenovic *et al.* 2000). Recent data show that anthropogenic greenhouse-gas emissions are already exceeding the higher SRES A1fi emission scenario. Therefore, our use of the A2 scenario should yield relatively conservative projections for changes to biodiversity and ecosystem services under a no-action scenario. In

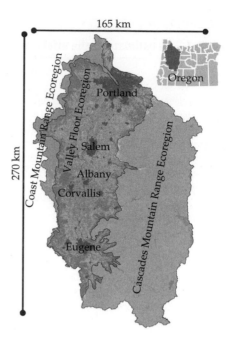

Figure 18.1 Map of the Willamette Basin, Oregon, USA.

addition, the UKMO-HadCM3 model tends to produce climate-change projections that lie just above the middle of the range of climate sensitivity defined by the other AOGCMs at the global scale.

Building the projected future climate dataset involved downscaling recent historic climate data and then applying anomalies from the future climate simulation to this historic climate dataset. Monthly climate data for 1901–2002 were created from the University of East Anglia Climatic Research Unit (CRU) CL 2.0 (New *et al.* 2002) and TS 2.1 (Mitchell and Jones 2005) climate data applied to a 30-second grid (~0.6 km^2 for the Willamette Valley) of the study area using a locally-weighted, lapse-rate-adjusted interpolation method developed by P. J. Bartlein (University of Oregon, personal communication). For future climate data, we used monthly climate data simulated for 2000–99 by the UKMO-HadCM3 coupled atmosphere–ocean general circulation model (AOGCM; Gordon *et al.* 2000) and obtained from the World Climate Research Programme's (WCRP's) Coupled Model Intercomparison Project phase 3 (CMIP3) multi-model dataset. Anomalies were calculated as differences (for temperature data) and ratios (for precipitation and percent cloud cover data) between each 2000–99 simulated monthly value and 1961–90 30-year mean monthly values calculated from the UKMO-HadCM3 simulated monthly data for the twentieth century. The anomalies were bi-linearly interpolated to the 30-second grid of the study area and applied to CRU-derived 1961–90 30-year mean data to create the future climate data used in this study.

Importantly, we were not simply interested in climate impacts, but climate impacts assuming different scenarios of land use and land cover. Projections of human land use were taken from Hulse *et al.* (2002) and represent alternative maps of agriculture, forest, and residential and industrial development assuming three different development scenarios. The three scenarios include a "development" trajectory that portrays an increase in development pressure, a "plan trend" trajectory that continues development at current rates and in current patterns, and a "conservation" trajectory that limits development in ways that promote the conservation of natural systems in the Basin. All three land-use scenarios assume that human population in the Basin will increase from 2.0 million in 1990 to 3.9 million people in 2050.

18.4.1 Water availability for agriculture in the Willamette Basin

At the heart of the Willamette Basin is a fertile river valley that produces a wide variety of agricultural goods, ranging from staple grains to wine grapes. Although the region receives significant amounts of precipitation, summers are dry and some of the crops grown in the valley must be irrigated to meet production goals. Here, we explore the potential impacts of future climate change on the provision of freshwater relative to the demand for irrigation water for berries and wine grapes, two of the valley's most valuable crops. We determine, all else being equal, how much irrigation water would be necessary to achieve current crop yields under a hypothesized future climate.

To estimate climate-driven changes in irrigation demand, we compared the volume of water needed for agricultural production under current conditions to the volume required under projected future

climatic conditions. For the future time period, we averaged the projected monthly precipitation totals and monthly average temperatures from 2041 to 2070 and used these 30-year monthly means to represent the average climatic conditions for the middle of the century. In this example, we used observed climate in 2000 to represent current (baseline) climate. Ideally, we would have used an averaged 30-year period (as we did with the future projections) to represent baseline conditions. However, our choice of a baseline period was limited by the availability of specific crop-yield information in the Basin.

In addition to climate projections, forecasting future irrigation demand requires an estimate of where vineyards and berry fields would be located by the middle of the 21st century. We use the projected "plan trend" land-use scenario to identify counties that will likely be involved in agricultural production in 2050 (Hulse *et al.* 2002). The allocation of vineyards and berry fields in a county in 2050 was based on current allocations of these crop types in the county. Of all the counties in the Basin, only Columbia County does not currently produce any of our focal agricultural products; therefore, we excluded it from our irrigation analyses.

By assuming the proportion of crop lands within a county dedicated to berry or grape production in 2050 would be the same as the proportion in 2000, we have likely overestimated the area dedicated to these crops in some regions and underestimated it in others. It is likely that some farmers will take various measures to adapt to changes in climate. Farmers will inevitably cease the cultivation of crops that demand too much irrigation or that fail to thrive under new climatic conditions. In some cases, new crops will be planted and, in more extreme circumstances, agriculture will be abandoned entirely. Although it would be more realistic to try to capture the effects of potential adaptation strategies in our analyses, doing so would introduce additional uncertainties. Early contributions of economic studies that addressed climate-change impacts on agricultural production attempted to allow for the mitigating impacts of adaptation (Mendelsohn *et al.* 1994). More recent efforts, however, have abandoned the focus on adaptation to avoid the potential biases associated with the functional form assumptions necessary to allow for adaptation (Deschenes and Greenstone, 2007; Schlenker *et al.* 2005). In the interest of simplicity, in our case studies, we did not model the potential effects of adaptation strategies.

We estimated crop-specific water requirements based on local climatic conditions. These requirements are outputs of CropWat, a tool developed by the Land and Water Development Division of FAO to aid in the design and management of irrigation schemes (FAO 1992). The inputs to this tool include climate data, crop coefficients (e.g., root depth and depletion fraction) and growth stages, and soil data for calculating plant water use. The growth-stage specific crop coefficients for the different agricultural products were taken from AgriMet, a satellite-based network of automated agricultural weather stations operated and maintained by the US Bureau of Reclamation. Although the AgriMet data provided the crop coefficients as a percentage of growth, the length of the growing season in CropWat is measured in days. The necessary crop coefficient values were interpolated assuming linear plant growth. The soil data for the counties in the region were derived from the SSURGO Soil Data Viewer (USDA 2008). Using the climate, land-use, and soils data, we ran CropWat for the five crops we focused on, raspberries, blackberries, strawberries, blueberries, and wine grapes.

We translated the crop-specific water requirements from CropWat into the total volume of irrigation water required within each county for year 2000 production of each of the five focus crops by multiplying each crop's water depth by the total area within each county in 2000 dedicated to the growth of that specific crop (crop area data are from the Oregon Agricultural Information Network (OAIN)).

We used the tier 1 water-availability model described in Chapter 4 to project average annual water depth across the landscape for current and predicted mid-century climatic conditions. The tier 1 water model requires temperature, precipitation, and land-cover inputs. Again, land cover was assumed to change based on the 2050 plan-trend scenario. Unlike the two other case studies below, our water-availability projections do not include the effects of potential climate-driven shifts in vegetation.

Most of the Basin, including the areas dedicated to wine grape and berry production, is projected to experience decreases in water yield (Figure 18.2; Plate 16). The largest decreases are along the crest of the Cascade Range, but decreases are also forecast for much of the basin floor. The largest increases are projected for the foothills of the Cascade Range and the northwestern corner of the basin. The decreases in water yields, in conjunction with increasing temperatures, led to increased irrigation demands in berry and grape fields across all counties expected to grow these products in 2050 (Table 18.1; recall that the spatial pattern of cropland in 2050 is based on 2050 projections but the fraction of these lands in the five focus crops is based on observed 2000 patterns). All counties were projected to require at least a 50% increase in irrigation water to maintain current berry and grape yields, with some counties expected to see as much as an 85% increase in irrigation needs.

These analyses make it clear that increased irrigation demand for berries and grapes is one of the potential impacts of climate change for the Willamette Basin. Of course, our analyses were done at relatively coarse temporal and spatial scales, whereas climatic extremes at fine temporal scales, such as frosts or water shortages, will significantly impact agricultural yields. Changes in the nature or frequency of climatic extremes, which are also associated with climate change (Alley *et al.* 2007) and specific weather events could significantly alter irrigation demands. Mismatches between precipitation events and water demand at a finer temporal resolution may increase irrigation demand above the levels projected by our models. Finally, as discussed above, some farmers will likely change cultivation practices and/or crops in response to changes in climate, thus, altering irrigation requirements.

18.4.2 Carbon storage in the Basin's upland forests

The uplands of the Willamette Basin are blanketed in forests, mostly conifer. The forests have contributed to the region's timber economy for over a century and include remnants of old-growth forests that provide important habitat for a variety of species. Here, we project the potential impacts of climate change on carbon storage in the Basin's upland forests. We used

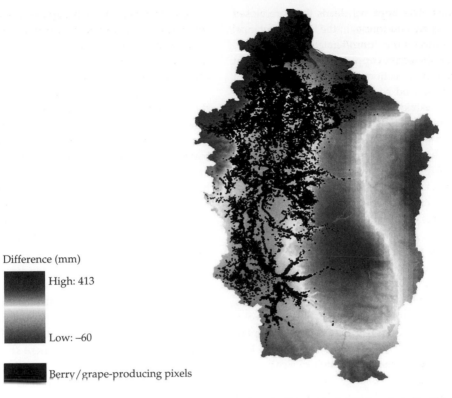

Difference (mm)

High: 413

Low: −60

Berry/grape-producing pixels

Figure 18.2   Change in annual average water availability (yield) between present day and mid-century for the Willamette Basin. (See Plate 16.)

Table 18.1   Projected mid-century climate-induced increases in water irrigation volumes required to maintain current yields for five different crops in the Willamette Basin

	Benton (%)	Clackamas (%)	Lane (%)	Linn (%)	Marion (%)	Multnomah (%)	Polk (%)	Washington (%)	Yamhill (%)
Raspberries	53.87	58.03	71.82	63.82	49.80	54.40	46.54	48.37	44.76%
Blackberries	53.87	58.03	71.82	63.82	49.80	54.40	46.54	48.37	44.76%
Strawberries	62.14	67.35	78.68	71.23	57.40	64.37	56.44	57.33	54.67%
Blueberries	53.31	57.25	70.88	63.04	49.33	53.72	31.81	47.85	44.35%
Wine grapes	67.60	n/a	97.55	n/a	61.38	n/a	56.21	62.70	55.55%
Total	58.61	58.82	84.89	65.00	51.92	55.07	50.99	52.30	52.16%

the tier 2 carbon-storage model documented in Chapter 7 and in Nelson *et al.* (2009) to estimate the carbon stored in both above and below ground biomass and soil on a given landscape.

To investigate the potential effect of climate-induced vegetation changes on carbon storage in the Basin's upland forests, we combined potential natural vegetation change data simulated for the Basin for 2050 with the three projections of land-use change for 2050 (the development, plan trend, and conservation scenarios described above). We integrated the climate-driven vegetation projection and the land-use change scenarios in two different ways. First, we assumed that climate-driven changes in vegetation would be the dominant driver of land cover in 2050. This alternative assumes that land owners and managers will allow their land to convert to the forest types as projected by the potential natural vegetation change maps. This approach leads to more closed mixed and hardwood forests in the Basin's uplands. We refer to this as the "unmitigated management" scenario. For the second approach, we assumed that forest management actions would be the dominant driver of forest land cover. In this scenario, timber companies, individual land owners, and natural resource management agencies would manage forests in such a way that the closed conifer forests that provide much of the economically valuable timber in the Basin's uplands today would be maintained regardless of changes in climate. In this scenario, forest management activities are allowed to mitigate the potential effects of climate change on closed canopy conifer forest, however, climate change is still assumed to be the driving factor for changes in all other land-cover types. We refer to this second scenario as the "mitigated management" scenario (see the chapter's online appendix for details on how these mitigated scenario maps were created).

The unmitigated and mitigated management scenarios represent two extreme cases—the actual contribution of land management to mitigating climate-driven land-cover change will likely lie somewhere between these two extremes. Table 18.2 summarizes the projected vegetation transitions for the six climate-affected mid-century land-cover scenarios (three land-use scenarios and two methods for combining land-use and climate-driven vegetation projections) relative to the mid-century land-use scenario maps that assume no climate-driven change in vegetation.

Before running the carbon-storage model, each of the six climate-change-affected and three baseline 2050 land-use maps, projected at a 30-m grid resolution, were summarized on a grid composed of 500-ha hexagons. We did this to reduce the number of spatial units that had to be run through the carbon-storage model. These hexagons merely served as spatial units over which the land-cover data were summarized. For example, a 500-ha hexagon could have 200 ha in 0- to 20-year closed conifer forest, 100 ha in 21- and 40-year closed conifer forest, and 300 ha in mixed conifer forest.

In general, climate change is projected to have a negative effect on carbon storage for the Willamette Basin (Table 18.3; see chapter's online appendix for calculation details). This trend is consistent across all six land-cover and management scenarios when compared to their relevant 2050 baseline maps (Figure 18.3). The 2050 landscapes that consider projected climate-driven changes in vegetation are estimated to store less carbon because the vegetation simulations project a shift from carbon rich,

Table 18.2 Modeled changes in mid-century vegetation cover in the Willamette Basin (in hectares) as a result of hypothesized climatic changes relative to projected mid-century land-cover maps that do not consider climate change

	2050 LULC Scenario					
	Plan trend		Development		Conservation	
	Unmitigated	Mitigated	Unmitigated	Mitigated	Unmitigated	Mitigated
Gain in closed mixed forest	147 420	97 458	147 394	96 029	147 314	124 293
Gain in closed hardwood forest	34 027	25 155	34 068	23 674	34 191	27 776
Loss in closed mixed forest	20 596	14 611	20 569	14 583	20 706	14 717
Loss in closed hardwood forest	1 783	1 783	1 777	1 777	1 784	1 784
Loss in closed conifer forest	159 067	106 218	159 116	103 343	159 014	135 569
Loss in semi-closed hardwood forest	0.18	0.18	0	0	0	0

The projected climate-driven vegetation maps were superimposed on three different potential mid-century land-use scenarios that do not consider climate change (the plan trend, development, and conservation scenarios). The projected vegetation maps were merged with each mid-century land-use scenario map twice. In the first merge, known as the unmitigated management approach, the vegetation map always determined land cover when there was a disagreement between the land-use scenario and vegetation maps. In the second merge, the mitigated management approach, closed conifer forest 0–60 years of age was managed for timber and continued to be closed conifer forest despite any projected climate-driven transitions to other forest types. Likewise, 50% of all closed mixed forest was assumed to remain in mixed forest despite any projected climate change due to management actions. Thus, the mitigated mid-century land-cover maps assume that land managers will be able to maintain a desired forest type despite climate change—an assumption that may not be valid in many cases. The unmitigated approach makes no such assumption. All changes are relative to the relevant baseline mid-century land-use scenario map.

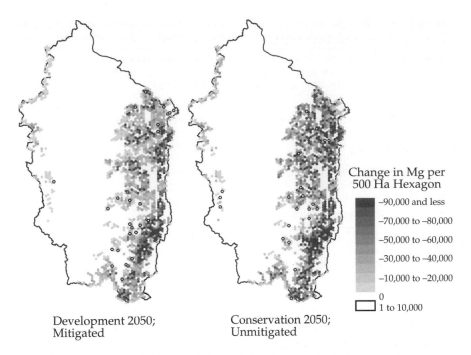

Development 2050;
Mitigated

Conservation 2050;
Unmitigated

Change in Mg per
500 Ha Hexagon

−90,000 and less

−70,000 to −80,000

−50,000 to −60,000

−30,000 to −40,000

−10,000 to −20,000

0
1 to 10,000

Figure 18.3   Difference in stored biomass and soil carbon in the Willamette Basin on two mid-century land-cover maps that incorporate projected climate-induced vegetation shifts relative to projected mid-century land-cover maps that do not consider climate change. The climate-change-affected development scenario map assumes mitigated land-cover change. The climate-change-affected conservation scenario map assumes unmitigated land-cover change (see description in legend for Table 18.2 for a brief description of the mitigated and unmitigated scenarios). When compared to their baselines, the development scenario map with mitigated change represents the lowest landscape-level loss in carbon storage of all the solutions presented in Table 18.3 (the 5th percentile solution of 5.6% (57.4 Tg)), whereas the conservation scenario with unmitigated change represents the greatest landscape-level loss in carbon storage of all the solutions presented in Table 18.3 (the 95th percentile solution of 7.0% (78.2 Tg)). Each spatial unit on the map is a 500-hectare hexagon.

older closed conifer forests, to relatively carbon poor, younger closed mixed forests and closed hardwood forests (Smith *et al.* 2006) (Table 18.3).

Despite the slightly larger negative effect of climate change on carbon storage under the conservation land-use scenario (Table 18.3), more carbon was stored under this scenario than under either the plan-trend or the development scenario. Given climate change, the conservation land-use scenario resulted in the median storage of 1040.5 Tg of carbon compared to 928.0 Tg stored under the plan-trend scenario and 944.5 Tg stored under the development scenario. We found that forest management strategies may also be able to mitigate some of the effects of climate change with respect to carbon storage. The "unmitigated" management scenarios resulted in less carbon storage than did

the corresponding "mitigated" management scenarios. The mitigated scenarios prevent larger losses of carbon because they restrict climate-driven changes in much of the established closed conifer and mixed conifer-hardwood forests from being manifested.

Our results highlight potential negative and positive feedback loops in the climate system. As atmospheric greenhouse-gas concentrations rise, the climate will change, driving shifts in vegetation. In some areas, vegetation may respond by sequestering more carbon and thus will help to mitigate emissions. In other areas, such as the Willamette Basin, vegetation responses to warming may result in reduced capacity for carbon storage and sequestration, thus amplifying the anthropogenic influence on climate. The models described in this book, when

Table 18.3   Percentage loss (and absolute loss in Tg) in biomass and soil carbon storage in the Willamette Basin due to hypothesized climatic changes relative to projected mid-century land-cover maps that do not consider climate change

Management approach	2050 LULC scenario		
	Plan trend	Development	Conservation
	Median [5th–95th Percentile]	Median [5th–95th Percentile]	Median [5th–95th Percentile]
Unmitigated	6.7% (66.2) [6.6 (65.7)–6.7% (66.4)]	6.6% (66.6) [6.5 (66.3)–6.6% (67.0)]	7.0% (77.8) [6.9 (77.5)–7.0% (78.2)]
Mitigated	6.0% (59.9) [6.0 (59.6)–6.1% (60.3)]	5.7% (57.5) [5.6 (57.4)–5.7% (58.0)]	6.7% (75.6) [6.7 (75.0)–6.8% (75.8)]

See the legend for Table 18.2 for a brief description of the various mid-century land-use scenario maps considered in this table. There is variability in results because we randomly simulated forest-age structure (this information is not provided by the mid-century land-cover maps) before calculating carbon-storage values. We repeated this process 1000 times. See the chapter's online appendix for simulation details. All losses are relative to the relevant baseline mid-century land-use use scenario map.

linked with ecosystem or vegetation models, provide a useful tool for investigating where these tradeoffs may occur and where management may be useful if increased carbon storage and sequestration is a goal.

18.4.3   Terrestrial vertebrate diversity in the Basin

The Willamette Basin hosts a rich biota that inhabits diverse environments including oak savannah, mixed hardwood and conifer forests, remnant riparian gallery forests, and young and mature temperate conifer rainforest. Here, we explore some of the potential effects of climate change on terrestrial vertebrate diversity in the Willamette Basin. Specifically, we investigate the effects of hypothesized clim-driven shifts in species distributions and vegetation change (i.e., habitat availability) on two measures of biodiversity status described in detail in Chapter 13, marginal biodiversity value (MBV) and countryside species area relationship (SAR). In short, MBV is a parcel-level measure of biodiversity based on both the number of species with geographic range in a parcel and the proportion of the parcel that likely serves as habitat for each of the species. SAR is a landscape-level estimate of biodiversity value based on multi-habitat species area relationships (Pereira and Daily 2006). The measure is referred to as a "countryside" metric because it was designed to

assess biodiversity value in diverse, human-dominated landscapes.

To model climate-driven changes in species distributions in the Willamette Basin, we used projected range shifts developed for a hemispheric assessment of species' geographic responses to climate change (Lawler *et al.* 2009). These range shifts are based on the same climate-change simulation used in the agriculture production and carbon-storage case studies above. Although the previous two case studies used the climate and vegetation projections at their native resolutions, (e.g., the 30-second grid), species range-shift projections were generated at a 50-km resolution. These projections were made for the average simulated climatic conditions for a period 2041–70. Our analyses use a set of bird, mammal, and amphibian species for which we have data on current and projected future geographic ranges as well as general habitat associations. This set includes 187 species (136 bird, 47 mammal, and 4 amphibian species). The modeled range shifts resulted in a potential loss of 8 of these species from the Basin by the middle of the century. At the same time, the climate model projects potential range expansions that could result in 17 new species moving into the Basin by the middle of the century, including 3 mammals and 14 birds (see Table 18.4 for a complete list of species with potential ranges that were projected to move in or out of the Basin).

As in the carbon-storage case study described above, we imposed projected climate-driven changes in vegetation (i.e., habitat availability) on the three 2050 land-use projections (Hulse *et al.* 2002) assuming two different land management responses (the unmitigated management and mitigated management schemes described in the carbon-storage example), thereby generating six 2050 maps affected to some degree or another by climate change. We also considered the three 2050 land-use projections that assume no climate-change effects (the baseline 2050 land covers). We also use a year 2000 land-cover map of the Basin for additional comparison purposes (this year 2000 map does not assume any climate-driven change in habitat availability). Finally, as we did in the carbon-storage case

study, land cover on all maps was summarized by 500-ha hexagon. Finally, unlike the carbon-storage case, we included the whole Basin in the biodiversity analysis, not just its upland forest areas.

Using these various scenarios, we generated several alternative landscapes across which we evaluated vertebrate diversity. First, we combined the 2000 land-cover map with the maps of *current* species distributions. This provided a map of current vertebrate biodiversity. Second, we combined each of the nine 2050 land-cover maps, including the three baseline maps, with maps of projected species distributions. This provided projections of how climate change and land-use change would likely affect vertebrate diversity values when both range shifts and climate-driven changes in habitat are taken into account and when range shifts alone are taken into account. Finally, we combined each of the nine 2050 land-cover maps, including the three baseline maps, with maps of *current* species distributions (i.e., the 2050 land cover assuming no climate-driven shifts in species ranges). This last combination provided an estimate of how both land-use and climate-driven changes in habitat would affect vertebrate diversity when range shifts were not taken into account.

For all projected vertebrate diversity maps, MBV values were calculated for each 500-ha hexagon and the SAR score was calculated for the entire landscape. As mentioned earlier, MBV measures the proportion of the modeled biodiversity supported by a hexagon (or, conversely, the proportion of the landscape's total biodiversity that would be lost if the hexagon were suddenly unable to support any species; see this chapter's online appendix for species-habitat compatibility scores for all modeled species).

One of the limitations of the MBV index is that it cannot be used to indicate the overall status of terrestrial biodiversity on the landscape (the summation of MBV values across all hexagons always equals one and does not allow for direct comparison of collective biodiversity status across scenarios). Instead, an MBV map indicates the relative provision of biodiversity across a landscape at some point in time. In contrast, the SAR scores allow for comparison across different landscapes, including the same landscape at different points in time. The SAR score for a species indicates the proportion of

Table 18.4 Species with geographic ranges projected to shift out of or into the Willamette Basin by mid-century

Emigrants	Immigrants
Gulo gulo (wolverine)	Myodes gapperi (Southern red-backed vole)
Anas acuta (Northern Pintail)	Onychomys leucogaster (Northern grasshopper mouse)
Histrionicus histrionicus (Harlequin Duck)	Ovis canadensis (Bighorn sheep)
Bucephala islandica (Barrow's Goldeneye)	Aeronautes saxatalis (White-throated swift)
Picoides dorsalis (Three-toed woodpecker)	Amphispiza bilineata (Black-throated sparrow)
Vireo olivaceus (Red-Eyed Vireo)	Carpodacus cassinii (Cassin's Finch)
Ammodramus savannarum (Grasshopper Sparrow)	Tringa semipalmata (Willet)
Xanthocephalus xanthocephalus (Yellow-Headed Blackbird)	Coccyzus americanus (Yellow-billed Cuckoo)
	Falco columbarius (Merlin)
	Phalaenoptilus nuttallii (Common Poorwill)
	Polioptila caerulea (Blue-gray Gnatcatcher)
	Recurvirostra americana (American Avocet)
	Sphyrapicus nuchalis (Red-naped Sapsucker)
	Tyrannus tyrannus (Eastern Kingbird)
	Anas crecca (Green-winged Teal)
	Aythya collaris (Ring-necked Duck)
	Melanerpes lewis (Lewis' Woodpecker)

its range space that contains compatible habitat raised to a power that is between 0 and 1 (the z score). A low z score is indicative of a species that only needs to find compatible habitat in a small portion of its range space to be relatively secure; increases beyond this threshold improve the condition for that species, but not dramatically. In contrast, the status of a species with a high z score is more affected by any changes in habitat in its range space. The landscape's SAR is the (weighted) average score for all species on the landscape. We calculated landscape SAR values in this case study assuming equal species weights and that each species had the same z score. Because we do not have information on the species' z scores, we calculated each landscape's SAR score four times, exploring a range of different z scores.

In general, simulated climate-driven changes in species ranges and habitat availability resulted in notable changes in the spatial distribution of hexagon MBVs from 2000 to 2050 under both the development and conservation land-use scenarios (Figure 18.4; we do not include the 2050 plan-trend maps in Figure 18.4 because these maps tend to lie between the extremes provided by the conservation and development scenarios). Climate impacts on MBV values appeared to be greatest along the crest of the Cascade Range and across portions of the basin floor. The distribution of MBVs on the maps that included climate-change effects, range shifts only, or both climate-driven range and habitat availability changes under "unmitigated" management (the last two 2050 maps for both scenarios in Figure 18.4), are significantly different than the distributions on the 2050 baseline maps (the first 2050 map for both scenarios in Figure 18.4) in several respects. First, climate change is projected to result in higher MBV values at higher elevations in the Cascade Range (the darker areas on the far right side of the maps) and some northern portions of the Coast Range. Second, climate change is projected to result in lower MBV values in the northern foothills of the Cascade Range and across portions of the Basin floor. These changes are a result of projected shifts in species distributions to higher elevations. In general, climate-driven changes in species ranges had a greater effect on the change in the spatial distribution of MBV values than did climate-driven changes in habitat—the differences in the leftmost

and middle of the three 2050 maps are more substantial than the differences between the middle and rightmost maps.

Climate change had a negative effect on overall biodiversity value as assessed by the SAR score (Table 18.5). Under the climate-change scenario, SAR scores decreased from the year 2000 to 2050 by 5.18 to 7.68% depending on the land-use scenario, the z score, and whether climate-driven range shifts only or climate-driven range shifts and climate-driven vegetation changes were considered. In contrast, in the absence of climate change, SAR scores decreased between 0.23 and 1.19% under the development scenario and increased between 0.12 and 0.48% under the conservation scenario. The pattern of land-cover change under the conservation scenario mitigated some of the deleterious effects of climate-driven changes in species ranges on SAR scores. However, this mitigating effect was relatively small. For example, with a z score of 0.64, climate-driven changes in species ranges resulted in a decrease in the SAR score of 6.18% under the conservation scenario, which was only slightly lower than the 7.25% decrease under the development scenario. When both climate-driven range shifts and climate-driven changes in vegetation were considered, the conservation scenario did little to mitigate the effects of climate change, particularly at higher z scores. Although these results indicate that general conservation strategies—such as those embodied in our conservation scenario—may be able to mitigate some of the potential effects of climate change on biodiversity, they ultimately suggest that conservation planning will need to directly address climate change with specific adaptation strategies for offsetting climate-driven habitat changes and range shifts.

Caution is needed when interpreting our analyses. First, it is not clear what a 5–8% decrease in landscape-level SAR scores means to the actual conservation status of the modeled species. Do these apparently modest decreases in SAR scores suggest that a species or two might go locally extinct? If so, the loss of even a single species from a region may have cascading ecological effects—a nuance not captured in our analyses. Second, although the vegetation models capture some of the basic changes in habitat availability, they likely fail to adequately describe changes in the habitats of many species

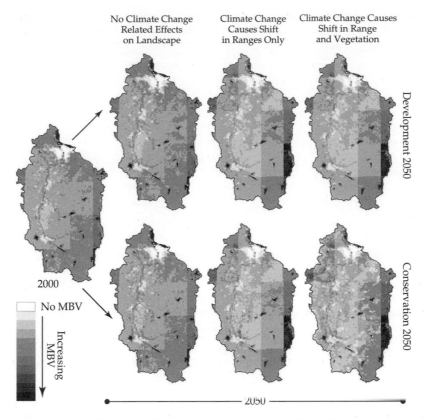

Figure 18.4   Marginal biodiversity values (MBVs) for the Willamette Basin for present day and mid-century. The six mid-century maps represent two different land-use scenarios (a development scenario and a conservation scenario) combined with three different levels of climate impact (none, climate-driven range shifts, and climate-driven range shifts and changes in habitat, i.e., vegetation). The first of the three mid-century maps assumes no climate change (species' range space does not change between present day and mid-century and no climate-driven vegetation changes occur). The second of the three mid-century maps assumes climate-change-induced species' ranges but no climate-driven vegetation changes occur. The last of the three mid-century maps assumes climate-driven shifts in species' ranges and climate-driven vegetation changes occur (in this case, we assume that vegetation transitions are *not* mitigated—see description in legend for Table 18.2 for a brief description of the mitigated and unmitigated scenarios). Given the coarse resolution of the range-shift projections, the effects of range changes are revealed by changes in the blocky patterns in the maps, whereas changes in habitat are much more finely resolved. White areas on the maps do not include any usable habitat.

that have critical relationships with specific plant species. In general, the biodiversity models presented in this book do not account for many of the factors that have been shown to affect species status on the landscape. Finally, the range-shift projections used in our analyses are based solely on climatic conditions. Shifts in species ranges will further be modified by changes in the distribution of predators, prey, competitors, and forage species. These limitations have likely led us to underestimate the potential impacts of climate change on biodiversity in the Basin. Nonetheless, the models presented here give us a first approximation of how climate change might affect terrestrial vertebrate diversity

in the Willamette Basin and where the climate impacts are likely to be the greatest.

18.5   Discussion and conclusions

Projected future climatic changes are likely to have significant effects on the functioning of ecological systems. These effects will in turn alter the degree to which natural and managed systems can provide many of the ecosystem services on which we currently depend. Here, we use several ecosystem service evaluation models and a biodiversity model to assess the potential impacts of climate change on irrigation demand, carbon storage, and biodiver-

Table 18.5 Landscape-level SAR scores in the Willamette Basin for the collection of modeled terrestrial vertebrates in the present day and mid-century with and without hypothesized climatic changes

	Z			
	0.11	**0.25**	**0.64**	**1**
2000[a]	0.855	0.737	0.533	0.421
2050				
Development LULC Scenario				
No climate-change effects[a]	0.853	0.734	0.529	0.416
	[−0.23%]	[−0.41%]	[−0.75%]	[−1.19%]
Climate change causes shifts in range only	0.790	0.680	0.494	0.393
(% change from 2000)[b]	[−7.57%]	[−7.68%]	[−7.25%]	[−6.56%]
Climate change causes shifts in range and vegetation	0.790	0.681	0.495	0.394
(% change from 2000)[b]	[−7.57%]	[−7.67%]	[−7.17%]	[−6.38%]
Conservation LULC Scenario				
No climate-change effects[a]	0.856	0.739	0.535	0.423
	[0.12%]	[0.27%]	[0.38%]	[0.48%]
Climate change causes shifts in range only	0.795	0.685	0.500	0.399
(% change from 2000)[b]	[−7.02%]	[−7.00%]	[−6.18%]	[−5.18%]
Climate change causes shifts in range and vegetation	0.792	0.682	0.493	0.390
(% change from 2000)[b]	[−7.39%]	[−7.46%]	[−7.48%]	[−7.38%]

[a] Species with geographic ranges overlapping the Willamette Basin in the present day are included.
[b] All species with geographic ranges overlapping the Basin in the present day and/or in the middle of the century are included. The species with range in the present day only contribute a SAR score of 0 to the mid-century SAR score.

sity. The three case studies provided here are meant to serve as examples of how the models described in this book can be applied to assess climate impacts. Our assessments were driven by a single future climate simulation based on the SRES mid-high A2 emissions scenario. Thus, our results should be seen as relatively conservative, and as a limited estimate of projected future climate-change impacts on ecosystem services. Our results are also likely to be conservative for two additional reasons. First, they do not consider the potentially multiplicative effects of climate change on a given sector or service through the influence of multiple climate drivers and interactions between services and sectors. And second, only mean temperature and precipitation changes were used to estimate impacts. Analyses show that many climate-change impacts, including ecosystem services, are likely to be just as (if not more) sensitive to changes in climate thresholds and extremes as they are to shifts in monthly, seasonal, or annual means. For all of these reasons, these results should be viewed as a demonstration of the potential application of the ecosystem-service

models for assessing potential climate-change impacts, rather than as providing quantitative estimates of how climate change is likely to affect the Willamette Basin.

To provide quantitative projections capable of being used by decision makers for long-term planning purposes, the analyses presented here should be modified in three important ways. First, it is important to quantify the extent to which combining the projections of multiple models can result in a compounding of model uncertainty. For example, there are uncertainties in the AOGCM projections, additional uncertainties in the dynamic global vegetation model and climate-envelope model projections, and additional uncertainties inherent in the ecosystem-service models themselves. All of these uncertainties need to be quantified to capture the full range of potential climate-change impacts on a given ecosystem service. A second limitation is that we used one future climate simulation from a single AOGCM in our analyses and a single emissions scenario. Given the wide variability in projections from different AOGCMs and the broad range of potential future emissions, it will be necessary to investigate the effects of alternative climate-change projections to begin to understand the range of possible climate impacts on ecosystem services. And lastly, capturing the potential impacts of climate change will require evaluating the effects of realistic adaptation strategies. Our analysis of irrigation demand did not take into account that farmers will likely shift crops or modify irrigation methods in response to climate change. Nor did we account for the multitude of ways in which forest managers might respond to climate-driven changes in tree-species composition.

Despite these limitations, the case studies investigated here demonstrate the importance of making even preliminary estimates of potential climate impacts on ecosystem services. For example, one can use the projected climate-induced change in each ecosystem service as a metric of "impact." The land-use scenario that yields the lowest impact score is the land-use scenario that provides the greatest potential for adaptation to climate change. In our assessments, this was generally the conservation scenario. Although, it is important to note that care must be taken in assessing the results of such scenario comparisons. For example, although we found a greater climate-driven loss in carbon storage under the conservation scenario, this differential resulted from the fact that the conservation scenario generated more carbon storage to begin with and thus there was more to lose by the middle of the century. More generally, given that climate change is expected to have profound effects on both human and natural systems, it will be critical to consider estimates of potential climate impacts when forecasting ecosystem services. As evidenced by the results of our simple case studies, climate change will alter many of the ecosystem services we rely on today. Both mitigating and adapting to climate impacts will require an understanding of how ecosystems and ecosystem services will respond to climate change. Models designed to assess ecosystem services will likely prove indispensable in our struggle to address climate change in the coming decades.

Acknowledgments

We thank Michelle Marvier, Katherine Hayhoe, Nathan Schumaker, and David Turner for useful comments and suggestions. We are also grateful to Peter Kareiva for his insightful editing and to Evan Girvetz for useful discussions. S. Shafer was supported by the US Geological Survey's Earth Surface Dynamics Program.

References

Alcamo, J., and Henrichs, T. (2002). Critical regions: A model-based estimation of world water resources sensitive to global changes. *Aquatic Sciences—Research across Boundaries*, **64**, 352–62.

Alcamo, J., Vuuren, D. v., Ringler, C., *et al.* (2005). Changes in nature's balance sheet: model-based estimates of future worldwide ecosystem services. *Ecology and Society*, **10**, Art. 19.

Alley, R., Berntsen, T., Bindoff, N. L., *et al.* (2007). *Climate change 2007: the physical science basis, summary for policymakers*. Working Group I contribution to the Intergovernmental Panel on Climate Change, Fourth Assessment Report, Geneva.

Arnell, N. W. (2004). Climate change and global water resources: SRES emissions and socio-economic scenarios. *Global Environmental Change*, **14**, 31–52.

Bachelet, D., Neilson, R. P., Lenihan, J. M., *et al.* (2001). Climate change effects on vegetation distribution and carbon budget in the United States. *Ecosystems*, **4**, 164–85.

Battin, J., Wiley, M. W., Ruckelshaus, M. H., *et al.* (2007). Projected impacts of climate change on salmon habitat restoration. *Proceedings of the National Academy of Sciences of the USA*, **104**, 6720–5.

Battisti, D. S., and Naylor, R. L. (2009). Historical warnings of future food insecurity with unprecedented seasonal heat. *Science*, **323**, 240–4.

Betts, R. A., Boucher, O., Collins, M., *et al.* (2007). Projected increase in continental runoff due to plant responses to increasing carbon dioxide. *Nature*, **448**, 1037.

Bouwman, A. F., Kram, T., and Goldewijk, K. K. (2006). *Integrated modelling of global environmental change. An overview of IMAGE 2.4*. Netherlands Environmental Assessment Agency (MPN), Bilthoven.

Carroll, C. (2007). Interacting effects of climate change, landscape conversion, and harvest on carnivore populations at the range margin: marten and lynx in the Northern Appalachians. *Conservation Biology*, **21**, 1092–104.

Cramer, W., Bondeau, A., Woodward, F. I., *et al.* (2001). Global response of terrestrial ecosystem structure and function to CO_2 and climate change: results from six dynamic global vegetation models. *Global Change Biology*, **7**, 357–73.

Deschenes, O., and Greenstone, M. (2007). The economic impacts of climate change: evidence from agricultural output and random fluctuations in weather. *American Economic Review*, **97**, 354–85.

de Wit, M., and Stankiewicz, J. (2006). Changes in surface water supply across Africa with predicted climate change. *Science*, **311**, 1917–21.

FAO. (1992). *CROPWAT, a computer program for irrigation planning and management by M. Smith*. FAO Irrigation and Drainage Paper 26, Rome.

Gordon, C., Cooper, C., Senior, C. A., *et al.* (2000). The simulation of SST, sea ice extents and ocean heat transports in a version of the Hadley Centre coupled model without flux adjustments. *Climate Dynamics*, **16**, 147–68.

Hulse, D., Gregory, S., and Baker, J. (2002). *Willamette River Basin planning atlas: trajectories of environmental and ecological change*. Oregon State University, Corvallis.

Lawler, J. J., Shafer, S. L., White, D., *et al.* (2009). Projected climate-induced faunal change in the western hemisphere. *Ecology*, **90**, 588–97.

Lobell, D. B., and Asner, G. P. (2003). Climate and management contributions to recent trends in U.S. agricultural yields. *Science*, **299**, 1032.

Lobell, D. B., and Field, C. B. (2007). Global scale climate–crop yield relationships and the impacts of recent warming. *Environmental Research Letters*, **2**, 014002.

Mendelsohn, R., Nordhaus, W. D., and Shaw, D. (1994). The impact of global warming on agriculture: A Ricardian analysis. *American Economic Review*, **84**, 753–71.

Milly, P. C. D., Dunne, K. A., and Vecchia, A. V. (2005). Global pattern of trends in streamflow and water availability in a changing climate. *Nature*, **438**, 347–50.

Mitchell, T. D., and Jones, P. D. (2005). An improved method of constructing a database of monthly climate observations and associated high-resolution grids. *International Journal of Climatology*, **25**, 693–712.

Nakicenovic, N., Alcamo, J., Davis, G., *et al.* (2000). *Special report on emissions scenarios: a special report of Working Group III of the Intergovernmental Panel on Climate Change*. Cambridge University Press, Cambridge.

Nelson, E., Mendoza, G., Regetz, J., *et al.* (2009). Modeling multiple ecosystem services, biodiversity conservation, commodity production, and tradeoffs at landscape scales. *Frontiers in Ecology and the Environment*, **7**, 4–11.

New, M., Lister, D., Hulme, M., *et al.* (2002). A high-resolution data set of surface climate over global land areas. *Climate Research*, **21**, 1–25.

Parmesan, C., and Yohe, G. (2003). A globally coherent fingerprint of climate change impacts across natural systems. *Nature*, **421**, 37–42.

Parmesan, C. (2006). Ecological and evolutionary responses to recent climate change. *Annual Review of Ecology and Systematics*, **37**, 637–69.

Pearson, R. G., Dawson, T. P., Berry, P. M., *et al.* (2002). SPECIES: A Spatial Evaluation of Climate Impact on the Envelope of Species. *Ecological Modelling*, **154**, 289–300.

Pereira, H. M., and Daily, G. C. (2006). Modeling biodiversity dynamics in countryside landscapes. *Ecology*, **87**, 1877–85.

Polasky, S., Nelson, E., Lonsdorf, E., *et al.* (2005). Conserving species in a working landscape: land use with biological and economic objectives. *Ecological Applications*, **15**, 1387–401.

Polasky, S., Nelson, E., Camm, J., *et al.* (2008). Where to put things? Spatial land management to sustain biodiversity and economic returns. *Biological Conservation*, **141**, 1505–24.

Ramankutty, N., Foley, J. A., Norman, J., *et al.* (2002). The global distribution of cultivable lands: current patterns and sensitivity to possible climate change. *Global Ecology and Biogeography*, **11**, 377–92.

Root, T. L., Price, J. T., Hall, K. R., *et al.* (2003). Fingerprints of global warming on wild animals and plants. *Nature*, **421**, 57–60.

Rosenzweig, C., and Parry, M. L. (1994). Potential impact of climate change on world food supply. *Nature*, **367**, 133–8.

Schlenker, W., Hanemann, W. M., and Fisher, A. C. (2005). Will U.S. agriculture really benefit from global warming? Accounting for irrigation in the hedonic approach. *American Economic Review*, **95**, 395–406.

Schumaker, N. H., Ernst, T., White, D., *et al.* (2004). Projecting wildlife responses to alternative future landscapes in Oregon's Willamette Basin. *Ecological Applications*, **14**, 381–400.

Sitch, S., Smith, B., Prentice, I. C., *et al.* (2003). Evaluation of ecosystem dynamics, plant geography and terrestrial carbon cycling in the LPJ dynamic global vegetation model. *Global Change Biology*, **9**, 161–85.

Smith, J. E., Heath, L. S., and Skog, K. E. (2006). *Methods for calculating forest ecosystem and harvested carbon with standard estimates for forest types of the United States. Gen Tech Rep NE-343*. US Department of Agriculture, Forest Service, Northeastern Research Station, Newtown Square.

Stewart, I. T., Cayan, D. R., and Dettinger, M. D. (2005). Changes toward earlier streamflow timing across western North America. *Journal of Climate*, **18**, 1136–55.

USDA. (2008). *Soil Survey Geographic (SSURGO) Database for Willamette Valley, Oregon*.

Vörösmarty, C. J., Green, P., Salisbury, J., *et al.* (2000). Global water resources: vulnerability from climate change and population growth. *Science*, **289**, 284–8.

Walther, G.-R., Post, E., Convey, P., *et al.* (2002). Ecological responses to recent climate change. *Nature*, **416**, 389–95.

CHAPTER 19

Incorporating ecosystem services in decisions

Emily McKenzie, Frances Irwin, Janet Ranganathan, Craig Hanson, Carolyn Kousky, Karen Bennett, Susan Ruffo, Marc Conte, James Salzman, and Jouni Paavola

19.1 Introduction

The world now faces unprecedented and interconnected challenges – reducing poverty, addressing climate change, and halting widespread environmental degradation. Scientific information on the links between humans and nature can help solve these problems (Lubchenco 1998; Bingham *et al.* 1995; Pielke 2007; MA 2005a). However, to seize this opportunity, scientific knowledge must be translated into action (NRC 2004; Daily *et al.* 2009; Cash *et al.* 2003). Advances in scientific modeling such as those described in this book are necessary but not sufficient. Here we provide an explicit examination of the channels through which the science of mapping and valuing ecosystem services can improve decisions.

Opportunities to use information on ecosystem services occur in all sectors of the economy and at all levels of decision-making: a mayor aims to increase flood protection for a city's citizens; a business requires a reliable supply of water for its manufacturing process; an international development agency attempts to reduce poverty by increasing small farm production. Although not always termed "ecosystem services," in each example, consideration of the benefits we gain from the environment is essential for making wise decisions. Information on ecosystem services can tell us how and which services are relevant to our goals, whether important services are at risk, where services are provided, who is affected, and the trade-offs of different choices; all key pieces of information for the design and implementation of a broad set of policy mechanisms.

19.1.1 Role of ecosystem service information in decisions

Scientific models move us from abstract, conceptual arguments about the importance of ecosystem services to specific quantification of the level, value and spatial distribution of ecosystem service benefits. This is of great relevance to the real world because such information can be applied to a range of policies and decisions, as in the examples from Oregon and Hawaii in Chapter 14 (see also Turner *et al.* 2000; Naidoo and Ricketts 2006; Nelson *et al.* 2009). A modeling framework that captures impacts on multiple ecosystem services over alternative scenarios enables stakeholders to weigh tradeoffs can serve as a basis for negotiation (Ghazoul 2007). Without such information, decision-makers tend to use intuitive or heuristic approaches that ignore ecosystem service values and distributional issues (Kiker *et al.* 2005). As noted by Ascher and Healy (1990), "information shapes many aspects of how resource issues are viewed and addressed: the focus of attention, the way problems come to be defined, and the ways that success or failure...is attributed to a project or policy."

There are several characteristics of ecosystem service models that are relevant to common policy contexts (Table 19.1). For example, the development of stakeholder-driven scenarios can ensure that ecosystem service valuations are aligned with the problems of interest to decision-makers, revealing sources of conflict and building consensus (Henrichs *et al.* 2008). This does not happen automatically, however, and typically requires active, iterative and

Table 19.1 Characteristics of ecosystem service models with relevance for decision-making

Model characteristics	Relevance to decisions
Integrated framework, including multiple ecosystem services	• Provides consistent standard for evaluating projects and policies • Encourages consideration of all ecosystem services including those that have not been emphasized in past policies • Enables analysis of trade-offs between different ecosystem services, stakeholders, and geographic areas • Informs and encourages coordinated, multi-sectoral, ecosystem-based management
Packaged as a "tool"	• Translates complex scientific results into policy relevant outputs, e.g., trade-off curves, efficiency frontiers (see Chapters 3 and 14) • Facilitates capacity building in ecosystem service modeling and valuation • Increases uptake due to lower costs and reduced difficulty of doing ecosystem service studies
Different levels of model complexity	• Tailors modeling to the level of certainty sufficient for different policy contexts (see Chapter 15)
Produces information on biophysical changes	• Enables use of biophysical ecosystem service units when they suffice for specific decision contexts • Can feed into decision-support tools such as multi-criteria analysis
Produces information on economic values	• Emphasizes connections between environmental sustainability and economic development • Provides common monetary metric, facilitating comparison of policy alternatives • Can feed into cost–benefit analysis
Scalable	• Enables selection of spatial scale most relevant to decision context and level of governance
Scenario based	• Provides structured way to consider implications of possible futures and policy alternatives • Tailors analysis to priority policy questions • Explores uncertain aspects of the future • Challenges assumptions about how systems operate
Stakeholder driven	• Encourages uptake and use of results • Enables local knowledge and policy priorities to shape the analysis
Spatially explicit	• Visual appeal for communication and advocacy • Determines where to target investments, policies, and payments • Determines locations where ecosystem services and biodiversity overlap • High-resolution maps enable policy responses to be targeted and context specific
Produces information on opportunity costs	• Determines the lower bound for payments for ecosystem services, and other incentive schemes
Possible to disaggregate to individual services	• Enables users to focus on individual services targeted by specific policies, such as payments for water yield
Enables analysis of distribution of ecosystem service costs and benefits	• Helps identify who bears benefits and costs of different policies, thereby identifying possible locations for payments for ecosystem services • Helps clarify how environmental policies affect social goals and priorities, such as poverty reduction

inclusive communication between scientific experts and stakeholders (Lubchenco 1998).

19.2 Putting ecosystem services on the agenda

People continue to degrade our environment in part because we simply do not appreciate the links between ecosystems and human well-being (Leopold 1949; Daly 1968; Wilson 1998; MA 2005c). When government ministries initiate programs or companies launch new ventures, they rarely consider how ecosystem services will affect their success, or the costs of replacing degraded services (Ranganathan *et al.* 2008).

19.2.1   Building better development policies

The Millennium Development Goals (MDGs) provide an example where the importance of ecosystem services to achieve poverty alleviation was not initially (but is now increasingly) recognized. In 2001, leaders adopted eight goals aimed at cutting global poverty by 50% by 2015. Although the MDGs originally included an environmental sustainability goal, it was narrowly defined, without consideration of the role of ecosystem services in achieving other goals (WRI 2005). Goals to increase incomes and reduce hunger did not account for how water, fertile soil and pollinators support agriculture. When countries made plans to achieve the MDGs, the links between development and nature were not addressed. Later analysis on the ecosystem service benefits of conservation found that significant environmental investment is required to achieve poverty reduction (Pearce 2005). Focusing targets on ecosystem services is now increasingly recognized as critical to achieve all eight MDGs (WRI 2005).

Some countries now explicitly consider ecosystem services in national poverty reduction strategies. Tanzania's 2005 National Strategy for Growth and Reduction of Poverty exemplifies this shift. It includes 15 environmental targets to protect and enhance ecosystem services (Assey *et al.* 2007). Distributional and social information is particularly important for policy formulation in developing countries such as Tanzania; investments that increase the supply of services and improve the welfare of certain groups of people may reduce services that support the livelihoods and well-being of others living in poverty (see Chapter 16). In Kenya, these types of trade-offs were explored by overlaying poverty maps—showing where poor people live and aspects of their well-being—with maps of ecosystems and their services. This identified where poverty indicators overlap with service supply and demand areas (Snel 2004; WRI 2007). For example, in the upper Tana region around Mt. Kenya, a large number of poor communities rely directly and exclusively on ecosystems to provide and filter drinking water. This type of information can be used to make sure that polices that impact ecosystems do not exacerbate the poverty of vulnerable communities.

Information on ecosystem services and their values can also influence decisions over major development projects. Ecosystem services that lack market prices are often not considered in project evaluations, enabling other interests to determine decisions (Balmford *et al.* 2002). Economic valuation of non-marketed ecosystem services makes their values clear in monetary terms, enabling comparison of all costs and benefits of proposed projects. This can help create political consensus around decisions that sustain ecosystem services. For example, in Borneo a rapid assessment of the economic value of ecosystem services provided by standing forests influenced conservation policy decisions over proposed oil palm plantations. The study assessed the ecosystem service benefits of carbon storage, the avoidance of health costs from forest fires, and the benefits of forest-agriculture mosaics. The information appears to have played a role in the policy decision, with the government representative declaring the oil palm development would not go forward because Borneo "is a resource of life for Kalimantan" (Naidoo *et al.* 2009).

It is possible to integrate ecosystem services into existing decision-support and environmental assessment tools (Le Quesne and McNally 2004). If mandated and enforced, these tools can ensure systematic, transparent evaluation of ecosystem service impacts. Examples include strategic environmental assessments (SEA) and regulatory impact assessments for evaluating policies and legislation, typically at the level of an economic sector or region, and cost–benefit analyses (CBA) and environmental impact assessments for individual projects. Although historically restricted to human health impacts of pollution, many CBAs and SEAs now consider a wider range of ecosystem services (DAC Network on Environment and Development Co-operation (ENVIRONET) 2008). For example, a recent SEA for a district planning process in Rwanda linked ecosystem service degradation to food, water, and fuel scarcity (UNDP 2007).

19.2.2   Building better business strategies

Businesses can use information on ecosystem services to shape their investments and strategies. Just as governments can find that ecosystem services are linked to the wellbeing of citizens, businesses can discover that ecosystem services contribute to profits. Although businesses can adversely impact ecosystems through consumption of natural resources, pollution, and land conversion, businesses also depend on ecosystems. Agribusiness, for example, depends on pollination, and control of pests and erosion (see Chapter 10). Approximately 75% of the world's 100 top agricultural crops rely on natural pollinators (Klein *et al.* 2007). Property developers benefit from the coastal protection provided by coral reefs, coastal forests and coastal wetlands (Turner *et al.* 1998; Cesar 2000). The tourism industry benefits from these ecosystems' aesthetic beauty (De Groot 1994).

Because of these impacts and dependencies, the degradation of ecosystem services presents significant risk to—and opportunities for—corporate performance (see Table 19.2 and Box 19.1). For example, industries relying on steady supplies of clean freshwater face operational risks when upstream deforestation increases sedimentation of rivers, disrupting business operations and increasing costs (see Chapter 6). Companies may face permit restrictions or damage brand image if they

pollute ecosystems with value to communities, leading to lawsuits and fines, and challenging their license to operate. Conversely, businesses that sustainably manage land or water resources may increase efficiency, differentiate their brand, reduce costs, and even generate new sources of revenue through markets for ecosystem services. Although these emerging markets face challenges in practice, they offer rewards to early business entrants (e.g., markets for carbon offsets or certified timber—see Section 19.3).

Most companies, however, fail to connect ecosystems with their business bottom line (Hanson *et al.* 2008). Many are not aware of the extent to which they impact or depend on ecosystem services, nor the ramifications of those impacts. Likewise, business tools such as environmental management systems and environmental impact assessments are often not attuned to the risks and opportunities arising from use and degradation of ecosystem services. Rather, they are suited to "traditional" issues of pollution and natural resource consumption. As a result, companies may be unprepared for ecosystem service risks or miss new sources of revenue associated with ecosystem change.

Table 19.2 Business risks and opportunities arising from dependencies and impacts on ecosystem services

Type	Risk	Opportunity
Operational	• Increased scarcity or cost of inputs • Reduced output or productivity • Disruption to business operations	• Increased efficiency • Low impact industrial processes
Regulatory and legal	• Extraction moratoria • Lower quotas • Fines • User fees • Permit or license suspension • Permit denial • Lawsuits	• Formal license to expand operations • New products to meet new regulations • Opportunity to shape government policy
Reputational	• Damage to brand or image • Challenge to social "license to operate"	• Improved or differentiated brand
Market and product	• Changes in customer preferences (public sector, private sector)	• New products or services • Markets for certified products • Markets for ecosystem services • New revenue streams from company-owned or managed ecosystems
Financing	• Higher cost of capital • More rigorous lending requirements	• Increased investment by progressive lenders and socially responsible investment funds

Box 19.1 An assessment of ecosystem services helps a paper and packaging business respond to emerging risks

Craig Hanson

One approach for businesses to identify connections between ecosystem services and their bottom line is to conduct a Corporate Ecosystem Services Review (ESR) (Hanson *et al.* 2008). The ESR is a structured methodology that helps managers proactively develop strategies to manage business risks and opportunities arising from their company's dependence and impact on ecosystems. Conducting an ESR involves identifying priority ecosystem services for a company (a facility, product or supply chain), analyzing trends in these services, and identifying business risks and opportunities. With this information, the company can develop response strategies.

Mondi is a leading international paper and packaging business. As of 2006, the company was Europe's largest producer of office paper, with operations in 35 countries. Much of the company's pulp comes from its plantations in South Africa. Mondi conducted an ESR to understand what business risks—and opportunities—might arise as changes in ecosystem services affect its plantations.

During its ESR, Mondi assessed the dependence and impact of these plantations on 24 different ecosystem services. The analysis identified six services as having the most impact on Mondi's corporate performance:

- *Freshwater.* Pine and eucalypt plantations significantly depend upon and affect the quantity of freshwater in their watersheds.
- *Water regulation.* Plantations rely on and impact the surrounding ecosystems that regulate the timing of water flows.
- *Biomass fuel.* As a byproduct, plantations generate biomass that can be used for energy by the company's mills and local villages.
- *Global climate regulation.* Plantations can sequester carbon dioxide thereby mitigating climate change (albeit dependent on the stage of the carbon cycle—see Chapter 7).
- *Recreation and ecotourism.* One of the plantations is located next to the isiMangaliso Wetland Park—a World Heritage Site—and contains wetlands and grasslands that could potentially provide new opportunities for recreation and ecotourism.
- *Livestock.* The plantations preclude surrounding villagers from using the land for large-scale livestock

grazing. Selective controlled grazing is, however, widely practiced.

A trends analysis of these six ecosystem services uncovered a number of emerging risks and opportunities facing Mondi. Freshwater in the three plantation watersheds is becoming increasingly scarce due to the proliferation of invasive alien plant species, increasing demand for water from nearby farmers, and climate change. This scarcity threatens to increase the cost of water, reduce the availability of wood fiber, and expose the company to reputational and regulatory risk. Meanwhile, new opportunities are arising as ecotourism grows in the region and new markets emerge for biomass fuel.

Through the ESR process, Mondi identified several strategies for managing these risks and opportunities:

- *Improve water-use efficiency.* To improve water efficiency, the company is now clearing invasive species more aggressively. It will also better match tree species to site conditions and more frequently burn grasslands. These strategies complement the company's past efforts to remove plantations from wetlands, thereby restoring natural hydrological systems.
- *Use invasive species as biomass fuel.* Mondi can tap into the growing market for biomass fuel by using the invasive species cleared from its plantations as feedstock for power and heat generation. Potential users of the feedstock are Mondi's own paper mills and a new biomass pellet manufacturer located close to one plantation.
- *Promote coppiced small-scale tree farms (woodlots) for biomass fuel.* Using the company's forestry expertise, Mondi can help local landowners and villages establish woodlots on degraded land for growing biomass fuel on coppiced rotations. Mondi could provide seedlings and offer extension services. Mondi could also purchase the wood to use in its mills or sell to nearby wood pellet manufacturers. These woodlots would provide additional income for villagers and thereby strengthen Mondi's reputation and stakeholder relationships.
- *Engage policy-makers to improve freshwater policies.* Mondi can support policies that encourage water-use efficiency in South Africa and, leveraging its expertise in water management, provide input into policy design.

Information on the value of ecosystem services can play a role in raising business awareness of the links between ecosystems and profits. Spatially explicit information on where ecosystem services are supplied can help businesses determine where to invest in ecosystem protection and restoration or where to avoid activity that degrades ecosystems.

19.2.3 Building public awareness

Because people often fail to connect their wellbeing to ecosystem conditions (Irwin and Ranganathan 2007), there is a need to build public awareness about this linkage before we can expect the public to hold decision-makers accountable (Pielke 2007). Education and awareness programs may be especially effective if they are directed at landowners or managers who lack information. For instance, if a farmer does not realize that an increase in habitat for native pollinators will raise yields, simply providing the type of information provided by the models outlined in Chapter 10 may induce landowners to set aside habitat. An example comes from a US Federal "Conservation Buffers" program, where training and advice is given to farmers on designing buffer zones to filter pollution along streams and wind barriers to reduce soil erosion (NRCS 2008). Some consider technical training to be more successful at altering land-use practices than direct payments (Daily and Ellison 2002; Salzman 2005).

When "externalities" exist—impacts that affect people other than the decision-maker—information alone will not suffice. For example, a farmer may learn that applying fertilizer causes water quality problems downstream but continue to do so because it leads to higher crop yields and the resulting water quality problem affects others. In these cases, education and awareness programs need to target the broader public to create demand for reform.

19.3 Instruments for sustaining and enhancing ecosystem services

When ecosystem service management becomes an institutional priority, several policy instruments can be used to influence the way people interact with the environment (MA 2005d). Some ecosystem services are private goods (e.g., agricultural production and timber), and as such, are easily traded in markets that regulate their provision. Other important services (e.g., pollination, climate regulation, and cultural values) are "public goods," which means private individuals cannot reap profits from providing them. Governments can develop regulations or incentives to alter the provision of these services. Here, we discuss a range of policy instruments for affecting the provision of both public and private goods.

19.3.1 Regulation

Government regulations have only recently begun to consider the full range of services provided by ecosystems, investing in them to meet human needs efficiently and effectively (Ruhl et al. 2007). Several kinds of regulations have recently been altered to include ecosystem services: licensing and permitting, zoning and land-use planning, and environmental standards.

Currently, most governments grant licenses and permits for extractive activities (e.g. mining, oil and gas) and infrastructure development (e.g. roads, developments) based on projected impacts to biodiversity and a subset of environmental conditions linked to human health. However, standard assessments usually fail to account for the full set of social impacts associated with proposed activities. Governments are slowly moving toward permitting and licensing procedures that more fully account for environmental and social damages. Models such as those discussed in this book can be informative for these new, expanded impact assessments. For example, the Colombian Ministry of Environment, Housing and Territorial Development is expanding the impacts considered in its licensing process for activities related to oil and gas, mining, and infrastructure to avoid, minimize, mitigate and compensate for biodiversity and ecosystem service damages. The Ministry plans to use information on where services are supplied and realized to identify risks from industrial activities, and to shape permit conditions and offset compensation programs.

Regulations of common pool resources, such as protected areas in marine, coastal and inland water ecosystems, are widespread. However, they are usually designed to maintain a single ecosystem service, such as fisheries or freshwater. Regulatory limits can be designed to supply a broader range of services such as water purification, waste treatment and recreation. For example, in the United Kingdom, the Department of Environment, Food and Rural Affairs (DEFRA) estimated the economic value of a range of ecosystem services affected by marine protected areas to inform the design of new regulations—Marine Conservation Zones—that are part of a national Marine Bill (McVittie and Moran 2008; Moran *et al.* 2007). The impact assessment supporting the Marine Bill legislation demonstrated that the benefits of Marine Conservation Zones outweigh the costs to industry. If the Marine Bill is passed, information on the value and location of ecosystem services and biodiversity is intended to inform selection of protected areas that provide multiple benefits matching policy priorities. A similar shift is happening in the United States, where a new ocean policy will likely require an ecosystem-based approach to marine zoning that explicitly considers a wide range of ecosystem services (Lubchenco and Sutley 2010).

Governments can also use information on the provision or value of ecosystem services to ban or set standards for activities that degrade the environment beyond a socially defined limit. In the Republic of the Marshall Islands, information on the value of coastal ecosystem services—protection from erosion, along with tourism, recreation and fisheries benefits—was used to advocate banning reef blasting and near-shore dredging (McKenzie *et al.* 2006). In addition, liability rules can hold those degrading ecosystem services responsible for the damage they cause, and enforce compensation for negligent actions, such as pollution (Thompson 2008b). In these cases, information on the economic value of degraded services can lead to more accurate estimates of environmental damages, and help build consensus about compensation among conflicting stakeholders (Van Beukering *et al.* 2007). For example, after the Exxon Valdez oil spill in Alaska in 1989, a study estimated the lower bound economic value of damages at $2.8 billion, based on the stated willingness to pay to prevent another similar spill (Carson *et al.* 1992).

19.3.2 Market-based approaches

To support regulation, governments may construct market-based instruments that create incentives for people to account for ecosystem services. The creation of property rights, typically enforced and regulated by governments, is common to many of these approaches, facilitating trade in activities that increase or maintain services. In theory, market-based approaches are cost effective as services can be provided at the lowest possible cost (Montgomery 1972). This is particularly the case when producers have flexibility—they can provide a service in differing ways and at differing costs (Revesz and Stavins 2004). In practice, efficiency advantages can be outweighed by the costs of operating these—often complex—mechanisms. Here we discuss several forms of market-based approaches that have been used to influence ecosystem service provision and detail how quantitative analyses can improve or facilitate the development of these policy instruments.

19.3.2.1 Cap and trade programs
Under "cap and trade" programs, the government sets a limit on an environmental externality, such as pollution. Permits to emit are issued to firms, who can then trade them. Cap and trade programs have reduced emissions at lower costs than regulations, a notable example being the market for SO_2 permits in the United States. Carbon markets are now emerging as a mechanism to achieve greenhouse gas emissions reductions. In this context, the scope of cap and trade programs may expand to encompass ecosystem services. For example, payments for Reduced Emissions from Deforestation and Forest Degradation (REDD) allow beneficiaries of the climate regulation service provided by forests to compensate those who conserve them (UNFCCC 2008). In this case, maps of carbon storage and projected estimates of future carbon change can be used to identify areas where carbon investments in natural or restored systems would be most profitable.

19.3.2.2 Voluntary markets
In addition to regulated markets, voluntary markets are emerging, such as the Chicago Climate Exchange for carbon. Given that most ecosystem services are

public goods or plagued by externalities, voluntary markets are only one part of the solution. In voluntary carbon markets, individuals or companies buy certified reductions in carbon emissions for moral reasons or to improve their reputation. As mentioned above, certification schemes exist to meet demand for "charismatic" carbon activities that deliver additional benefits (CCBA 2008b). For example, the Climate, Community and Biodiversity Alliance recently awarded the Juma Reserve project in Brazil a Gold rating, verifying additional benefits through income generation and the promotion of local business and biodiversity (CCBA 2008a). Integrated models of multiple ecosystem services can estimate how activities to reduce greenhouse gas emissions will change a larger suite of ecosystem services. Such estimates might be used to grant certification to projects whose projected impacts benefit multiple services. Follow up monitoring will be essential in determining whether these projected benefits are realized, but modeling can help minimize negative impacts of early carbon investments.

The Voluntary Carbon Standard (VCS—a widely accepted certification scheme for carbon offsets) is adding the need for a different kind of analysis. Under VCS, projects must demonstrate that their emissions reductions are additional to what would have happened anyway. To pass, a project must be neither legally mandated nor common practice, and face unique barriers to implementation that carbon finance could help overcome (VCS 2008). Modeling the consequences of different scenarios can help estimate additional ecosystem service benefits, by comparing a "baseline" scenario (what is assumed to happen without the project), with a "project" scenario. The difference in service delivery helps identify the truly additional quantity, or that eligible for compensation. When framed by baseline scenarios, the models described in the rest of this book can make this type of calculation, estimating both additional carbon sequestration benefits (see Chapter 7) and the additional benefits of other services.

19.3.2.3 *Payments for ecosystem services (PES)*
Payments for ecosystem services (PES) represent an increasingly popular method for creating incentives to sustain services. PES involve contracts in which one party agrees to compensate another party in

exchange for a guarantee that a specified amount of an ecosystem service, or a particular land use or practice believed to provide that service, will be delivered (Wunder 2005). A famous example of a PES program involves payments by Nestle (the owner of the Vittel mineral water company) to farmers in north eastern France to safeguard the water supply. Upstream farmers had replaced natural grasslands—which filtered and cleaned the water—with corn and cattle, resulting in nitrate contamination of the aquifer. After calculating the higher costs of water purification plants, Nestle decided to use a PES approach. The company signed long-term contracts with farmers to manage their animal waste, graze cattle sustainably, and reforest water filtration areas (Perrot-Maitre 2006).

The majority of PES schemes are government-funded. For example, in the well-known Catskills example, the city of New York decided to protect the Delaware and Catskills watersheds, which provide 90% of its drinking water (Heal 2000, NRC 2004) . Regulations were important for driving the scheme; the US Safe Drinking Water Act required that water must be filtered unless watersheds are sufficiently protected. The costs of a new filtration plant vastly outweighed the costs of restoring and protecting the watersheds (Postel and Thompson 2005). To protect the watersheds, a set of policy measures were introduced alongside standard payments, including land purchase and easements. Within the first five years of the policy initiative, the city purchased approximately 14 000 ha of land and another 1000 ha of conservation easements, thereby doubling the buffer zones around important reservoirs. The scheme had to be carefully negotiated with diverse landowners; in this case there were 477 property owners (Postel and Thompson 2005). Those providing water quality services beyond the regulated minimum are rewarded with a payment package, including additional income, but also in-kind benefits, such as farm management consultations (Appleton 2002).

Beyond the New York City watershed case, most existing water-related PES programs have been established without information on where investments will provide the greatest ecosystem service returns, and payments are made based on implementation of activities, not on actual changes in ecosystem service levels. This approach can lead to inefficient programs that fail to produce desired ecosystem service returns because the marginal changes in ecosystem services associated with a given management change are not constant (Jack et al. 2008). For example, a kilometer of fencing installed along a riverbank where cattle graze will usually reduce erosion much more than a kilometer of fencing installed around croplands in the upper reaches of the same watershed. Although policy creation in such contexts is difficult, it can be dramatically simplified with the use of models that reveal how management changes in different regions affect desired ecosystem services. More complex, but efficient payment designs (e.g. differential taxes, trading zones) can be established with such information. In addition, models such as those described in this book can help locate areas for new PES schemes by identifying where there are sources of service supply and beneficiaries who may be willing to pay for service delivery. This is particularly powerful when combined with information on land tenure, to determine who can influence service delivery and should receive payments. For further discussion on additional conditions under which PES can be most effective, see Jack et al. (2008).

19.3.2.4   Government fiscal incentives

Governments can create further financial incentives to supply ecosystem services using subsidies, subsidy reform and taxes. Subsidies differ from PES schemes in that payments are not conditional on service delivery, or some proxy for it. Subsidies that account for the full economic value of ecosystem services can help to maintain or increase their provision. Many existing subsidies, however, actually encourage activities that degrade services, by focusing on a narrowly defined outcome and ignoring ecosystem service losses (Myers 2001). Significant cases exist from agricultural subsidies. Farmers in Europe and the United States have long received payments linked to crop productivity (OECD 1999). Such payments promote land conversion, leading to erosion of topsoil, and increasing use of fertilizers and pesticides that affect water quality. Recent reforms have attempted to replace agricultural subsidies with payment programs (often called "agri-environment" schemes), which reward farm-

ers for managing the land sustainably and providing ecosystem services, such as regulating nutrient runoff and soil erosion, and maintaining the spiritual and symbolic values of agricultural landscapes (Baylis et al. 2008, Ilbery and Bowler 1998).

Although growing in scale, agri-environment programs nevertheless remain a small proportion of agricultural expenditure (OECD 2003). Reducing agricultural subsidies has proved particularly difficult in Europe and the United States due to perceived impacts on farmers' livelihoods and lifestyles (Myers 2001). As subsidies are often maintained by rent-seeking by special interest groups who lose out from subsidy reform, reform policies are more popular if done gradually and combined with public awareness campaigns (Pearce and Finck von Finckenstein 1999). Such campaigns can draw on valuation studies that quantify the effects of subsidies on ecosystem services to build political support, and create transparency and accountability about who wins and who loses from policy change.

19.3.2.5   Offsets

Offset schemes attempt to ensure that degradation of ecosystem services in one location is compensated through activities elsewhere. For example, wetland mitigation banking is a consolidated offset scheme in the United States to encourage compliance with the Clean Water Act. This regulation requests that developers avoid dredging wetlands connected to other water bodies and offset any unavoidable damages through creation of wetlands elsewhere. Landowners that restore, establish, enhance, and (in exceptional circumstances) preserve wetlands gain permits that can be sold to developers filling wetlands. However, mitigation banking has been controversial because of the scheme's potential to redistribute wetland ecosystem services between human populations and the mixed quality of offsets relative to the wetlands destroyed (Ruhl et al. 2007). Offsets work best when there is a well-defined, scalable unit that can be exchanged, whether it is an acre of wetland or a pound of CO_2, and when exchanges are fungible—an acre of wetland in one location produces the same services as an acre elsewhere (Salzman and Ruhl 2000). Since an acre of wetland seldom provides the same service in different watershed contexts, standard, reproducible, and relatively easy ways of measuring ecosystem services are essential for establishing trades that actually avoid the loss of ecosystem services. Combining biophysical models with information on beneficiaries can determine how offsets affect the distribution of services among stakeholder groups, allowing agencies to preempt equity issues associated with developments and their offsets.

19.4   Choosing the right instrument

Decision-makers need to select the policy mechanism most appropriate to their local context if they are to succeed in delivering ecosystem services. How to choose? Some approaches are appropriate to certain institutions—governments are the only entity with authority to levy taxes, for example— and some will be constrained by the nature of the service (its "production function"). Others require strong systems of governance and institutions that may not exist in developing countries, such as clear property rights and competitive markets (Pearce 2005). To choose the best approach, information is required on a range of issues, many of which are summarized in Table 19.3. For further details within the conservation context, see McKenney *et al.* (2008) who describe a set of enabling conditions affecting the success of ecosystem service projects.

A summary of the main conditions affecting successful use of each policy mechanism is given in Table 19.4.

Table 19.3   Information needed to select a policy instrument to provide ecosystem services

The service	• How do interventions alter the level of service being produced?
	• Can these changes be measured?
	• What (if any) other policies and interventions are compatible with providing the service?
	• Is a particular spatial pattern of land uses required?
	• Can the service be broken down into discrete, fungible units?
	• At what scale is the service produced?
	• Are other services co-produced with the desired one?
	• Is there uncertainty regarding the link between actions and service delivery?
	• Are there unpredictable shocks that could alter service levels?
Producers	• If other land uses are compatible with service provision, are behaviors that increase or decrease service levels observable and enforceable?
	• Are there any relevant information asymmetries between producers, the decision-maker, and beneficiaries?
Beneficiaries	• Who benefits from the ecosystem service? Where are they located?
	• Will this change over time?
	• Can others be prevented from benefiting from the service (in the language of economics, is the service a public good)?
Costs and funding	• Who is bearing the costs? Where are they located?
	• What are the costs of various approaches for providing the service? (This should include direct costs, indirect costs such as reductions in the supply of other ecosystem services, and transaction costs.)
	• How much are people willing to pay? How much funding is available (and over what time frame) and where is it coming from?
Goals	• What are the goals of the policy? Goals could be improving livelihoods, biodiversity conservation, or simply cost-effectiveness.
Institutional context	• What institutions are needed for an approach and are they present (for example, contracts enforced by a court of law)?
	• Are there clear property rights?
	• What other laws are in existence that might affect implementation of the approach? Do these conflict with or enhance any of the approaches to providing the service?
Views of stakeholders	• Who wins and who loses with each approach?
	• Which stakeholder groups could prevent implementation?
	• Whose support is needed and what are their views?
Performance over time	• How will each approach respond to changing conditions over time, such as price changes, changes in technology, changes in funding levels, or changes in drivers of service degradation?

Table 19.4 Enabling conditions for policy instruments

	Cap and trade schemes, voluntary markets, offsets	Public and private payments	Taxes, subsidies and liability rules	Regulations	Information provision
Ecosystem service— biophysical attributes	• Well-defined, fungible units	• Alternative land uses compatible with supplying service	• Changes in service levels can be attributed to individual behavior or specific actions (liability)	• No flexibility in how to supply service • Specific spatial arrangements of land uses required • Ecosystem service at or near tipping point	• Alternative land uses compatible with supplying service
Producers of ecosystem service	• Heterogeneity in supply costs among providers • Flexibility among providers on how to increase services • Numerous point sources	• Small number of large producers • Land-use changes would not happen without payment • Opportunity costs of land-use change not prohibitively high	• Small number of producers (liability) • Action clearly correlated with supply or degradation of service that is amenable to a fiscal incentive (tax or subsidy)	• Producers already subject to other forms of regulation • Large point sources	• Producers unsure how to increase service levels • Producers unable to capture demand for service due to information asymmetry
Beneficiaries	• Willing and able buyers	• Beneficiaries can organize to make sufficient payments	• No collective action problems restricting use of liability rules	• A certain level of service must be guaranteed	• Beneficiaries would pay for service if they knew it was being provided
Costs and benefits	• Cost of trading is low	• Transaction costs not prohibitive • Funding available to sustain program over time	• Revenue needed (taxes) • Insufficient funding for payments (liability/taxes)	• Minimal transaction costs to regulator • Landowners face similar costs for achieving service level	• Costs of providing information of sufficient quality not prohibitive

Quantified, mapped information on ecosystem services and their values can answer a number of these questions, particularly those relating to services and their biophysical attributes—the first rows in Tables 19.3 and 19.4. Ecosystem service maps can identify where services are supplied. Distributional and land tenure information can identify who bears the costs and who benefits from service delivery. This underlines the fact that ecosystem service modeling can help select policy instruments appropriate to local contexts. But the tables also highlight that biophysical and economic models will usually need to be complemented by additional information, looking at the institutional context, property rights, and other social and economic conditions.

19.5 Building stronger organizations

Effective actions to provide ecosystem services require strong and legitimate organizations and processes for making decisions (Pearce and Seccombe-Hett 2000). The Millennium Ecosystem Sub-global Assessments found that "effective" responses to ecosystem degradation tended to involve collaboration across governance levels. Organizations must be equipped to implement coordinated, cross-sectoral policies that consider multiple services and trade-offs. There are three key prerequisites that need to be met for organizations to become effective implementers of ecosystem service policy.

First, ecosystem services must move from the periphery to the center of environmental and natural resource management organizations' agendas. Currently, natural resource ministries typically focus on provisioning services, such as fish, timber and water, and neglect regulating services. Fortuantely, there are indications that governments are giving regulating services—and ecosystem services in general—a more central role. For example, the US Forest Service has recently reformulated its mission to focus on conserving ecosystem services. In the United Kingdom, DEFRA has adopted an action plan to take a more holistic approach to policy-making to maintain ecosystem services and ensure their values are reflected in decisions (DEFRA 2007).

Second, organizations must establish mechanisms to distribute benefits and costs of ecosystem service change equitably, both at the local level and between local, regional and global levels (see Chapter 16). Ecosystem degradation is a significant factor—sometimes the principal one—exacerbating poverty (MA 2005b) because the rural poor depend on ecosystems for their livelihoods (Paavola 2008). Yet local communities have often been left out of environmental management decisions, typically dominated by national or international interests. One positive example of a step in the right direction entails Namibia, which in 1996, adopted the Nature Conservation Act giving rural communities rights to form conservancies in communal lands to build tourism and wildlife industries. By 2007, 50 conservancies benefited over 230 000 rural people and covered 14% of the country's area. Improved management has increased wildlife populations and reduced overgrazing (WRI 2008). In this case, decentralizing authority to representative local institutions and establishing clear property rights strengthened environmental management. Equity is not only about the distribution of costs and benefits: participation is also important (Paavola 2007). If boundaries of participation are set locally, a forest assessment might focus on timber and non-tim-

---

Box 19.2 Cultural evolution as an enabling condition for the use of ecosystem services in decisions

Paul R. Ehrlich, Lee D. Ross, and Gretchen C. Daily

The idea of "ecosystem services" originated through cultural evolution, and this idea holds the secret to their continuation. Most human evolution *does not* involve changes in our DNA, but rather alterations in the gigantic library of non-genetic information—the culture—of *Homo sapiens*. Culture comprises such evolving entities as languages, beliefs, scientific theories, and systems of justice and oppression, in addition to the information embodied in books, songs, computer disks, and artifacts ranging from potsherds to jumbo jets. The decisions we make and the policies we devise to manage our ecosystem services are shaped by our culture. Improving those decisions and policies thus depends on cultural evolution.

The idea of ecosystem services entered the modern mainstream of cultural evolution about three decades ago,

as it became clear that the norm of "putting aside a nature reserve to save a charismatic species" was failing to stem the loss of biodiversity. The "ecosystem service norm" now spreading makes plain the dependence of human wellbeing on investment in natural capital (e.g., Daily and Matson 2008; Goldman *et al*. 2008; Kareiva and Marvier 2007; MA, 2003). Policy efforts to reward investments in natural capital must tap into, and reinforce, this emerging norm.

Numerous failed civilizations attest to the difficulty of directing cultural evolution toward environmentally sustainable practices (Diamond 2005; Tainter 1988). On Easter Island, early inhabitants surely knew that chopping down the entire palm forest would not be a good thing. They depended heavily on porpoise meat, hunted from palm wood canoes. Nonetheless, they cut down every last tree. They then wiped out coastal food resources, resorted in desperation to eating rats, and finally turned on each other. Possession of "scientific" understanding—the importance and declining availability of palms would have been evident to all—did not prevent disaster. Other islands followed equally grim paths. But on others, truly sustainable economies emerged. What accounted for the difference? It appears that size matters. Tikopia, a model of success, is only about 1.7 mi^2. Kirch (1997) proposes that where everyone knew everyone else, ecological limits to human activities were more likely to be accepted, and major "policy changes" (like giving up pork) and new institutions (regulating fertility) adopted. Conversely, the Easter-scale (64 m^2) islands were prone to dividing into "them" and "us" in a race to the bottom.

Two island lessons are particularly relevant to today. First, like the islanders, we know enough scientifically to recognize trouble and start moving in the right direction. Second, human beings evolved as a small-group animal. Our future prospects depend on whether, in a population of 6 billion-plus, we can design and implement cooperative solutions.

Social psychology is central to understanding cultural evolution, and has important, often counter intuitive lessons for fostering change. Consider a popular policy option: payments for ecosystem service (PES). Three key issues arise: *what* should people pay for; *how much* can and should they pay; and what *time frame* for payment is best?

On the question of *what* people should pay for, social psychologists stress that overselling, as a way of achieving greater public "buy-in" for the measures being advocated, should be avoided as it may be both counterproductive in the long run and unnecessary. School lunch programs in the United States, for example, were promoted as a means to achieve a range of social goals (kids with full bellies, it was claimed, could learn better, leave school with better job prospects, and be less likely to turn to crime). American taxpayers did not see these effects, and resistance to the program increased. The irony is that Americans would probably have supported lunch programs for disadvantaged students on the basis of decency norms that no child should go hungry in school.

The ecosystem service outcomes that are promised, whenever feasible, should reinforce existing norms and show results in a satisfying time frame. In cases where we know a lot scientifically, as with carbon sequestration, we might do well to promise outcomes that are specific, attainable within a generation (or two), and congruent with shared values (for example preserving forests). Where we know less or have less local/regional control, as with provision of water quantity, it may prove detrimental to make specific promises. Our scientific understanding is sure to evolve greatly; policy instruments, and the institutions that shape them, need to evolve with this.

On the question of *how much* to pay, relying on financial incentives or disincentives can be tricky. Payments for ecosystem services may actually reduce their supply if they remove cultural norm violation as a factor influencing behavior. Yet carefully tuned payments are important for establishing new norms, especially where suppliers and beneficiaries of services are widely separated.

On the question of *how long* payments should be made, evidence from social psychology argues for shorter-term (one generation, say) payments to achieve greater support. Once a program is adopted and becomes the status quo, people are unlikely to violate group norms by not "doing their share." They are unlikely to exert political pressure to "opt out" (even if they had not initially favored "opting in").

Underlying these tactical questions are deeper ones. What combination of services should be targeted in policy, given that there are trade-offs among services? Who decides? The answers may emerge quite differently in different places. Yet, our culture is evolving rapidly in the way people think about the environment, and we have a window of opportunity to foster this change productively around the world. This will require cooperative efforts to develop and deploy innovative policy approaches worldwide. A *Millennium Assessment of Human Behavior* may be the best shot at achieving the "small group" communication necessary to do this (Ehrlich and Kennedy 2005). With luck, it might keep us from eating each other.

ber forest products. If set more broadly, water regulation and filtration, and climate regulation may receive greater emphasis (Reid *et al*. 2006). If local voices are excluded, important ecosystem services may continue to be degraded. This emphasizes the need for participatory stakeholder processes for framing ecosystem service modeling assessments.

Third, there is a need to coordinate across sectors, levels and timeframes. Organizations, such as a national forestry agency or a city council, follow political boundaries and jurisdictions, rather than the ecosystem service boundaries. Yet effective management requires partnerships and networks across *ecosystem* boundaries. "Bridging organizations" can lower the costs of collaboration and conflict resolution (Folke 2005). An increasingly popular approach, co-management, involves government agencies working with local stakeholders and organizations. A national agency's authority to adopt regulations, provide scientific data, and obtain funding and political support complements local organizations' on the ground understanding of the natural resource and management capability (Irwin and Ranganathan 2007; WRI 2005). For example, locally managed marine areas in Fiji and elsewhere in the Pacific use traditional practices to set aside portions of fishing grounds as restricted areas, allowing fisheries to recover. Local communities lead in designating restricted areas, monitoring, and enforcement. The national government used its regulatory authority to introduce a 12-mile limit on fishing by foreign trawlers (WRI 2008).

19.6 Future directions

Our scientific understanding of ecosystem services is growing increasingly sophisticated—both in the natural and social sciences. However, to improve outcomes on the ground, information must be salient to the world's problems, and channeled into behavior change. We have seen that information on ecosystem services is already making a difference. Governments, businesses and the public increasingly realize the significance of ecosystem services for achieving their ends. A variety of policy instruments and business practices are being implemented more effectively to maintain important services. And organizations are evolving to manage ecosystem services effectively and equitably, using approaches such as adaptive co-management. Models with the characteristics described in this book can be powerful forces for change. They can alter perceptions about the importance of the environment for people. They can also answer many of the questions needed to select, design and implement different policy instruments.

While progress is evident, it is difficult to change the way people perceive and value ecosystems; one may have to fight fierce political battles, build new laws, and wrestle with powerful vested interests. As a first step, there is much to learn from the existing understanding of best practice. Objective evaluations are required in two areas: first, to evaluate how ecosystem service interventions perform over time and in different contexts, and thereby learn how to be most effective (Berkes *et al.* 2003; Ostrom 2005; Carpenter *et al.* 2009); second, to assess continually the information, and level of certainty of that information, required for different interventions, so that scientific models and tools can be refined in light of user needs.

Our constant challenge is to ensure that science and practice are effectively integrated, by working across disciplines and political boundaries, and sharing ideas and experiences. Ecosystem service science needs to be grounded in sound theory but, to be most effective, it must always keep the final application—the "practice"—firmly in mind.

References

Appleton, A. F. (2002). How New York City used an ecosystem services strategy carried out through an urban-rural partnership to preserve the pristine quality of its drinking water and save billions of dollars. A paper for Forest Trends. Tokyo.

Ascher, W. and Healy, R. (1990). The policy process and complexity in natural resource policy-making. *Natural resource policy-making in developing countries: Environment, economic growth, and income distribution.* Duke University Press, Durham.

Assey, P., Bass, S., Cheche, B., *et al.* (2007). Environment at the heart of Tanzania's development: lessons from Tanzania's National Strategy for Growth and Reduction of Poverty—MKUKUTA *National Resource Issues Series 6.* International Institute for Environment and Development, London.

Balmford, A., Bruner, A., Cooper, P., *et al.* (2002). Economic reasons for conserving wild nature. *Science,* **297,** 950–3.

Baylis, K., Peplow, S., Rausser, G., *et al.* (2008). Agri-environmental policies in the EU and United States: A comparison. *Ecological Economics,* **65,** 753–64.

Berkes, F., Colding, J., and Folke, C. (2003). *Navigating social-ecological systems: Building resilience for complexity and change.* Cambridge University Press, Cambridge.

Bingham, G., Bishop, R., Brody, M., *et al.* (1995). Issues in ecosystem valuation: Improving information for decision making. *Ecological Economics,* **14,** 73–90.

Carpenter, S. R., Mooney, H. A., Agard, J., *et al.* (2009). Science for managing ecosystem services: Beyond the Millennium Ecosystem Assessment. *Proceedings of National Academy of Sciences,* **106,** 1305–12.

Carson, R. T., Mitchell, R. C., Hanemann, W. M., *et al.* (1992). A contingent valuation study of lost passive use values resulting from the Exxon Valdez oil spill. Attorney General of the State of Alaska.

Climate Community and Biodiversity Alliance (CCBA). (2008a). *The Juma Sustainable Development Reserve Project: reducing greenhouse gas emissions from deforestation in the States of Amazonas, Brazil. CCB—Validation Report.* TUV SUD Industrie Service GmbH.

Climate Community and Biodiversity Alliance (CCBA). (2008b). *Climate, Community and Biodiversity Project Design Standards,* second edi. CCBA, Arlington VA.

Cesar, H. S. J. (2000). *Collected essays on the economics of coral reefs,* CORDIO, Kalmar University, Sweden.

DAC Network on Environment and Development Co-operation (ENVIRONET). (2008). *Strategic environmental assessment and ecosystem services.* OECD.

Daily, G. C., and Ellison, K. (2002). *The new economy of nature: The quest to make conservation profitable.* Island Press, Washington, DC.

Daily, G. C., and Matson, P. A. (2008). Ecosystem services: from theory to implementation. *Proceedings of National Academy of Sciences,* **105,** 2455–6.

Daily, G. C., Polasky, S., Goldstein, J., *et al.* (2009). Ecosystem services in decision-making: Time to deliver. *Frontiers in Ecology and the Environment,* **7,** 21–8.

Daly, H. E. (1968). On economics as a life science. *Journal of Political Economy,* **76,** 392–406.

De Groot, R. (1994). Environmental functions and the economic value of natural systems. In: A. M. Jansson, M. Hammer, C. Folke, and R. Costanza, Eds., *Investing in natural capital: the ecological economics approach to sustainability.* Island Press, Washington, DC.

DEFRA. (2007). *Securing a healthy natural environment: An Action Plan for embedding an ecosystems approach.* Department for Environment Food and Rural Affairs, London.

Diamond, J. (2005). *Collapse: how societies choose to fail or succeed.* Viking, New York

Ehrlich, P. R., and Kennedy, D. (2005). Millennium assessment of human behavior. *Science,* **309,** 562–3.

Folke, C. (2005). Adaptive governance of social-ecological systems. *Annual Review of Environment and Resources,* **30,** 441–73.

Ghazoul, J. (2007). Recognising the complexities of ecosystem management and the ecosystem service concept. *GAIA—Ecological Perspectives for Science and Society,* **16,** 215–21.

Goldman, R. L., Tallis, H., Kareiva, P., *et al.* (2008). Field evidence that ecosystem service projects support biodiversity and diversify options. *Proceedings of National Academy of Sciences,* **105,** 9445–8.

Hanson, C., Ranganathan, J., Iceland, C., *et al.* (2008). *The Corporate Ecosystem Services Review: guidelines for identifying business risks and opportunties arising from ecosystem change.* World Resources Institute, Washington, DC.

Heal, G. (2000). *Nature and the market-place: capturing the value of ecosystem services,* Island Press, Washington, DC.

Henrichs, T., Zurek, M., Eickhout, B., *et al.* (2008). Scenario development and analysis for forward-looking ecosystem assessments. *Ecosystems and human well-being: A manual for assessment practitioners.* Draft for external review.

Ilbery, B. and Bowler, I. (1998). From agricultural productivism to post-productivism. In: B. Ilbery, Ed., *The geography of rural change.* Prentice Hall, London.

Irwin, F., and Ranganathan, J. (2007). *Restoring nature's capital,* World Resources Institute, Washington, DC.

Jack, B., Kousky, C. and Sims, K. E. (2008). Designing payments for ecosystem services: Lessons from previous experience with incentive-based mechanisms. *Proceedings of National Academy of Sciences,* **105,** 9465–70.

Kareiva, P., and Marvier, M. (2007). Conservation for the people. *Scientific American,* **297,** 50–7.

Kiker, G. A., Bridges, T. S., Varghese, A., *et al.* (2005). Application of multicriteria decision analysis in environmental decision making. *Integrated Environmental Assessment and Management,* **1,** 95–108.

Kirch, P. V. (1997). Microcosmic histories: Island perspectives on "global" change. *American Anthropologist,* **99,** 3–42.

Klein, A.-M., Vaissiere, B. E., Cane, J. H., *et al.* (2007). Importance of pollinators in changing landscapes for world crops. *Proceedings of the Royal Society Biological Sciences,* **274,** 303–13.

Le Quesne, T. and McNally, R. (2004). *The green buck—using economic tools to deliver conservation goals: a WWF field guide,* WWF-UK.

Leopold, A. (1949). *A Sand County Almanac, and sketches here and there.* Oxford University Press, Oxford.

Lubchenco, J., and Sutley, N. (2010). Proposed ocean policy for ocean, coast and great lakes stewardship. *Science,* **328,** 1485–6.

Lubchenco, J. (1998). Entering the century of the environment: A new social contract for science. *Science,* **279,** 491–7.

Millennium Ecosystem Assessment (MA). (2003). *Ecosystems and human well-being: a framework for assessment.* Island Press, Washington, DC.

Millennium Ecosystem Assessment (MA). (2005b). *Ecosystems and human well-being: synthesis.* Island Press, Washington, DC.

Millennium Ecosystem Assessment (MA). (2005c). *Living beyond our means: natural assets and human well-being: Statement from the Board.* World Resources Institute, Washington, DC.

Millennium Ecosystem Assessment (MA). (2005d). *Ecosystems and human well-being: policy responses.* Island Press, Washington DC.

McKenney, B., Morris, B. and McKenzie, E. (2008). *Framework for assessing the viability of an ecosystem service approach to conservation: the top 10 screening criteria.* Nature Conservancy, Natural Capital Project, and WWF-US.

McVittie, A., and Moran, D. (2008). Determining monetary values for use and non-use goods and services: Marine biodiversity—primary valuation. Department for Environment, Food and Rural Affairs, London.

Montgomery, D. W. (1972). Markets in licenses and efficient pollution control programs. *Journal of Economic Theory,* **5,** 395–418.

Moran, D., Hussain, S., and Fofana, A. (2007). Marine Bill marine nature conservation proposals: Valuing the benefits. Department for Environment, Food and Rural Affairs, London.

Myers, N. (2001). *Perverse subsidies: How tax dollars can undercut the environment and the economy.* Island Press, Washington, DC.

Naidoo, R., Malcolm, T., and Tomasek, A. (2009). Economic benefits of standing forests in highland areas of Borneo: quantification and policy impacts. *Conservation Letters,* **2,** 35–44.

Naidoo, R., and Ricketts, T. H. (2006). Mapping the economic costs and benefits of conservation. *PLoS Biology,* **4,** 2153–64.

Nelson, E., Mensoza, G. M., Regetz, J., *et al.* (2009). Modeling multiple ecosystem services, biodiversity conservation, commodity production and tradeoffs at landscape scales. *Frontiers in Ecology and the Environment,* **7,** 4–11.

NRC. (2004). *Valuing ecosystem services: toward better environmental decision-making.* National Academies Press, Washington, DC.

NRCS. (2008) Buffer strips: common sense conservation. United States Department of Agriculture, Natural Resources Conservation Science.

OECD (1999). *Agricultural policies in OECD countries: Monitoring and evaluation.* Organisation for Economic Co-operation and Development, Paris.

OECD (2003). *The greening of the WTO Green Box.* Organisation for Economic Co-operation and Development, Paris.

Ostrom, E. (2005). *Understanding institutional diversity.* Princeton University Press, Princeton.

Paavola, J. (2007). Institutions and environmental governance: a reconceptualisation. *Ecological Economics,* **63,** 93–103.

Paavola, J. (2008). Livelihoods, vulnerability and adaptation to climate change: lessons from Morogoro, Tanzania. *Environmental Science & Policy,* **11,** 642–54.

Pearce, D. W. (2005). *Investing in environmental wealth for poverty reduction: environment for the MDGs.* United Nations Development Programme, United Nations Environment Programme, International Institute for Environment and Development, IUCN, World Resources Institute, New York.

Pearce, D. W., and Finck Von Finckenstein, D. (1999). Advancing subsidy reform: towards a viable policy package. *Finance for sustainable development: Testing new policy approaches.* United Nations Division for Sustainable Development.

Pearce, D. W., and Seccombe-Hett, T. (2000). Economic valuation and environmental decision-making in Europe. *Environmental Science & Technology,* **34,** 1419–25.

Perrot-Maitre, D. (2006). The Vittel payments for ecosystem services: A perfect "PES" case?, International Institute for Environment and Development.

Pielke, R. A., Jr. (2007). *The honest broker: making sense of science in policy and politics.* Cambridge University Press, Cambridge.

Postel, S. L., and Thompson, B. (2005). Watershed protection: capturing the benefits of nature's water supply services. *Natural Resources Forum,* **29,** 104–5.

Ranganathan, J., Raudsepp-Hearne, C., Lucas, N., *et al.* (2008). *Ecosystem services: a guide for decision makers.* World Resources Institute, Washington, DC.

Reid, W., Berkes, F., Wilbanks, T. J., *et al.* (2006). *Bridging scales and knowledge systems: Concepts and applications in ecosystem assessment.* World Resources Institute, Washington, DC.

Revesz, R. L., and Stavins, R. (2004). Environmental law and policy. In: A. M. Polinsky and S. Shavell, Eds., *The handbook of law and economics.* Elsevier Science, Amsterdam.

Ruhl, J. B., Kraft, S. E., and Lant, S. L. (2007). *The law and policy of ecosystem services.* Island Press, Washington, DC.

Salzman, J. (2005). Creating markets for ecosystem services: Notes from the field. *NYU Law Review,* **80,** 870–961.

Salzman, J., and Ruhl, J. B. (2000). Currencies and the commodification of environmental law. *Stanford Law Review,* **53,** 607–94.

Snel, M. (2004). Poverty-conservation mapping applications. *IUCN World Conservation Congress.* Cooperazione Italiana, IUCN, UNEP and GRID Arendal.

Tainter, J. (1988). *The collapse of complex societies.* Cambridge University Press, Cambridge.

Thompson, D. (2008b) Union Pacific to pay $102 million for forest fire. *Associated Press.*

Turner, R. K., Lorenzoni, I., Beaumont, N., *et al.* (1998). Coastal management for sustainable development: analysing environmental and socio-economic changes on the UK coast. *The Geographical Journal,* **164,** 269(1).

Turner, R. K., Van den Bergh, J. C. J. M., Soderqvist, T., *et al.* (2000). Ecological-economic analysis of wetlands: Scientific integration for management and policy. *Ecological Economics,* **35,** 7–23.

UNDP (2007). *Pilot integrated ecosystem assessment of Bugesera.* United Nations Environment Programme, Repubulika y'u Rwanda and United Nations Environment Programme.

UNFCCC (2008). United Nations Climate Change Conference in Bali.

Van Beukering, P., Brander, L., Thompson, E., *et al.* (2007). *Valuing the environment in small islands: An environmental economics toolkit.* Joint Nature Conservation Committee, Peterborough, UK.

Voluntary Carbon Standard (VCS). (2008). Voluntary Carbon Standard 2007.1: Specification for the project-level quantification, monitoring and reporting as well as validation and verification of greenhouse gas emission reductions or removals.

Wilson, E. O. (1998). *Consilience: the unity of knowledge.* Kopf, New York.

WRI (2005). *The wealth of the poor—Managing ecosystems to fight poverty.* World Resources Institute, Washington, DC.

WRI (2007). *Nature's benefits in Kenya: An atlas of ecosystems and human well-being.* World Resources Institute, Washington, DC; Department of Resource Surveys and Remote Sensing, Ministry of Environment and Natural Resources, Kenya; Central Bureau of Statistics, Ministry of Planning and National Development, Kenya; International Livestock Research Institute; Nairobi.

WRI (2008). Roots of resilience—growing the wealth of the poor. World Resources Institute, Washington, DC.

Wunder, S. (2005). Payments for environmental services: Some nuts and bolts. *Occasional Paper 42.* Center for International Forestry Research, Jakarta.

Index